This book is
Dedicated to Dr. Ann Oaks, FRSc for
her outstanding contribution in the field

NITROGEN NUTRITION
IN
HIGHER PLANTS

NITROGEN NUTRITION IN HIGHER PLANTS

Foreword by
Peter J. Lea
University of Lancaster

Edited by
Prof. H.S. Srivastava
Dean, Faculty of Life Sciences
Rohilkhand University, Bareilly (India)

&

Dr. R.P. Singh
Department of Biosciences
M.D. University, Rohtak (India)

2016
Associated Publishing Company
A Division of
Astral International Pvt. Ltd.
New Delhi - 110 002

© EDITOR
First Published, 1995
Reprinted, 2016

ISBN: 978-93-5130-985-7(International Edition)

Published by   : **Associated Publishing Company**
                            A Division of
                            Astral International Pvt. Ltd.
                            – ISO 9001:2008 Certified Company –
                            4760-61/23, Ansari Road, Darya Ganj
                            New Delhi-110 002
                            Ph. 011-43549197, 23278134
                            E-mail: info@astralint.com
                            Website: www.astralint.com

Typesetting   : **Divya Computers, Meerut-250 002**

Printed at   : **Replika Press Pvt. Ltd.**

Contents

List of Contributors

Amonkar, D.V.
Department of Botany, R.J. College, Ghatkopar, Bombay-400 086, India

Banowetz, G.M.
USDA, ARS, NFSPR Centre 3450, S.W. Campus Way, Corvallis, Oregan, 97331, U.S.A.

Bhadula, S.K.
High Altitude Plant Physiology Research Centre, P.B. 14, Srinagar-246 174, Garhwal, U.P., India

Dadarwal, K.R.
Department of Microbiology, CCS Haryana Agricultural University, Hisar-125 004, India

Fletcher, R.A.
Department of Environmental Biology, University of Guelph, Guelph, Ontario, Canada, N1G 2W1

Fernandes, M.S.
UFRRJ-KM47-23851-Itaguai-RJ, Brazil

Ghosh, S.
Department of Biology, University of Waterloo, Waterloo, Ontario, Canada, N2L 3G1

Glass, A.D.M.
Department of Botany, 3529-6270 University Boulevard, Vancouver, B.C., Canada V6T 1Z4

Griffith, S.M.
USDA, ARS, NFSPR Centre, 3450, S.W. Campus Way, Corvallis, Oregan, 97331, U.S.A.

Haynes, R.J.
New Zealand Institute for Crop and Food Research, Canterbury Agriculture and Science Research Centre, Private Bag 4704, Christchurch, New Zealand

Hale, B.
Department of Horticultural Sciences, University of Guelph, Ontario, Canada, N1G 2W1

Jaiwal, P.K.
Department of Biosciences, M.D. University, Rohtak-124 001, India

Kao, C.H.
Department of Agronomy, National Taiwan University, Taipei, Taiwan, Republic of China

Karmarkar, S.M.
Department of Botany, R.J. College, Ghatkopar, Bombay-400 086,। India

Lara, C.
Instituto de Bioquimica Vegetal Y Fotosintesis, Universidad de Sevilla, CSIC, Apdo. 1113, 41080-Sevilla, Spain

Magalhaes, J.R.
EMBRAPA/CENARGEN, SAIN-Parque Rural, 70770-900-Brasilia-DF-Brazil

Ormrod, D.P.
Office of the Graduate Studies, University of Guelph, Ontario, Canada N1G 2W1

Paliyath, G.
Department of Horticultural Sciences, University of Guelph, Guelph, Ontario, Canada N1G 2W1

Puranik, R.M.
School of Biochemistry, Devi Ahilya University, Indore-452 001, M.P., India

Peirson, D.
Department of Biology, Wilfrid Laurier University, Waterloo, Ontario, Canada, N2L 3C5

Sindhu, S.S.
Department of Microbiology, CCS Haryana Agricultural University, Hisar-125 004, India

Singh, R.P.
Department of Biosciences, M.D. University, Rohtak-124 001, India

Srivastava, H.S.
Department of Plant Sciences, Rohilkhand University, Bareilly-243 005, India

Shargool, P.D.
Department of Biochemistry, University of Saskatchewan, Saskatoon, Canada, S7N 0W0

Sawhney, S.K.
Department of Chemistry and Biochemistry, CCS Haryana Agricultural University, Hisar-125 004, India

Singh, B.K.
American Cyanamid Company, P.O. Box 400, Princeton, NJ 08543-0400, U.S.A.

Singh, R.
Department of Chemistry and Biochemistry, CCS Haryana Agricultural University, Hisar-125 004, India

Stulen, I.
Department of Plant Biology, University Groningen, P.O. Box 14, 9750 AA. Haren, The Netherlands

Siddiqui, M.Y.
Department of Botany, 3529-6270 University Boulevard, Vancouver, B.C., Canada, V6T 1Z4

ter Steege, M.
Department of Plant Biology, University Groningen, P.O. Box 14, 9750 AA. Haren, The Netherlands

Williams, P.H.
New Zealand Institute for Crop and Food Research, Canterbury Agriculture and Science Research Centre, Private Bag 4704, Christchurch, New Zealand

Yamaya, T.
Department of Applied Biological Chemistry, Faculty of Agriculture, Tohoku University, 1-1, Tsutsumidori, Amamiyamachi, Aoba-Ku, Sendai, 981, Japan

Abbreviations Used

N,	nitrogen
LATS,	low affinity transport systems
HATS,	high affinity transport systems
ENDOR,	electron nuclear double resonance spectroscopy
EPR,	electron paramagnetic resonance
NiR,	nitrite reductase
NR,	nitrate reductase
NH_3,	ammonia
Fd red,	ferrodoxin reduced
GDH,	l-glutamate dehydrogenase
GS,	glutamine synthetase
GOGAT,	glutamate synthase
GS-GOGAT,	glutamine synthetase-glutamate synthase
MSO,	methionine sulfoximine
AHAS,	acetohydroxy acid synthase
AK,	aspartate kinase
AS,	anthranilate synthase
CM,	chorismate mutase
DAHP,	3-deoxyarabinoheputlosonate-7-phosphate
DHAD,	dihydroxyacid dehydratase
DHQ,	dehydroquinate
DHPS,	dihydrodipicolinate synthase
EPSP,	5-enolpyruvylshikimate-3-phosphate
GSA,	glutamate semialdehyde
HK,	homoserine kinase
HSDH,	homoserine dehydgenase
IAA,	indole acetic acid
IPA,	indole pyruvic acid
IGP,	indole glycerol phosphate
KARI,	ketoacid reductoisomerase
5-MA,	5-methyl anthranilate
PAT,	anthranilate phosphoibosyltransferase
P-5-C,	pyrroline-5-carboxylate
PEP,	phosphoenolpyruvate
SAM,	s-adenosylmethionine
SDS,	sodium dodecyl sulfate

S-3-P,	shikimate-3-phosphate
TD,	threonine dehydratase
TS,	trytophan synthase
SDP,	short day plants
LDP,	long day plants
ABA,	abscisic acid
GAS,	gibberellins
NAA,	napthalene acetic acid
PGR,	plant growth regulator
SA,	saliaylic acid
PGA,	phosphoglyceric acid
RUBP,	ribulose
1-5	biphosphate
ESP,	exchangeable sodium percentage
SAR,	sodium absorption ratio

Foreword

Nitrogen is a constituent of a large number of important compounds found in all living cells. Particular notable examples are amino acids, proteins and nucleic acids, while others e.g. polyamines and chlorophyll may play a major role in some organisms. Most animals do not have the capacity to assimilate inorganic nitrogen or to synthesise half the amino acids found in proteins, unless assisted by bacteria (e.g. in the rumen of sheep and cattle). Plants therefore act as the primary source of all organic nitrogen for animal and human nutrition.

Of the nitrogen present in the biosphere, 99.95% (4×10^{15} tonnes) exists as atmospheric or dissolved nitrogen gas, which is not directly available to plants. A relatively small number of bacteria are able to convert (fix) nitrogen to ammonia using the enzyme nitrogenase. The rhizobium-legume symbiosis in the root nodule is a particularly important mechanism of obtaining nitrogen in deficient soils. If available, plants are able to take up either ammonium or nitrate ions through the roots, although normally ammonia is converted to nitrate by soil microorganisms. Nitrate is converted to ammonia by a two step process involving the enzymes nitrate and nitrite reductase. Nitrate reductase has been the subject of intense study and is now known to be regulated by a number of factors including the availability of nitrate, light and carbohydrate.

Ultimately all inorganic nitrogen is converted to ammonia, which may also be formed internally in the plant by a number of metabolic reactions including photorespiration, lignin synthesis, transport compound breakdown and proteolysis. Ammonia is initially assimilated into the amide position of glutamine by glutamine synthetase. At least four genes are known to regulate the formation of glutamine synthetase. Specific functions of the enzyme have now been identified in the leaf chloroplast, legume root nodule and the vascular system of plants. Following the synthesis of glutamine, the amide nitrogen may be transferred to the transport and storage compound asparagine. In a limited number of tropical legumes the ureides, allantoin and allantoic acid act as transport compounds.

The majority of the ammonia assimilated into glutamine is eventually transferred to the α-amino position of glutamate by glutamate synthase, which may use either NADH or reduced ferredoxin as a source of reductant. The two isoenzymes are regulated by different genes with the ferredoxin dependent enzyme playing a major role in the leaf chloroplast, whilst the NADH-dependent enzyme is present in the root and vascular system. A third enzyme glutamate dehydrogenase was originally thought to be involved in ammonia assimilation, it is now proposed that the enzyme catalyses the deamination of glutamate to yield 2-oxoglutarate. Following the synthesis of glutamate, the α-amino group is rapidly transaminated to other amino acids in particular alanine and aspartate.

The remainder of the amino acids are grouped into "families" depending upon the precursor of amino acid. The "aspartate family" of lysine, threonine, methionine and isoleucine have received particular attention as lysine and methionine are the limiting amino acids in most plant feedstuffs. Mutant selection and genetic engineering has now produced plants that contain elevated levels of these essential amino acids. A number of new low dose herbicides are now known to inhibit the enzyme acetolactate synthase that is involved in the synthesis of isoleucine, leucine and valine.

The aromatic amino acids tryptophan, phenylalanine and tyrosine are synthesised via the shikimate pathway, with chorismate being a common substrate for each of the amino acids. A range of secondary compounds including lignin, flavonoids, phytoallexins, alkaloids and the plant growth regulator indole acetic acid are also derived from the aromatic amino acid pathway. The herbicide glyphosate is a potent inhibitor of the synthesis of aromatic amino acids.

The polyamines, spermine, spermidine and putrescine are synthesised from arginine and S-adenosylmethionine. The compounds are now known to act as growth regulators and may play an important role in drought resistance, flowering and senescence.

The gas nitrogen dioxide often occurs as an atmospheric pollutant and may have an inhibitory effect on plant growth particularly in the presence of sulphur dioxide. However under certain conditions plants are able to take up nitrogen dioxide and use it as a nitrogen source. Stimulatory effects on the enzymes nitrate and nitrite reductase have frequently been demonstrated.

An important fact that has arisen from the studies on nitrogen metabolism is the major role of the chloroplast. Following the conversion of nitrate to nitrite all the processes of ammonia assimilation and amino acid biosynthesis take place in the chloroplast. In the leaf the ATP and reductant required for the metabolic pathways are derived from the light reactions. Thus amino acid biosynthesis can be considered photosynthesis in the same way as CO_2 assimilation.

Edited by H.S. Srivastava and R.P. Singh, the book 'Nitrogen Nutrition in Higher Plants' is a collection of valuable articles on these aspects, and is expected to meet the aspirations of researchers, scientists and teachers in this and related areas.

Peter J. Lea
University of Lancaster
July, 1994

Editorial

Nitrogen is the most abundant mineral nutrient of plants, and is essential component of important bio-molecules such as amino acids, proteins, nucleic acids, chlorophylls, alkaloids, several vitamins and growth regulators etc. Although it is present abundantly as N_2 gas in the atmosphere, only a few plants with the symbiotic association of bacteria and actinomycetes and perhaps to a limited extent with the association of free living microbes, are able to utilise this gaseous nitrogen. Most others acquire this nutrient from the soil in the form of inorganic ions and then assimilate them to synthesize nitrogenous organic molecules. Thus, the acquisition and the assimilation of inorganic nitrogen is often a critical determinant of plant growth and productivity. Some of these aspects have been reviewed periodically through various articles in research journals. A few books have also been published covering one or more aspects of nitrogen metabolism in higher plants. But, it has to be realized that the aspects of nitrogen metabolism have been more dynamic and open fields than most other aspects of Plant Physiology and Biochemistry; and every year hundreds or perhaps thousands of new research publications have been surfacing up, some giving altogether a new dimension to the established postulates and some opening up new avenues. As in other physiological and biochemical areas, molecular biology has been providing a strong theoretical and technical background for investigating various aspects of nitrogen nutrition also. At the other end, global concern over ever increasing load of chemical fertilizers, agro-chemicals and other wastes in the environment has compelled plant and soil scientists to ponder over the problems of nitrate pollution and nitrogen cycling with a new vision and vigour. The possible use of immobilized plant nitrate reductase for the alleviation of nitrate pollution problem, will be a good example of integrating knowledges from physiology, bio-chemistry and molecular biology to solve environmental problems. The study of nitrogen uptake and assimilation by plants has been providing the background and enthusiastic support to other cognate and equally important disciplines as well; and thus a review of these aspects will be never out of date.

Our main objective in producing this book was to present an overview of nitrogen uptake and assimilation by the higher plants, which could be useful for both beginners as well as advanced readers in this field. Obviously, it was a very difficult task to identify the contributors and more so to limit their views within a preplanned frame, as they had been well known and experienced researchers. Nevertheless, most of the articles in this book, with a strong introductory paragraph, a good discussion on contemporary status and with a concluding paragraph, will be able to provide something to both beginners as well as advanced learners.

The understanding of many of the problems in nitrogen nutrition in plants is more clear today than it was in late 80's. Some of the basic postulates evolved as a result of studies in micro-organisms have been tested and found to be relevant in higher plants as well. For example, plants appear to be sharing many pathways in amino acid biosynthesis with *E. coli*. The knowledge gained from the amino acid biosynthetic pathway may be utilized in evolving herbicide resistant crop plants, as convenient techniques of cloning, insertion and integration of genes are becoming available now. New concepts are emerging in relation to uptake and intracellular mobilization of inorganic and organic nitrogen; the osmotic potential across the plasmalemma and tonoplast is being considered as an important factor in the process. The reduction and assimilation of nitrate processes are also better understood today than a few years back. NAD(P)H specific isoform of nitrate reductase, discovered originally in soybean has been shown to be present in many other species as well. More than one isoform of nitrite reductase are also known. The polymorphism in these enzymes indicates that the 'bottleneck' in nitrate assimilation process is perhaps not as constricted as it was though to be, onetime. Although, the importance of GS-GOGAT pathway in ammonium assimilation is well established, there is need to examine the significance of high level of GDH protein and its induction by ammonium in plants. It would be also interesting to study the nitrogen assimilation and nitrogen productivity in GDH mutants. The molecular biology of symbiotic nitrogen fixation, the ureide metabolism, the role of polyamines, the role of nitrogen in flowering and nitrogen mobilization during senescence have also been the subject matter for extensive investigation in the recent past and some definite informations are available now, on these aspects. Light is a critical factor in plant life and its effect on nitrogen assimilation is also better understood now. Light-dark mediated phosphorylation-dephosphorylation of nitrate reductase is a significant discovery of the nineties. The study of the hormonal regulation of nitrogen assimilation is to be extended beyond nitrate reduction step. Among various factors of the environment affecting plant life, pollutants have acquired an importance in the recent past. The nitrogenous air pollutants seem to be having both positive as well as negative influence on nitrogen assimilation in plants. This book focusses on these aspects and we do hope that it will be useful for physiologists, biochemists, molecular biologists, agriculturists and all those who are engaged in the task of improving plant life by manipulating nitrogen status.

It will be rather difficult to list all those who have provided moral and material support in the production of this book. Nevertheless, we wish to record our deep appreciation for Dr. Ann Oaks for her inspiration and for help in finalizing the names of contributors in many cases. We also wish to acknowledge the inspiring support from Drs. D.P. Ormrod, P.J. Lea, A.R. Wellburn, P.A. Jolliffe, V.C. Runeckles, A.N. Purohit, R.C. Pant and many others. We also acknowledge the moral support of our students, colleagues and family members. Lastly, we are thankful to Mr. S.K. Dutta, our chief publisher for taking all the pains and interest for a timely and excellent production.

Bareilly, August 1, 1994   H.S. Srivastava & R.P. Singh

Nitrogen Nutrition in Higher Plants, 1995
Editors: H.S. Srivastava & R.P. Singh
Associated Publishing Co., New Delhi, India
pp. 1-20.

Nitrogen in the Plant Environment

P.H. WILLIAMS and R.J. HAYNES

I. Introduction

Nitrogen is widely distributed throughout nature. The major reservoir of nitrogen (N) is the lithosphere where it is found mainly in primary igneous rocks (Table 1). Only a small proportion of this N is present in the soil in plant available forms (NH_4^+ and NO_3^-). Weathering of the primary rocks to release N is very slow, consequently little of the lithosphere N is supplied to the biosphere. Instead the major source of biosphere N is the atmosphere.

Table 1. Quantities of nitrogen in various spheres (Stevenson, 1986).

Sphere	Tg N
Lithosphere	
Igneous rocks	1.9×10^{11}
Sediments	4×10^8
Terrestrial soils	
Organic matter	3×10^5
Inorganic	1.6×10^4
Atmosphere	3.9×10^9
Hydrosphere	2.2×10^7
Biosphere	2.4×10^5

Since N is a very mobile element it readily moves between spheres and from one chemical form to another as a result of biological, chemical and physical processes (Fig. 1). Thus the availability of N for plants depends not only on the amount present in the soil but also on the rate at which N cycles occur through the soil-plant system. This chapter reviews the forms of N in the soil and how these forms

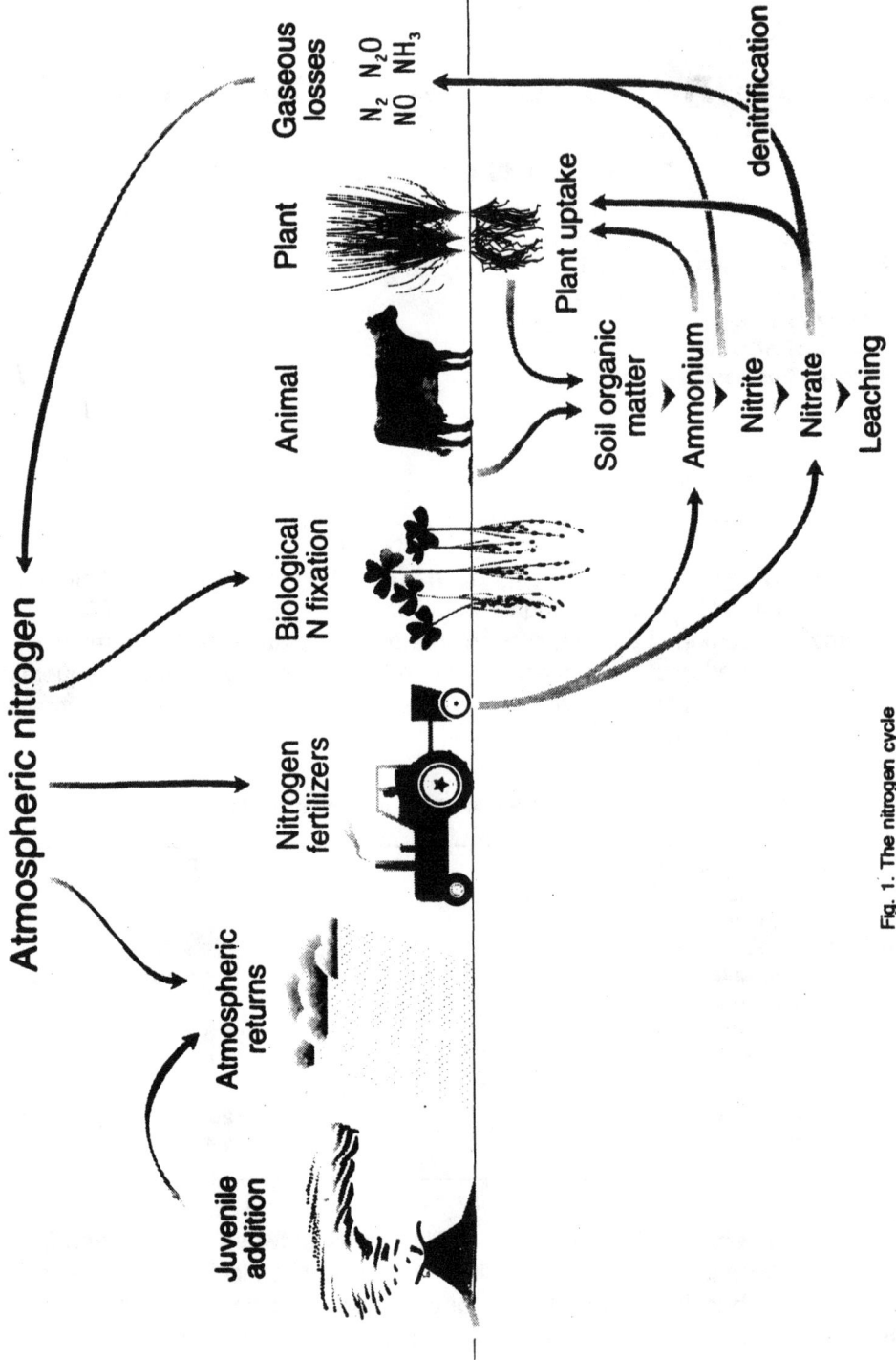

Fig. 1. The nitrogen cycle

are inter-related. The gains to and losses from the soil are also considered as is the practical role of N fertilizers in plant production.

II. Forms of Nitrogen in the Soil

A fertile mineral soil contains approximately 7000 kg N ha^{-1} in the plant rooting depth (Mengel and Kirkby, 1987). At least 90% of this N is organically bound mainly as NH_2 groups in highly complex compounds. Between 20 and 50% of the N is thought to be in amino acid forms, 5-10% as amino sugars and the rest is largely unidentified (Stevenson, 1986). Organic N originates from the microbial decomposition of plant and animal residues. This decomposition process can result in the release of organic N as NH_4^+ and NO_3^- (mineralization) and it can also lead to incorporation of N into microbial tissue or into humic substances which are relatively resistant to microbial attack (immobilization). These two processes of mineralization and immobilization occur simultaneously and form an important link in the N cycle of most natural and agricultural ecosystems.

Mineral N in the soil exists as NH_4^+, NO_3^-, and sometimes NO_2^- in solution and on exchange sites of soil particles or as NH_4^+ held by the clay minerals. Some gaseous N may also be found in the soils atmosphere and dissolved in soil solution.

The N content of soils varies from less than 0.1% in desert soils to 2% in highly organic soils (Haynes, 1986a). Since most of the N is organically bound, the N content closely relates to soil organic matter content. Soil development is influenced by 5 factors (parent material, climate, vegetation, topography and time; Jenny, 1961). Of these, climate and vegetation are considered to have the most influence on soil organic activity and vegetation type (Stevenson, 1982). For example, N contents are low in desert areas where vegetation and consequently organic residues are sparse. Similarly N contents are low in tropical soils where high temperatures ensure that microbial activity is high and organic residues are rapidly decomposed. In contrast, grassland soils in a temperate environment tend to accumulate high levels of organic matter and organic N. The exception to this pattern is poorly drained soils in which organic matter accumulates regardless of climate and vegetation as the decomposition rate is restricted by the poor oxygen supply (e.g. peat soils).

During soil development N accumulates in the soil due to biological fixation of N_2 mainly with some contribution from wet and dry deposition. With time soil N levels increase sufficiently to support nonfixing plants and these ultimately dominate the vegetation. Eventually the soil N content reaches a stable equilibrium value as dictated by the soil forming factors. Should the ecosystem be disturbed (e.g. by fire, cultivation or felling of forest vegetation) this equilibrium value can suddenly be changed. For example, cultivating a grassland ecosystem results in a dramatic reduction in organic matter content which can continue over a period of many years. This loss occurs through the encouragement of organic matter oxidation, organic

N mineralization and nitrification by cultivation. Wind and water erosion may also be accelerated resulting in a loss of soil particles and therefore N. Under continuous cropping a new equilibrium level may be reached at a much lower N content than before. The addition of fertilizers and manures or use of legume crops may help to increase soil N content above this equilibrium level. If cultivation ceases then succession may begin again and soil organic matter and N content may increase. This is the basis for the practice of shifting cultivation carried out in many tropical areas.

III. Transformations of Soil Nitrogen

Soil N undergoes many transformations within the soil-plant system. Nitrogen is taken up by plants and returned to the soil in the form of plant litter and crop residues. Within the soil nitrogen is continuously being interchanged between organic and inorganic forms by the microbial biomass (e.g. fungi, bacteria, actinomycetes, algae and protozoa) which itself is part of the soil organic matter. The length of time the N remains in any one form varies from a few days for some elements of the biomass to over a thousand years for stable humus (Stevenson, 1986). In addition to the biochemical transformations of N, the clay minerals present in some soils can fix NH_4^+ thus transforming nitrogen into a form which is not available to the plant.

A. Plant Uptake

Nitrogen is taken up by plants from the soil as NO_3^- or NH_4^+. Plant species vary in their ability to absorb and utilize these N sources, but mostly prefer a mixture of high NO_3^- and low NH_4^+. In acid soils NH_4^+ is the predominant source of soil inorganic N and the plants have adapted to use NH_4^+ in preference to NO_3^- (Haynes, 1986a).

The amount of N taken up by field crops varies during the growing season (Goh and Haynes, 1986; Mengel and Kirkby, 1987). Initially the emerging crop has a small demand for N but the requirement increases rapidly as the plant grows. In many situations soil N can not match plant requirements during this phase and fertilizer may be required to maintain maximum crop growth rates. As the crop reaches maturity, uptake of N diminishes and internal N is mobilized from vegetative parts to the developing fruits and seeds. Thus throughout the season there are constant changes in N concentration and distribution in the plant. By the time the crop is harvested there can be more N present in the harvested produce than in the crop residues (e.g., cereal crops; Table 2).

There is also a difference between species in the amount of N present in plant tissues (Table 2). Forage legumes tend to have a higher N content compared with forage grasses, cereal grain crops or other arable crops due to the supply of N from their root nodules (see section on nitrogen fixation).

Table 2. Amount of nitrogen present in various crops (Mengel and Kirkby, 1987).

Crop	Yield (t ha^{-1})	N content (kg N ha^{-1})
Barley (grain)	2.2	40
Barley (straw)	2.5	17
Maize (grain)	9.5	150
Maize (straw)	11.0	110
Lucerne hay	10.0	200
Red clover hay	6.0	110
Timothy hay	6.0	66
Sugarbeet	50	20
Sugarcane	75	110
Cotton (seed and lint)	1.7	45
Cotton (stalks, leaves and burs)	2.2	39
Potatoes (tubers)	27	90
Cabbage	50	145

B. Mineralization and Immobilization

The biological interchange of organic N with inorganic N occurs through the simultaneous processes of mineralization and immobilization (Fig. 2). The first phase of mineralization is the breakdown of organic N into amino acids and then ammonification to NH_3 by energy yielding enzymatic reactions:-

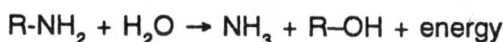

$$R\text{-}NH_2 + H_2O \rightarrow NH_3 + R\text{–}OH + energy$$

The NH_3 is rapidly hydrolysed to NH_4^+ and the latter can then be nitrified in a two-step process to NO_3^-. In most soils nitrification is carried out by a small group of autotrophic soil bacteria which derive their energy from oxidation of N compounds rather than C ones. In the first step NH_4^+ is oxidised to NO_2^- by *Nitrosomonas* and a limited number of other organisms such as *Nitrosolubus* and *Nitrospira*:-

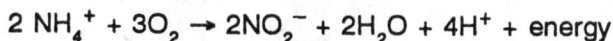

$$2\,NH_4^+ + 3O_2 \rightarrow 2NO_2^- + 2H_2O + 4H^+ + energy$$

Then the NO_2^- is rapidly converted to NO_3^- by *Nitrobacter*

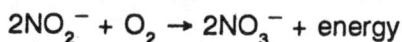

$$2NO_2^- + O_2 \rightarrow 2NO_3^- + energy$$

While ammonification can be carried out in both aerobic and anaerobic conditions, nitrification requires an aerobic state. So in situations where O_2 is restricted there can be an accumulation of NH_4^+ e.g. in flooded rice soils.

Although mineralization readily occurs in most soil systems NO_3^- seldom accumulates in the soil. This is due to the rapid removal of NO_3^- by the plants as soon

$$\text{Organic N} \longrightarrow NH_3 \underset{-H^+}{\overset{+H^+}{\rightleftharpoons}} NH_4^+ \xrightarrow{\text{Nitrosomonas}} NO_2^- \xrightarrow{\text{Nitrobacter}} NO_3$$

—ammonification →   ———————nitrification————→

———————————— Mineralization ————————————→

←———————————— Immobilization ————————————

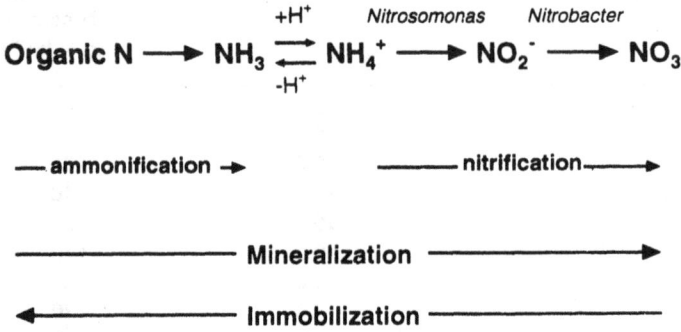

Fig. 2. The processes of mineralization and immobilization in soils (Stevenson, 1986)

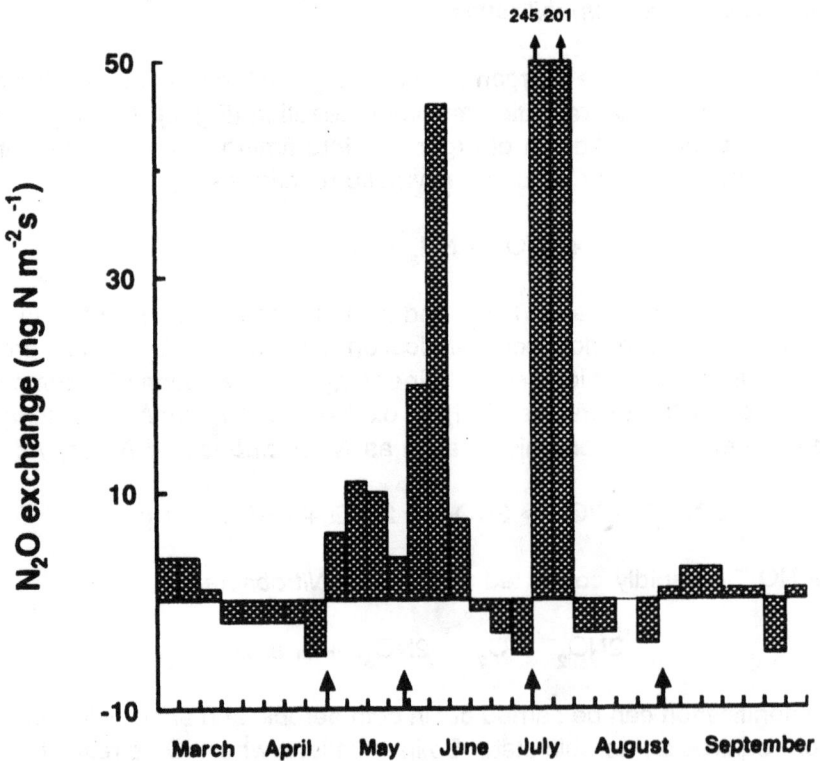

Fig. 3. Nitrous oxide exchange between the atmosphere and cut grassland receiving 4 applications of 62.5 kg N ha^{-1} as ammonium nitrate. Times of fertilizer application are marked (†) (Ryden, 1981)

as it is formed. The amount of NH_4^+ available for nitrification may also be limited due to competition for NH_4^+ by the microbial biomass and plants. When environmental conditions favour mineralization and there are no plants present or plant uptake is restricted by some factor NO_3^- can accumulate in cultivated soil (e.g. during autumn and winter). In such circumstances nitrification is often undesirable as the NO_3^- may be lost due to leaching or denitrification. Alternatively, nitrification can be desirable in some circumstances as it reduces the build up of NH_4^+ in the soils which can have a phytotoxic effect or result in volatilization losses of NH_3.

Since microbial activity is involved in these transformations, they are influenced by temperature, moisture and pH. The N content and the quality of organic residues (the C/N ratio, lignin and polyphenol content) can also influence the rate of mineralization (Haynes', 1986b). Organic materials have a wide range of C/N ratios being low in microbial tissues and high in cereal straw and forest wastes (Table 3). In general if the N content is low (C/N ratio >30) then the amount of N assimilated into organic forms to meet the requirements of the microorganisms is greater than that released during decomposition and so net immobilization occurs. Thus incorporating cereal residues and straw into the soil can result in net immobilization of N. In this situation fertilizer N may need to be applied to overcome N starvation by the subsequent crop. If the N content of the residues is high (C/N ratio is <20) then net mineralization can occur. So addition of material with a low C/N ratio such as animal manure can increase the soil NO_3^- content. However if the residues contain high concentrations of lignin or phenols (e.g. legumes) then despite a high N content there may not be a net mineralization of this N (Fox et al., 1990). This is because the lignin and polyphenols and their decomposition products can combine with the plant proteins and amino acids to form humic polymers that resist decomposition (Haynes, 1986b).

Table 3. Typical C/N ratios of various organic materials (Stevenson, 1986).

Material	C/N
Microbial tissues	6-12
Sewage sludge	5-14
Soil humus	10-12
Animal manures	13-25
Legume residues and green manures	13-25
Cereal residues and straw	60-80
Forest wastes	150-500

C. Nitrogen from Plant Litter and Crop Residues

Decomposition of plant roots and above ground herbage is a major pathway for returning N, often in large quantities, to the soil. For example, in a high producing pasture grazed by dairy cows 550kg N ha^{-1} may be returned to the soil annually in decaying roots and herbage (Steele, 1982).

In cultivated soils, plant roots and crop residues (e.g. straw and stover) are incorporated into the soil. Legume species grown as a crop or as a green manure can contribute significant amounts of N in this manner, the total amount varying according to species. Measurements carried out under New Zealand conditions showed that residues from a pea crop contained 8 kg N ha^{-1} in the roots and 36 kg N ha^{-1} in the stover while field bean residues contained 30 and 49 kg N ha^{-1}, respectively (Haynes et al., 1993). The amount of biomass incorporated into the soil will also affect N supply. For example, alfalfa may contribute 50-80 kg N ha^{-1} when fall regrowth, roots and plant crowns are incorporated into the soil, but if the herbage is mown several times retained on the soil surface and cultivated into the soil along with the other residues then 300 kg N ha^{-1} may be supplied (Hesterman et al., 1987).

The availability of N in the plant litter and residues to subsequent crops is determined by the rate at which they are decomposed and mineralized as previously discussed. It has been shown that about 10-35% of the N in legume residues can be recovered in the first succeeding crop (Prasad and Power, 1992) and 4% in the second crop (Ladd and Amado, 1986). This contribution may be sufficient to reduce fertilizer requirements of the succeeding crops. Using *Sesbania aculeata* as a summer legume green manure can save 60 kg N ha^{-1} off the fertilizer requirements of a subsequent rice crop in a cropping rotation in northern India (Goswami et al., 1988).

D. Adsorption and Fixation of Nitrogen

Soil clay particles and organic matter have a predominantly negative charge which is capable of attracting and holding positively charged cations e.g. NH_4^+. These charges are referred to as cation exchange sites and they arise in several ways. On micaceous clays they occur through substitution of one structural cation for another of a lower valency. For amorphous clay minerals additional negative charges are created by the presence of weak acids on the clay surfaces. The negative charges of organic matter are derived from carboxyl and phenolic groups which behave like weak acids (Nommik and Vahtras, 1982). Cation exchange is a reversible process with the cations in solution in equilibrium with those on the exchange sites. Therefore NH_4^+ held in the soil on the exchange sites is protected from leaching but is still in a plant available form.

Ammonium ions can be held between layers of micaceous clays in positions not readily accessible for exchange with solution cations. Such positions occur between structural layers of micas, intergrade hydrous micas and vermiculites or in the wedge zones formed as mica is weathered and the layers start to expand. Due to the affinity of the clay surface charges for NH_4^+ and the position of these ions, they are not easily exchanged with other cations and so they are referred to as fixed or nonexchangeable (Nommik and Vahtras, 1982). This form of NH_4^+ can be released slowly into solution and so become available to the plants if the concentration in solution decreases. Fixation of NH_4^+ may be of significance following the application of fertilizers to soils containing micaceous clays but with time this N can be recovered by crop plants (Preston, 1982).

Anion exchange can also occur where NO_3^- ions are held electrostatically to positive charges which occur on some soil colloids (e.g. iron and aluminium oxides and hydroxides on kaolinite and allophane). This is less common and is important only in some tropical and volcanic soils (Cameron and Haynes, 1986).

IV. Gains of Nitrogen

Nitrogen can be added to the soil via biological fixation of N_2 or through the deposition of N gases and salts from the atmosphere. The application of N fertilizers to soil is another major gain of N to the soil.

A. Biological Nitrogen Fixation

Biological fixation of N occurs when atmospheric N is reduced to NH_3 by soil microorganisms. The process requires a supply of energy and the nitrogenase enzyme complex;

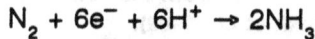

$$N_2 + 6e^- + 6H^+ \rightarrow 2NH_3$$

Only a few genera of bacteria and actinomycetes are capable of N_2 fixation. In agriculture the greatest contribution is made by the *Rhizobium* species in association with leguminous plants. These bacteria invade the root hair cells of the legumes and induce the formation of root nodules. Bacteria within the nodules receive C substrates from the plant and the NH_3 from the fixation process is exported to the plant via a specialised transport system (Havelka et al., 1982). It has been estimated that N fixation rates by this symbiotic relationship may be as high as 600 kg N ha^{-1} y^{-1} (Table 4) with rates of 100-300 kg N ha^{-1} y^{-1} common in temperate grass/legume pastures (Hoglund and Brock, 1987). The rate of fixation is variable as it is affected by light, temperature, water availability and soil pH. Nitrogen supply in the soil is another factor and fixation is greatest when soil N is low (Stevenson, 1986).

Nitrogen can also be fixed by a variety of free living microorganisms (e.g. blue green algae, *Azotobacter*) and microorganisms living in association with non-legume plants (e.g. *Actinomyces* and alder trees). These sources of N are important in forests, woodlands and aquatic environments (Evans and Barber, 1977). Fixation rates by free living microorganisms are generally low compared with the *Rhizobia* systems (Table 4).

Table 4. Rates of biological N$_2$ fixation (Evans and Barber, 1977)

	N Fixed (kg ha^{-1} y^{-1})
Legumes	
Clover	100-160
Alfalfa	125-600
Lupins	150-170
Soybeans	55-100
Cowpeas	85
Free-living microorganisms	
Blue-green algae	25
Azotobacter	0.3

The total amount of N fixed by microorganisms is approximately 100-200 Tg N y^{-1} (Table 5). This N is released into the soil upon decomposition of the microbes and associated plant roots, root nodules and above ground herbage. In past years fertilizers rather than legumes have been the main form of N supply to agricultural systems. However, increased costs of fertilizer N and concern for environmental pollution from fertilizer manufacture and use is currently encouraging greater reliance on the use of legumes as a N source than in the past.

Table 5. Gains and losses of N to and from the soil-plant environment (Haynes, 1986; Stevenson, 1986; Aulakh et al., 1992).

Soil-plant environment	Tg N y^{-1}
Gains	
Biological fixation	100-200
Wet and dry deposition	120-280
Fertilizers	80
Losses	
Ammonia volatilization	35-250
Denitrification	100-190
Chemodenitrification	1-15
Leaching	10-40
Erosion and runoff	5

B. Wet and Dry Deposition

Wet deposition involves the precipitation of gaseous and particulate matter from the atmosphere, while dry deposition refers to gravitational settling and gaseous absorption directly by soils and plants. It has been estimated that the combination of wet and dry deposition results in the addition of 0.8-22 kg N ha^{-1} y^{-1} as NO$_3$$^-$

and NH_4^+ (Stevenson, 1982). Higher rates than this may be deposited near to industrial areas due to the burning of fossil fuels and in agricultural areas due to volatilization of NH_3 from animal excreta.

C. Fertilizer Nitrogen

Throughout the world there is a reliance on N fertilizers to maintain high yields of arable crops. Consequently 80 Tg N are applied annually (FAO, 1989) to 11% of the earth's land surface area (Hauck, 1981). Most of this fertilizer is derived from NH_3 manufactured synthetically by the Haber-Bosch process. This involves reacting N_2 with H_2 at high pressure and temperature in the presence of an iron catalyst (Haynes, 1986a).

The efficiency with which fertilizer N is used is not high as the crop plants recover only 25-70% of the applied N. The rest of the N can be incorporated into soil organic forms or lost via leaching, surface runoff, leaching, volatilization and denitrification (Williams, 1992).

V. Losses of Nitrogen

The main pathways for loss of N from the soil are gaseous emissions, leaching and erosion. Nitrogen losses are of concern not only because they result in the removal of N from the farming system but also because many nitrogen compounds cause environmental pollution. The N gases N_2O, NO and NO_2 are involved in ozone depletion in the stratosphere which shields the biosphere from harmful exposure to UV radiation. N_2O is also one of the gases implicated in the greenhouse effect and global warming. The NH_3 gas is a pollutant both directly due its corrosive nature and indirectly through the formation of ammonium salts (Bouwman, 1990).

The loss of soil N to groundwater by leaching and surface water by runoff can lead to environmental problems. Groundwater is an important source of drinking water for much of the worlds population and its quality and purity affect human and animal health. High NO_3^- levels in drinking water have been linked with haemoglobinaemia (blue baby syndrome) particularly in infants (Keeney, 1982). In surface waters high N contents can increase the growth of algae and plants thus accelerating eutrophication and affecting water quality and usage.

A. Gaseous Loss

Nitrogen can be emitted from soils as NH_3, N_2, N_2O, NO_2 and NO during volatilization, nitrification, biological denitrification and chemodenitrification (Haynes and Sherlock, 1986). The extent to which these processes occur depends on the ecosystem, soil characteristics, cropping procedure, fertilizer techniques and the prevailing weather. For example, while NH_3 volatilization may be important when urea is broadcast on to the soil surface of a row crop, biological denitrification may be the major process when fertilizer is drilled into an irrigated soil (Freney, 1992). Both volatilization and denitrification losses of N may occur when urea is broadcast onto flooded soils (Freney et al., 1990).

B. Volatilization

Volatilization occurs when free NH_3 present in the soil atmosphere or the soil solution is released into the atmosphere. Estimates of the amount of N volatilized from land varies from 36 to 250 Tg N y^{-1} (Haynes, 1986a), thus it represents a significant loss of N to the atmosphere (Table 5).

The source of the NH_3 is mainly NH_4^+ originating from the decomposition of soil organic matter, plant residues and animal excreta. Fertilizers like urea, ammonium salts, anhydrous or aqueous NH_3 can also provide a source of NH_3/NH_4^+ but contribute only 3-4 Tg N y^{-1} (Bouwman, 1990).

The amount of NH_3 volatilized is affected by a combination of biological, chemical and physical factors which affect the concentration of NH_3/NH_4^+ in the soil, the ability of the soil to retain NH_4^+ (e.g., texture, cation exchange capacity) and the weather conditions which affect evapotranspiration. The conversion of NH_4^+ to NH_3 is dependent on pH, being enhanced at high pH;

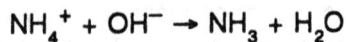

$$NH_4^+ + OH^- \rightarrow NH_3 + H_2O$$

Temporary increases in soil pH occur following the application of fertilizer urea or animal urine to soil due to the hydrolysis of the urea to NH_4^+ and this can encourage volatilization losses;

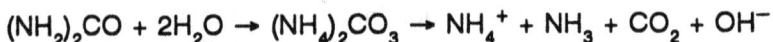

$$(NH_2)_2CO + 2H_2O \rightarrow (NH_4)_2CO_3 \rightarrow NH_4^+ + NH_3 + CO_2 + OH^-$$

Similarly losses are greater from soils with naturally high pH compared with acidic soil. Losses are accentuated by wind and high temperatures which favour drying of the soil. Ammonia losses can readily occur from wetland rice fields as the pH of the submerged soils is often near neutral and turbulence at the water-air interface readily occurs.

C. Biological Denitrification

Denitrification occurs when certain bacteria utilise the oxygen of NO_3^- as a substitute for O_2 under anaerobic conditions (Stevenson, 1986). During this process NO_3^- is sequentially reduced to N_2O and eventually N_2 by the following pathway;

$$\uparrow NO_3^- \rightarrow NO_2^- \rightarrow NO \rightarrow N_2O \rightarrow N_2 + H_2O$$

The energy for these reactions is derived from the decomposition of carbohydrates. While NO, N_2O and N_2 are all capable of being lost from the soil as gases, N_2O is the main form. It has been estimated that denitrification contributes 100-190 Tg N y^{-1} to the atmosphere as N_2O, and this amount is similar to the amount of N fixed annually (Table 5). Thus on a global scale denitrification is an important pathway for returning N to the atmosphere.

Denitrification is enhanced under conditions of high soil pH (>5), the presence of a C substrate, temperatures >5°C, high soil NO_3^- content and high soil moisture content/low O_2 availability (Aulakh et al., 1992). Thus losses form organic soils tend to be much higher than from mineral soils (>100 kg N ha^{-1} y^{-1} from organic soils compared with <10 kg N ha^{-1} y^{-1} from mineral soils; Bouwman, 1990). Losses can also be high from small localised areas of soil such as anaerobic microsites in well drained soils or following heavy rain onto soil with a high NO_3^- content. Thus losses tend to occur as pulses when conditions are right rather than as a continuous emission. A typical pattern of N_2O production from a fertilized mown sward is shown in Fig. 3. Emission occurred soon after application of the N fertilizer and was greater in July when soil temperatures were higher.

Flooded rice soil systems can favour denitrification especially if high concentrations of C and NO_3^- are present and losses of up to 50% of applied fertilizer N have been recorded (Freney, 1992). Where urea is used as a N fertilizer the N may be present as NH_4^+ rather than NO_3^- due to restricted nitrification in a flooded soil. However, rice grown on coarse textured soils may not be continuously flooded hence there can be "dry" spells when nitrification can occur followed by denitrification when the soil is flooded again. If nitrification does not occur then high concentrations of NH_4^+ can build up in the floodwater which may result in volatilization losses. Thus in flooded rice systems denitrification and volatilization can be interdependent (Freney et al., 1990).

D. Chemodenitrification

Gaseous losses on N (as N_2, NO, NO_2 and N_2O) which are independent of microbial activity can occur from soils due to chemical reactions of NO_2^- in the soil (Haynes and Sherlock, 1986). The amount of N lost in this way is thought to be in the order of 1-15 Tg N y^{-1} which is small compared with losses due to denitrification and volatilization (Table 5). Chemodenitrification is likely to be significant only when NO_2^- accumulates in the soil. Since NO_2^- is rapidly converted to NO_3^- in well aerated soils by *Nitrobacter* the accumulation of NO_2^- occurs only under specific conditions e.g. when high concentrations of NH_4^+ in the soil from fertilizer or animal urine inhibit *Nitrobacter* activity. Several reaction pathways have been proposed by which NO_2^- can lead to gaseous emission of N. The most significant of these are the decomposition of NO_2^-/HNO_2 at low soil pH and the nitrosation of various components of organic compounds like phenols and amino acids (Haynes and Sherlock, 1986).

E. Nitrification

During nitrification of NH_4^+ to NO_3^- gaseous losses of N_2O can occur (Bouwman, 1990). The exact mechanism is unclear but under field conditions nitrification and denitrification are considered to occur simultaneously and both contribute to gaseous losses of N (Haynes, 1986a).

F. Leaching Losses

Leaching can result in large losses of N from soils with the removal of >100 kg N ha^{-1} y^{-1} in some situations (Table 6). Nitrogen is leached mainly as NO_3^- due to its high solubility and low retention in the soil. While NH_4^+ can be leached in some circumstances, in most soils NH_4^+ is strongly retained by cation exchange, fixed by clay lattices and microbial immobilization and/or readily nitrified hence there is usually little NH_4^+ present in the soil that is susceptible to leaching.

Nitrate is leached by mass flow of water through the soil profile (convection), movement along a concentration gradient in the soil solution due to the uneven distribution of nitrate in the soil (diffusion) and/or mechanical mixing with the soil solution as it flows through the soil (dispersion; Cameron and Haynes, 1986). Hence NO_3^- leaching is a physical process affected by the quantity of water passing through the soil profile. Drainage occurs when there is an excess of rainfall over evapotranspiration. Thus it is more common in the winter than in the summer but can occur following excess irrigation.

Table 6.   Leaching losses of nitrate from different land use systems (Cameron and Haynes, 1986; Juergens-Gschwind, 1989).

Land use	Soil type	N applied (kg N ha^{-1})	N leached (kg N ha^{-1})
Bare fallow	Sandy loam	0	100
Bare fallow	Sand	0	144
Bare fallow	Sandy loam	0	84
		224	231
Bare fallow	Heavy loam	0	34
Grass	Sandy loam	0	7
Grass	Clay	3 x 50	5
Crop rotation	Sand	0	52
Barley	Loamy sand	0	54
		113	61
Winter rye	Sandy loam	0	61
		80	74
Corn, carrots	Sandy loam	396	155
Potatoes	Sandy soil	0	43
		80	47

The other factor affecting NO_3^- leaching is the concentration of NO_3^- in the soil profile at the time of drainage. The NO_3^- can originate from many sources including mineralization of organic matter, natural soil NO_3^-, sewage effluent and animal excreta. Of these sources mineralization of organic matter is thought to be a major source of N, particularly in cultivated soil as the incorporation of plant residues into the soil during cultivation can result in large quantities of NO_3^- being released into the soil. For example, ploughing in a 3 year old grass/legume pasture in the autumn resulted in the incorporation of 176 kg N ha^{-1} into the soil from plant residues. These residues were readily mineralized and prior to the onset of leaching in the winter soil NO_3^- levels were 139 kg N ha^{-1}. Over the following 4 months 80 kg N ha^{-1} were leached (Francis et al., 1992). In temperate climates a seasonal pattern of NO_3^- leaching is often observed with highest losses during the winter and lowest losses in summer (Fig. 4). This is due to the combination of higher drainage coinciding with significant quantities of NO_3^- in the soil profile as previously described.

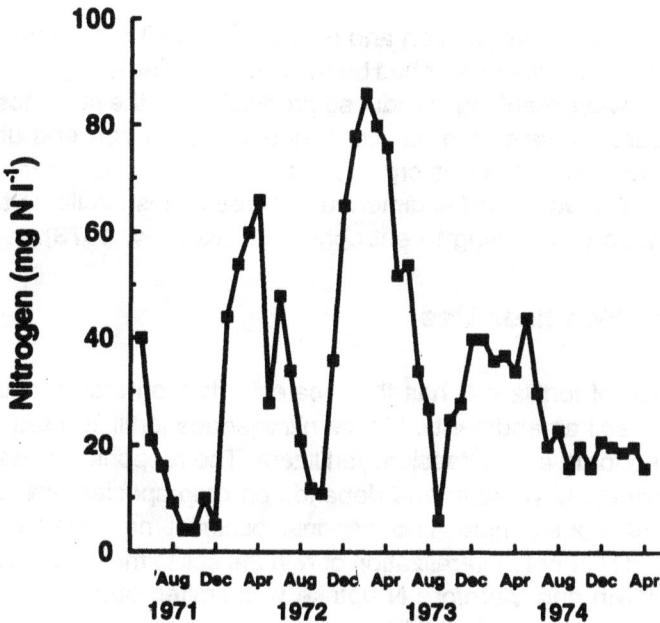

Fig. 4. Seasonal changes of nitrate concentration in percolate from a tilled corn-cropped lysimeter (Chichester, 1977).

Fertilizer N can also contribute to leaching losses particularly when the application rate is higher than crop requirements. For instance, the application of 800 kg N ha^{-1} to vegetable crops which have a requirement of 380 kg N ha^{-1} can lead to very high soil nitrate contents in the autumn after harvest followed by leaching losses of 100 kg N ha^{-1} over the winter (Wehrmann and Scharpf, 1989). Where fertilizer N is applied at rates appropriate for crop needs and at a time when the crops are actively taking up N then leaching losses are usually small (Powlson, 1988).

Leaching losses can also be affected by land use, losses being highest from bare fallow periods when no plants are present to take up NO_3^- from the soil and least from grassland systems (Table 6). However, if the grassland systems are grazed then nitrate leaching can increase due to the presence of small areas of highly concentrated N in urine patches which provide a source of leachable NO_3^- (Haynes and Williams, 1993). Spreading organic manures and wastes onto land can also result in leaching losses of N if application rates exceed plant nutrient requirements. Thus disposal of effluent from intensive animal production systems (poultry, pig and beef) can lead to NO_3^- leaching (also NH_3^- volatilization and surface run off).

G. Runoff and Erosion

The removal of natural vegetation and cultivation of soil can expose the topsoil to runoff and erosion losses of N. It has been estimated that 4.5 Tg N are lost annually by these pathways resulting in reduced productivity of the soil. Most of the erosion (80%) occurs as water erosion consequently the N can end up in surface waterways. While most of the N is organic it is readily converted to inorganic N via mineralization in the water and sediment and so becomes available to algae and aquatic plants thus contributing to eutrophication (Keeney, 1973).

VI. Nitrogen Fertilizer Use

The application of fertilizer N has the greatest effect on crop productivity compared with other soil amendments. Hence nitrogenous fertilizers are more widely used than phosphorus and potassium fertilizers. The responsiveness of crops to N fertilizer is extremely variable and depends on crop species, soil and environmental conditions. For example, if no response occurs it may mean that there was sufficient residual N or net mineralization of N in the soil to meet plant requirements or that plant growth and therefore N uptake was limited due to low soil temperatures or limited soil water availability.

The amount of N required by a particular crop depends on the species as different crops remove different amounts of N (Table 2). For instance, 100-200 kg N ha^{-1} may be removed in hay crops compared with 40 kg N ha^{-1} in cereal grains. The expected yield will also affect requirements with greater rates required for higher yields. The amount of N needed may be offset by mineralization of soil organic matter and any residual fertilizer N from previous crops. Any losses expected to occur from leaching and denitrification also need to be compensated for. Since the amount of fertilizer required is dependent on so many factors, most of which are environmentally controlled, it is not surprising that the ability to predict plant N requirements is unreliable. Often fertilizer recommendations are made simply by taking into account previous cropping history and residual N in the soil profile.

Fertilizer recommendations would be greatly improved if it were possible to predict soil N availability throughout the growing season. A number of tests have been proposed to do this. They vary from measurements of mineral N in the soil profile to incubation and chemical extractions which attempt to predict potentially mineralizable N. As yet no widely acceptable test has been developed to predict N mineralization. Some progress, however, has been made towards developing mathematical models to predict mineralization for a particular site (e.g., Scholefield et al., 1991).

The application of excessive amounts of fertilizer N is undesirable. Excess N may be lost thereby causing an economic loss of N and environmental problems. As well excess N in the soil can affect the quality of the crops, result in high NO_3^- concentrations that affect human and animal health, cause lodging in cereals which may affect harvesting and quality and increase disease susceptibility.

Table 7. Chemical formula and nitrogen content of common fertilizers (Goh and Haynes, 1986; Mengel and Kirkby, 1987).

Fertilizer material	Chemical formula	% N
Anhydrous ammonia	NH_3	82
Aqueous ammonia	NH_3, NH_4OH	24
Ammonium sulphate	$(NH_4)_2SO_4$	21
Ammonium chloride	NH_4Cl	26
Ammonium nitrate	NH_4NO_3	35
Ammonium nitrate sulphate	$NH_4NO_3 \cdot (NH_4)_2SO_4$	26
Calcium ammonium nitrate	$NH_4NO_3 + CaCO_3$	21
Calcium cyanide	$CaCN_2$	21
Diammonium phosphate	$(NH_4)_2HPO_4$	18
Isobutylidene di-urea (IBDU)	$(CH_3)_2C_2H_2(NHCONH_2)_2$	32
Potassium nitrate	KNO_3	14
Sodium nitrate	$NaNO_3$	16
Urea	$CO(NH_2)_2$	46
Blood and bone		7
Cattle dung		2

There are a range of N fertilizers available and these supply N either as NO_3^- or NH_3/NH_4^+ (Table 7). Urea is becoming the major fertilizer used due to its high N content and ease of handling. Ammonium sulphate is also a popular choice if sulphur is required. In some countries organic N sources such as animal manure are used as fertilizers. These generally have a lower N content (<10%) and are slower to release N for plant use compared to manufactured fertilizer. In recent years slow release fertilizers have been developed in an effort to match N supply with crop requirements and reduce losses. These products either have limited solubility (e.g. IBDU) or are made by coating the fertilizer particles with resin or S. The use of urease and nitrification inhibitors has also been shown to be beneficial under experimental conditions (Keeney, 1982).

Literature Cited

Aulakh, M.S., Doran, J.W., and Mosier, A.R. 1992. Soil denitrification - significance, measurement and effects of management. *Adv. Soil Sc.* **18**: 1-57.

Bouwman, A.F. 1990. *Soils and the Greenhouse Effect.* Wiley, Chichester. 575 p.

Cameron, K.C., and Haynes, R.J. 1986. Retention and movement of nitrogen in soils. *In Mineral Nitrogen in the Plant-Soil System.* (Haynes R.J. ed), Academic Press, Florida. pp. 166-241.

Chichester, F.W. 1977. Effects of increased fertilizer rates on nitrogen content of runoff and percolate from monolith lysimeters. *J. Environ. Qual.* **6**: 211-217.

Evans, H.J., and Barber, L.E. 1977. Biological nitrogen fixation for food and fibre production. *Science* **197**: 332-339.

FAO. 1989. *Food and Agriculture Organisation. Fertilizer Yearbook* **39**.

Fox, R.H., Myers, R.J.K., and Vallis, I. 1990. The nitrogen mineralization rate of legume residues in soil as influenced by their polyphenol, lignin and nitrogen contents. *Plant Soil* **129**: 251-259.

Francis, G.S., Haynes, R.J., Sparling, G.P., Ross, D.J., and Williams, P.H. 1992. Nitrogen mineralization, nitrate leaching and crop growth following cultivation of a temporary leguminous pasture in autumn and winter. *Fertilizer Research*: **33**: 59-70.

Freney, J.R. 1992. Transformations and transfers of nitrogen in the plant-soil system. *In Proceedings of the International Symposium on Nutrient Management for Sustained Productivity,* Department of Soils, Punjab Agricultural University, Ludhiana, India. pp. 334-343.

Freney, J.R., Trevitt, A.C.F., De Datta, S.K., Obcemea, W.N., and Real, J.G. 1990. The interdependence of ammonia volatilization and denitrification as nitrogen loss processes in flooded rice fields in the Philippines. *Bio. Fert. Soils* **9**: 31-36.

Goh, K.M., and Haynes, R.J. 1986. Nitrogen and agronomic practice. *In Mineral Nitrogen in the Plant-Soil System.* (Haynes, R.J. ed.). Academic Press, Florida. pp. 379-468.

Goswami, N.N., Prasad, R., Sarkar, M.C., and Singh, S. 1988. Studies on the effect of green manuring in nitrogen economy in a rice-wheat rotation using a ^{15}N technique. *J. Agric. Sci. Camb.* **111**: 413-417.

Hauck, R.D. 1981. Nitrogen fertilizer effects on nitrogen cycle processes. *In Terrestrial Nitrogen Cycles.* (Clark, F.E. and Ruswall T., eds) Ecological Bulletins, Stockholm. pp. 551-562.

Havelka, U.D., and Boyle, M.G., Hardy, R.W.F. 1982. Biological nitrogen fixation. *In Nitrogen in Agricultural Soils.* (Stevenson, F.J. ed.). American Society of Agronomy, Madison, Wisconsin, pp. 365-422.

Haynes, R.J. 1986a. Origin, distribution, and cycling of nitrogen in terrestrial ecosystems. *In Mineral Nitrogen in the Plant-Soil System*. (Haynes, R.J. ed.) Academic Press, Florida. pp. 1-51.

Haynes, R.J. 1986b. The decomposition process: mineralization, immobilization, humus formation, and degradation. *In Mineral Nitrogen in the Plant-Soil System*. (Haynes, R.J. ed.), Academic Press, Florida. pp. 52-126.

Haynes, R.J., and Sherlock, R.R. 1986. Gaseous losses of nitrogen. *Mineral Nitrogen in the Plant-Soil System*. (Haynes, R.J. ed.) Academic Press, Florida. pp. 242-302.

Haynes, R.J., and Williams, P.H. 1993. Nutrient cycling and soil fertility in the grazed pasture ecosystem. *Adv. Agron.* **49**: 119-199.

Haynes, R.J., Martin, R.J., and Goh, K.M. 1993. Nitrogen fixation, accumulation of soil N and N balance for some field-grown legume crops. *Field Crops:* **35**: 85-92.

Hesterman, O.B., Russelle, M.P., Sheaffer, C.C., and Heichel, G.H. 1987. Nitrogen utilization from fertilizer and legume residues in legume-corn rotations. *Agron. J.* **79**: 726-731.

Hoglund, J.H., and Brock, J.L. 1987. Nitrogen fixation in managed grasslands. *In: Managed Grasslands*. (Snaydon, R.W. ed.) Elsevier, Oxford. pp. 187-196.

Jenny, H. 1961. Derivation of state factor equations of soils and ecosystems. *Soil Sci. Soc. America Proc.* **25**: 385-388.

Juergens-Gschwind, S. 1989. Ground water nitrates in other developed countries (Europe)- relationships to land use patterns. *In Nitrogen Management and Groundwater Protection.* (Follett, R.F., ed.) Elsevier, Amsterdam. pp. 75-138.

Keeney, D.R. 1973. The nitrogen cycle in sediment-water systems. *J. Environ. Qual.* **2**: 15-29.

Keeney, D.R. 1982. Nitrogen management for maximum efficiency and minimum pollution. *In Nitrogen in Agricultural Soils*. (Stevenson, F.J. ed.) American Society of Agronomy, Madison, Wisconsin. pp. 605-649.

Ladd, J.N., and Amado, M. 1986. The fate of nitrogen from legume and fertilizer sources in soils successively cropped with wheat under field conditions. *Soil Biol. Biochem.* **18**: 417-425.

Mengel, K., and Kirkby, E.A. 1987. *Principles of Plant Nutrition*. International Potash Institute, Bern.

Nommik, H., and Vahtras, K. 1982. Retention and fixation of ammonium and ammonia in soils. *In Nitrogen in Agricultural Soils*. (Stevenson, F.J., ed.), American Society of Agronomy, Madison, Wisconsin. pp. 123-171.

Powlson, D. 1988. Measuring and minimising losses of fertilizer nitrogen in arable agriculture. *In Nitrogen Efficiency in Agricultural Soils.* (Jenkinson, D.S., and Smith, K.A. eds.), Elsevier, London. pp. 231-245.

Prasad, R., and Power, J.F. 1992. Crop residue management. *Adv. Soil Sc.* **18**: 205-251.

Preston, C.M. 1982. The availability of residual fertilizer nitrogen immobilized as clayfixed ammonium and organic N. *Can. J. Soil Sci.* **62**; 479-486.

Ryden, J.C. 1981. N_2O exchange between a grassland soil and the atmosphere. *Nature* **292**: 235-237.

Scholefield, D., Lockyer, D.R., Whitehead, D.C., and Tyson K.C. 1991. A model to predict transformations and losses of nitrogen in UK pastures grazed by beef cattle. *Plant Soil*. **132**: 165-177.

Steele, K.W. 1982. Nitrogen in grassland soils. *In Nitrogen Fertilisers in New Zealand Agriculture.* (Lynch, P.B., ed.), New Zealand Institute of Agricultural Science, Wellington, New Zealand. pp. 29-44.

Stevenson. F.W. 1982. Origin and distribution of nitrogen in soils. *In Nitrogen in Agricultural Soils.* (Stevenson F.J., ed.), American Society of Agronomy. Madison, Wisconsin. pp. 1-42.

Stevenson, F.W. 1986. *Cycles of Soil C, N, P, S and micronutrients*. John Wiley, New York. p. 380.

Wehrmann, J., and Scharpf, H.C. 1989. Reduction of nitrate leaching in a vegetable farm-fertilization, crop rotation, plant residues. *In Nitrogen Management and Groundwater Protection.* (Follett R.F., ed), Elsevier, Amsterdam. pp. 147-157.

Williams, P.H. 1992. The role of fertilizers in environmental pollution. *In Proceedings of the International Symposium on Nutrient Management for Sustained Productivity,* Department of Soils, Punjab Agricultural University, Ludhiana, India. pp. 195-215.

Nitrogen Nutrition in Higher Plants, 1995
Editors : H.S. Srivastava & R.P. Singh
Associated Publishing Co., New Delhi, India
pp. 21-56.

Nitrogen Absorption by Plant Roots

A.D.M. GLASS and M.Y. SIDDIQI

Abbreviation: N, nitrogen; LATS, low affinity transport systems; HATS, high affinity transport systems.

I. Introduction

Elemental analyses of plant tissues typically reveal that nitrogen comprises 2-5% of dry weight. In quantitative terms, therefore, after carbon, hydrogen and oxygen, N is the most important of the essential elements. Indeed, it most commonly limits plant growth under natural conditions on land or in water. Using the more conservative figure of 2% N by dry weight, a corn crop yielding e.g. 10 tonnes of grain ha^{-1} should remove at least 200 kg N ha^{-1} from soil solution simply to satisfy the requirements of the grain. In fact, maximum grain yields for corn exceed 20 tonnes ha^{-1}. In order to sustain such demands on the N reserves of agricultural soils, rates of N application should at least match these rates of withdrawal by crop species. On a global basis rates of N application are currently estimated to be 10^{11} kg per annum. The U.S. alone consumes roughly 10% of this fertilizer at an annual cost of around 3 billion dollars.

Considering the above, it is not surprising that the absorption of N by plant roots and its assimilation into amino acids and proteins has been the subject of intensive investigation for a number of years. During this time there have been regular reviews of this subject (Haynes and Goh, 1978; Kleiner, 1981; Clarkson, 1986; Bloom, 1988; Glass, 1988; Clarkson and Lüttge, 1991). The present review will focus principally on soil systems and root absorption of N. The emphasis will be on the kinetics, energetics and regulation of N uptake.

Limitations of space and time do not allow us to examine many important and exciting whole plant phenomena such as the heterogeneity of N uptake along roots (Lazoff et al., 1992; Henriksen et al., 1992), the importance of photosynthesis (Rufty et al., 1984) and photoperiod (Raper et al., 1991) and the role of mycorrhizae in NO_3^- and NH_4^+ uptake (Johansen et al., 1993).

Many reviews of the topic of N absorption have begun by stating that nitrate is the preferred N source for plants. Yet it is well known that for many plant species, including those of the forest and the Arctic tundra, this is simply not the case. Even if we limit our discussion to agricultural species, it is commonly observed that rates of ammonium absorption are typically higher than those of nitrate from equimolar concentration (Bloom, 1988; Macklon et al., 1990), especially at low $[N]_o$ (Bloom, 1988). Moreover, in energetic terms, NH_4^+ is less costly to absorb and to assimilate than NO_3^-, and since Q_{10} values for NO_3^- uptake are commonly significantly higher than those for NH_4^+, uptake of the latter is significantly more important at lower temperatures. In many species NO_3^- uptake is rapidly and reversibly inhibited by the presence of NH_4^+ in the external medium (Haynes and Goh, 1978; Glass et al., 1985; Lee and Drew, 1989), but even more interesting is that long term growth rates and total N uptake are typically enhanced by growth on equimolar mixture of NO_3^- and NH_4^+, compared to growth on a single N source (Minotti et al., 1969; Cox and Reisenauer, 1973; Zhang and Mackown, 1993). Thus, the question of "preferred N source" may require much more careful definition with due consideration to species and environmental conditions. Another important question we wish to address in this review is the importance of amino acids as a direct source of N from soils.

II. Availability of Nitrogen to Plant Roots in Soils

In soil solution, it is generally agreed that NO_3^- and NH_4^+ represent the most readily available forms of N for root absorption; the former typically predominating in well aerated agricultural soils at modest temperature and at neutral pH. By contrast in forest soils or in Arctic tundra or even in agricultural soils during colder seasons of the year, NH_4^+ may assume significantly greater importance. Estimates of the concentrations of NO_3^- and NH_4^+ in soil solutions indicate that both of these N species may be present in the 0.1 – 10 mM range in agricultural soils (Reisenauer, 1966; Barber, 1974; Nye and Tinker, 1977). However, the variation among natural soils is considerable, ranging over 3-4 orders of magnitude (Jackson and Caldwell, 1993). In part this is due to potent influences of environmental variables e.g. temperature, pH, and water status, on components of the N cycle (Haynes and Goh, 1978; Bloom, 1988). The concentration of NH_4^+ tends to be buffered by the adsorption of this ion to negatively charged soil particles. By contrast, NO_3^- is highly mobile in soils and can readily be leached.

Nevertheless, the major part of soil N is present in organic form, of which 20 to 50% may be amino N (Barber, 1974). Significant quantities of this amino N are in the form of amino acids which can readily be released from the organic fraction into soil solution. Concentrations of various amino acids are typically in the μ molar range, depending upon soil type, but total amino acid concentration may exceed that of NH_4^+ nitrogen in some soils (Schobert and Komor, 1987; Monreal and McGill, 1989; Kielland, 1993). In such cases, therefore amino acids could represent a significant source of nitrogen for plant uptake.

The uptake of NO_2^- by plant roots is probably of limited significance in nature (Lee, 1979). Nevertheless, NO_2^- uptake has been the subject of much interest (Lee, 1979; Breteler and Luczak, 1982) because this ion appears to share a common transport system with NO_3^- (but see also Breteler and Luczak, 1982; Cordoba et al., 1986). Aslam et al. (1993) have recently completed a comprehensive study of this topic in barley, concluding that either NO_3^- or NO_2^- were capable of inducing the transporters for NO_3^- and NO_2^- and their corresponding reductases. In the opinion of the latter group, NO_2^- and NO_3^- share a common transport system. In the remainder of this review the main focus of attention will therefore be upon the absorption of NO_3^-, NH_4^+ and amino acids.

III. The Kinetics of Nitrate and Ammonium Uptake

Despite the spectacular advances in our understanding of ion transport processes which have emerged from the application of biochemical, biophysical and molecular techniques to plant systems, the principal evidence for the existence of specific transport systems for NO_3^- and NH_4^+ in plant roots remains kinetic; in particular the patterns of concentration-dependence.

At this point in the review it is convenient to define a limited number of terms in order to simplify subsequent discussions. The term induction is used to describe the process involved in the up-regulation of NO_3^- uptake observed in the period following first exposures to NO_3^- or NO_2^--containing solutions. Constitutive transport systems are those which are expressed without the necessity for induction and appear to be insensitive to repression. The term negative feedback is used to denote the down-regulation of N uptake which occurs when N uptake exceeds assimilation, leading to significant accumulation of N. The use of this term does not imply a specific mechanism e.g. direct negative feedback inhibition as opposed to transcriptional repression.

The earlier literature (Van den Honert and Hooymans, 1955; Lycklama, 1963; Neyra and Hageman, 1975; Rao and Rains, 1976; Pilbeam and Kirkby, 1990) established that net uptake could be described by Michaelis-Menten kinetics with K_m values typically in the range from ~10 to ~150 µM for roots of cereal crops. Thus, Van den Honert and Hooymans (1955), Lycklama (1963), and Rao and Rains (1976) established the cases for NO_3^- uptake, and Becking (1956) and Lycklama (1963) for NH_4^+ uptake. At the same time Rao and Rains (1976) and Neyra and Hageman (1975) recognized that the NO_3^- uptake data could not be resolved by a single Michaelis-Menten isotherm.

In recent years, the development of automated systems (Goyal and Huffaker, 1986; Bloom, 1989) has considerably improved techniques for measuring net uptake of N by plant roots. The use of various isotopes (^{15}N, and ^{13}N) as well as analogues of NO_3^- and NH_4^+ ($^{36}ClO_3^-$ and $^{14}CH_3NH_3^+$, respectively) has enabled researchers to dissect uptake into its component processes of influx and efflux, as well as subsequent events such as assimilation and translocation to the shoot. Moreover, the considerably increased sensitivity provided by isotopes has allowed short-term influx measurements to be made even at high external ion concentration. Although extensive use has been made of $^{36}ClO_3^-$ and $^{14}CH_3NH_3^+$ in the past, their use should be approached with caution because in some cases they have proven not to be entirely faithful analogues.

At least 3 transport systems for NO_3^- have been identified by kinetic measurements in plant roots. Two of these display saturable kinetics. One is low capacity constitutive system (Lee and Drew, 1986; Behl et al., 1988; Aslam et al., 1992), the other a high capacity inducible system. In addition, a non-saturating low affinity, high capacity system becomes apparent only at higher external NO_3^- concentrations (Siddiqi et al., 1990 and references therein). According to Breteler and Nissen (1982), NO_3^- uptake by *Phaseolus* was mediated by a single (multiphasic) transport system which underwent 4 concentration-dependent phase changes in the concentration range from 0.004 to 5 mM $[NO_3^-]_o$. Recently, H. Kronzucker (unpublished) has observed two saturable phases for influx of $^{13}NO_3^-$ and $^{13}NH_4^+$ in the concentration range from 2.5 to 350 µM in spruce seedlings.

For NH_4^+ uptake, at least two systems have been characterized. These are saturable and non-saturable operating at low and high external $[NH_4^+]$, respectively (Ulrich et al., 1984; M.Y. Wang unpublished). In addition, net uptake of NH_4^+ by soybean roots was considered to be multiphasic (Joseph et al., 1975), with K_m values of 0.6, 1.9 and 3.5×10^{-4} M.

A. High Affinity Nitrate Transporters

At low external nitrate concentrations a high affinity high capacity transport system is induced only after plants have been exposed to NO_3^-. For example Behl et al. (1988), reported that net NO_3^- uptake by barley roots grown without NO_3^- was ~1 μmol g^{-1} f.w. h^{-1} on initial exposure to NO_3^-. Net uptake of K^+ by these plants was roughly 10 times the NO_3^- flux. By 4 h after initial exposure to NO_3^-, net NO_3^- uptake had increased to a rate comparable with K^+ uptake. Behl et al., (1988) suggested that the initial low rate of uptake was mediated by a "constitutive low capacity" carrier system. Clearly, even though cytoplasmic $[NO_3^-]$ is extremely low in NO_3^--starved roots, the large electrical component of $\Delta\mu NO_3^-$ would preclude a diffusive entry of NO_3^- unless external NO_3^- is raised above 1 mM. At the very least, cytoplasmic $[NO_3^-]$ would rise rapidly following exposure to external NO_3^- thus halting a diffusive entry of NO_3^-. To take a specific example, with cytoplasmic $[NO_3^-]$ set at even 1 μM, with external NO_3^- at 1 mM and an electrical potential difference of ~200 mV, predictions of the Nernst equation indicate that NO_3^- entry is energetically uphill, with an energy requirement of 2.5 kJ mol^{-1}. Thus a constitutive low capacity carrier system appears to be necessary in order to actively transport NO_3^- into the roots; this inturn leads to induction of the high capacity system. Without having characterized the constitutive system, Behl et al. (1988) were unable to evaluate whether this transport system was different in character from the induced system. The simplest explanation of the data would be that the constitutive system represented a low level expression of the same gene product responsible for the induced transport. Siddiqi et al. (1990) referred to this as a 'genetic leak'. Lee and Drew (1986) used $^{13}NO_3^-$ to characterize NO_3^- influx in barley plants grown with NH_4^+ as sole source of N and then N-starved for 2 days. This study revealed that the constitutive transport system had a significantly lower K_m (~7 μM) than the induced system (K_m = 13 μM). On this basis, Lee and Drew (1986) suggested that the two transport systems were distinct. Recently, Aslam et al. (1992) measured net NO_3^- uptake in uninduced and induced barley plants using HPLC. They too observed different K_m values, 7 μM and ~35 μM, respectively for the two transport systems and distinguished between the constitutive and inducible systems by means of a Lineweaver Burk plot, which was bimodal. The change of slope occurred at a $[NO_3^-]$ ~10 μM. Such a finding, it was argued, was inconsistent with the proposal of a single carrier system expressed in greater abundance following induction.

Using $^{13}NO_3^-$ to measure NO_3^- influx in barley roots, Lee and Drew (1986), Hole et al. (1990) and Siddiqi et al. (1990) observed significant changes in V_{max} and K_m when external N sources were manipulated. V_{max} values for induced plants increased substantially when NO_3^- availability was reduced, but in uninduced plants, V_{max} values were quite low. Siddiqi et al. (1990) reported V_{max} values of 0.344 μmol g^{-1} h^{-1} and 9.41 μmol $g^{-1}h^{-1}$ respectively, for barley plants grown without NO_3^- or exposed to 100 μM NO_3^- for 24 h. Corresponding values in Lee and Drew (1986) for plants grown without N and with 1.5 mM NO_3^- were 1.3 and

2.8, respectively. The differences in V_{max} appear to correspond to inherent differences in the constitutive system among different barley varieties (Siddiqi et al., unpublished) and the duration/concentration of NO_3^- exposure (Siddiqi et al., 1989). In the experiments reported by Siddiqi et al. (1990) the V_{max} for influx declined to 3.63 and 1.69 μmol g^{-1}h^{-1}, respectively, after 4 days exposure to 100 μM NO_3^- and 24 h in 1 mM NO_3^-.

K_m values of uninduced plants were also different from those of induced plants (Lee and Drew, 1986), increasing from 7 (Lee and Drew, 1986), 7 (Aslam et al., 1992) and 20 μM (Siddiqi et al., 1990) to 13, 35 and 79 μM respectively, following NO_3^- provision. Notwithstanding this initial increase associated with NO_3^- provision, Siddiqi et al. (1990) observed that K_m values declined from 79, through 45 to 30 μM as tissue $[NO_3^-]$ increased.

B. Low Affinity Nitrate Transporters

At concentrations beyond the range of saturation for the inducible HATS, NO_3^- uptake appears to increase linearly with increasing $[NO_3^-]_o$ (see Siddiqi et al., 1990 for references), with no indication of saturation even at 100 mM (Omata et al., 1989). In the diatom *Skeletonema* the linear phase became apparent at concentrations as low as 10 μM (Serra et al., 1978). On the basis of the effects of cycloheximide and p-hydroxymercuribenzoate, which abolished the saturable phase of uptake, but not the linear phase, the authors concluded that this linear phase was diffusion-mediated. Using $^{13}NO_3^-$, Colios et al. (1992) also observed linear kinetics at quite low $[NO_3^-]_o$ in two marine diatoms. Clearly, unless these organisms have no $\Delta\psi$ across the plasmalemma, it is inconceivable that NO_3^- uptake from such low $[NO_3^-]_o$ could be diffusive (see Section IV). The observations of linear kinetics in single-celled organisms indicate that it is not necessary to invoke a heterogeneous tissue organization to rationalize the (LATS). Indeed the observation that 5 mM NO_3^- strongly depolarized $\Delta\psi$ in roots of uninduced barley (see below) clearly identifies the site of the LATS at the plasmalemma of epidermal and cortical cells (Glass et al., 1992).

The linear concentration dependence of the LATS has been reported in a wide range of organisms (Glass, 1988; Siddiqi et al., 1990). In their early discussions of the LATS for NO_3^- in barley, Siddiqi et al. (1990; 1991) referred to this system as constitutive. However, there are clear indications from long-term studies (Clement et al., 1978) that this system is subject to negative feedback (see Section VI of this Chapter) and hence it is inappropriate to use the term constitutive. Rather the LATS appears to be repressible but not subject to induction.

A final word regarding the much discussed multiphasic patterns of ion uptake; two observations arising from electrophysiological and molecular studies, respectively, deserve consideration. In uninduced barley, exposure to 100 μM NO_3^- caused no depolarization of $\Delta\psi$ even though this concentration was strongly depolarizing in induced plants. However, on first exposure to 5 mM NO_3^- there was a large depolarization of $\Delta\psi$ (Glass et al., 1992). In Omata's study of a genetically engineered mutant of *Synechococcus* (Omata et al., 1989), rendered defective in the HATS, there was no NO_3^- uptake <1 mM. Yet between 1 and 100 mM there was substantial NO_3^- uptake. These two observations appear to be incompatible with a single transporter functioning differentially at low and high $[NO_3^-]_o$.

C. High Affinity Ammonium Transporters

At low external $[NH_4^+]$ (<1mM) plant roots possess a saturable high affinity mechanism to transport NH_4^+ across the plasma membrane. First characterized kinetically by Becking (1956) in maize roots, its existence has subsequently been demonstrated in a wide range of species. These include barley (Bloom and Chapin, 1981), wheat (Goyal and Huffaker, 1993b), ryegrass (Lycklama, 1963), rice (M.Y. Wang et al., 1993b), soybean (Joseph et al., 1975), tomato (Smart and Bloom, 1988), *Chara* (Deane - Drummond, 1984b), *Chlorella* (Schlee and Komor, 1986) and *Lemna* (Ullrich et al., 1984). K_m values for this transporter range from ~14 to 167 μM. Earlier studies typically measured net absorption of $^{14}NH_4^+$ by depletion of uptake solutions, but more recently $^{15}NH_4^+$ uptake and $^{13}NH_4^+$ influx have seen wider use, as well as ^{14}C-methylamine, an analogue of NH_4^+ (Deane - Drummond, 1984b). In place of a single saturable transporter, a multiphasic system for NH_4^+ uptake, with K_m values of 0.6, 1.9 and 3.5 x 10^{-4} M was proposed for soybean roots (Joseph et al., 1975).

Genetic evidence for the existence of a saturable HATS capable of accumulating high concentrations of NH_4^+ from low external $[NH_4^+]_o$ has been provided by the isolation of bacterial, fungal and algal mutants (Kleiner, 1981; Jayakumar et al., 1981; Franco et al., 1987). These appear to be incapable of surviving at low $[NH_4^+]_o$ and are tolerant of concentrations of methylamine which normally prove toxic. In *E. coli* a structural gene (amt A) has been cloned by complementation and a 27 kD peripheral protein, responsible for NH_4^+ uptake, proposed (Fabiny et al., 1991).

Ammonium uptake by this HATS has been demonstrated to be strongly depolarizing, consistent with NH_4^+ (rather than NH_3) transport across the plasmalemma (Ullrich et al., 1984; Wang, et al., 1993b). Likewise, several studies have failed to confirm the increase of N uptake anticipated of NH_3 entry at elevated pH (e.g. see also Glass, 1988 for references). Nevertheless, there are a limited number of cases (Lycklama, 1963) in which N uptake increased dramatically above pH 7, suggestive of NH_3 uptake.

A commonly observed phenomenon associated with the introduction of NH_4^+ into growth media has been the inhibition of NO_3^- uptake. In some cases e.g. barley (Lee and Drew, 1989) this is apparent within 3 min. Removal of NH_4^+ caused NO_3^- uptake to be restored as quickly. The mechanism of this effect is unclear. At least in barley, Lee and Drew (1989) have examined this phenomenon intensively and provide evidence that the site of action of NH_4^+ is at the plasmalemma. The inhibition increased approximately as the log of $[NH_4]_o$, leading to the suggestion that an electrical effect e.g. membrane depolarization, as proposed earlier by Ullrich et al. (1984), was responsible. Lee and Drew however, point to the lack of agreement between the concentration-dependence of membrane depolarization and inhibition of NO_3^- uptake. Also, the lack of a similar inhibitory effect of K^+, which also strongly depolarizes, $\Delta\psi$ appears to make the above hypothesis untenable. Despite the rapidity of this inhibitory effect of NH_4^+ in barley and other species, in corn, the inhibitory effect was not significant until at least 2 h had elapsed (Mackown et al., 1982). This appears to indicate that a direct effect of NH_4^+ at the plasma membrane is not universal. Based upon the apparent lack of effect of NH_4^+ on $^{36}ClO_3^-$-influx in barley, it was suggested by Deane-Drummond and Glass (1983) that NH_4^+ effects on NO_3^- uptake resulted from a stimulated efflux. Direct measurements of NO_3^- efflux using ^{13}N have failed to confirm this (see detailed discussion in Lee and Drew, 1989). In addition to the $NH_4^+:NO_3^-$ interactions, NH_4^+ has commonly been found to inhibit K^+ uptake (Deane - Drummond and Glass, 1983; Scherer et al., 1984).

D. Low Affinity Ammonium Transport Systems

Beyond ~0.5 mM $[NH_4^+]_o$ net uptake of NH_4^+ and $^{13}NH_4^+$ influx have been demonstrated to conform to linear kinetics (Ullrich et al., 1984; Wang et al., 1993b). According to Ullrich et al. (1984) this LATS mediated diffusive flux of NH_4^+, or even NH_3. The use of $^{13}NH_4^+$ to measure influx in rice has demonstrated that the LATS is energy-dependent but passive, while the pH profiles are of odds with NH_3 fluxes, even at elevated concentrations (Wang et al., 1993b). Likewise the strong depolarization of $\Delta\psi$ by NH_4^+ at high $[NH_4^+]_o$ argues against NH_3 entry. Nevertheless, in *Lemna* depolarization of $\Delta\psi$ due to $[NH_4^+]_o$ was saturated at ~0.5 mM NH_4^+, corresponding to saturation of the HATS (Ullrich et al., 1984). Despite a linear uptake of N at higher concentrations of NH_4^+ there was no further depolarization of $\Delta\psi$. This observation is consistent with NH_3 uptake at high $[NH_4^+]_o$ in this organism. As discussed below, the K^+ channel (KAT1) is also permeable to NH_4^+ (Schachtman et al., 1992). Perhaps this channel is the pathway for NH_4^+ entry through the LATS.

IV. Energetics: Thermodynamic and Mechanistic Considerations of Nitrate and Ammonium Fluxes

A. Nitrate Fluxes Across the Plasmalemma

The results of early experiments with various metabolic inhibitors, anoxia or low temperature (Lycklama, 1963; Rao and Rains, 1976; Clarkson and Warner, 1979; see also Haynes and Goh, 1978; Clarkson, 1986; Glass, 1988; Bloom, 1988 for reviews) established that NO_3^- uptake by plant roots was an energy-dependent process. More recently, the use of metabolic inhibitors as well as Q_{10} determinations for $^{13}NO_3^-$ influx (see Siddiqi et al., 1991 and references therein) in barley have confirmed this conclusion. In a study of root respiration, Bloom et al. (1992) estimated that NO_3^- absorption by barley roots consumed 5% of the energy derivedfrom respiration. A further 15% and 3%, respectively, were expended on the reduction of $[NO_3^-]_o$ and NH_4^+ assimilation. These observations are all consistent with an energy-dependent absorption of $[NO_3^-]_o$. However, a definitive analysis of the nature of the fluxes between external solution and the cytosol and between various other compartments (e.g. cytosol/vacuole, cytosol/stele) demands as a minimum that the nitrate concentrations of these compartments be known, as well as the extent of their electrical potential differences ($\Delta\psi$). As early as 1967, Higinbotham et al. (1967) concluded that NO_3^- absorption was an active process based upon their analysis of average tissue $[NO_3^-]$ and $\Delta\psi$ values of oat and pea roots. The availability of a single estimate of cytoplasmic $[NO_3^-]$ ($[NO_3^-]_c$) derived from compartmental analysis of barley roots using $^{13}NO_3^-$ (Lee and Clarkson, 1986) enabled Glass (1988) to calculate $\Delta\mu NO_3^-$ to be ~17 kJ mol^{-1} when external $[NO_3]$ ($[NO_3^-]_o$) was 10 mM and $\Delta\psi$ was - 150 mV. Clarkson and Hanson (1980) used this same estimate of $[NO_3^-]_c$ in barley roots to predict values of 28 and 9 kJ mol^{-1}, respectively for NO_3^- uptake in ryegrass from $[NO_3^-]_o$ of 1.48 μM and 14.8 mM. The value assumed for $\Delta\psi$ was -75 mV. Since that time estimates of $[NO_3^-]_c$ in roots have been obtained by several methods including compartmental analysis using ^{13}N and ^{15}N (Macklon et al., 1990; Siddiqi et al., 1991), ion specific micro-electrodes (Miller and Zhen, 1991), and *in vivo* nitrate reductase rates (King et al., 1992). All of these methods provided estimates in the milli molar concentration range. The study by Siddiqi et al. (1991) showed that $[NO_3^-]_c$ could vary from ~1 mM to 36.5 mM in Klondike barley depending on $[NO_3^-]_o$. Using the same barley variety, Glass et al. (1992) recorded $\Delta\psi$ values in epidermal and cortical cells in the range from -170 to -246 mV in dilute (1/80th strength) inorganic nutrient solution. In $CaSO_4$ solution $\Delta\psi$ values as low as -300 mV were recorded. Thus it is now possible in the same cultivar to determine $\Delta\mu NO_3^-$ with some certainty under different conditions of $[NO_3^-]_o$ supply. Using a quite conservative estimate of $[NO_3^-]_c$ at 5 mM, and varying $\Delta\psi$ between -50 and -200 mV, with $[NO_3^-]_o$ set at values ranging from 10^{-6} to 10^{-1} M, calculations using the Nernst equation indicate that only at an unreasonably high value for $\Delta\psi$ (-50 mV) with $[NO_3^-]_o$ set at 100 mM can the entry of $[NO_3^-]_o$ be passive. This finding clearly contradicts earlier suggestions that the LATS for NO_3^- might be channel-mediated (Siddiqi et al., 1990). In addition it makes untenable the interpretation of the linear LATS as due to diffusion (Omata et al., 1989; Clarkson and Lüttge, 1991; Wieneke, 1992).

Electrophysiological evidence from studies with *Lemna*, *Zea* and *Hordeum* indicates that NO_3^- uptake across the plasma membrane occurs by means of proton/nitrate symporters (Ullrich and Novacky, 1981; McClaure et al., 1990; Glass et al., 1992). Energetic considerations (Glass, 1988) led to the conclusion that there was ample free energy associated with the proton gradient to drive NO_3^- through a $2H^+/1NO_3^-$ symport into the cytosol (with $[NO_3^-]_c$ set at 26 mM and $\Delta\psi$ at —150mV) except at high pH (>7) and low external $[NO_3^-]$. These calculations were made using a constant value for $[NO_3^-]_c$. The demonstration that $[NO_3^-]_c$ declined significantly in barley roots as $[NO_3^-]_o$ declined from 1.0 to 0.01 mM (Siddiqi et al., 1991) indicates that at low $[NO_3^-]_o$, the gradient against which NO_3^- must be transported is less than originally predicted. Hence the available proton gradient is competent to drive uphill transport of NO_3^- over an even greater range of conditions.

As discussed above, initial suggestions that the LATS might be downhill and channel mediated even in cells actively involved in NO_3^- reduction (Siddiqi et al., 1991) were contradicted by estimates of $[NO_3^-]_c$ by *in vivo* nitrate reductase assays (King et al., 1992) and by the demonstration that NO_3^- uptake by the LATS is electrically depolarizing (Glass et al., 1992). This was apparent in barley roots which had experienced no previous exposure to NO_3^- to ensure that depolarization was not due to the inducible HATS. These observations are most simply interpreted to indicate that both the HATS and the LATS are powered by proton/NO_3^- symports with minimum stoichiometries of $2H^+$: $1\ NO_3^-$.

Considering the above, it follows that efflux of NO_3^- across the plasmalemma should be downhill and hence might occur via NO_3^--specific channels. Despite frequent reference in the literature to NO_3^- as a highly permeant anion, several studies have demonstrated that $^{13}NO_3^-$ efflux is virtually zero in intact roots (Ingemarsson et al., 1987a; Wieneke, 1992), even when these have been grown at $[NO_3^-]_o$ which would result in relatively high $[NO_3^-]_c$. This would seem to indicate that a purely diffusive release from the cytosol is unlikely. Indeed, the permeability coefficients ($P_{NO_3}^-$ and $P_{NH_4}^+$) for plasma membrane and tonoplast (Macklon et al., 1990) in onion roots were estimated to be ~0.5 to 5×10^{-11} m s^{-1} and ~1 to 16 $\times 10^{-11}$ m s^{-1}, respectively for NO_3^- and NH_4^+. These values are quite low compared to P_{Cl}^- and P_K^+, reported to be ~1×10^{-10} m s^{-1} and ~1 to 30×10^{-10} m s^{-1}, respectively, in barley roots (Pitman, 1969). Thus channel-mediated efflux appears to be more reasonable than pure diffusion. Notwithstanding the considerable progress in the characterization of anion channels in plant membranes (Tyerman, 1992) no detailed characterization of NO_3^- specific channels has been forthcoming to date.

B. Nitrate Fluxes Across the Tonoplast

While the situation vis a vis NO_3^- transport at the plasmalemma is reasonably clear, there is lack of agreement concerning fluxes across the tonoplast. Glass (1988) and others (Martinoia et al., 1986; Zhen et al., 1991) have conclued that the accumulation of NO_3^- within the vacuole may involve active transport as vacuolar accumulation of NO_3^- becomes appreciable. As a corollary, NO_3^- release to the cytosol should then be passive. By contrast, others have argued for passive nitrate entry via an electrogenic uniport driven by the positive value of $\Delta\psi$ across the tonoplast, and a NO_3^-/H^+ symport for release of NO_3^- (Glass, 1988; Deane-Drummond, 1990; Pope and Leigh, 1990).

There can be little doubt that NO_3^- is accumulated to levels approaching 100 μmol g^{-1} in both root and leaf tissues when external nitrate is supplied at levels exceeding 1 mM (Cram, 1973; Belton et al., 1985; Martinoia et al., 1986; Siddiqi et al., 1990; 1991). Notwithstanding what has been stated concerning the $[NO_3^-]_c$ in the preceding section, the bulk of tissue NO_3^- is localized within the vacuole. Hence, Belton et al. (1985) failed to detect a cytoplasmic NO_3^- pool using NMR, although a single peak corresponding to the vacuolar NO_3^- pool was evident. Similar conclusions were arrived at by Martinoia et al. (1986) who estimated that 98% of the NO_3^- contained in barley leaf protoplasts was within the vacuole. Direct measurement of vacuolar $[NO_3^-]$ by single cell sampling of barley root epidermal and cortical cells gave values of 53 and 101 mM, respectively (Zhen et al., 1991). Using NO_3^--specific microelectrodes to measure vacuolar and cytosolic $[NO_3^-]$, the same authors obtained values of 72.8 and 3.2 mM, respectively in cortical cells. Using these values, $\Delta\psi$ across the tonoplast would need to be ~ +79 mV to sustain a passive distribution of NO_3^-. In the same laboratory, $\Delta\psi$ across the tonoplast was determined to be only +12.3 mV. Such small (positive) values of across the tonoplast are not atypical (Macklon, 1975) and clearly represent serious difficulties for a passive accumulation of NO_3^- within the vacuole.

Using vacuoles isolated from barley leaves, Martinoia et al. (1986), demonstrated that ATP stimulated Cl^- influx, while reducing NO_3^- efflux. The pattern of ion efflux from these isolated vacuoles ($NO_3^- > Cl^- > Pi$), led the authors to conclude that anion efflux was passive; hence anion uptake should be active. At an earlier date, Cram (1968) concluded on the basis of electrical measurements and [ion], derived from compartmental analysis, that Cl^- influx across both the plasmalemma and tonoplast were probably active in carrot root tissue. The use of valinomycin to generate a negative value of $\Delta\psi$ across the tonoplast strongly increased NO_3^- efflux but had little effect on Cl^- or Pi efflux (Martinoia et al., 1986). These observations are consistent with active uptake of NO_3^- across the tonoplast and passive release, although, clearly, the ATP effects could be explained as the result of stimulating H^+ pumping across the tonoplast and thus hyperpolarizing (increasingly positive) $\Delta\psi$. Active NO_3^- uptake across the tonoplast might occur via a proton-nitrate antiporter or by means of a NO_3^- specific ATPase. Such mechanisms are, however, entirely speculative.

At the other extreme, Siddiqi et al. (1991) estimated $[NO_3^-]_v$ and $[NO_3^-]_c$, on the basis of compartmental analysis of $^{13}NO_3^-$ efflux from barley roots grown in 1 mM NO_3^-, to be 100 and 37 mM, respectively. This distribution of NO_3^- would indicate a requirement for $\Delta\psi$ across the tonoplast of +25 mV, a value which is less remote from those typically observed for tonoplast potentials than the values required by the data of Zhen et al. (1991) discussed above. Evidence in support of an electrogenic (passive) uniport has also been derived from studies of NO_3^-/Cl^- interactions in tonoplast vesicles. Both NO_3^- and Cl^- have been demonstrated to stimulate H^+ pumping in tonoplast vesicles, replacing $\Delta\psi$ by ΔpH (Blumwald and Poole, 1985; Kaestner and Sze, 1987). Pope and Leigh (1990) demonstrated that Cl^- entry into tonoplast vesicles from red beet was stimulated by the generation of an inside positive $\Delta\psi$ and competitively inhibited by NO_3^-. These results were interpreted to indicate that NO_3^- and Cl^- shared the same transporter, a passive uniport driven by $\Delta\psi$. In the *Chara* tonoplast the anion channel appears to show a greater permeability for NO_3^- than Cl^- but remains open for a shorter proportion of time in the presence of NO_3^- than Cl^- (Tyerman and Findlay, 1989). If the electrical gradient is indeed sufficient to drive Cl^- or NO_3^- across the tonoplast, then release of NO_3^- when NO_3^- supplies are interrupted might occur through a proton/-NO_3^- symport (Blumwald and Poole, 1985; Schumaker and Sze, 1987; Deane-Drummond, 1990). It is feasible, even, that a single channel might mediate NO_3^- entry into the vacuole by electrogenic uniport under conditions of adequate NO_3^- supply and serve also to release NO_3^- from the vacuole as metabolism and translocation of NO_3^- reduce $[NO_3^-]_c$ following interruption or exhaustion of $[NO_3^-]_o$. In onion roots, Macklon et al. (1990) concluded on the basis of $^{15}NO_3^-$ efflux analysis and (tonoplast) that NO_3^- fluxes across the tonoplast were passive in both directions. In summary, there is evidence consistent both with passive and active NO_3^- entry across the tonoplast.

C. Fluxes to the Xylem

The greater part of absorbed NO_3^- is exported to the shoot for reduction and assimilation even in those plant species which reduce significant quantities of NO_3^- in the roots. Hence the energetics of this translocation step between root symplasm and xylem tracheary elements is another important consideration. Realistic estimates of xylem $[NO_3^-]$ are rare, since virtually all available data are derived from analyzing the composition of xylem sap exuding from decapitated plants. Values obtained typically place $[NO_3^-]$ in the range from ~1 to 20 mM (Davis and Higinbottham, 1969; Ezeta and Jackson, 1975; Triplett et al., 1980). Because the driving force for sap exudation is root pressure rather than transpiration, these are probably overestimates of $[NO_3^-]$. It is also evident that $[NO_3^-]_x$ may vary according to prevailing light conditions (Margolis and Vezina, 1988). Thus, in *Betula* the $[NO_3^-]_x$ increased tenfold as plants were switched to low light conditions (Margolis and Vezina, 1988). Nevertheless, accepting that these values (1-20 mM) may represent overestimates, it is possible to determine the direction of the driving force for NO_3^- transport. Using the same analysis as was used for the uptake of NO_3^- (see Section IVA) it is evident that only if $\Delta\psi$ were extremely high (-50 mV) and the xylem sap extremely concentrated (~100 mM) could the flux between root symplasm and xylem be uphill. This is not to suggest that this flux is unregulated. Indeed this flux appears to be much more sensitive to inhibitors of protein synthesis, e.g., paraflurophenylalanine, than the uptake step (Behl et al., 1988). This was also true of K^+ translocation compared to K^+ uptake (Schaefer et al., 1975). The NO_3^- flux is also extremely sensitive to carbohydrate supply to the roots (Pate, 1980). Rufty et al. (1987) demonstrated that during a 12h dark period NO_3^- translocation and NO_3^- uptake by soybeans were reduced to 27 and 62%, respectively, of that reported during an equivalent light period. More than twice as much $^{15}NO_3^-$ was accumulated in roots in the 12 h dark period as in an equivalent 12 h light period. This NO_3^- was translocated to leaves and assimilated in the following light period (Rufty et al., 1984). Recently, a Pi translocation mutant was identified in *Arabidopsis* which appeared to be defective in the transport of Pi to the stele although normal in Pi uptake (Poirier et al., 1992). In contrast to the assumption that anion transport to the stele is passive, however, Pitman (1972) concluded on the basis of the effects of the inhibitor CCCP that transport of Cl^- to the stele involved a second active transport step. Clearly, channel-mediated transport might account for a passive transport of NO_3^- to the stele. However, there is virtually no information available on this topic.

Notwithstanding the relatively high $[NO_3^-]$ of the xylem sap, NO_3^- is virtually absent from the phloem (Pate, 1980) and hence cycling of N and redistributions associated with localized N supplies or split-root experiments must depend upon the translocation of organic N (Pate, 1980). In summary, NO_3^- transport across the plasmalemma is virtually always active, even at concentrations characteristic of the LATS. Entry to the xylem is probably always passive, but clearly regulated, while entry into the vacuole could be active or passive according to the $[NO_3^-]_c$.

D. Ammonium Fluxes Across the Plasmalemma

Estimates of $[NH_4^+]_c$ have been obtained using compartmental analysis (Presland and Mc Naughton, 1986; Macklon et al., 1990, Wang et al. 1993a) and by means of NMR (Lee and Ratcliffe, 1991). The latter method indicated that $[NH_4^+]_c$ in corn roots is typically in the low millimolar range (8-12 mM). Values obtained by efflux analysis are sometimes higher, and estimates as high as 76 mM were calculated by Macklon et al. (1990) for onion roots. However, estimates of $[NH_4^+]_c$ and $\Delta\psi$ have only rarely been obtained in the same plant system. Macklon et al. (1990) concluded that plasmalemma NH_4^+ influx was active at 2 mM $[NH_4^+]_o$. Wang et al. (1993a) found that $[NH_4^+]_c$ varied from 3.72 to 38 mM in roots of rice plants, depending on $[NH_4^+]_o$ in the range from 2 to 1000 μM. In the same strain of rice, plasmalemma $\Delta\psi$ values ranged from ~ - 140 to -89 mV, depending on $[NH_4^+]_o$. External NH_4^+ was strongly depolarizing and plots of depolarization versus $[NH_4^+]_o$ conformed to the influx isotherms for $^{13}NH_4^+$, indicating that NH_4^+ rather than NH_3 was the major ion transported, even at high external $[NH_4^+]$. Using the values obtained from this study, it was concluded that only below 40 μM was it necessary to invoke active transport (Wang et al., 1994). This is similar to the conclusion of Ullrich et al. (1984), whose equivalent figure for *Lemna* was 67 μM. However, although such thermodynamic evaluations are critical in order to make definitive statements concerning the driving forces for transport, they need to be evaluated in light of the kinetic data. For example, in the data from rice (Wang et al., 1994) there appear to be no discontinuities in the $^{13}NH_4^+$ influx isotherm corresponding to a change of transport mechanism at ~40 μM, the putative switch point from active to passive transport. The mechanism of this active NH_4^+ flux has received little attention. Glass (1988) postulated that it might occur via a H^+/NH_4^+

symport, analogous to the postulated H^+/K^+ symport of *Neurospora* (Rodriguez-Navarro, 1986). At higher concentrations of NH_4^+ a linear LATS is evident and fluxes through this system are correctly viewed as passive, though clearly energy-dependent (Macklon et al., 1990). It is tempting, considering the competitive interactions between K^+ and NH_4^+ at the uptake step, (Deane-Drummond and Glass, 1983; Scherer et al., 1984), as well as the recent demonstration that the NH_4^+ conductance of one of the K^+ channels of *Arabidopsis* is 30 ± 12% that of its K^+ conductance (Schactman et al., 1992), to suggest that NH_4^+ influx by the LATS might be channel-mediated, via the K^+ channel.

E. Ammonium Fluxes Across the Tonoplast

Based upon their analyses of $^{15}NH_4^+$ efflux and electrical measurements of tonoplast $\Delta\psi$, Macklon et al. (1990) concluded that NH_4^+ influx to the vacuole was passive, while the flux back to the cytosol was uphill. Passive entry of NH_4^+ might proceed via a tonoplast channel, whereas return of NH_4^+ to the cytosol could be driven by a H^+/NH_4^+ symport. Clearly, much more information is required before these fairly fundamental questions can be resolved. Alkalinization of the vacuole during exposure of corn root tips to NH_4^+ solutions led Roberts and Pang (1992) to conclude that NH_3 was the permeant species, although no changes of cytoplasmic pH were reported. If this were the case, the N species transported back to the cytoplasm would necessarily be NH_4^+, because of the low pH typical of the vacuole.

F. Ammonium Fluxes to the Stele.

Virtually no NH_4^+ is found in xylem or phloem sap (Lewis and Chadwick, 1983; Wang, et al., 1993a). Rather, when NH_4^+ serves as the N source for root absorption, nitrogen is translocated to the shoot in the form of amino acids.

V. Amino Acid Uptake

Most reviews of N absorption by plants have focussed almost exclusively on nitrate and ammonium absorption. There are valid reasons, however, to include amino acids as potential sources of N for plants in soil. As detailed in the introduction, amino acids are present in significant concentrations in soil solution. Moreover, reported concentrations (Paul and Schmidt, 1961; Kuprevich and Scherbakova, 1971; Monreal and McGill, 1985; Schobert and Komor, 1987) are in the range of reported apparent K_m values for amino acid uptake (Schobert and Komor, 1987). Perhaps more important than their absolute concentrations in soil solution are the rates of turnover of amino acid pools. According to Monreal and McGill (1989) turnover times for soluble cystine in Chernozemic and Luvisolic soils are but a few seconds.

Extensive studies of amino acid uptake by plant roots (Aldag and Young, 1970; Soldal and Nissen, 1978; Reinhold and Kaplan, 1984; Schobert and Komor, 1987; Burton, 1989), leaf slices (Lien and Rognes, 1977), coleoptiles (Etherton and

Rubinstein, 1978), suspension cultures (Wyse and Komor, 1984) and even proto-plasts (Reinhold and Kaplan, 1984) have demonstrated the existence of amino acid transport systems. A limited number of studies have made use of subcellular pre-parations such as plasma membrane vesicles (Li and Bush, 1990) and vacuoles (Goerlach and Willms-Hoff, 1992) to investigate processes involved in regulating amino acid fluxes. Furthermore, many investigators have demonstrated that amino acids can serve very effectively as the sole source of N for plant growth. Such studies include the use of entire plants maintained in sterile culture (Brozozowska et al., 1974; Mori and Nishimura, 1979; Datko and Mudd, 1985; Schobert et al., 1988) as well as excised plant organs and suspension cultures (Wyse and Komor, 1984). In a recent study at the University of Alberta, Burton (1989) compared the growth of maize plants on different sources of N in sterile solution culture. A mixture of alanine, aspartate, glutamate and glycine increased total plant dry weight by 37% compared to equimolar ammonium nitrate. Thus, not only has the existence of amino acid transport systems been clearly demonstrated in plants, but the latter are able to supply N at sufficient rates to satisfy the N requirements of normal plant growth.

Considering the above, it is not unreasonable to expect that plant roots might access significant quantities of soil amino acids under field conditions. Unfortunately only a limited number of investigators have addressed this question directly. Scho-bert et al. (1988) concluded, on the basis of the absorption of ^{14}C-labelled proline, injected into the rhizosphere of castor bean seedlings, that roots can successfully compete with microorganisms for the acquisition of this amino N. Based on their earlier estimates of the kinetics of amino acid uptake by roots of *Ricinus* and deter-minations of the amino acid concentrations of soil solution, Schobert and Komor (1987) estimated that 10-20% of the nitrogen acquisition of these plants might be derived from amino acid uptake. Even higher values (10-80%) were suggested by Kielland (1993) for arctic tundra plants. The results of Burton's investigations with barley, canola and maize (Burton, 1989) were less encouraging. The three species examined failed to retain more than 2% of the amino acid mixtures introduced into the rhizosphere of these plants. However, in this study, like that of Schobert et al. (1988) the amino acids were ^{14}C-labelled; recovery was based upon 14 or 23 days exposures. Hence the capacity to transaminate the absorbed amino acids and release the ^{14}C label as CO_2 via respiration must certainly have biased the findings of this study. For all three species, $^{14}CO_2$ released from the amino acids and retained in CO_2 traps amounted to as much as 35% of ^{14}C supplied as amino acids. Clearly, much more work is warranted in this area, particularly in natural ecosys-tems where inorganic N may be in limited supply and where mycorrhizal associa-tions may significantly enhance plant acquisition of amino N.

A. Transport Systems

Prior to the late 1970's, relatively few studies of amino acid transport had been reported. Yet, by 1984 in a major review of sugar and amino acid transport in plants (Reinhold and Kaplan, 1984), over 50 papers dealing with amino acid transport were cited. In their thorough review of this topic, Reinhold and Kaplan (1984) detail many of the fundamental characteristics of amino acid transport in plants. These include the many documented cases of active transport of amino acids, the overwhelming evidence for proton symport of amino acids at the plasmalemma and the complexity of published kinetic data. What remained unclear were the mechanisms of transport processes at the tonoplast, the role of K^+ in amino acid absorption and the interpretation of the complex kinetic data. Evidence summarized in the same review (Reinhold and Kaplan, 1984) was also equivocal about the existence of a common carrier for amino acids as opposed to several carriers with limited specificity for different classes of amino acids. For example Kinraide (1981) and Kinraide and Etherton (1980) proposed the existence of basic, neutral and acidic amino acid transport systems on the basis of work undertaken with oat coleoptiles.

Clearly, there is a need for amino acid transport at several membrane barriers within cells of higher plants. These include the plasma membranes of root and leaf mesophyll cells where amino acids may be withdrawn from soil solution or from xylem sap, respectively. Particularly when plants are absorbing significant amounts of NH_4^+ from external media, or when they establish symbiotic relations with N fixing bacteria, the major N form translocated, via xylem, to the shoot will be amino acids. Thus, at the root symplasm/xylem interface amino acids must leave the root symplasm to enter the xylem sap. When the bulk of incoming N is in the form of NO_3^- and reduction is mainly in the shoot, amino acids will circulate within the phloem to rapidly expanding metabolic sinks. This applies equally to the provision of amino acids derived from stored protein in cotyledons or other protein reserves. Indeed, cycling of amino acids from shoot to root and back to the shoot via the phloem and xylem, respectively, may be the *status quo*. According to Cooper and Clarkson (1989) this N cycling may represent an important signalling process, serving to regulate N uptake by plant roots and thereby co-ordinate plant demands at the whole plant level. Finally, according to Schobert and Komor (1987), amino acid uptake systems in roots may serve to retrieve amino acids which are lost from phloem offloading into the apoplasm.

B. Kinetics of Uptake

The kinetics of amino acid uptake have been investigated in a wide range of plant systems. These include both single celled and multicellular organisms (Aldag and Young, 1970; Jung and Luttge, 1980; Datko and Mudd, 1985; Sauer and Tanner, 1985; Borstlap et al., 1986; Schobert and Komor, 1987; Burton, 1989), plant roots (Wright, 1962; Soldal and Nissen, 1978; Shobert and Komor, 1987), coleoptiles (Etherton and Rubinstein, 1978; Kinraide and Etherton, 1980), leaf tissue (Reinhold et al., 1970; Lien and Rognes, 1977; Despeghel and Delrot, 1983), phloem tissue (Servailes et al., 1979; Riens et al., 1991), isolated mesophyll cells (Chervel and Jullien, 1979; McCutcheon et al., 1988), suspension cells (Paszkowski et al., 1980; Harrington and Henke, 1981; McDaniel et al., 1982; Wyse and Komor, 1984), protoplasts (Guy et al., 1978; Paszkowski et al., 1980; Suzuki, 1981; uer Nooy and

and Lin, 1986), vacuoles (Dietz et al., 1990; Martinoia et al., 1991; Goerlach and Willms-Hoff, 1992; Winter et al., 1992) and vesicles isolated from plasma membrane preparations (Bush and Langston-Unkefer, 1988; Li and Bush, 1990; 1991). As with many solute transport systems defined by kinetic studies, amino acid uptake kinetics are rarely simple. At low external concentrations of amino acids (<1 mM) uptake is usually saturable, conforming to Michaelis-Menten kinetics (Chervel and Jullien, 1979; Harrington and Henke, 1981; uer Nooy and Lin, 1986; Li and Bush, 1990) with apparent K_m values ranging from ~5 x 10^{-16} M, as in the case of lysine transport by tobacco callus cultures (Harrington and Henke, 1981), to ~2 mM for glutamate uptake by plasma membrane vesicl₋s isolated from sugar beet leaves (Li and Bush, 1990). When wider limits of concentration have been examined, biphasic (Harrington and Henke, 1981; Borstlap et al., 1986; Li and Bush, 1990; 1991) or multiphasic (Reinhold et al., 1970; Lien and Rognes, 1977; Soldal and Nissen, 1978; Blackman and McDaniel, 1980) transport kinetics have commonly been reported. In those cases of biphasic kinetics, the second (higher concentration) system may be saturable (Li and Bush, 1990; 1991) or even linear (uer Nooy and Lin, 1986; Schobert and Komor, 1987; Goerlach and Willims-Hoff, 1992).

Classical methods of kinetic analysis have been widely used to examine for competition between individual amino acids. Data arising from such experiments have formed the basis of the recognition of 1 (Chervel and Jullien, 1979; McDaniel et al., 1982), 2 (Harrington and Henke., 1981), 3 (Kinraide and Etherton, 1980; Sauer and Tanner, 1985, Martinoia et al., 1991), or 4 (Li and Bush, 1990; 1991) different amino acid transporters. Using plasma membrane vesicles isolated from sugar beet leaves, Li and Bush (1991) provided kinetic evidence for an acidic, a basic and 2 neutral amino acid transporters. Encouraging support for the concept of discrt multiple transporters for amino acids has come from the identification of amino acid transport mutants in *Chlorella* (Sauer and Tanner, 1985). Using the toxic analogues, L-canavanine (for arginine) or L-azetidine-2-carboxylic acid (for proline) both single and double mutants were isolated. The authors, Sauer and Tanner (1985) proposed the existence of three separate uptake systems, responsible for the absorption of basic amino acids, for the absorption of proline, alanine, serine and glycine, and a general system capable of transporting a number of neutral and acidic amino acids.

C. Energetics

The literature dealing with amino acid uptake by plant systems is replete with data documenting the inhibition of amino acid fluxes by inhibitors of metabolism such as azide (Chervel and Jullien, 1979; Harrington and Henke, 1981), cyanide (Kinraide and Etherton, 1980), dinitrophenol (Chervel and Jullien, 1979; Kinraide and Etherton, 1980; Harrington and Henke, 1981; McDaniel et al., 1982), CCCP (Harrington and Henke, 1981; McDaniel et al., 1982), and DCCD (Harrington and Henke, 1981), and the stimulation of amino acid uptake by low pH (Etherton and Rubinstein, 1978; Harrington and Henke, 1981; Li and Bush, 1990). Likewise many reports confirm the alkalinization of external media (Jung et al., 1982; Despeghel and Delrot, 1983; Kinraide et al., 1984), the depolarization of membrane electrical potentials (Etherton and Rubinstein, 1978; Harrington and Henke, 1981; Lüttge et al., 1981; Jung et al., 1982; Kinraide et al., 1984; Felle and Johannes, 1990) associated with amino acid uptake, as well as the spontaneous repolarization of membrane electrical potentials which typically follows, within minutes, of depolarization (Kinraide and Etherton, 1980). This repolarization failed to occur in the presence of metabolic inhibitors, e.g. cyanide or dinitrophenol (Kinraide and Etherton, 1980). Taken together, this information provides strong evidence for active (or at least energy-dependent) transport of amino acids which is coupled to the proton motive force via proton symports (Reinhold and Kaplan, 1984). Clearly, with cytoplasmic amino acid concentrations in the millimolar range and soil solution strength in the μmolar range, uptake of negatively charged and neutral amino acids is necessarily active. The case for positively charged amino acids is less clear, particularly since their concentrations are typically much lower than the negatively charged amino acids.

Investigations of the distribution of amino acids between cytosol and vacuole have typically revealed that vacuolar concentrations are significantly lower than those of the cytosol (Riens et al., 1991; Winter et al., 1992). In spinach leaves, for example, total amino acid concentration of the cytosol was 121 mM, while that of the vacuole was an order of magnitude lower (Riens et al., 1991). Similar results were documented for the leaves of barley (Winter et al., 1992). Nevertheless, the considerably larger volume of the vacuole, compared to that of the cytosol of differentiated mesophyll cells would indicate that vacuoles still represent a significant store of N.

Given the large gradient of amino acid concentration between cytosol and vacuole, amino acid entry into the vacuole might be anticipated to occur passively. Experiments with vacuoles isolated from mesophyll protoplasts have recently been used quite extensively to demonstrate the uptake of various amino acids by these organelles (Dietz et al., 1990; Martinoia et al., 1991; Goerlach and Willms-Hoff, 1992). These studies indicate a dependence upon ATP for transport which appears to represent a direct regulatory role, rather than provision of an energy source. Thus, using barley mesophyll vacuoles, for example, Dietz et al. (1990) demonstrated that the uptake of alanine, leucine and glutamine was stimulated 3 to 6 fold by the presence of ATP, MgATP or even adenylylimidodiphosphate, a non-hydrolyzable analogue of ATP. Moreover, this stimulated amino acid uptake was independent of ΔpH or $\Delta\psi$ across the tonoplast. The authors proposed that these amino acids were transported into the vacuole by an ATP regulated importer. Similar conclusions were arrived at for the transport of glycine (Goerlach and Willms-Hoff, 1992) and for arginine and aspartate (Martinoia et al., 1991) by barley mesophyll vacuoles. If this is the case, the retrieval of this N source from the vacuole would demand "uphill" transport; perhaps by a H^+: amino acid symport.

Nevertheless, earlier studies by Horneyer et al. (1989) had indicated that the absorption of phenylalanine by barley mesophyll vacuoles depends upon the activity of both the ATPase and the PPase of the tonoplast. According to these authors a lipophilic amino acid carrier system is energized by those hydrolases. According to

Martinoia et al. (1991), at least 3 transport systems mediate amino acid transport across the tonoplast: (1) an aromatic carrier, dependent upon pmf generated by the tonoplast ATPase and PPase; (2) a basic amino acid porter; and (3) a third importer (channel) of broad specificity. The latter two systems appear to be regulated but not energized by ATP.

In concluding this section on amino acid uptake by plants, it must be stressed that physiological and biochemical studies of amino acid uptake and metabolism have far outreached those at the ecophysiological level. A limited number of studies have indicated the potential importance of amino acids as a source of N (Schobert and Komor, 1987; Schobert et al., 1988; Burton, 1989; Kielland, 1993) but much more of a definitive nature needs to be done, especially with mycorrhizal plants grown under natural conditions.

VI. Regulation of Nitrogen Uptake

The uptake, translocation and metabolism of the various forms of absorbed N appear to be subject to precise regulation. At the whole plant level, convincing proof of this was provided by the work of Clement et al. (1978) who grew ryegrass for 8 weeks in solutions varying in $[NO_3^-]$ from 1.4 μM to 142 mM. Between 14.2 μM and 14.2 mM there was virtually no effect of $[NO_3^-]_o$ on rates of NO_3^- uptake, plant growth or plant N status. Likewise, Wang et al. (1993b) have documented similar acclimation to NH_4^+ supply in rice, from 2 μM to 1000 μM $[NH_4^+]_o$. Thus, notwithstanding the earlier discussion of the complex kinetics observed in short-term influx measurements (Section III) the up/down regulation of N fluxes has the effect of rendering the whole plant independent of external concentration over a wide range of $[N]_o$. There was evidence of this feedback as early as 1906, when Brezeale (1906) showed that withholding NO_3^- from hydroponically-grown wheat plants for 18 h caused > 3 fold increase of NO_3^- uptake in a subsequent uptake period. In addition to this negative feedback (repression?) and its relief (derepression?) when N supplies are interrupted, which is common for many nutrients, NO_3^- uptake appears to be unique in being subject to induction by provision of NO_3^- (Section III).

A. Induction of Nitrate Uptake

Plants grown in a NO_3^-; free medium (uninduced plants) exhibit low rates of NO_3^- uptake. Upon exposure to NO_3^-, following a lag of ~ 1-2 h (Clarkson, 1986), the rates gradually increase, reaching a maximum which is many times higher than the constitutive level. The duration of exposure to NO_3^- required for maximum induction varies greatly with the genotype e.g. 4-6 h in maize (Jackson et al., 1973), ~ 24 h in barley (Siddiqi et al., 1989; King et al., 1993), 3d in spruce (H. Kronzucker, pers. comm.) and is also a function of $[NO_3^-]_o$ (Neyra and Hageman, 1975. Deane - Drummond, 1982; Maeck and Tischner, 1986; Mackown and McClure, 1988; Siddiqi et al., 1989; Hole et al., 1990). In beans, however, induction was independent of $[NO_3^-]_o$ in the range 0.01 - 10 mM (Breteler and Nissen, 1982). There is also a high degree of genetic variation in the constitutive rates of uptake; e.g. in barley, cultivars Klondike and Steptoe showed a ~ 5-10 fold difference (Siddiqi et al., 1989; King et al., 1993).

(i) *Substrate Specificity*

There is general agreement that NO_3^- *per se,* rather than a product of NO_3^- assimilation, induces NO_3^+ uptake. This conclusion is supported by several independent lines of evidence: (a) Neither NH_4^- (Jackson et al., 1972; Omata et al., 1989; King et al., 1993; Aslam et al., 1993) and, by implication, nor any subsequent product of NH_4^- assimilation is capable of inducing NO_3^- uptake. (b) Exposure to NO_3^- in the presence of WO_4^{2+} prevented the induction of nitrate reductase activity (NRA) whereas induction of NO_3^- uptake proceeded normally (Heimer and Filner, 1971; Miyagi et al., 1992; unpublished data). (c) In nitrate reductase (NR) defective mutants of *Arabidopsis thaliana* (Doddema et al., 1978; Doddema and Telkamp, 1979), pea (Deane-Drummond and Jacobsen, 1986) and barley (Warner and Huffaker, 1989; King et al., 1993) NO_3^- fully induced the uptake of NO_3^- while NRA was undectable.

In suspension culture cells of *Chenopodium rubrum* (Beck and Renner, 1989) and intact barley roots (Bloom and Sukrapanna, 1990; Bloom et al., 1992), however, it has been claimed that provision of NH_4^+ induced NO_3^- uptake. Beck and Renner (1989) suggest that the effect of NH_4^+ may be indirect; NH_4^+ caused release of NO_3^- from vacuale to the cytoplasm which then induced NRA and NO_3^- uptake. In any event there is a great preponderance of evidence available from a variety of organisms ranging from microbes to higher plants, showing that provision of NH_4^+ either inhibits or has no effect on the induction of NO_3^- uptake (Doddema and Otten, 1979; Mackown et al., 1982; Glass et al., 1985; Madueno et al., 1988; Sivak et al., 1989; Unkles et al., 1991; Miyagi et al., 1992; King et al., 1993; Aslam et al., 1993). Finally, indirect positive effects of the provision of nutrients resulting from the alleviation of general deficiency should be distinguished from the true induction phenomena (Lefebvre and Glass, 1982; Siddiqi and Glass, 1987. Clarkson et al., 1989; Rufty et al., 1991).

In most cases, ClO_3^- and NO_2^- utilize the same transport system as NO_3^- (Deane-Dummond and Glass, 1982; Deane-Drummond, 1984a; Criddle et al., 1988; Chodera and Briskin, 1990; Aslam et al., 1992; Siddiqi et al., 1992). At the induction step, however, genotypes differ in whether these transport analogs can substitute for NO_3^-. In barley (Siddiqi et al., 1992) and maize (McClure et al., 1986) ClO_3^- failed to induce NO_3^- or ClO_3^- uptake and NRA (LaBrie et al., 1991; Siddiqi et al., 1992). By contrast, ClO_3^- induced NO_3^- uptake in *Chara corallina* (Deane-Drummond, 1984) and ClO_3^- uptake in *Arabidopsis thaliana* (Doddema and Telkamp, 1979). In the study by La Brie et al. (1991), NR-mRNA was induced by

ClO_3^- as well as by NO_3^- but NR protein was absent in ClO_3^- pretreated plants. The same authors suggested that ClO_2^-, formed by the reduction of ClO_3^-, inactivated NR and that in this inactivated state NR protein was readily degraded. Alternately, ClO_3^- may interfere with the synthesis of NR protein in some yet unknown manner.

In higher plants as well as in algae, NO_2^- is a competitive inhibitor of NO_3^- uptake (Ullrich, 1987; Fuggi, 1989; Siddiqi et al., 1992; Aslam et al., 1992). In barley (Siddiqi et al., 1992; King et al., 1993) and in *Aspergillus nidulans* (Unkles et al., 1991) NO_2^- fully induced NO_3^- uptake but not NRA. In barley leaves, however, provision of NO_2^- induced NRA but only as a result of accumulation of NO_3^- formed from oxidation of NO_2^- (Aslam et al., 1987) which was not detected in barley roots (Siddiqi et al., 1992). Recently, however, Aslam et al. (1993) have reported that provision of NO_2^- induced both, NO_3^- uptake as well as NRA in barley roots. According to these authors induction of NO_3^- uptake as well as NRA was independent of NO_3^- accumulation in the roots.

(ii) Biochemistry and Molecular Biology

Using inhibitors of protein synthesis, it has long been known that induction of NO_3^- uptake involves *de novo* protein synthesis (Jackson et al., 1973; Neyra and Hageman, 1976; Clarkson, 1986). From these studies and from the time taken for induction upon NO_3^- supply, or deinduction upon NO_3^- withdrawal, the rate of turnover of the inducible NO_3^- carrier has been estimated to be 2-3 h (Clarkson, 1986; Siddiqi et al., 1989).

Since these early studies, SDS-PAGE analyses of microsomes and plasma membrane-enriched fractions have shown the appearance or enhancement of specific polypeptides following the exposure of uninduced plants to NO_3^- (see Table 1). The appearance and responses of these polypeptides paralleled the NO_3^- uptake activity, e.g. ~ 3 hour lag in appearance, induction by NO_3^-, inhibition by NH_4^+ (Table 1). This information is, of itself, insufficient to warrant the conclusion that these polypeptides constitute the NO_3^- transport system.

Detailed genetic/molecular analyses have lead to the identification of genes encoding NO_3^- transport proteins in *Synechococcus* (= *Anacystis nidulans*) (Omata et al., 1989; Omata, 1991), *Aspergillus nidulans* (Brownlee and Arst, 1983; Unkles et al., 1991), *Escherichia coli* (Noji et al., 1989), and most recently in *Arabidopsis thaliana* (Tsay et al., 1993). The corresponding putative NO_3^- transport proteins these genes encode are: 45, 52, 50, and 65 kD, respectively (Table 1). Interestingly, even amino acid sequences of these proteins show no significant homologies (Tsay et al., 1993). It may be that these transporters correspond to kinetically different systems. In *Arabidopsis thaliana*, for instance, the NO_3^- transporter in question is a low affinity system and, more importantly, it is non-specific: it transports Cl^- and K^+ as well (Tsay et al., 1993).

The aforementioned studies have clearly established that induction of NO_3^- uptake (*de novo* carrier synthesis) is controlled at the level of transcription. Provision of NO_3^- induces the synthesis of mRNA encoding NO_3^- carrier protein(s) (Omata et al., 1989; Unkles et al., 1991; Tsay et al., 1993). The corresponding mRNA's are lacking in NH_4^+ grown cells and plants; in fact, NH_4^+ suppresses the synthesis of the corresponding mRNA even in the presence of NO_3^- (Unkles et al., 1991).

Table 1.   Polypeptides induced by NO_3^- pretreatment

Organism	Mr (kD)	Location	Notes	References
Barley	31	TP/ER	NO_3^- induced. Regulatory	McClure et al., 1987.
Maize	30,47	MIC	NO_3^- induced	Dhugga et al., 1988a,b.
	*40		NO_3^- induced	
Synechococcus R2	48	PM	NO_3^- induced, NH_4^+ inhibited	Madueno et al., 1988.
Synechococcus	47	PM	NO_3^- induced, 3 h lag, NH_4^+ inhibited	Sivak et al., 1989.
Synechococcus	45	PM	Gene sequenced. NO_3^- induced	Omata et al., 1989; Omata, 1991.
Escherichia coli	50	PM	Gene (narK) sequenced. NO_3^- induced	Noji et al., 1989.
Aspergillus nidulans	52	PM	Gene (crnA) sequenced. NO_3^-/ NO_2^- induced, N-metabolite repressed	Unkles et al., 1991.
Heterosigma akashiwo	26	PM	NO_3^- induced, 3 h lag. NH_4^+ inhibited	Miyagi et al., 1992.
Chlamydomonas reinhardtii	21	PM	NO_3^- induced	Watt et al., 1992.
Arabidopsis thaliana	65	PM	Gene (CHL1) cloned. NO_3^- or low pH induced, N-metabolite inhibited. Also transports Cl^- and K^+	Tsay et al., 1993.

*a group of peptides ~ 40 kDa. TP: tonoplast, ER: endoplasmic reticulum, MIC: microsome, PM: Plasma membrane.

B.   Induction of Ammonium Uptake by Ammonium

As far as we are aware, three groups have demonstrated an apparent induction of NH_4^+ uptake in N starved plants when NH_4^+ was resupplied (Goyal and Huffaker, 1986a; Morgan and Jackson, 1988, M.Y. Wang, unpublished). In all cases, NH_4^+ influx or net uptake increased for 2-5 h and then declined. This situation is different from most cases of repressible transport systems where provision of the limiting nutrient usually causes influx to decline. It is also quantitatively different from the classical induction of NO_3^- uptake by NO_3^- in so far as ammonium uptake rates in the N-starved plants were already high and increased by –0.2 - 0.3 fold. Induction of NO_3^- uptake by NO_3^- provision may increase V_{max} for influx by as much as 30 fold (Siddiqi et al., 1990). Nitrogen uptake by N-starved plants may represent a special case because the derepression signal (amino N) is also the substrate for carrier synthesis. Thus it may not be possible to respond fully to the signal because of N-limitation. Hence, when N is provided exogenously, the first response may be to increase carrier synthesis. This N-limitation appears to be a general effect on carrier synthesis because in rice plants grown on low N and low K^+, provision of NH_4^+ caused K^+ influx to increase during the first 6 h in parallel with the increases of NH_4^+ uptake (M.Y. Wang, unpublished). Thus we view this as a non-specific effect arising from N-limitation rather than true induction of NH_4^+ uptake.

C. Repression of Nitrate and Ammonium Uptake

Plants deprived of N for any appreciable length of time demonstrate increased uptake of NO_3^-, NH_4^+ or even amino acids when these are resupplied. This pheno- menon, first documented by Brezeale (1906), has subsequently been investigated extensively (Jackson et al., 1976, Lee, 1982; Lee et al., 1982; Mackown et al., 1982; Lee and Drew, 1986; Lee and Rudge, 1986; Siddiqi et al., 1989, 1990; see Clarkson and Luttge, 1991 for review). Siddiqi et al. (1989) demonstrated that when NO_3^- was first provided to barley seedlings grown for 5 days without exogenous N, $^{13}NO_3^-$ influx increased rapidly, peaking as early as 12 h at the highest $[NO_3^-]$ pro- vided and then declined to a considerably lower value which varied inversely accord- ing to the $[NO_3^-]$ supplied. A plot of $^{13}NO_3^-$ influx against root $[NO_3^-]$ therefore took the form of a parabola. Clearly, the concentrations of many N pools are altered during NO_3^- provision. Given the possibility of satisfying a plant's N demand from sev- eral different forms of N it would not be unreasonable if a common feedback signal arising at some biochemical distance from the entry point (NO_3^-)served to regu- late N uptake. This argument has been forcefully made by Lee and Rudge (1986), and by Clarkson and Lüttge (1991). In prokaryotes and fungi the case seems clear that glutamine is the signal for negative feedback (Clarkson and Lüttge, 1991). In microalgae, also, there is evidence that glutamine is involved in regulating NO_3^- and NH_4^+ uptake (Syrett and Peplinska, 1988). Yet in higher plants evidence from feeding specific amino acids to plant roots indicates that a wide range of amino acids are capable of down regulating NO_3^- uptake (Ferguson, 1970; Oaks et al., 1977; Doddema and Otten, 1979; Breteler and Arnozis, 1985; Lee et al., 1992) and/or NH_4^+ uptake (Lee et al., 1982; M.Y. Wang, unpublished). The obvious shortcom- ing of this strategy is that feeding of one amino acid may rapidly lead to conversion to another (Lee et al., 1992, and references therein). Moreover, there are differ- ences in the compartmentation of NO_3^- and NH_4^+ assimilation. Thus NO_3^- is assimilated to NH_4^+ and beyond within plastids, whereas absorbed NH_4^+ is assimi- lated to glutamine by cytosolic glutamine synthetase. Thus exogenously fed amino acids may result in abnormal subcellular distributions. Nevertheless, Lee et al. (1992) established that in maize, conditions leading to elevated glutamine or asparagine levels (NH_4^+ or amino acid feeding) led to reduced rates of NH_4^+ and NO_3^- influx. Glutamine feeding of NO_3^- grown plants e.g. increased tissue glutamine from

1 to 23 μmol g$^-$ and reduced ^{15}NO$_3^-$ uptake. The critical question here is whether NO$_3^-$ grown plants ever accumulate such high glutamine levels. By contrast, with-olding N or treating NH$_4^+$-fed plants with Methionine sulfoximine led to a decline of cytosolic glutamine and asparagine and concomitant increases of NO$_3^-$ and NH$_4^+$ uptake. However, there are anomalies to be explained regarding the role of glutamine. For example, during the up-regulation of NH$_4^+$ uptake associated with provision of NH$_4^+$ to N-starved plants.(Section VI d) glutamine levels increased sig-nificantly (Lee et al., 1992, M.Y. Wang, unpublished). Also, following transfer of rice plants from 1 mM to 2 μM NH$_4^+$ solutions tissue NH$_4^+$ declined, glutamine levels declined and ^{13}NH$_4^+$ influx increased. However, when the 2 μM NH$_4^+$ solution con-tained MS, to block glutamine synthesis, [NH$_4^+$] increased, glutamine levels decreased and ^{13}NH$_4^+$ influx failed to increase (M.Y. Wang, unpublished). Finally, in the study by Lee et al. (1992) 1.5 mM NO$_3^-$-grown plants contained only 1 μmol glutamine per g compared to 17 μmol g^{-1} in 1.5 mM NH$_4^+$-grown plants. Yet, ^{15}NO$_3^-$ and ^{15}NH$_4^+$ fluxes were rather similar, despite the considerable differences of glutamine concentrations.

While the concept of a reduced N pool serving to regulate N fluxes is attractive from the viewpoint of coordinating N uptake and is supported by some evidence in higher plants, it certainly does not preclude feedback from other sources. In this context, attention has been directed to the effects of NH$_4^+$ application on N fluxes (Clarkson and Lüttge, 1991). In many cases (Glass et al., 1985; Ingemarsson et al., 1987; Lee and Drew, 1989) NH$_4^+$ has been shown to cause an almost immediate inhibition of NO$_3^-$ influx. This topic was discussed in section III and should probably not be confused with regulaton of influx because in long-term studies using NH$_4$NO$_3$, NO$_3^-$ uptake is not diminished compared to growth on NO$_3^-$ or NH$_4^+$ as sole N sources (Zhang and Mackown, 1993) and growth is signi-ficantly enhanced.

Siddiqi et al. (1989) proposed that accumulated (vacuolar) NO$_3^-$ might feedback on NO$_3^-$ influx by its effect on the fluxes from cytoplasm to vacuole and fluxes into reduced N pools. King et al. (1993) used barley NR mutants to examine the pat-terns of down-regulation of ^{13}NO$_3$ influx after the peak influx associated with induc-tion. In the mutants, both induction (see also Warner and Huffaker, 1989) and the subsequent down-regulation were similar to wild type barley even though rates of NO$_3^-$ reduction in the mutant are but 5 - 10% of wild type. It was concluded that NO$_3^-$ itself might play a role in regulating NO$_3^-$ influx. This is not unreasonable since tissue NO$_3^-$ accumulation can be high (100 μmol g^{-1}) and if feedback only arose from an amino N pool, vacuolar pools of NO$_3^-$ or NH$_4^+$ would be circum-vented and not accounted for in regulating influx.

D.   Regulation of Amino Acid Transport

Notwithstanding the importance of the proton motive force as the source of energy for amino acid transport, Felle and Johannes (1990) have stressed the importance of other more critical factors in regulating the fluxes of amino acids. In their study of aminoisobutyric acid (AIB) uptake in *Riccia fluitans*, amino acid uptake was much more sensitive to small changes of cytosolic pH than to changes of ΔpH.

In addition, evidence from several sources indicates that amino acid uptake is sensitive to regulation in response to the N or C status of the system. For example, studies with *Chlamydomonas* (Malhotra and Glass, unpublished), isolated soybean embryos (Bennett and Spanswick, 1983) and with suspension cultures (King and Olenuik, 1973; McDaniel and Wozniak, 1982) demonstrated increased rates of amino acid uptake some hours after removal or reduction of exogenous N supply. Transfer of *Chlamydomonas* from normal growth medium, containing 9 mM NH_4^+, to 100 μM NH_3^+ for 24 h increased rates of ^{14}C-arginine uptake 4 fold in a subsequent 6 h labelling period. Even more dramatic were the increases reported by King and Oleniuk (1973), who observed that ^{14}C-alanine uptake by soybean suspension cells increased by 100 fold following growth on a N-free medium. Interestingly, amino acid uptake by high N cells showed no indications of saturation. By contrast, N-deprived cells exhibited concentration-dependent saturation with apparent K_m value equal to 2.6×10^{-6} M. Similar results were reported by McDaniel and Wozniak (1982); L-leucine uptake by suspension cultures of tobacco increased 3 fold, peaking 3 days after removal of N. These workers also demonstrated stimulation of amino acid uptake in response to C deprivation. Thus, amino acid uptake peaked after 1 day of growth on C-free media. Clearly amino acids in the growth medium provide both C and N to heterotrophically growing cells. In a study of amino acid uptake by *Chlorella vulgaris*, Sauer et al. (1983) demonstrated that N deprivation or provision of glucose caused induction (or derepression?) of two amino acid uptake systems, for arginine and proline, respectively. One hour after provision of glucose, arginine uptake had increased ~10 fold with virtually no change of cellular sugar or amino acid concentration. Growth on -N media appeared to take much longer (5 h) to bring about a similar increase of arginine uptake. This increase appeared to be correlated with decreased cellular amino acid concentration.

Bennett and Spanswick (1983), working with isolated soybean embryos also observed non-saturating amino acid uptake by N-replete embryos. The authors concluded that this flux was not due to diffusion because of its sensitivity to pH and apparent energy dependence. Transfer to -N media caused increased uptake of AIB and glutamine. Simultaneous measurements of total free amino acid concentrations of the embryos during a 24 h period of N-deprivation revealed a lack of correlation between amino acid uptake rate and cellular amino acid concentration. While the latter fell rapidly in the first 4-8 h and then remained constant, amino acid uptake rates climbed steadily after a lag period of 4-6 h. The authors favoured the hypothesis of derepression of carrier synthesis rather than trans-inhibition of uptake by cellular amino acids.

It is evident that amino acid uptake may be subject to regulation by cellular C or N pools. Moreover, down-regulation of amino acid uptake in response to elevated cellular N levels may arise from nitrate, ammonium or amino acid feeding. This may indicate a common negative feedback from an amino acid pool which regulates all forms of N absorption (NO_3^-, NH_4^+, amino acids) and/or feedback from separate cellular N pools, as discussed earlier.

VII. Conclusions and Future Prospects

In conclusion, it may be stated that very significant progress has been achieved in characterizing the kinetics and energetics of N transport by plant roots in recent years. The cloning and sequencing of genes involved in NO_3^- and NH_4^+ transport in micro-organisms and, most recently, the CHL1 nitrate transport system of *Arabidopsis* (Tsay et al., 1993), are encouraging developments which will ultimately allow definitive testing of hypotheses derived from kinetic studies. Having said this, it remains true that the diversity of N sources for plant absorption and, in particular, their ecological significance, is worthy of considerably more attention. The complexity of higher plant compartmentation, translocation and recycling, and, especially, the integration of these physiological processes, provide major challenges to those scientists who would capitalize on the rewards of the very fruitful reductionist approaches to attempt the more daunting task of putting the parts together again.

Literature Cited

Aldag, R.W., and Young, J.L. 1970. D-amino acids in soils. I. Uptake and metabolism by seedling maize and ryegrass. *Agron. J.* **62**: 184-188.

Aslam, M., Rosichan, J.L., and Huffaker, R.C. 1987. Comparative induction of nitrate reductase by nitrate and nitrite in barley leaves. *Plant Physiol.* **83**: 579-584.

Aslam, M., Travis, R.L., and Huffaker, R.C. 1992. Comparative kinetics and reciprocal inhibition of nitrate and nitrite uptake in roots of uninduced and induced barley (*Hordeum vulgare* L.) seedlings. *Plant Physiol.* **99**: 1124-1133.

Aslam, M, Travis, R.L., and Huffaker, R.C. 1993. Comparative induciton of nitrate and nitrite uptake and reduction systems by ambient nitrate and nitrite in intact roots of barley (*Hordeum vulgare* L.) seedlings. *Plant Physiol.* **102**: 811-819.

Barber, S.A. 1974. Soil Nutrient Bioavailability: A Mechanistic Approach. John Wiley & Sons Inc. New York.

Beck, E., and Renner, U. 1989. Ammonium triggers uptake of nitrate by *Chenopodium rubrum* suspension culture cells and remobilization of their vacuolar nitrate pool. *Plant Cell Physiol.* **30**: 487-495.

Becking, J.H. 1956. On the mechanism of ammonium uptake by maize roots. *Acta. Bot. Neerl.* **5**: 2-79.

Behl, R., Tischner, R., and Raschke, K. 1988. Induction of a high capacity nitrate uptake mechanism in barley roots prompted by nitrate uptake through a constitutive low-capacity mechanism. *Planta.* **176**: 235-240.

Belton, P.S., Lee, R.B., and Ratcliffe, R.G. 1985. A ^{14}N nuclear magnetic resonance study of inorganic nitrogen metabolism in barley, maize and pea roots. *J. Exp. Bot.* **36**: 190-210.

Bennett, A.B., and Spanswick, R.M. 1983. Derepression of amino acid - H^+ cotransport in developing soybean embroys. *Plant Physiol.* **72**: 781-786.

Blackman, M.S., and McDaniel, C.N. 1980. Amino acid transport in suspension-cultured plant cells. II. Characterization of L-leucine uptake. *Plant Physiol.* **66**: 261-266.

Bloom, A.J. 1988. Ammonium and nitrate as nitrogen sources for plant growth. ISI Atlas. *Animal and Plant Scienes* 1: 55-59.

Bloom, A.J. 1989. Continuous and steady-state nutrient absorption by intact plants. *In Applications of Continuous and Steady-State Methods to Root Biology.* (Torrey J.G., and Winship L.J., eds.), Kluwer Academic Publications, Dordrecht, pp. 147-163.

Bloom, A.J., and Chapin, F.S. 1981. Differences in steady-state net ammonium and nitrate influx by cold and warm adapted barley varieties. *Plant Physiol.* 68: 1064-1067.

Bloom, A.J., and Sukrapanna, S.S. 1990. Effects of exposure to ammonium and transplant shock upon the induction of nitrate absorption. *Plant Physiol.* 94: 85-90.

Bloom, A.J., Sukrapanna, S.S., and Warner, R.L. 1992. Root respiration associated with ammonium and nitrate absorption and assimilation by barley. *Plant Physiol.* 99: 1294-1301.

Blumwald, E., and Poole, R.J. 1985. Nitrate storage and retrieval in *Beta vulgaris*: effects of nitrate and chloride on proton gradients in tonoplast vesicles. *Proc. Natl. Acad. Sci. U.S.A.* 82: 3683-3687.

Borstlap, A.C., Meenks, J.L.D., van Eck, W.F., and Briker, J.J.E. 1986. Kinetics and specificity of amino acid uptake by the duckweed *Spirodela polyrhiza. J. Exp. Bot.* 37: 1020-1035.

Breteler, H., and Arnozis, P.A. 1985. Effect of amino compounds on nitrate utilization by roots of dwarf bean. *Phytochemistry.* 24: 653-657.

Breteler, H., and Nissen, P. 1982. Effect of exogenous and endogenous nitrate concentration on nitrate utilization by dwarf beans. *Plant Physiol.* 70: 754-759.

Breteler, H., and Luczak, W. 1982. Utilization of nitrite and nitrate by dwarf bean. *Planta.* 156: 226-232.

Brezeale, J.F. 1906. The relation of sodium to potassium in soil and solution cultures. *J. Amer. Chem. Soc.* 28: 1013-1025.

Brownlee, A.G., and Arst, H.N. Jr. 1983. Nitrate uptake in *Aspergillus nidulans* and involvement of the third gene of the nitrate assimilation gene cluster. *J. Bacteriol.* 155: 1138-1146.

Brozozowska, J., Hanower, P., and Chezeau, R. 1974. Free amino acids of *Hevea braziliensis* ψ *Experientia* 30: 894-896.

Burton, D.L. 1989. Control of amino acid catabolism in soil and direct assimilation by plants. *Ph.D. Thesis.* ψ.

Bush, D.R., and Langston-Unkefer, P.J. 1988. Amino acid transport into membrane vesicles isolated from zucchini: evidence of a proton-amino acid symport in the plasmalemma. *Plant Physiol.* 88: 487-490.

Chervel, J., and Jullien, M. 1979. Amino acid uptake into cultivated mesophyll cells from *Asparagus officinalis* L. *Plant Physiol.* 63: 621-626.

Chodera, A.J., and Briskin, D.P. 1990. Chlorate transport in isolated tonoplast vesicles from red beet (*Beta vulgaris* L.) storage tissue. *Plant Sci.* 67: 151-160.

Clarkson, D.T., Saker, L.R., and Purves, J.V. 1989. Depression of nitrate and ammonium transport in barley plants with diminished sulfate status. Evidence of coregulation of nitrogen and sulfate intake. *J. Exp. Bot.* 40: 953-963.

Clarkson, D.T. 1986. Regulation of the absorption and release of nitrate by plant cells: A review of current ideas and methodology. ψ *Fundamental, Ecological and Agricultural Aspects of Nitrogen Metabolism in Higher Plants.* (Lambers, H., Neteson, J.J. and Stulen, I, eds.), Martinus Nijhoff, Boston, pp 3-27.

Clarkson, D.T., and Hanson, J.B. 1980. The mineral nutrition of higher plants. *Annu. Rev. Plant. Physiol.* 31: 239-298.

Clarkson, D.T., and Luttge, U. 1991. Mineral nutrition: Inducible and repressible nutrient transport systems. *Progress in Botany* 52: 61-83.

Clarkson, D.T., and Warner, A. 1979. Relationships between root temperature and transport of ammonium and nitrate ions by Italian and perennial ryegrass *Lolium multiflorum* and *Lolium perenne. Plant Physiol.* 64: 557-561.

Clement, C.R., Hopper, M.J., and Jonnes, L.H.P. 1978. The uptake of nitrate by *Lolium perenne* from flowing nutrient solutions. I. Effect of NO_3^- concentration. *J. Exp. Bot.* **29**: 453-464.

Collos, Y., Siddiqi, M.Y., Wang M.Y., Glass, A.D.M., and Harrison, P.J. 1992. Nitrate uptake kinetics by two marine diatoms using the radioactive tracer [13]N. *J. Exp. Mar. Biol. Ecol.* **163**: 251-260.

Cooper, H.D., and Clarkson, D.T. 1989. Cycling of amino-nitrogen and other nutrients between shoots and roots in cereals - A possible mechanism integrating shoot and root in the regulation of nutrient uptake. *J. Exp. Bot.* **40**: 753-762.

Córdoba, F., Cardenas, J., and Fernandez, E. 1986. Kinetic characterization of nitrite uptake and reduction by *Chlamydomonas reinhardtii. Plant Physiol.* **82**: 904-908.

Cox, W.J., and Reisenauer, H.M. 1973. Growth and ion uptake by wheat supplied nitrogen as nitrate, or ammonium, or both. *Plant Soil* **38**: 363-380.

Cram, W.J. 1968. Compartmentation and exchange of chloride in carrot root tissue. *Biochim. Biophys. Acta* **163**: 339-353.

Cram, W.J. 1973. Internal factors regulating nitrate and chloride influx across the plasmalemma. *J. Exp. Bot.* **24**: 328-341.

Criddle, R.S., Ward, M.R., and Huffaker, R.C. 1988. Nitrogen uptake by wheat seedlings, interactive effects of four nitrogen sources: NO_3^-, NO_2^-, NH_4^+ and urea. *Plant Physiol.* **86**: 166-175.

Datko, A.H., and Mudd, S.H. 1985. Uptake of amino acids and other organic compounds by *Lemna paucicostata* Hegelm 6746. *Plant Physiol.* **77**: 770-778.

Davis, R.F., and Higinbotham, N. 1969. Effects of external cations and respiratory inhibitors on electrical potential of the xylem exudate of excised corn roots. *Plant Physiol.* **44**: 1383-1392.

Deane-Drummond, C.E. 1982. Mechanisms for nitrate uptake into barley (*Hordeum vulgare* L. cv Fergus) seedlings grown at controlled nitrate concentrations in the nutrient medium. *Plant Sci. Lett.* **24**: 79-89.

Deane-Drummond, C.E. 1984a. The apparent induction of nitrate uptake by *Chara corallina* cells following pretreatment with or without nitrate and chlorate. *J. Exp. Bot.* **35**: 1182-1193.

Deane-Drummond, C.E. 1984b. Nitrate transport into *Chara corallina* cells using ClO_3^- as an analog for nitrate. II Comparison with ^{14}C methylamine fluxes at different pH and NH_4^+/NO_3^- interactions. *J. Exp. Bot.* **35**: 1299-1308.

Deane-Drummond, C.E. 1990. Biochemical and biophysical aspects of nitrate uptake and its regulation. ψ *Nitrogen in Higher Plants.* (Abrol. Y.P., ed.), John Wiley & Sons Inc. New York. pp. 1-37.

Deane-Drummond, C.E., and Glass, A.D.M. 1982. Nitrate uptake into barley (*Hordeum vulgare*) plants. A new approach using [36]ClO_3^- as an analog for NO_3^-. *Plant Physiol.* **70**: 50-54.

Deane-Drummond, C.E., and Glass, A.D.M. 1983. Short-term studies of nitrate uptake into barley plants using ion-specific electrodes and [36]ClO_3^-. *Plant Physiol.* **73**: 105-110.

Deane-Drummond, C.E., and Jacobsen, E. 1986. Characteristics of [36]ClO_3^- influx into nitrate reductase deficient mutant E_1 *Pisum sativum* seedlings: Evidence for restricted 'induction' by nitrate compared with wild type. *Plant Sci.* **46**: 169-173.

Despeghel, J.P., and Delrot, S. 1983. Energetics of amino acid uptake by *Vicia faba* leaf tissue. *Plant Physiol.* **71**: 1-6.

Dhugga, K.S., Waines, J.G., and Leonard, R.T. 1988a. Nitrate absorption by corn roots: Inhibition by phenylglyoxal. *Plant Physiol.* **86**: 759-763.

Dhugga, K.S., Waines, J.G., and Leonard, R.T. 1988b. Correlated induction of nitrate uptake and membrane polypeptides in corn roots. *Plant Physiol.* **87**: 120-125.

Dietz, K.J., Jager, R., Kaiser, G., and Martinoia, E. 1990. Amino acid transport across the tonoplast of vacuoles isolated from barley mesophyll protoplasts. *Plant Physiol.* **92**: 123-129.

Doddema, H., and Telkamp, G.P. 1979. Uptake of nitrate by mutants of *Arabidopsis thaliana*, disturbed in uptake or reduction of nitrate. II. Kinetics. *Physiol. Plant.* **45**: 332-338.

Doddema, H., and Otten, H. 1979a. Uptake of nitrate by mutants of *Arabidopsis thaliana*, disturbed in uptake or reduction of nitrate. III. Regulation. *Physiol. Plant.* **45**: 339-346.

Doddema, H, Hofstra, J.J., and Feenstra, W.J. 1978. Uptake of nitrate by mutants of *Arabidopsis thaliana*, disturbed in uptake or reduction of nitrate. I. Effect of nitrogen source during growth on uptake of nitrate and chlorate. *Physiol. Plant.* **43**: 343-350

Epstein, E. 1976. Kinetics of ion transport and the carrier concept. In Encylopedia of Plant Physiology, New Series. (Luttge, U. and Pitman, M.G. eds.), Berlin/Heidelberg/New York: Springer. **2B**: 70-94.

Etherton, B., and Rubinstein, B. 1978. Evidence for amino acid-H^+ co-transport in oat coleoptiles. *Plant Physiol.* **61**: 933-937.

Ezeta, F.W., and Jackson, W.A. 1975. Nitrate translocation by detopped corn seedlings. *Plant Physiol.* **56**: 148-156.

Fabiny, J.M., Jayakumar, A., Chinault, C., and Barnes, E.M. Jr. 1991. Ammonium transport in *Escherischia coli*: location and nucleotide sequence of the amt gene. *J. Gen. Microbiol.* **137**: 983-989.

Felle, H., and Johannes, E. 1990. The regulation of proton/amino acid symport in *Riccia fluitans* L. by cytosolic pH and proton pump activity. *J. Exp. Bot.* **41**: 587-592.

Ferguson, A.R. 1970. Nitrogen metabolism of *Spirodela oligorrhizao* III. Amino acids and the utilization of nitrate. *Planta.* **90**: 365-369.

Franco, A.R., Cárdenas, J., and Fernández, F.A. 1987. A mutant of *Chlamydomonas reinhardtii* altered in the transport of ammonium and methylammonium. *Mol. Gen. Genet.* **206**: 414-418.

Fuggi, A. 1989. Competition between nitrate and nitrite as a tool to study the regulation of nitrate metabolism by ammonium in the microalga *Cyanidium caldarium*. *Plant Physiol. Biochem.* **27**: 563-568.

Glass, A.D.M. 1988. Nitrogen uptake by plant roots. *ISI Atlas Animal and Plant Sciences* **1**: 151-156.

Glass, A.D.M., Thompson, R.G., and Bordeleau, L. 1985. Regulation of NO_3^- influx in barley. Studies using $^{13}NO_3^-$. *Plant Physiol.* **77**: 379-381.

Glass, A.D.M., Shaff, J.E., and Kochian, L.V. 1992. Studies of nitrate uptake in barley. IV. Electrophysiology. *Plant Physiol.* **99**: 456-463.

Goerlach, J., and Willms-Hoff, I. 1992. Glycine uptake into barley mesophyll vacuoles is regulated but not energized by ATP. *Plant Physiol.* **99**: 134-139.

Goyal, S.S., and Huffaker, R.C. 1986a. The uptake of NO_3^-, NO_2^- and NH_4^+ by intact wheat (*Triticum aestivum*) seedlings. *Plant Physiol.* **82**: 1051-1056.

Goyal, S.S., and Huffaker, R.C. 1986b. A novel approach and a fully automated microcomputer-based system to study kinetics of NO_3^-, NO_2^- and NH_4^+ transport simultaneously by intact wheat seedlings. *Plant Cell Environ.* **9**: 209-215.

Guy, M., Reinhold, L., and Laties, G.G. 1978. Membrane transport of sugars and amino acids in isolated protoplasts. *Plant Physiol.* **61**: 593-596.

Harrington, H.M., and Henke, R.R. 1981. Amino acid transport into cultured tobacco cells. I. Lysine transport. *Plant Physiol.* **67**: 373-378.

Haynes, R.J., and Goh, K.M. 1978. Ammonium and nitrate nutrition of plants. *Biol. Rev.* **53**: 465-510.

Heimer, Y.M., and Filner, P. 1971. Regulation of the nitrate assimilation pathway in cultured tobacco cells. III. The nitrate uptake system. *Biochim. Biophys. Acta* **230**: 363-372.

Henriksen, G.H., Raman, D.R., Walker, L.P., and Spanswick, R.M. 1992. Measurement of net fluxes of ammonium and nitrate using ion-selective microelectrodes. II. Patterns of uptake along the root axis and evaluation of the microelectrode flux estimation technique. *Plant Physiol.* **99**: 734-747.

Higinbotham, N., Etherton, B., and Foster, R.J. 1967. Mineral ion content and cell transmembrane electropotentials of pea and oat seedling tissue. *Plant Physiol.* **42**: 37-46.

Hole, D.J., Emran, A., Farres, Y., and Drew, M.C. 1990. Induction of nitrate transport in maize roots and kinetics of influx measured with nitrogen-13. *Plant Physiol.* **93**: 642-647.

Horneyer, U., Litek, K, Huchzermeyer, B., and Schultz, G. 1989. Uptake of phenylalanine into iso-lated barley vacuoles is driven by both tonoplast adenosine triphosphatase and pyrophos-phatase. *Plant Physiol.* **89**: 1388-1393.

Ingemarsson, B., Oscarson, P., af Ugglas, M., and Larsson, C.-M. 1987a. Nitrogen utilization in *Lemna* II. Studies of nitrate uptake using $^{13}NO_3^-$. *Plant Physiol.* **85**: 860-864.

Ingemarsson, B., Oscarson, P., Ugglas, M., and Larsson, C.-M. 1987b. Nitrogen utilization in *Lemna* III. Short-term effects of ammonium on nitrate uptake and nitrate reduction. *Plant Physiol.* **85**: 865-867.

Jackson, R.B., and Caldwell, M.M. 1993. The scale of nutrient heterogeneity around individual plants and its quantification with geostatics. *Ecology* **74**: 612-614.

Jackson, W.A., Flesher, D., and Hageman, R.H. 1973. Nitrate uptake by dark-grown corn seed-lings. Some characteristics of apparent induction. *Plant Physiol.* **51**: 120-127.

Jackson, W.A., Kwik, K.D., Volk, R.J., and Butz, R.G. 1976. Nitrate influx and efflux by intact wheat seedlings: Effects of prior nitrate nutrition. *Planta* **132**: 149-156.

Jackson, W.A., Pan, W.L., Moll, R.H., and Kamprath, E.J. 1986. Uptake, translocation, and reduc-tion of nitrate. In *Biochemical Basis of Plant Breeding. Vol. II. Nitrogen Metabolism.* (Neyra C.A. ed.), CRC Press, Cleveland, Ohio. pp. 63-108.

Jackson, W.A., Volk, R.J., and Tucker, T.C. 1972. Apparent induction of nitrate uptake in nitrate-depleted plants. *Agron. J.* **64**: 518-521.

Jayakumar, A., Hwang, S.J., Fabiny, J.M., Chinault, A.C., and Barnes, E.M., Jr. 1981. Isolation of an ammonium or methylamine ion transport mutant of *Escherischia coli* and complementa-tion by the cloned gene. *J. Bacteriol.* **171**: 996-1001.

Johansen, A., Jakobsen, I., and Jensen, E.S. 1993. Hyphal transport by a vesicular-arbuscular fungus of N applied to the soil as ammonium or nitrate. *Biol. Fert. Soils* **16**: 66-67.

Joseph, R.A., Van Hai, T., and Lambert, J. 1975. Multiphasic uptake of ammonium by soybean roots. *Physiol. Plant.* **34**: 321-325.

Jung, K.D., and Lüttge, U. 1980. Amino acid uptake by *Lemna gibba* by a mechanism with affinity to neutral L- and D- amino acids. *Planta.* **150** : 230-235.

Jung, K.D., Lüttge, U., and Fischer, E. 1982. Uptake of neutral and acidic amino acids by *Lemna gibba* correlated with the H^+-electrochemical gradient at the plasmalemma. *Physiol. Plant.* **55**: 351-355.

Kaestner, K.H., and Sze, H. 1987. Potential-dependent anion transport in tonoplast vesicles from oat roots. *Plant Physiol.* **83**: 483-489.

Kielland, K. 1994. Amino acid absorption by arctic plants: implications for plant nutrition and nitro-gen cycling. *Ecology*, in Press.

King, B.J., Siddiqi, M.Y., and Glass, A.D.M. 1992. Studies of the uptake of nitrate in barley. V. Esti-mation of root cytoplasmic nitrate concentration using nitrate reductase activity - implications for nitrate influx. *Plant Physiol.* **99**: 1582-1589.

King, B.J. Siddiqi, M.Y., Ruth, T.J., Warner, R.L., and Glass, A.D.M. 1993. Feedback regulation of nitrate influx in barley roots by nitrate, nitrite and ammonium. *Plant Physiol.* **102**: 1279-1286.

King J., and Olenuik, F.H. 1973. The uptake of alanine-^{14}C by soybean root cells grown in sterile suspension culture. *Can. J. Bot.* **51**: 1109-1114.

Kleiner, D. 1981. The transport of NH_3 and NH_4^+ across biological membranes. *Biochim. Biophys. Acta* **639**: 41-52.

Kinraide, T.B. 1981. Interamino acid inhibition of transport in higher plants: Evidence for two trans-port channels with ascertainable affinities for amino acids. *Plant Physiol.* **68**: 1327-1333.

Kinraide, T.B., and Etherton, B. 1980. Electrical evidence for different mechanisms of uptake of basic neutral and acidic amino acids in oat coleoptile. *Plant Physiol.* **65**: 1085-1089.

Kinraide, T.B., Newman, T.A., and Etherton, B. 1984. A quantitative simulation model for H^+-amino acid cotransport to interpret the effects of amino acids on membrane potential and extra-cellular pH. *Plant Physiol.* **76**: 806-813.

Kuprevich, V.F., and Shcherbakova, J.A., 1971. Comparative enzymatic activity in diverse types of soil. In *Soil Biochemistry.* Vol 2. (McLaren, A.D. and Skujins, J. eds.), Marcel Dekker Inc. New York. pp. 167-201.

LaBrie, S.T., Wilkinson, J.Q., and Crawford, N.M. 1991. Effect of chlorate treatment on nitrate reductase and nitrite reductase gene expression in *Arabidopsis thaliana. Plant Physiol.* **97**: 873-879.

Lazoff, D.B., Rufty, T.W., and Redinbaugh, M.G. 1992. Localization of nitrate absorption and translocation within morphological regions of the corn root. *Plant Physiol.* **100**: 1251-1258.

Lee, R.B. 1979. The effect of nitrate on root growth of barley and maize. *New Phytol.* **83**: 615-622.

Lee, R.B. 1982. Selectivity and kinetics of ion uptake by barley plants following nutrient deficiency. *Ann. Bot.* **50**: 429-449.

Lee, R.B., and Clarkson, D.T. 1986. Nitrogen-13 studies of nitrate fluxes in barley roots. I. Compartmental analysis from measurements of ^{13}N efflux. *J. Exp. Bot.* **37**: 1753-1767.

Lee, R.B., and Drew, M.C. 1986. Nitrogen-13 studies of nitrate fluxes in barley roots. II. Effect of plant N-status on the kinetic parameters of nitrate influx. *J. Exp. Bot.* **185**: 1768-1779.

Lee, R.B., and Drew, M.C. 1989. Rapid, reversible inhibition of nitrate influx in barley by ammonium. *J. Exp. Bot.* **40**: 741-752.

Lee, R.B., and Ratcliffe, R.G. 1991. Observations on the subcellular distribution of the ammonium ion in maize root tissue using in-vivo ^{14}N-nuclear magnetic resonance spectroscopy. *Planta* **183**: 359-367.

Lee, R.B., and Rudge, K.A. 1986. Effects of nitrogen deficiency on the absorption of nitrate and ammonium by barley plants. *Ann. Bot.* **57**: 471-486.

Lee, R.B., Purves, J.V., Ratcliffe, R.G., and Saker, L.R. 1992. Nitrogen assimilation and the control of ammonium and nitrate absorption by maize roots. *J. Exp. Bot.* **43**: 1385-1396.

Lefebvre, D.D., and Glass, A.D.M. 1982. Regulation of phosphate influx in barley roots: Effects of phosphate deprivation and reduction of influx with provision of orthophosphate. *Physiol. Plant.* **54**: 199-206.

Lewis, O.A.M., and Chadwick, S. 1983. An ^{15}N investigation into nitrogen assimilation in hydroponically-grown barley (*Hordeum vulgare* L. cv. clipper) in response to nitrate, ammonium and mixed nitrate and ammonium nutrition. *New Phytol.* **95**: 635-646.

Li, Z-C., and Bush, D.R. 1990. pH-dependent amino acid transport into plasma membrane vesicles isolated from sugar beet leaves. Evidence for carrier-mediated electrogenic flux through multiple transport systems. *Plant Physiol.* **94**: 268-277.

Li, Z.C., and Bush, D.R. 1991.Δ pH-dependent amino acid transport into plasma membrane vesicles isolated from sugar beet (*Beta vulgaris*) leaves. II. Evidence for multiple aliphatic, neutral amino acid symports. *Plant Physiol.* **96**: 1338-1344.

Lien, R., and Rognes, S.E. 1977. Uptake of amino acids by barley leaf slices: Kinetics, specificity and energetics. *Physiol. Plant.* **41**: 175-183.

Lüttge, U., Jung, K-D., and U-Eberius, C.J. 1981. Evidence for amino acid-H^+ cotransport in *Lemna gibba* given by effects of fusicoccin. *Z. Pflanzenphysiol.* **102**: 117-125.

Lycklama, J.C. 1963. The absorption of ammonium and nitrate by perennial ryegrass. *Acta. Bot. Neerl.* **12**: 361-423.

Macklon, A.E.S. 1975. Cortical cell fluxes and transport to the stele in excised root segments of *Allium cepa* L. I. Potassium, sodium and chloride. *Planta* **122**: 109-130.

Macklon, A.E.S., Ron, M.M., and Sim. A. 1990. Cortical cell fluxes of ammonium and nitrate in excised root segments of *Allium cepa* L.; Studies using [15]N. *J. Exp. Bot.* **41**: 359-370.

Madueno, F., Vega-Palas, M.A., Flores, E., and Herrero, A. 1988. A cytoplasmic membrane protein repressible by ammonium in *Synechococcus* R2: altered expression in nitrate-assimilation mutants. *FEBS Lett.* **239**: 289-291.

MacKown, C.T., Jackson, W.A., and Volk, R.J. 1982. Restricted nitrate influx and reduction in corn seedlings exposed to ammonium. *Plant Physiol.* **69**: 353-359.

MacKown, C.T., and McClure, P.R. 1988. Development of accelerated net nitrate uptake: Effects of nitrate concentration and exposure time. *Plant Physiol.* **87**: 162-166.

MacKown, C.T., Volk, R.J., and Jackson, W.A. 1981. Nitrate accumulation, assimilation and transport by decapitated corn roots. *Plant Physiol.* **68**: 133-138.

Maeck, G., and Tischner, R. 1986. Nitrate uptake and reduction in sugarbeet seedlings. In Fundamental, Ecological and Agricultural Aspects of Nitrogen Metabolism in Higher Plants. (Lambers, H., Neeteson, J.J. and Stulen, I. eds), Martinus Nijhoff, Boston, pp 33-36.

Margolis, H.A., and Vezina, L.P. 1988. Nitrate content, amino acid composition and growth of yellow birch seedlings in response to light and nitrogen source. *Tree Physiol.* **4**: 245-253.

Martinoia, E., Schramm, M.J., Kaiser, G., Kaiser, W.M., and Heber, U. 1986. Transport of anions in isolated barley vacuoles. I. Permeability to anions and evidence for a Cl⁻ uptake system. *Plant Physiol.* **80**: 895-901.

Martinoia, E., Thume, M., Vogt, E., Rentsch, D., and Dietz, K-J. 1991. Transport of arginine and aspartic acid into isolated barley mesophyll vacuoles. *Plant Physiol.* **97**: 644-650.

McClure, P.R., Kochian, L.V., Spanswick, R.M., and Shaff, J.E. 1990. Evidence for cotransport of nitrate and protons in maize roots. I. Effects of nitrate on the membrane potential. *Plant Physiol.* **93**: 281-289.

McClure, P.R., Ohmolt, T.E., and Pace, G.M. 1986. Anion uptake in maize roots: Interactions between chlorate and nitrate. *Physiol. Plant.* **68**: 107-112.

McClure, P.R., Ohmolt, T.E., Pace, G.M., and Bouthyette P-Y. 1987. Nitrate-induced changes in protein synthesis and translation of RNA in maize roots. *Plant Physiol.* **84**: 52-57.

McCutcheon, S.L., Ciccarelli, B.W., Chung, I., Shelp, B., and Bown, A.W. 1988. L-Glutamate-dependent medium alkalinization by *Asparagus* mesophyll cells. *Plant Physiol.* **88**: 1042-1047.

McDaniel, C.N., Holterman, R.K., Bone, R.F., and Wozniak, P.M. 1982. Amino acid transport in suspension-cultured plant cells. III. Common carrier system for the uptake of L-arginine, L-aspartic acid, L-histidine, L-leucine and L-phenylalanine. *Plant Physiol.* **69**: 246-249.

McDaniel, C.N., and Wozniak, P.M. 1982. Amino acid transport in suspension-cultured cells. V. Influence of L-leucine, carbon-free and nitrogen-free media on L-leucine uptake. *Planta.* **154**: 110-114.

Miller, A.J., and Zhen, R.G. 1991. Measurement of intracellular nitrate concentrations in *Chara* using nitrate-selective microelectrodes. *Planta* **184**: 47-52.

Miller, A.J., Zhen, R.G., and Smith, S.J. 1992. Compartmentation of nitrate. *J. Exp. Bot.* **43**: p. 49 (Supplement).

Minotti, P.L., Williams, D.C., and Jackson, W.A. 1969. The influence of ammonium on nitrate reduction in wheat seedlings. *Planta* **86**: 267-271.

Miyagi, N., Satoh, S., and Fujii, T. 1992. A nitrate-inducible plasma membrane protein of a marine alga, *Heterosigma akashiwo. Plant Cell Physiol.* **33**: 971-976.

Monreal, C.M., and McGill, W.B. 1985. Centrifugal extraction and determination of free amio acids in soil solutions by TLC using tritiated 1-Fluoro-2,4-dinitrobenzene. *Soil Biol. Biochem.* **17**: 533-539.

Monreal, C.M., and McGill, W.B. 1989. The dynamics of free cystine cycling at steady-state through the solutions of selected cultivated and unclutivated chernozemic and luvisolic soils. *Soil Biol. Biochem.* **21**: 689-694.

Morgan, M.A., and Jackson, W.A. 1988. Inward and outward movement of ammonium in root systems: transient responses during recovery from nitrogen deprivation in presence of ammonium. *J. Exp. Bot.* **39**: 179-191.

Mori, S., and Nishimura, Y. 1979. Nitrogen absorption by plant roots from culture mediums where organic and inorganic nitrogen coexist. II. Which nitrogen is preferentially absorbed from among [U-^{14}C] GluNH$_2$, [2, 3-^{3}H] Arg and Na^{15}NO$_3$? *Soil Sci. Pl. Nutrit.* **25**: 51-58.

Neyra, C.A., and Hageman, R.H. 1975. Nitrate uptake and induction of nitrate reduction in excised corn roots. *Plant Physiol.* **56**: 692-695.

Neyra, C.A., and Hageman, R.H. 1976. Relationship between carbon dioxide, malate and nitrate accumulation and reduction in corn (*Zea mays* L.) seedlings. *Plant Physiol.* **58**: 726-730.

Noji, S., Nolmo, T., Saito, T., and Taniguchi, S. 1989. The narK gene product participates in nitrate transport induced in *Escherichia coli* nitrate-respiring cells. *FEBS Lett.* **252**: 139-143.

Nye, P.H., and Tinker, P.B. 1977. Solute movement in the soil-root system. *Studies in Ecology. Vol. 4.* Blackwell. Oxford. 34 p.

Oaks, A., Aslam, M., and Boesel, I. 1977. Ammonium and amino acids as regulators of nitrate reductase in corn roots. *Plant Physiol.* **59**: 391-394.

Omata, T. 1991. Cloning and characterization of the nrtA gene that encodes a 45-kDa protein involved in nitrate transport in the cyanobacterium *Synechococcus* PCC 7942. *Plant Cell Physiol.* **32**: 151-157.

Omata, T., Ohmori, M., Arai, N., and Ogawa, T. 1989. Genetically engineered mutant of the cyanobacterium (Synechococcus) PCC 7942 defective in nitrate transport. *Proc. Natl. Acad. Sci. U.S.A.* **86**: 6612-6616.

Oscarson, P., Ingemarsson, B., af Ugglas, M., and Larsson, C.-M. 1987. Short-term studies of NO$_3^-$ uptake in *Pisum* using ^{13}NO$_3^-$. *Planta* **170**: 550-555.

Paszkowski, J., Lorz, H., Potrykus, I., and D-Ventling, C. 1980. Amino acid uptake and protein synthesis in cultured cells and protoplasts of *Zea mays* L. *Z. Pflanzenphysiol.* **99**: 251-259.

Pate, J.S. 1980. Transport and partitioning of nitrogenous solutes. *Annu. Rev. Plant Physiol.* **31**. 313-340.

Paul. E.A., and Schmidt, E.L. 1961. Formation of free amino acids in rhizosphere and nonrhizosphere soil. *Soil Sci. Soc. Amer. Proc.* **25**: 359-362.

Pilbeam, D.J., and Kirkby, E.A. 1990. The physiology of nitrate uptake. In *Nitrogen in Higher Plants.* (Abrol. Y.P. ed.), John Wiley & Sons Inc. New York. pp. 39-64

Pitman, M.G. 1969. Simulation of Cl$^-$ uptake by low-salt barley roots as a test of models of salt uptake. *Plant Physiol.* **44**: 1417-1427

Pitman, M.G. 1972. Uptake and transport of ions in barley seedlings. II. Evidence for two active transport stages in transport to the shoot. *Aust. J. Plant. Physiol.* **25**: 243-257.

Poirier, Y., Thoma, S., Sommerville, C., and Schiefelbein. J. 1992. A mutant of *Arabidopsis* deficient in xylem loading of phosphate. *Plant Physiol.* **97**: 1087-1093.

Pope, A.J., and Leigh, R.A. 1990. Characterization of chloride transport at the tonoplast of higher plants using a chloride-sensitive fluorescent probe. *Planta.* **181**: 406-413.

Presland, M.R., and McNaughton, G.S. 1986. Whole plant studies using radioactive 13-nitrogen. IV. A compartmental model for the uptake and transport of ammonium ions by *Zea mays*. *J. Exp. Bot.* **37**: 1619-1632.

Rao, K.P., and Rains, D.W. 1976. Nitrate absorption by barley. I. Kinetics and energetics. *Plant Physiol.* **57**: 55-58.

Raper, C.D., Vessey, J.K., and Henry, L.T. 1991. Increase in nitrate uptake by soybean plants during interruption of the dark period with low intensity light. *Physiol. Plant.* **81**: 183-189

Reinhold, L., and Kaplan, A. 1984. Membrane transport of sugars and amino acids. *Annu. Rev. Plant Physiol.* **35**: 45-83.

Reinhold, L., Shtarkshall, R.A., and Ganot, D. 1970. Transport of amino acids in barley leaf tissue. II. The kinetics of uptake of an unnatural analogue. *J. Exp. Bot.* **21**: 926-932.

Reisenauer, H.M. 1966. Mineral nutrients in soil solution. In *Environmental Biology.* (Altman, P.L. and Dittmer, D.S., eds.), *Fed. Amer. Soc. Exp. Biol.* Bethesda. pp 507-508.

Riens, B., Lohaus, G., Heineke, D., and Heldt, W.H. 1991. Amino acid and sucrose content determined in the cytosolic, chloroplastic, and vacuolar compartments and in the phloem sap of spinach leaves. *Plant Physiol.* **97**: 227-233.

Roberts, J.K.M., and Pang, M.K.L. 1992. Estimation of ammonium ion distribution between cytoplasm and vacuole using nuclear magnetic resonance spectroscopy. *Plant Physiol.* **100**: 1571-1574.

Rodriguez-Navarro, A., Blatt, M.R., and Slayman, C.L. 1986.. A potassium-proton symport in *Neurospora crassa*. *J. Gen. Physiol.* **87**: 649-674.

Rufty, T.W. Jr., Israel, D.W., and Volk, R.J. 1984. Assimilation of $^{15}NO_3^-$ taken up by plants in the light and in the dark. *Plant Physiol.* **76**: 769-775.

Rufty, T.W. Jr., Volk, R.J., and MacKown, C.T. 1987. Endogenous NO_3^- in the root as a source of substrate for reduction in the light. *Plant Physiol.* **84**: 1421-1426.

Rufty, T.W. Jr., Siddiqi, M.Y., Glass, A.D.M., and Ruth, T.J. 1991. Altered $^{13}NO_3^-$ influx in phosphorus limited plants. *Plant Sci.* **76**: 43-48.

Sauer, N., and Tanner, W. 1985. Selection and characterization of *Chlorella* mutants deficient in amino acid transport: further evidence for three independent systems. *Plant Physiol.* **79**: 760-764.

Sauer, N., Komor, E., and Tanner, W. 1983. Regulation and characterization of two inducible amino-acid transport systems in *Chlorella vulgaris*. *Planta.* **159**: 404-410.

Schachtman, D.P., Schroeder, J.I., Lucas, W.J., Anderson, J.A., and Gaber, R.F. 1992. Expression of an inwardly-rectifying potassium channel by the *Arabidopsis* KAT1 cDNA. *Science* **258**: 1654-1658.

Schaefer, N., Wildes, R.A., and Pitman, M.G. 1975. Inhibition by p-flourophenylalanine of protein synthesis and of ion transport across the roots in barley seedlings. *Aust. J. Plant Physiol.* **2**: 61-73.

Scherer, H.W., MacKown, C.T., and Leggett, J.E. 1984. Potassium-ammonium uptake interactions in tobacco seedlings. *J. Exp. Bot.* **156**: 1060-1070.

Schlee, J., and Komor, E. 1986. Ammonium uptake by *Chlorella*. *Planta.* **168**: 232-238.

Schobert, D., and Komor, E. 1987. Amino acid uptake by *Ricinus communis*: Characterization and physiological significance. *Plant Cell Environ.* **10**: 493-500.

Schobert, D., Köckenberger, W., and Komor, E. 1988. Uptake of amino acids by plants from the soil: A comparative study with castor bean seedlings grown under natural and axenic soil conditions. *Plant Soil* **109**: 181-188.

Schumaker, K.S., and Sze, H. 1987. Decrease of pH gradients in tonoplast vesicles by NO_3^- and Cl^-: Evidence for a H^+-coupled anion transport. *Plant Physiol.* **83**: 490-496.

Serra, J.L., Llama, M.J., and Cadenas, E. 1978. Nitrate utilization by the diatom *Skeletonema costatum*. I. Kinetics of nitrate uptake. *Plant Physiol.* **62**: 987-990.

Servailes, J.C., Schrader, L.E., and Jung, D.J. 1979. Energy dependent loading of amino acids and sucrose into the phloem of soybean. *Plant Physiol.* **64**: 546-560.

Siddiqi, M.Y., and Glass, A.D.M. 1987. Regulation of K^+ influx in barley: Evidence for the direct control of influx by K^+ concentration of root cells. *J. Exp. Bot.* **38**: 935-947.

Siddiqi, M.Y., Glass, A.D.M., Ruth, T.J., and Fernando, M. 1989. Studies of the regulation of nitrate influx by barley seedlings using $^{13}NO_3^-$. *Plant Physiol.* **90**: 806-813.

Siddiqi, M.Y., Glass, A.D.M., Ruth, T.J., and Rufty, T.W. 1990. Studies of the uptake of nitrate in barley. I. Kinetics of $^{13}NO_3^-$ influx. Plant Physiol. **93**: 1426-1432.

Siddiqi, M.Y., Glass, A.D.M., and Ruth, T.J. 1991. Studies of the uptake of nitrate in barley III. Compartmentation of NO_3^-. *J. Exp. Bot.* **42**: 1455-1463.

Siddiqi, M.Y., King, B.J., and Glass, A.D.M. 1992. Effects of nitrate, chlorate, and chlorite on nitrate uptake and nitrate reductase activity. *Plant Physiol.* **100**: 644-650.

Sivak, M.N., Lara, C., Romero, J.M., Rodriguez, R., and Guerrero, M.G. 1989. Relationship between a 47-kDa cytoplasmic membrane polypeptide and nitrate transport in *Anacystis nidulans*. *Biochem. Biophys. Res. Comm.* **158**: 257-262.

Smart, D.R., and Bloom, A.J. 1988. Kinetics of ammonium and nitrate uptake among wild and cultivated tomatoes. *Oecologia* **76**: 336-340.

Soldal, T., and Nissen, P. 1978. Multiphasic uptake of amino acids by barley roots. *Physiol. Plant* **43**: 181-188.

Suzuki, M. 1981. L-leucine transport in isolated protoplasts of *Vinca* suspension culture: Characterization of uptake. *Plant Cell Physiol.* **22**: 1269-1278.

Syrett, P.J., and Peplinska, A.M. 1988. Effects of nitrogen-deprivation, and recovery from it, on the metabolism of microalgae. *New Phytol.* **109**: 289-296.

Triplett, E.W., Barnett, N.M., and Blevins, D.G. 1980. Organic acids and ionic balance in xylem exudate of wheat during nitrate or sulfate absorption. *Plant Physiol.* **65**: 610-613.

Tsay, Y-F., Schroeder, J.I., Feldmann, K.A., and Crawford N.M. 1993. The herbicide sensitivity gene CHL1 of *Arabidopsis* encodes a nitrate-inducible nitrate transporter. *Plant Cell* **72**: 705-713.

Tyerman, S.D. 1992. Anion channels in plants. *Annu. Rev. Plant Physiol.* **43**: 351-373.

Tyerman, S.D., and Findlay, G.P. 1989. Current-voltage curves of single Cl⁻ channels which coexist with two types of K$^+$ channel in the tonoplast of *Chara corallina*. *J. Exp. Bot.* **40**: 105-117.

Ullrich, W.R. 1987. Nitrate and ammonium uptake in green algae and higher plants: Mechanism and relationship with nitrate metabolism. In *Inorganic Nitrogen Metabolism*. (Ullrich, W.R., Aparicio, P.J., Syrett, P.J. and Castillo, F., eds.), Springer-Verlag, Berlin, pp. 32-38.

Ullrich, W.R., and Novacky, A. 1981. Nitrate-dependent membrane potential changes and their induction in *Lemna gibba*. *Plant Sci. Lett.* **22**: 211-217.

Ullrich, W.R., Larsson, M., Larsson, C-M., Lesch, S., and Novacky, A. 1984. Ammonium uptake in *Lemna gibba* G1, related membrane potential change and inhibition of anion uptake. *Physiol. Plant.* **61**: 369-376.

Unkles, S.E., Hawker, K.L., Grieve, C., Campbell, E.I., Montague, P., and Kinghorn, J.R. 1991. crnA encodes a nitrate transporter in *Aspergilus nidulans*. *Proc. Natl. Acad. Sci. USA* **88**: 204-208.

van den Honert, T.H., and Hooymans, J.J.M. 1955. On the absorption of nitrate by maize in water culture. *Acta. Bot. Neerl.* **4**: 376-384.

ver Nooy, C.D., and Lin, W. 1986. Amino acid transport in protoplasts isolated from soybean leaves. *Plant Physiol.* **81**: 8-11.

Wang, M.Y., Siddiqi, M.Y., Ruth, T.J., and Glass, A.D.M. 1993a. Ammonium uptake by rice roots. I. Fluxes and subcellular distribution of $^{13}NH_4^+$. *Plant Physiol.* **103**: 1249-1258.

Wang, M.Y., Siddiqi, M.Y., Ruth, T.J., and Glass, A.D.M. 1993 b. Ammonium uptake by rice roots. II. Kinetics of $^{13}NH_4^+$ influx across the plasmalemma. *Plant Physiol.* **103**: 1259-1267.

Wang, M.Y., Glass, A.D.M., Shaff, J.E., and Kochian, L.V. 1994. Ammonium uptake by rice roots. III. Electrophysiology. *Plant Physiol.* **104**: 899-906.

Warner, R.L., and Huffaker, R.C. 1989. Nitrate transport is independent of NADH and NAD(P)H nitrate reductases in barley seedlings. *Plant Physiol.* **91**: 947-953.

Watt, D.A., Anory, A.M., Watt, M.P., and Creswell, C.F. 1993. Appearance of nitrate concentration-dependent polypeptides in N-Limited *Chlamydomonas reinhardtii* cells. *J. Exp. Box.* **44**: 447-455.

Werner-Washburne, M., and Keegstra, K. 1985. L-aspartate transport into pea chloroplasts. *Plant Physiol.* **78**: 221-227.

Wieneke, J. 1992. Nitrate fluxes in squash seedlings measured with ^{13}N. *J. Plant Nutriti.* **15**: 99-124.

Winter, H., Lohaus, G., and Heldt, H.W. 1992. Phloem transport of amino acids in relation to their cytosolic levels in barley leaves. *Plant Physiol.* **99**: 996-1004.

Wright, D.E. 1962. Amino acid uptake by plant roots. *Arch. Biochem. Biophys.* **97**: 174-180.

Wyse, R., and Komor, E. 1984. Mechanisms of amino acid uptake by sugarcane-suspension cells. *Plant Physiol.* **76**: 865-870.

Zhang, N., and MacKown, C.T. 1993. Nitrate fluxes and nitrate reductase activity of suspension cultured tobacco cells: Effects of internal and external nitrate concentrations. *Plant Physiol.* **102**: 851-857.

Zhen, R-G., Smith, S.J., and Miller, A.J. 1992. A comparison of nitrate-selective microelectodes made with different nitrate sensors and the measurement of intracellular nitrate activities in cells of excised barley roots. *J. Exp. Bot.* **43**: 131-138.

Zhen, R-G., Koyro, H-W., Leigh, R.A., Tomos, A.D., and Miller, A.J. 1991. Compartmental nitrate concentrations in barley root cells measured with nitrate-selective microelectrode and by single-cell sap sampling. *Planta.* **185**: 356-361.

Nitrogen Nutrition in Higher Plants, 1995
Editors : H.S. Srivastava & R.P. Singh
Associated Publishing Co., New Delhi, India
pp. 57-129.

Molecular Biology of Nodule Development and Nitrogen Fixation in *Rhizobium*—Legume Symbiosis

S.S. SINDHU and K.R. DADARWAL

I. Introduction

Nitrogen, an element essential to all life, makes up approximately 80% of the elements present in the earth's atmosphere. Among plant nutrients, this element constitutes one of the three major plant nutrients and successful crop production to a very great extent depends on the availability of this nutrient in adequate amounts. The plants and other eukaryotes are unable to directly utilize atmospheric nitrogen, or dinitrogen to meet their biological requirements for this element. However, plants can utilize combined forms of nitrogen such as ammonia and nitrate as sources of nitrogen. The increased cost of fertilizer production coupled with progressively increasing use of chemical fertilizers, particularly needed by high yielding varieties are adding to the cost of cultivation of crops. Recent energy crisis, rapid depletion of non-renewable sources like naphtha and natural gas and release of pollutants during fertilizer production has necessitated the development of alternate or supplemental technologies to obtain this plant nutrient. In this context, biological nitrogen fixation (BNF) has drawn much attention and in this process, dinitrogen gas is reduced to ammonia using the bacterial enzyme nitrogenase.

During the past two decades, there has been intense interest in nitrogen-fixing bacteria that are capable of forming symbiotic relationships with plants. Three important bacterial plant symbiotic systems i.e. *Rhizobium* - leguminuous plants, *Frankia*-actinorhizal plants and *Anabaena*-Azolla symbiosis have been studied in great detail which contribute a major share of biological nitrogen fixation. In *Rhizobium*-legume symbiosis, gram negative, rod shaped soil bacteria-rhizobia, induce the formation of nitrogen-fixing root nodules on the leguminuous plants (except the genus *Parasponia* of the *Ulmaceae* family). The Leguminosae family comprises three subfamilies, Caesalpiniodeae, Mimosoideae and Papilionoideae, each of which contains genera able to form nodules (Allen and Allen, 1981; Polhill and Raven, 1981). Leguminous plants (approximately 15,000 species) exhibit very diverse morphology, habitat and ecology ranging from arctic annuals to tropical trees. The symbiosis in majority of legumes with rhizobia is apparently not an adaptation to a specialized ecological niche but rather depends on some complex genetic particularity of legumes, that has rarely evolved elsewhere in the plant kingdom (Young and Johnston, 1989).

An important feature of the symbiosis is that it exhibits specificity, particular legumes being infected only by a limited range of rhizobial strains or species. Host plants, which are infected by one species of *Rhizobium*, belong to a so called "cross-inoculation group" and this has formed the basis of classification system for *Rhizobium* bacteria. The use of modern methods of bacterial systematics such as nume-rical taxonomy, nucleic acid (DNA/DNA or DNA/RNA) hybridization and DNA sequencing led to the definition of three genera *Rhizobium, Bradyrhizobium* and *Azorhizobium* (Jarvis et al., 1986; Dreyfus et al., 1988; Young and Johnston, 1989) (Table 1). *Rhizobium* includes the fast-growing species, *Bradyrhizobium* includes slow-growing species and *Azorhizobium* includes the fast-growing species capable for forming both stem and root nodules on tropical water logged legume *Sesbania*. These genera are quite distinct and are much closer to non-symbiotic relatives than they are to each other. For example, *Rhizobium* is closely related to the plant-associated *Agrobacterium* while *Rhodopseudomonas palustris*, a soil phototroph and *Xanthobacter* spp. are the closest relatives of *Bradyrhizobium* and *Azorhizo-bium* respectively (Hennecke et al., 1985; Jarvis et al., 1986; Dreyfus et al., 1988). The reason for grouping of these diverse bacteria into a single family, the *Rhizobia-ceae,* is their common ability to establish a nitrogen-fixing symbiosis ·with legumes.

Table 1. Classification of rhizobia

Genus	Members	Some of the host plants nodulated
Rhizobium	R. meliloti	Alfalfa (*Medicago sativa*) Sweet clover (*Melilotus alba*) Fenugreek (*Trigonella foenumgraecum*)
	R. leguminosarum	
	Biovar viciae	Pea (*Pisum sativum*) Vetch (*Vicia sativa*) Sweet pea (*Lathyrus sativus*)
	Biovar phaseoli	Bean (*Phaseolus vulgaris*)
	Biovar trifolii	Clovers (*Trifolium* spp.)
	R. loti	Trefoil (*Lotus* spp.)
	R. fredii	Soybean (*Glycine max*)
	R. tropici	Bean (*Phaseolus vulgaris*) Leucaena (*Leucaena leucocephala*)
	R. sp. NGR 234	26 legumes species including Leucaena (*Leucaena leucocephala*) Cowpea (*Vigna unguicultata*) Siratro (*Macroptilium atropurpureum*) Soybean (*Glycine max*) Parasponia (non-legume)
Bradyrhizobium	B. japonicum	Soybean (*Glycine max*)
	B.sp. "Cowpea"	Cowpea (*Vigna unguiculata*) Mungbean (*Vigna radiata*) Pigeonpea (*Cajanus cajan*) Chickpea (*Cicer arietinum*) Peanut (*Arachis hypogaea*) Sunhemp (*Crotolaria juncea*)
Azorhizobium	A. caulinodans	Root and stem nodules on Sesbania (*Sesbania rostrata,* *Sesbania bispinosa*)

The *Rhizobium*-legume symbiosis is the most significant in terms of global nitrogen fixation with representatives in tropical and temperate zones, in pasteurs, aerable land and forests. On a global basis, this symbiotic association may reduce 20 million tonnes of atmospheric nitrogen to ammonia per annum which accounts for about 70-80% of all biologically fixed nitrogen per year (Burris, 1980) and one-third of total nitrogen input needed for world agriculture (Gallon and Chaplin, 1987). In *Rhizobium*-legume symbiosis, rhizobial cell transforms into an enlarged spherical or 'Y' shaped bacteroid into the interior of the nodule that has different membrane and physiological properties from free-living cells (reviewed by Appleby, 1984; Brewin, 1991; Hirsch, 1992). *Rhizobium* and legumes jointly synthesize the oxygen-binding compound, leghemoglobin which modulate the supply of oxygen to the bacteroids (Appleby, 1984). In this symbiotic association, the plant provides a microaerobic environment for the functioning of oxygen sensitive enzyme nitrogenase and carbohydrates for the bacterial endosymbionts to support their metabolism. In return the bacteria fix atmospheric nitrogen that is used by the plant for the synthesis of organic nitrogenous compounds to meet its biological needs for this element. Due to this relationships, plants capable of forming these symbiosis are generally able to grow in nitrogen-poor soils that would not otherwise support vigorous plant growth. Therefore, the process of symbiosis is important in providing high protein diet and fodder.

With the introduction of molecular biology techniques to the study of symbiotic nitrogen fixation in the mid-70's, there were many predictions of a rapid improvement in the symbiotic performance of legumes and perhaps more importantly, of the development of important non-legumes with nitrogen-fixing symbiotic capability. The results of decades of research in this area has greatly enhanced understanding of the biochemistry, genetics and regulation of biological nitrogen fixation. Recent studies have clearly demonstrated that genes in both the macro - and micro-symbiont are specifically expressed when each is exposed to the other (Long, 1989b). Therefore. *Rhizobium*- legume symbiosis has been a useful model for understanding the process of symbiotic nitrogen fixation. The goal of this chapter is to review the current state of our understanding of the genetics and regulation of biological nitrogen fixation in the important *Rhizobium-Bradyrhizobium*-leguminous plants symbiotic system and to suggest avenues for future approaches that may lead to enhancement of nitrogen fixation in this symbiotic association.

II.   Development of Nitrogen-fixing Legume Root Nodules

The overall mechanism for the induction of nodule formation appears to be broadly similar for all forms of rhizobia and involves a controlled and coordinated expression of both bacterial and plant genes (Long, 1989 a,b; Caetano-Anolles and Gresshoff, 1991; Denarie et al., 1992; Fisher and Long, 1992; Hirsch, 1992; Verma et al., 1992). The morphology of legume root nodules shows much diversity (Corby, 1988), reflecting considerable variation in the details of the symbiotic interaction for different species (de Faria et al., 1989). Nevertheless, four anatomical features are generally characteristic of legume nodules : induction of new plant meristem normally involving root cortical cells (Rolfe and Gresshoff, 1988); tissue and cell invasion by *Rhizobium* (Sprent, 1989); development of a central tissue with reduced access to atmospheric oxygen within which nitrogen fixation reaction can take place (Bergersen, 1982) and development of peripheral vascular tissue located outside the central mass of infected tissue, but inside the nodule endodermis (Truchet et al., 1989b). Combined efforts of cytologists, plant physiologists, geneticists and molecular biologists have given insight into the process of nodule formation and function and nodule development can be arbitrarily divided into stages of nodule initiation, nodule invasion and bacteroid differentiation followed by nodule senescence (Long, 1989b; Brewin, 1991; Hirsch, 1992).

(i) Nodule Initiation

Recent results show that the initial nodulation steps can be characterized as a two way molecular conversation. The host legume releases signal compounds that stimulate the coordinate expression of bacterial nodulation (*nod*) genes. These *nod* genes, in turn, encode enzymes involved in the synthesis of Nod factors (substituted oligosaccharides) that cause morphological changes in the plant root.

A. Early Signals from Plant to Bacterium

A given leugme releases several flavonoids and other specific compounds with or without inducing properties for its compatible *Rhizobium*. The nature and amount of the compounds exuded depend on the plant and its stage of development. For alfalfa, bean or soybean, the spectrum of flavonoids present in seed exudates is different from that present in root exudates (Peters et al., 1986 Hartwig et al., 1989; Hungaria et al., 1991a,b; Graham, 1991). The different flavonoids present in exudates can interfere with each other's ability to induce *nod* gene expression and some secreted flavonoids can act as competitive inhibitors (anti-inducers) rather than transcriptional activators (Fig.1) (Firmin et al., 1986; Djordjevic et al., 1987; Peter and Long, 1988; Kosslak et al., 1990; Gyorgypal et al., 1991). Rhizobia respond by positive chemotaxis to plant root exudates and move towards localized sites on legume roots (Gulash et al., 1984; Caetano-Anolles et al., 1988b; Kape et al., 1991). The flavonoids in root exudates also induce the transcription of an important set of nodulation genes of *Rhizobium* (Peters et al., 1986) and *Bradyrhizobium* (Kosslak et al., 1987) which are not expressed or expressed at very low levels in free-living rhizobia in the absence of a plant (Long, 1989a). Recently, betaines (i.e., trigonelline and stachydrine), which are structurally distinct from flavonoids and are excreted in significant quantities from alfalfa seeds, were found to induce *nod* gene expression in *R. meliloti* (Phillips et al., 1992).

After chemotaxis, rhizobia attach to root hairs all over the root, but the hairs that are most responsive to *Rhizobium* infection are just behind the apical meristem at the site of emergence of root hairs (Bhuvaneswari et al.,1981). In the infectible zone of roots, rhizobia attaches to susceptible root hairs via a two step attachment process (Dazzo et al., 1984; Smit et al., 1987). First, they loosely attach to a plant receptor through an acidic extracellular polysaccharide (Halverson and Stacey, 1986) or via a protein on the bacterial surface known as rhicadesin (Smit et al., 1987). Rhicadesin is a calcium-binding protein that appears to be common among ·*Rhizobiaceae*. Then tighter adherence occurs either by means of cellulose fibrils (Smit et al., 1987) or fimbriae (Vesper and Bauer, 1986). Entry of bacteria appears to

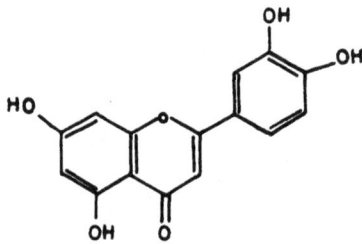

Luteolin
(flavone)

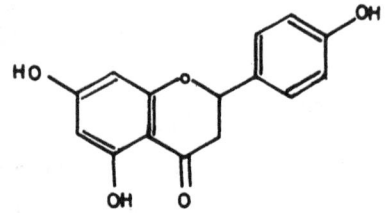

Naringenin
(flavanone)

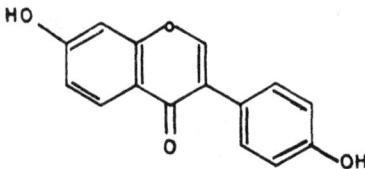

Daidzein
(Isoflavone)

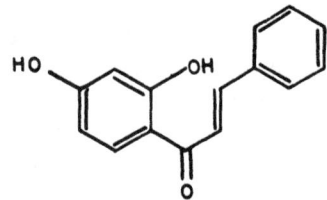

Chalcones

Fig. 1.   Chemical structure of the plant-derived flavonoids and chalcone, active as *nod* gene inducers in different *Rhizobium* and *Bradyrhizobium* strains. Luteolin and chalcone are the most
active inducers of *R.meliloti nod* genes. Naringenin is active in *R. leguminosarum* bv. *viciae*
and diadzein is the *nod* gene inducer of *B. japonicum*.

occur at the root hair tip, probably because the cell wall is thinner and less cross
linked. Depending upon the host, root hair deformation takes place 6-18 h after inoculation. The deformed root hairs in various legumes may form different structures,
including corkscrews, branches, twists, spirals and Shepherd's crooks. Root hair
deformation is dependent on the presence of functional *Rhizobium nod* genes.

B.   Structure and Function of Rhizobial Nodulation Genes

The nodulation genes of *R. leguminosarum* and *R. meliloti* are located on large plasmids of about 200 to 1500 kb, so called symbiotic plasmids (*psym*), whereas in *Bradyrhizobium*, such symbiotic genes are located on the chromosome (Beynon et al., 1980; Long, 1989a; Fisher and Long, 1992; Denarie et al., 1992). The use of Tn5 mutagenesis, DNA sequencing and *in vitro* transcription/translation, has allowed the identification and characterization of a large number of *nod* and *nol* genes (Downie et al., 1985; Long, 1989 a,b; Denarie et al., 1992). So far, 44 bacterial *nod* and *nol* genes have been reported to be involved in nodulation in various *Rhizobium* species (Fig. 2). Despite all efforts, our knowledge of the role of nodulation genes in symbiosis is limited to sequence homology and a few biochemical studies that point out to the possible involvement of genes in hormone metabolism, membrane transport, and lipid and polysaccharide biosynthesis.

These nodulation genes may be divided into regulatory, common and host specificity genes. The *nod* D is a regulatory gene, which is constitutively expressed. Its product is essential for the expression of other *nod* genes in conjunction with host signal molecules, i.e. flavones, flavanones and isoflavones, secreted by legume roots. The *nod* D exists as a single gene in *R. leguminosarum* and *R. trifolii* (Kondorosi, 1992) while *R. meliloti* (Gottfert et al., 1986), *R. phaseoli* (Davis and Johnston, 1990), *R. fredii* (Appelbaum et al., 1988) and *B. japonicum* (Gottfert et al., 1989) carry two or three copies of *nod* D. The different *nod* D genes are conserved at the nucleotide sequence level and share homology with the *lys* R family of prokaryotic regulatory genes (Henikoff et al., 1988). In common with the Lys R-family proteins, the N-terminal end of *nod* D contains a helix-turn-helix motif characteristic of DNA-binding proteins (Henikoff et al., 1988; Gyorgypal et al., 1991). Point mutations in *nod* D results in *nod* gene activation with a broader range of inducers (Burn et al., 1987, 1989; Mc Iver et al., 1989) and chimaeric *nod* Ds proteins exhibit the flavonoid specificity of the *nod* D product constituting their C-terminal end (Horvath et al., 1987) suggesting that flavonoid specificity is largely determined by the C-terminal portion of *nod* D.

Upstream of all the inducible *nod* operons, there is a highly conserved DNA sequence of about 47bp nucleotides. These have been called *nod* boxes (Rostas et al., 1986; Shearman et al., 1986; Schofied and Watson, 1986). They are essential for inducible promoter activity and the *nod* D gene product binds directly to these *nod* box sequences (Hong et al., 1987; Fischer et al., 1988). In the genetically distant *Azorhizobium caulinodans* and *B. japonicum* less conserved *nod* boxes have been identified and new shorter consensus sequences have been proposed to account for these divergent *nod* boxes (Wang and Stacey, 1991; Goethals et al., 1992). Some *nod* D genes are subject to negative autogenous regulation (Rossen et al., 1985), others are unregulated (Mulligan and Long, 1985) or their expression is repressed by nol R (Kondorosi et al., 1989), or subjected to inducible expression (Banfalvi et al., 1988).

In *R. meliloti*, the *syr* M gene, another member of the lys R family, activates the expression of the *nod* D3 gene, and *nod* D3 activates in turn the expression of *syr* M (Kondorosi et al., 1991; Rushing et al., 1991). When both *nod* D3 and *syr* M are carried on a multicopy plasmid, they can induce *nod* gene expression at a high level

Rhizobium leguminosarum bv viciae

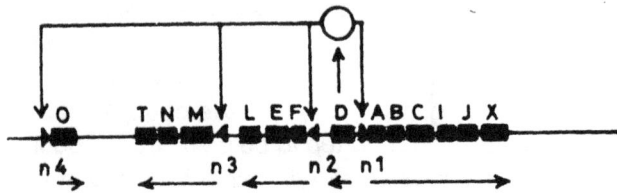

Rhizobium meliloti

Bradyrhizobium Japonicum

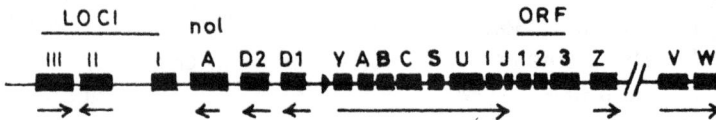

Fig. 2.   Organization of the *nod* genes in *Rhizobium leguminosarum* bv. *viciae, R. meliloti* and *Bradyrhizobium japonicum*. The capital letters represent *nod* genes and the arrows indicate the direction of gene transcription. The black triangles indicate location and orientation of *nod* boxes, while the broken vertical lines show large interruptions in the map. The fig. is modified from Economou et al. (1990) and Denarie et al. (1992).

even without plant inducers (Gyorgypal et al., 1988; Mulligan and Long, 1989; Honma et al., 1990; Maillet et al., 1990; Kondorosi et al., 1991). In several strains of *R. meliloti, nol* R has been found to regulate *nod* gene expression by inhibiting the transcription of *nod* DI and *nod* D2 (Kondorosi, 1992). A mutation in *nol* R results in a slight delay in nodulation, which suggests that *nol* R is required for optimizing *nod* gene expression during infection and nodulation (Kondorosi et al., 1989). Mutation of *R. meliloti gro* EL prevents expression of *nod* D3 and reduces the activity of Nod D1 and Nod D3 when they are expressed constitutively (Long et al., 1991). Indeed, *R. meliloti* and *B. japonicum* have multiple copies of *gro* EL (*gro* EL=chaperonin protein), one or more of which are symbiotically regulated (Stacey et al., 1992). The functions of *nod* Ds may also be diverse. In *R. meliloti*, two inducer-requiring *nod* Ds (*nod* DI and *nod* D2) are important for optimal responses to different hosts (Honma and Ausubel, 1987; Horvath et al., 1987). The inducer-independent circuit controlled by *nod* D3 (Mulligan and Long, 1989) may be important for late *nod* gene expression (Sharma and Signer, 1990; Kondorosi et al., 1991) and/or may respond to nitrogen, growth and/or inducers (Dusha et al., 1989; Wang and Stacey, 1991).

Common nod Genes

Mutations in the *nod* D gene as well as mutations in the common *nod* ABC genes completely block nodulation, including root hair curling and meristem induction (Rossen et al., 1984; Torok et al., 1984; Long, 1985; Egelhoff and Schofield and Watson, 1986). The common *nod* ABC genes have been found in all *Rhizobium, Bradyrhizobium* and *Azorhizobium* isolates studied so far (Goethals et al., 1989; Martinez et al., 1990; Stacey, 1990). These genes are functionally interchangeable between *Rhizobium* and *Bradyrhizobium* species without altering the host range (Martinez et al., 1990; Barbour et al., 1992). These common *nod* ABC genes have been shown to produce a basic common signal, Nod factor, for eliciting root-hair deformation and root's cortical cell division which is then modified to host plant specific signals through the action of host-specificity genes (Faucher et al., 1989; Lerouge et al., 1990; Fisher and Long, 1992; Denarie et al., 1992).

The *nod* IJ genes are present in *R. leguminosarum, R. trifolii* and *B. japonicum* (Evans and Downie, 1986; Göttfert et al., 1990; Surin et al., 1990) and partial sequence data suggest their presence in *R. meliloti* and *A. caulinodans* (Jacobs et al., 1985; Goethals et al., 1989). Mutations in *nod* IJ result in nodulation delay with *R. leguminosarum*, but have no detectable effect with *B. japonicum* (Evans and Downie, 1986; Gottfert et al., 1990b). The presence of the *nod* ABC IJ genes in the genetically distant *Rhizobium, Bradyrhizobium* and *Azorhizobium* isolates suggest a common origin for these genes. Recently, the *nod* M and *nod* N genes of *R. leguminosarum* bv *viciae* have been found to complement the *R. meliloti nod* M and *nod* N mutations and these genes have been described as common *nod* genes (Baev et al., 1992).

Host-specific nod Genes

The host-specific nodulation genes are required for the correct induction of root hair curling and cell division in the appropriate legume host and are pesent in one species and not in others. The *nod* P, Q and H genes which determine alfalfa specificity, are present in *R. meliloti* but not in *R. leguminosarum, R. trifolii* and *B. japonicum* (Debelle et al., 1986; Horvath et al., 1986; Rodrigues-Quinones et al., 1987).

The *nod* SU genes, present in *B. japonicum* and in the broad-host-range tropical *Rhizobium* sp. NGR 234 and *nod* T genes present in *R. leguminosarum* and *R. trifolii* have not been identified in *R. meliloti* (Gottfert et al., 1990a; Lewin et al., 1990; Surin et al., 1990). In *R. leguminosarum* bv *Viciae* a gene. *nod* X, confers the ability to nodulate Afghanistan peas (Davis et al., 1988). Some host specific *nod* genes can be present only in some biovars, for example, the *nod* O gene is found in *R. leguminosarum* bv *viciae* but not in the closely related biovar *trifolii* (Economou et al.,

1990). Similarly, the *nol* A gene of *B. japonicum* is essential for nodulation of soybean genotypes restrictive to a particular *Bradyrhizobium* serocluster (Sadowsky et al., 1991). A single dominant plant gene appears to be involved in this nodulation restriction (Barbour et al., 1992). Thus, the rhizobial *nod* X and *nol* A genes might represent the bacterial counterpart of gene-to-gene interaction systems analogous to those found in many pathogen-plant relationships (Keen and Staskawicz, 1988; Barbour et al., 1992).

Mutations in the host-specific nodulation genes cannot be fully complemented by DNA from other rhizobial species or biovars (Kondorosi et al., 1984; Djordjevic et al., 1985; Debelle et al., 1986; Horvath et al., 1986; Lewin et al., 1990). In *B. japonicum, nod* Z and *nod* VW mutants lose the ability to nodulate siratro but not soybean (Gottfert et al., 1990a; Barbour et al., 1992) and *Rhizobium* sp. NGR 234 *nod* SU mutants are Nod⁻ on *Leucaena leucocephala* but Nod⁺ on siratro (Lewin et al., 1990). Mutations in some host-specific *nod* genes may result in broadening of the host range. In *R. trifolii, nod* FE⁻ mutants poorly nodulate white and red clovers but have acquired the ability to infect and nodulate peas, in contrast to wild-type strain (Djordjevic et al., 1985). In *R. meliloti, nod* H⁻ mutants have lost the ability to infect and nodulate the homologous host alfalfa but have acquired the ability to infect and nodulate heterologous hosts such as vetch species, while *nod* Q mutants can infect both alfalfa and vetch (Debelle et al., 1986; Horvath et al., 1986; Faucher et al., 1989).

Sequence homology of the *nod* gene products and known proteins in other organisms has led to the rapid identification of the biochemical functions of a number of the *nod* gene products (Table 2). Sequence comparisons suggest enzymatic functions for particular host-range proteins and attention is currently focused on the roles that individual *nod* genes have in the synthesis of the plant morphogenetic factors or in the modification of surface macromolecular bacterial components.

C. Nod Factor Synthesis and Signal Exchange from Rhizobium to Plant

Both the common and host specific nodulation genes are involved in the production of a low molecular weight diffusible signaling substance, identified as a lipo-oligosaccharide (glycolipid). This glycolipid molecule of *R. meliloti* causes root hair deformations at 10^{-11} M concentration and stimulates cortical cell division at 10^{-7} M concentration in alfalfa roots (Truchet et al., 1991). The chemical structure of the root hair deformation factor of *R. meliloti* (termed Nod Rm-I) is a sulphated β-1, 4-tetra-D-glucosamine with three acetylated amino groups (Lerouge et al., 1990). A C_{16} unsaturated fatty acid occupies the non-reducing end of the molecule, while the reducing end contains a sulphate group. Other lipo-oligosaccharides besides Nod Rm-I have been identified from *R. meliloti*. Truchet et al. (1991) have found a Nod Rm-I Like factor with an O-acetyl group at carbon 6 of the terminal sugar at the non-reducing end. The mutations in *nod* H and *nod* Q genes of *R. meliloti* lead to the production of another compound, designated Nod Rm-2 which is structurally related to Nod Rm-I, but lacks the sulphate group at the reducing end of the tetrasaccharide (Lerouge et al., 1990). Other Nod factor molecules containing five instead of four glucosamine residues and a trisaccharide-containing Nod factor has been isolated from *R. meliloti* (Schultze et al., 1992). Roche et al. (1991) proposed a uniform method for naming of Nod factors. According to this Nod Rm-I is now described as Nod Rm-IV(s); Rm signifies *R. meliloti,* IV the four glucosamine residues and S the sulphate on the reducing end of the molecule.

Recently, Nod factors from *Rhizobium leguminosarum* species have been chemically characterized (Spaink et al., 1991 a,b). In *R. leguminosarum,* four Nod factors, namely Nod Rlv-IV (Ac, C 18:1), Nod Rlv-V (Ac, C 18:1), Nod Rlv-IV (Ac, C 18:4) and Nod Rlv-V (Ac, C 18:4) have been purified (Spaink et al., 1991b). *R. leguminosarum* Nod factors lack the sulphate group and the fatty acid side chains contains 18 carbons with either one or four double bonds whereas Nod Rm-I contains 16 carbons and two double bonds (Spaink et al., 1991b). Another difference from Nod Rm-I is the presence of an O-acetyl group on the non-reducing end of *R. leguminosarum* bv *viciae* Nod factor (Spaink et al., 1991b). Sometimes Nod Rm-I is also acetylated in this position (Truchet et al., 1991). *R. leguminosarum* bv *viciae* *nod* L mutants make a Nod factor that lacks the O-acetyl group.

The extracellular Nod factors produced by *Rhizobium* sp. NGR234 are also N-acetylated chitin oligomers having five N-acetyl glucosamine residues. The terminal nonreducing glucosamine residue is substituted by one or two carbomyl groups located at various positions and the reducing glucosamine residue carries a methyl-fucose sugar substituted by a sulphate or an acetyl group (Denarie et al., 1992). The Nod factor secreted by *Bradyrhizobium japonicum* strains is a pentamer of ß-1-4-linked N-acetyl-D-glucosamine which is N-acetylated on the terminal nonreducing residue and it carries a sugar substituent on the reducing end (Denarie et al., 1992). The particular decorations present on the chitosan backbone may determine the host specificity of the Nod factor (Stacey et al., 1992). For example, the lipo-oligosaccharide signals produced by *R. meliloti* and *R.* sp. strain NGR234 are capable for alfalfa nodulation. Roche et al. (1991) have shown that *R. meliloti* signal must be sulfated on the reducing-end sugar in order to be biologically active on alfalfa. The Nod factor from *R.* sp. strain NGR 234 was also found to have sulphate substitution except that the sulphate substitution was found on the fucosyl residue. Similarly, the mutation in *nod* H gene of *R. meliloti* lead to the production of Nod Rm-2 (without a sulphate group) that elicits root hair deformation on vetch (*Vicia*), but not on alfalfa (Lerouge et al., 1990).

The synthesis of the Nod factors is mediated by specific *nod* genes (Fig. 3). The DNA sequence of *nod* M is homologous to the *glm* S gene in *E. coli* which encodes D-glucosamine synthetase (Baev et al., 1991). The *nod* C with a sequence similarity to yeast chitin synthetase (Hirsch, 1992) may be involved with linking individual glucosamine units synthesized by the activity of *nod* M. The *nod* L gene product is thought to acetylate the glucosamine residues (Downie, 1989) while *nod* FE-encoded proteins are involved in synthesizing the fatty acid side chain (Horvath et al., 1986; Shearman et al., 1986). A non-sulphated Nod factor is produced by

Table 2.   Functions and properties of the nodulation genes of rhizobia

Gene	Species biovar[*]	Cellular localization	Homologies	Characteristics/functions	References
nod A	Common	Cytoplasmic	Unknown	Required for Nod factor production	Rossen et al., 1984; Torok et al., 1984.
nod B	Common	Cytoplasmic	Unknown	Required for Nod factor production	Rossen et al., 1984; Torok et al., 1984.
nod C	Common	Outer membrane	Chitin synthases	Proposed to form β-1, 4-glycosyl bond	Rossen et al., 1984; Jacobs et al., 1985; John et al., 1988.
nod D	Common	Cytoplasmic membrane	Lys R-family proteins	Transcriptional activator of inducible *nod* genes	Mulligan and Long, 1985, 1989; Honma and Ausubel, 1987.
nod E	Rm, Rl, Rt	Cytoplasmic membrane	β-ketoacyl synthases	Proposed to synthesize Nod factor acyl chain	Horvath et al., 1986; Spaink et al., 1989b.
nod F	Rm, Rl Rt	Cytoplasm	Acyl carrier proteins	Proposed to synthesize Nod factor fatty acyl chain	Shearman et al., 1986; Spaink et al., 1991a.
nod G	Rm		Ribitol or glucose dehydrogenases	Proposed to modify Nod factor fatty acyl side chain	Debelle et al., 1986; Horvath et al., 1986.
nod H	Rm	Outer membrane	Sulphotransferases	Proposed to transfer activated sulphate to Nod factor	Kondorosi et al., 1984; Horvath et al., 1986; Roche et al., 1991.
nod I	Rl, Rt, Bj	Cytoplasmic membrane	ATP-binding transport proteins	Proposed to form membrane transport complex with *nod* J	Evans and Downie, 1986; Surin et al., 1990.
nod J	Rl, Rt, Bj	Cytoplasmic membrane	Transmembrane proteins	Proposed to form membrane transport complex with *nod* I	Evans and Downie, 1986; Surin et al., 1990.
nod L	Rm, Rl, Rt	Cytoplasmic membrane	Acetyl transferases	Proposed to add O-acetyl group to Nod factor	Surin and Downie, 1988; Downie, 1989.
nod M	Rm, Rl, Rt		D-glucosamine synthetase	Proposed to provide sugar moiety for the synthesis of Nod factor	Surin and Downie, 1988; Baev et al., 1991, 1992.

Gene	Species biovar*	Cellular localization	Homologies	Characteristics/functions	References
nod N	Rm, Rl, Rt			Proposed to provide substrate(s) for the anabolic pathway leading to synthesize of Nod factor.	Surin and Downie, 1988; Baev et al., 1991, 1992.
nod O	Rl	secreted	haemolysin, Ca^{2+} binding	Proposed to form Ca^{2+}-regulated ion channels in membranes for transport of K^+ and Na^+	Economou et al., 1990; Stacey et al., 1992.
nod P	Rm		ATP-sulfurylase	Proposed to provide activated sulphate for transfer to Nod factor	Roche et al., 1991; Schwedock and Long, 1990, 1992.
nod Q	Rm		ATP-sulfurylase and APS kinase	Proposed to provide activated sulphate for transfer to Nod factor	Fisher and Long, 1992. Schwedock and Long, 1990, 1992. Gottfert et al., 1990b; Lewin et al., 1990.
nod SU	BJ, NGR				
nod T	Rl, Rt		Transit sequence	Unknown	Surin et al., 1990;
nod V	BJ	Cytoplasmic membrane	Sensor, two component regulatory family	Required for isoflavone-mediated induction of the *nod* ABC genes	Gottfert et al., 1990a; Stacey et al., 1992.
nod W	BJ	Cytoplasmic membrane	Regulator, two component regulatory family	Required for isoflavone-mediated induction of the *nod* ABC genes	Gottfert et al., 1990a; Stacey et al., 1992;
nod X *nol* A	Rl (TOM) BJ		DNA-binding proteins	Unknown	Davis et al., 1988; Sadowsky et al., 1991.
nol R	Rm			Repressor of *nod* D	Kondorosi et al., 1989.

* Rhizobia species and biovars : Rm, *R. meliloti*; Rl, *R. leguminosarum bv. viciae*; Rt, *R. leguminosarum bv. trifolii*; BJ, *B. japonicum*; NGR, *Rhizobium* sp. NGR 234.
Other genes including *nod* K, *nod* R, *nod* Y, *nod* Z, *nol* D, *nol* E and *nol* G have been identified (compiled by Stacey, 1990) but their sequence homology or possible functions are unknown.

either *nod* Q or *nod* H mutants of *R. meliloti. nod* H is homologous to a sulpho-transferase (Roche et al., 1991), while *nod* P and *nod* Q genes specify ATP sulphurylase activity (Schwedock and Long, 1990, 1992).

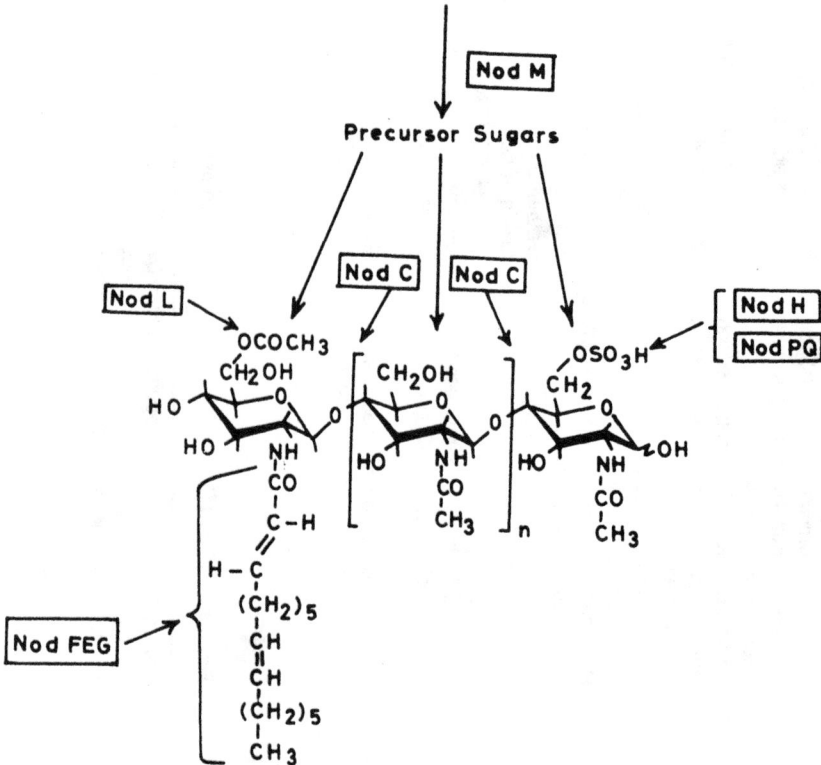

Fig. 3.　Hypothetical pathway of Nod factor biosynthesis. The lipid moiety shown in this figure is typical of Nod factor secreted by *R. meliloti* and *n* refers to the number of glucosamine residues. The proposed roles of the *nod* gene products in the synthesis of Nod factor are indicated by arrows. Other Nod factor modifications seen in different bacteria include the C-6 sulphate, added by the action of *nod* H and *nod* PQ genes in *R. meliloti* and the C-6 acetyl group that requires *nod* L in *R. leguminosarum* bv *viciae.* Modified from Hirsch (1992).

D.　Induction of the Nodule Meristem

Prior to the entry of bacteria into the host, initiation of cell division occurs in the root cortex of infectible zone of roots. In this region, there is induction of mitotic activity within 12-24 h after inoculation with rhizobia (Calvert et al., 1984). Initially, the plane of cell divisions is oriented so that the axis of the new wall deposition is perpendicular to the longitudinal axis of the root (anticlinal). Subsequently, this nodule primordium gives rise to an organized meristem consisting of a mass of small cells dividing in all planes. Besides, signalling effect of Nod factors, these glycolipid molecules also cause morphological effects. *R. meliloti Nod* factor has been reported to induce plant cell membrane depolarization (Ehrhardt et al., 1992). In alfalfa, Nod factors elicit the deformation of root hairs and the formation of nodule-like structures (Truchet et al., 1991). In vetch, the Nod factor initiates pre infection thread structures as well as nodule primordia. Cytoplasmic bridges resembling those preceding infection threads were observed in the outer root cortex and at the same time cell divisions were evident in the inner cortex (van Brussel et al., 1992; Clarke et al., 1992). These signals may interact with the cell cycle control mechanism so that resting cortical cells enter mitosis (Verma, 1992). Although many infections occur on the root, only a small fraction of them give rise to cortical cell divisions, suggesting that further modification of some factor(s) alongwith the path of infection process is essential for eventual success in generating nodule primordia.

Recently, many plant and bacterial mutants have been isolated in which root hair deformation, cortical cell divisions and infection thread formation are uncoupled. Miller et al. (1991) have described five loci in non-nodulating *Melilotus albus* Desr. mutants. In some of these mutants, root hair deformations occur, but no subepidermal cell divisions take place in response to inoculation with *R. meliloti*. Similarly, two soybean mutants exhibit an uncoupling of root hair deformation and cortical cell divisions (Carroll et al., 1986) and two non-nodulating soybean lines belonging to the same complementation group exhibited subepidermal cell divisions but no root hair curling (Mathews et al., 1987, 1989). A mutation in a loci downstream of *nod* made *B. japonicum* incapable of inducing subepidermal cell division without affecting the ability to curl root hairs (Deshmane and Stacey, 1989). Similarly, Nod factor from *R. leguminosarum* bv *viciae nod* E mutants, which produce a more saturated fatty acid, or from *R. meliloti nod* PQ mutant, which has 16 carbons in its fatty acid tail but no sulphate on the glucosamine backbone, elicit root hair deformation, but no cortical cell divisions on vetch (Lerouge et al., 1990; Spaink et al., 1991b).

A number of bacterial factors (BF) have been identified in *R. leguminosarum* bv *trifolii* which promote root hair proliferation or work synergistically to elicit hair deformation and cortical cell divisions on clover (Hollingsworth et al., 1990). One of these factors, BF-5, which is dependent on *nod* gene induction by flavonoids, has been identified as N-acetylglutamic acid (Philip-Hollingsworth et al., 1991). BF-5 does not elicit these responses on alfalfa and *Lotus*. Moreover, the empty nodules could be induced by the application of naphthyl phthalamic acid which inhibits polar transport of auxins (Hirsch et al., 1989) or by the application of cytokinins (Arora et al., 1959; Libbenga et al., 1973). Kondorosi and his colleagues reported a correlation between auxin sensitivities of alfalfa cultivars and their tendencies to form nodules (Clarke et al., 1992). These results suggest that Nod factors presumably act by altering phytohormone (auxin/cytokinin ratio) levels in some way. Local cell divisions may in turn lead to further changes in translocation of additional growth substances which may result in the induction of a region of cell division, the nodule primordium. A cyclin B cDNA (*cdc* 2) has also been cloned from soybean and this gene has been shown to be expressed at a very high level in early nodule development, which correlates with cell proliferation (Hata et al., 1991; Verma, 1992).

Although rhizobia do not invade root and nodule meristems, they do invade dividing cells in the cortex, indicating that the dividing cells of the nodule primordium differ from those of root and nodule meristems. The type of nodule that develops depends on the host plant and not on the rhizobial strain (Dart, 1977; Newcomb, 1981). Because the same bacteria can form determinate or indeterminate nodules in different hosts, the persistence of the meristematic activity depends upon the host. The cessation of meristematic activity gives rise to determinate (spherical) nodules and final form of the nodule results from cell enlargement rather than cell division. Nodules of soybean, *Phaseolus* bean and *Lotus* are examples of determinate nodules. In roots of alfalfa, clover, vetch and pea, the nodule primordium arises in the uninfected inner layers of the root cortex adjacent to the pericycle near the xylem pole. The persistent meristem causes indeterminate nodules to be elongate and club-shaped because new cells are constantly being added to the distal end of the nodule. All stages of nodule development are represented in one nodule because an age gradient occurs from the distal meristem to the proximal point of attachment to the parent root.

(ii) Nodule Invasion

A. Infection Thread Formation

The root hair cells that are most susceptible to *Rhizobium* induced deformation and infection thread development are those that are rapidly expanding (Bhuvaneswari et al., 1981). Upon root hair curling, rhizobia become entrapped in the curls. A hyaline spot is usually the first sign of infection thread penetration. Marked cytoplasmic streaming occurs in response to the attachment of rhizobia to root hair. The nucleus migrates towards the refractile spot. Following dissolution of the cell wall by bacterial-produced hydrolytic enzymes, the plasma membrane of the root hair invaginates and cell wall is deposited around it (Verma and Long, 1983). The invagination with the newly formed cell wall forms the infection thread (Kijne, 1992) (Fig. 4). The host cell nucleus, attached by microtubules to the infection thread tip, precedes the infection thread as it passes through the root hair cell (Lloyd et al., 1987; Bakhuizen, 1988). This inwardly growing tubular tunnel-structure (infection thread) contains rhizobia embedded in a mucigel composed of cell wall polysaccharides, a plant-derived matrix glycoprotein and their own extracellular polysaccharides (Bauer, 1981; Callaham and Torrey, 1981; Robertson and Lyttleton, 1982; Vanden Bosch et al., 1989). The expression of two early nodulins: hydroxy proline-rich glycoprotein, ENOD12 and arabinogalactan protein, ENOD5 has also been suggested to be involved with the development of infection threads, both in root hair cells and during sbsequent stages of invasion (Scheres et al., 1990 a, b).

Although numerous root hairs deform, very few of them form Shepherd's crooks. Moreover, infection thread formation and penetration events resulting in the formation of N_2-fixing nodules are very low (Wood and Newcomb, 1989). Infection threads often abort in the root hair cells, although some penetrate into the root cortical cells (Libbenga and Harkes, 1973; Dart, 1977). Several studies have suggested that infection thread abortion is associated with a hypersensitive (HR)-like response by plants which causes localized necrosis and accumulation of phenolic compounds at abortive site. The aborted infection threads in alfalfa terminate their growth within a cell that is positioned in the mid-cortex. This cell is observed to contain hydroxyproline-rich glycoproteins and increased level of phenylalanine ammonia lyase (PAL) and chalcone synthase (CHS) (Hirsch, 1992). Estabrook and Sengupta-Gopalan (1990) also observed rapid increases in CHS and PAL transcripts in soybean following *B. japonicum* inoculation. These two enzymes CHS and PAL are the enzymes of phenylpropanoid biosynthetic pathway by which phytoalexin are produced. These results suggest that flavonoid biochemistry and physiology are important in nodule induction and morphogenesis. Other studies with *exo* mutants of *R. meliloti* have indicated that cells containing aborted infection threads, and adjacent cells, autofluoresce, indicating the presence of phenolic compounds in their cell walls (Puhler et al., 1991). Phenolic accumulation in cell walls usually foreshadows the lignification of the responding cells. Lignification as well as phytoalexin accumulation are symptoms of HR response.

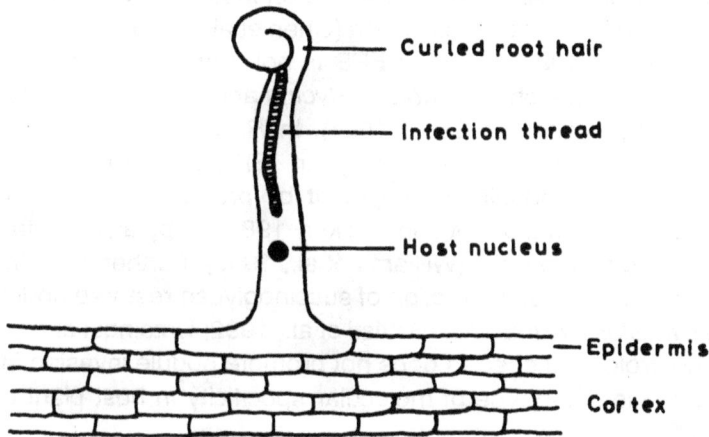

Fig. 4.   The curling of infected root hairs and formation of the infection thread. In the presence of appropriate rhizobia, an infection thread develops following a redirection of the root hair cell wall growth. The bacteria grow along this intracellular tunnel which elongates towards the developing nodule meristem.

B.   Role of the Bacterial Surface Polysaccharides in Nodule Invasion

The synthesis of surface and extracellular polysaccharides such as exopolysaccharides (EPS), capsular polysaccharides (CPS), the periplasmic cyclic oligosaccharide ß (1 → 2) - glucan and lipopolysaccharides (LPS) appear to be essential for successful infection thread formation. Mutants deficient in the production of acidic EPS (*exo* mutants) have been described for several *Rhizobium* strains (reviewed by Gray and Rolfe, 1990; Reuber et al., 1991; Leigh and Coplin, 1992; Sindhu and Dadarwal, 1993a) and several functions have been proposed for the role of EPS in *Rhizobium*-legume interactions (Darvill et al., 1989; Gray and Rolfe, 1990; Reuber et al., 1991). These include masking of the bacterial surface to avoid the elicitation of host defenses, encapsulation of the bacterium as a protection against physiological stresses encountered in infection thread, recognition of the bacterial surface by a host plant receptor, elicitation of host responses as a result of the release of oligosaccharide signalling molecules and activation of critical enzymes such as polysaccharase that may be involved in root hair curling or infection thread growth.

In the nodules induced by *exo* mutants, the root hair curling is delayed (Leigh et al., 1987), infection threads are formed but do not penetrate the nodule meristem (Finan et al., 1985), compounds containing phenolic structures accumulate around bacteria and infection threads abort (Puhler et al., 1991) and nodules lack a discrete, persistent meristem (Yang et al., 1992). Several recent genetic and biochemical evidences suggest that acidic exopolysaccharide is required for the establishment of effective N_2-fixing nodules on indeterminate nodulating legumes such as *Pisum, Medicago* and *Leucaena* (Chen et al., 1985; Leigh et al., 1985; Borthakur et al., 1986; 1987) whereas EPS is not found essential for development of determinate nodules on *Phaseolus, Glycine* and *Lotus* (Borthakur et al., 1986; Kim et al., 1989; Hotter and Scott, 1991). In *R. meliloti exo* mutants that fail to synthesize the normal acidic succinoglycan (EPS I.) can be complemented for the phenotype of full nodule development by production of a second exopolysaccharide EPS II (Glazebrook and Walker, 1989) or by a particular lipopolysaccharide specified by the *lps* Z (Williams et al., 1990). Furthermore, the application of only low-molecular-weight fraction of succinoglycan restored nodule invasion in alfalfa by *R. meliloti exo* mutants (Battisti et al., 1992; Urzainqui and Walker, 1992). EPS from heterologous species does not promote nodule invasion in alfalfa, implying that there is some type of molecular specificity in host plant recognition of rhizobial EPS.

The synthesis of cyclic (ß-1 → 2)-glucans by *Rhizobium* is also important for infection thread development. The chromosomal mutants in *ndv* genes produced normal amounts of exopolysaccharides but no cyclic oligosaccharide ß-(1 → 2)-glucan (Dylan et al., 1986). The *ndv* mutants rendered the bacteria capable of inducing nodules without infection threads and bacteroids (Dylan et al., 1986). The *ndv* pseudorevertants selected for their ability to induce normal N_2-fixing nodules on alfalfa, do not recover the ability to synthesize ß-(1 → 2)-glucan (Dylan et al., 1990). Thus the effects of the *ndv* locus on nodule development, particularly with regard to facilitating rhizobial entry into plant tissues, are unclear. One interesting suggestion is that the centres of these cyclic macromolecules (sometimes termed cyclosophorans) may be important for interaction with plant cells (Dudman, 1984). Brewin

(1991) proposed that such a structure might entrap signalling molecules, facilitating the transfer of flavonoid compounds to the bacterial inner membrane, or might act as carrier molecules to facilitate the transport of excreted glycolipids like Nod Rm-I to the plant cell surface.

Lipopolysaccharides are major immunogenic components of the outermembrane of rhizobia (Carlson, 1984; de Maagd et al., 1989a; Sindhu et al., 1990) and mutants with major defects in the structure of the carbohydrate components of LPS cannot establish a nitrogen-fixing symbiosis in legume root nodules (Noel, 1991; Kannenberg et al., 1992). *R. leguminosarum* bv *viciae* mutants deficient in O-antigen are incapable of complete infection of the legume host pea because the rhizobia are not released from the infection threads (de Maagd et al., 1989b; Brewin et al., 1990; Kannenberg et al., 1992). Infections on *Phaseolus* beans by such *lps* mutants abort early in nodule development and small, bump like nodules, devoid of bacteria are formed (Noel et al., 1986; Cava et al., 1989). Also, *lps* mutants lacking O-antigen in *B. japonicum* have a severe inf⁻ phenotype on *Glycine* (Puvanesarajah et al., 1987; Stacey et al., 1991). On *Vicia* and *Trifolium* hosts, infections by such mutants break down at later stages of nodule development in which bacteria are being released to become bacteroids (Priefer, 1989; Brink et al., 1990). However, *R. meliloti lps* mutants induce N_2-fixing nodules on alfalfa (Clover et al., 1989).

Several important functions have been proposed for LPS macromolecules which might be relevant to their role in the development of nitrogen-fixing bacteroids. Through their effects on membrane stability, LPS macromolecules may contribute to resistance of the bacteria to a variety of physiological stresses encountered within the nodule such as high or low pH, salt stress or low oxygen (Kannenberg and Brewin, 1989). LPS macromolecules may be involved in plant-microbe surface interactions, perhaps involving general 'compatibility' effect dependent on surface charge or hydrophobicity (Bradley et al., 1986; de Maagd et al., 1989b). LPS could also be involved in specific molecular recognition (Maiti and Podder, 1989) acting either positively as an inducer of nodule morphogenesis or may function in the avoidance of a plant defence response (Kannenberg et al.,1992).

In summary, the *Rhizobium* cell membrane and cell surface components encoded by *lps*, *exo* and *ndv* genes are in some way involved with the entry of rhizobia into plant cells. The extracellular polysaccharides are found to be more important for infection thread development and the precise structure of lipopolysaccharide is more important for endocytosis and bacteroid differentiation.

(iii) Nodule Maturation and Differentiation

In the development of both determinate and indeterminate nodules, the infection threads grow towards the nodule primordia. In indeterminate nodules, anticlinal cell divisions take place in the inner cortex of the root (Libbenga and Harkes, 1973). The branches of the infection threads invade cells of the nodule primordium and these cells stop dividing and begin differentiating (Newcomb et al., 1979). Mitoses in cells adjacent to the nodule primordium, near the middle of root cortex, generate the nodule meristem. The nodule meristem gives rise to all the tissues of the nodule, except the cells of the nodule cortex and the tissues at the base of nodule.

The nodule cortex results from enlargement as well as division of the cells of the outer cortical layers of the root (Bond, 1948) and consists of cells which are highly vacuolated and separated from each other by intercellular spaces. Immediately proximal to the nodule meristem, the earliest infection threads spread from cell to cell along the longitudinal axis of the nodule. A subsequent stage of tissue invasion by rhizobia involves release into the plant cell cytoplasm, and this occurs a few cell diameters behind the earliest infection threads. A cylindrical infection structure, the infection droplet, is formed in which naked bacteria come into contact with naked plant cell plasma membrane for the first time (Brewin and Vanden Bosch, 1988) (Fig. 5). This infection droplet generally is 10-25 μm in diameter and may contain 10-100 bacteria. The bacteria are released in the plant cell cytoplasm from the infection droplets by a process that resembles phagocytosis (Kijne, 1992) and an organelle-like structure, termed symbiosome is formed (Roth and Stacey, 1989 a,b). After release from the infection thread, the bacteroids actively divide with a concommitant synthesis and division of the peribacteroid membrane (PBM), which results in the enclosure of separate bacteroids in individual membrane envelopes. The PBM, which is initially derived from the plant cytoplasmic membrane is formed by the fusion with the membrane vesicles derived from Golgi bodies (Brewin et al., 1985) and the endoplasmic reticulum (Newcomb and Mc Intrye, 1981). The synchronization between bacteroid division and PBM synthesis is maintained until the bacteroids cease dividing. After the division has stopped, the bacteroids increase in volume by factors estimated from 10 to 40 (Sutton, 1983) and undergo morphological changes.

The nodule meristem gives rise to cells that differentiate into the central tissue, which consists of infected and uninfected cells and the peripheral tissue, which consists of the nodule cortex and the nodule endodermis in addition to the nodule parenchyama (inner cortex) containing the vascular bundles (Newcomb, 1981). In the central tissue, three developmental zones can be distinguished from the distal nodule meristem to the older, proximal tissues attached to the parent root (Fig. 6). In the invasion zone immediately adjacent to the apical meristem, release of bacteria from the infection threads continues to establish newly infected cells. Approximately half of the cells are not penetrated by infection threads and remain uninfected. The invasion zone is followed by the early symbiotic zone, in which plant cells elongate and bacteria proliferate. In the late symbiotic zone, the infected cells are completely filled with bacteria that have differentiated into their pleiomorphic endosymbiotic bacteroid form. Nitrogen fixation takes place in this late symbiotic zone. In older nodules, a fourth zone is the senescent zone, in which both plant cells and bacteroids degenerate.

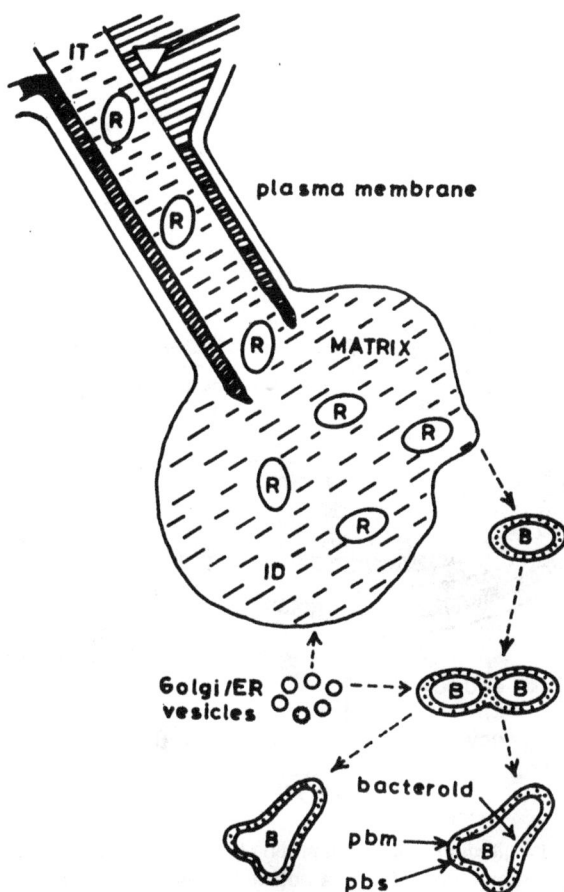

Fig. 5. Endocytosis and the development of bacteroids in the pea plant cell forming indeterminate nodules. Rhizobia (R) are released from the walled infection thread (IT) into the unwalled infection droplet (ID) which contains plant-derived matrix glycoprotein. Individual bacteria are engulfed by the naked plant membrane surrounding this droplet. Within the plant cytoplasm, the bacteria grow, divide and differentiate into nitrogen-fixing bacteria (B) enclosed by a plant derived peribacteroid membrane (pbm) and peribacteroid space (pbs). (Courtesy of N.J. Brewin, John Innes Institute, Norwich, U.K.).

In determinate nodules, the first cell divisions that occur in response to rhizobial infection are anticlinal and hypodermal (Newcomb et al., 1979; Rolfe and Gresshoff, 1988). Later cell divisions occur in the pericycle and inner cortex and eventually, the two meristematically active regions coalesce and give rise to the incipient nodule. Like the indeterminate nodules, the determinate nodule can be broadly subdivided into central and peripheral tissues. The peripheral tissues consist of nodule parenchyma and nodule cortex separated from each other by the nodule endodermis.

ZONES

Nitrogen fixation

Senescence   Development
of
symbiosome

Infection
thread

Meristem

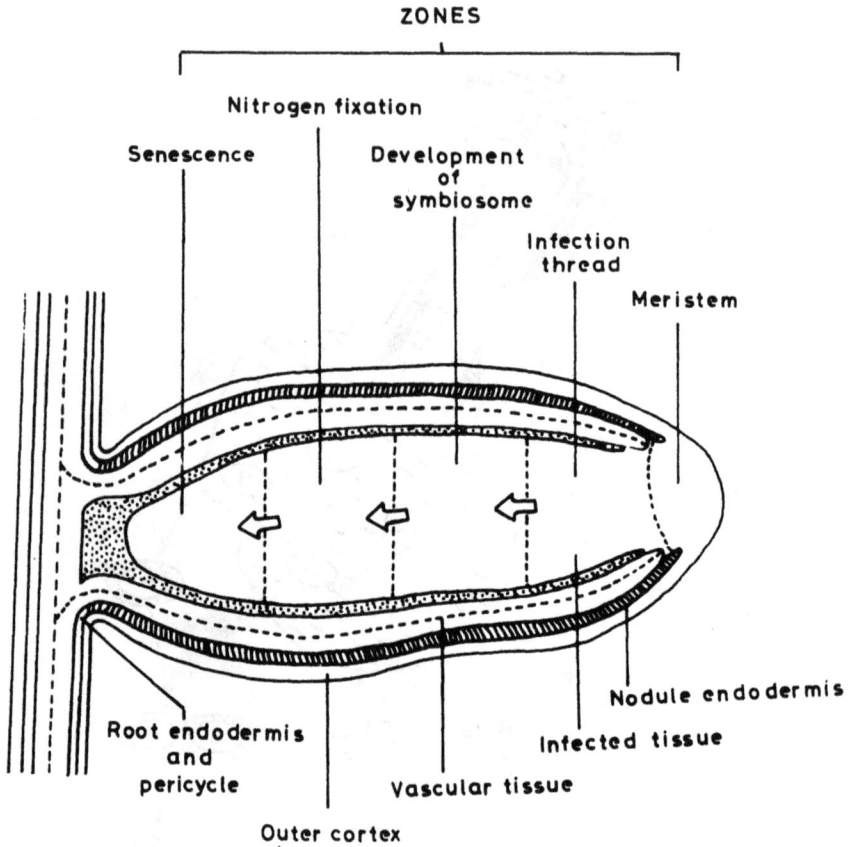

Root endodermis
and
pericycle

Nodule endodermis

Infected tissue

Vascular tissue

Outer cortex

Fig. 6.   Diagrammatic longitudinal section illustrating the organization and development of an indeterminate nodule meristem. The pea nodule has an apical meristem which is uninfected
and gives rise both to specialized uninfected tissues and a central mass of tissues which
shows successive stages of host cell invasion and differentiation by *Rhizobium*. Intracellular bacteria (bacteroids) differentiate within organelle-like structures termed "symbiosomes". Bacteroids induce the nitrogen-fixing (nitrogenase) enzyme system and ancillary
enzymes and then ultimately senesce. Modified from Brewin (1991).

Vascular bundles are embedded in the nodule parenchyma. The nodule cortex which
is to the outside of the nodule endodermis, is derived from the root cortical cells that
surround the nodule primordium. The persistent nodule meristem is found in indeterminate nodules whereas determinate nodules lack such a meri-stem. In determinate
nodules, like *Phaseolus* bean and soybean, the proliferation of the PBM is arrested
at an earlier stage of nodule development than the division of rhizobia, resulting
in the enclosure of 5 to 20 bacteroids within one single membrane envelope. The
bacteroids in these determinate nodules do not exhibit major morphological changes
and resemble free-living bacteria (Sutton, 1983). In a fully developed infected nodule
cell, packed with PBM enclosed bacteroids, the amount of membrane synthesized
is 20 to 40 times as much as in uninfected cell (Verma et al., 1978).

The peribacteroid space (PBS) surrounding the bacteroids, enclosed between the bacteroid membrane and the PBM, is the lumen of symbiosome. The PBS is shown to contain a specific set of polypeptides, different from both bacteroid and plant cytoplasm (Robertson et al., 1978; Fortin et al., 1985; Katinakis et al. 1988a). Katinakis et al., (1988a) showed that approximately 90% of the 40 major PBS proteins isolated from *R. leguminosarum* bacteroids are of bacterial origin. However, the plant host seems to play a role in the level of accumulation of the different bacteroid encoded PBS proteins (Katinakis et al., 1988b). Since most of the metabolites designated to or derived from the bacteroid have to pass through the PBS, specific carriers are supposed to be present in the PBS and Nodulin-26 has been proposed to form a specific channel to transport the various metabolites (Verma, 1992). The contents of the peribacteroid space have been found to be highly susceptible to autohydrolysis at low pH (Brewin et al., 1990), indicating that acid hydrolases may be present in the peribacteroid space.

In the central tissues of the nodule, relatively small uninfected cells are interspersed with the enlarged cells that harbor rhizobia. It is not known what factors govern cell infectibility (Joshi et al., 1991), but infected and uninfected cells assume different roles to meet the metabolic needs of the nodule (Scheres et al., 1990b). The symbiotic zone is surrounded by nodule parenchyma and endodermis. The nodule parenchyma controls diffusion of oxygen (Witty et al., 1986) and the ENOD2 gene product is localized in this region (van de Weil et al., 1990a,b). The endodermis seems to prevent diffusion of ammonia (Miao et al., 1991). The infected cells produce leghemoglobin, which allows the maintenance of oxygen flux under low oxygen tension (Appleby, 1984). The organization of nodule tissue into infected and uninfected cells represent a distribution network between the vascular tissue of the inner cortex and the infected cells, which conduct carbon substrates towards the N_2-fixing cells and organic nitrogen compounds away from this area (Fig. 7). In determinate nodules, the uninfected interstitial cells harbor peroxisomes containing a nodule-specific form of uric oxidase (Vanden Bosch and Newcomb, 1986) for translocation of ureides from these nodules (Larsen and Jochimsen, 1986; Kaneko and Newcomb, 1990). Plant cells adapt in relation to metabolism of oxygen, carbon and nitrogen compounds (Verma and Delauney, 1988) leading to leghemoglobin synthesis for oxygen regulation, sucrose synthetase production for carbon metabolism (Thummler and Verma, 1987) and glutamine synthetase production for nitrogen assimilation (Forde and Cullimore, 1989). The peribacteroid membrane develops a key role in the exchange of metabolites. Dissolved gases diffuse passively across the PBM: oxygen and nitrogen diffuse inwards, while CO_2, NH_3 and hydrogen diffuse outwards (Day et al., 1990). The major carbon sources for bacteroids such as malate and succinate are transported across the PBM via a dicarboxylate transporter (Yang et al., 1990). The PBM also appears to have a plasma-membrane type ATPase (Day et al., 1990). The intracellular bacteria progressively differentiate into pleiomorphic bacteroid forms (Hennecke, 1990) and nitrogen fixation occurs in the late symbiotic zone.

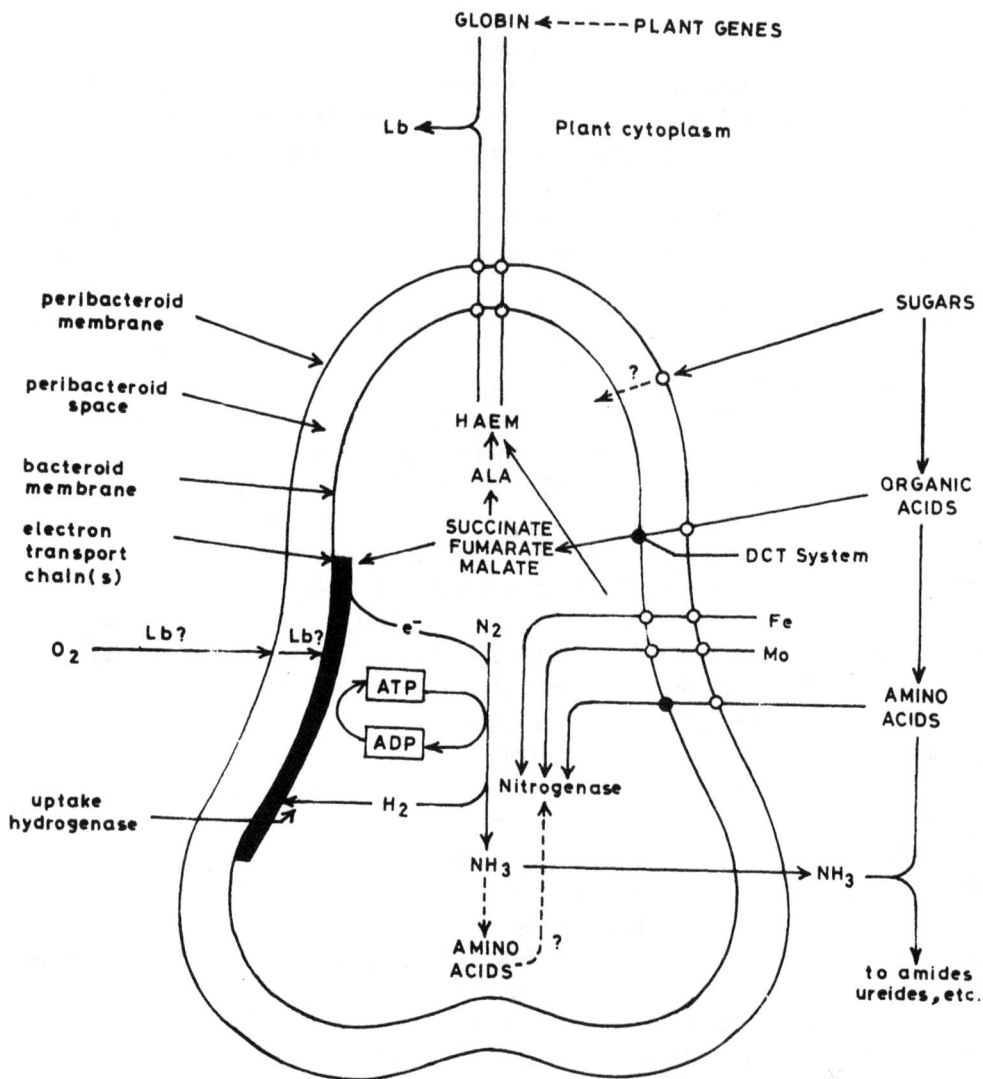

Fig. 7. Schematic diagram of metabolic exchanges between plant and bacterial cells in a N_2-fixing nodule. *-represents a known transport system and O—a potential transport system. Lb=leghemoglobin; ALA=δ aminolaevulinic acid. The organic acids are transported by the dicarboxylate transport system and specific carrier proteins are supposed to be present in the peribacteroid space for transport of various metabolites to and from the bacteroids. (Modified from Dillworth and Glenn, 1984).

(iv) Nodule Senescence

The programmed senescence of nitrogen-fixing bacteroids is an integral part of the developmental sequence in indeterminate nodules (Vasse et al., 1990). In this process, bacteroids growth and division is stopped and lysis of N_2-fixing bacteroids and the host cells occur (Kneen et al., 1990). Brewin (1991) proposed that the physiological stresses experienced by intracellular bacteria within the symbiosome compartment is one of the factors leading to nodule senescence. Bacteroids of *B. japonicum* synthesize stress proteins (Brewin, 1991) resembling GroEL and GroES of *Salmonella* (Buchmeier and Heffron, 1990). Moreover, the LPS of *Rhizobium* also appears to be adapted to survive physiological stress because it carries 27-hydroxyoctacosanoic acid as a membrane spanning component of lipid A (Bhat et al., 1991) and because epitope modification of the O-antigen, which occurs during nodule development, can be simulated by culturing free-living bacteria under conditions of low oxygen and pH (Kannenberg and Brewin, 1989).

The symbiosome can be considered as a form of lysosomal compartment within the plant endomembrane system (Werner et al., 1984; Mellor, 1989), and this may be the basis for the physiological stress encountered by the bacteroids. Two characteristic activities of the PBM would tend to acidify the peribacteroid compartment through the action of proton-ATPases (Udvardi and Day, 1989) and transport systems for dicarboxylic acids (Yang et al., 1990). Also, Golgi-derived acid hydrolases are present in the peribacteroid fluid (Brewin, 1990). Hence the symbiosome apparently has the potential to develop into a lytic vesicle if the internal pH drop significantly leading directly or indirectly, to the degradation of enclosed bacteroids. However, the endosymbiotic bacteroids can counter balance the plant-induced acidification of the symbiosome compartment (Kannenberg and Brewin, 1989), firstly by uptake of dicarboxylic acids for respiration (Ronson et al., 1984) and secondly by the excretion of ammonia as the product of nitrogen fixation (Brewin et al., 1990). One obvious prediction of this model is that the pH of the peribacteroid space should decrease following growth of nitrogen-fixing nodules in an argon/oxygen atmosphere where the nitrogenase-dependent production of ammonia would be impossible. An interesting consequence of this model is that bacterial mutants with fail to fix nitrogen (Regensburger et al., 1986; Hirsch and Smith, 1987) or to metabolize dicarboxylic acids (Finan et al., 1988) would be unable to prevent the acidification of the peribacteroid compartment, which would lead to the premature lysis of these bacteria (Werner et al., 1984).

III. *Rhizobial* Genes Involved in Nitrogen Fixation and their Regulation

The *Rhizobium* genes involved in the reduction of dinitrogen are generally divided into two groups: *nif* genes which exhibit a structural and/or functional homology to the *nif* gene cluster of free-living nitrogen fixer *Klebsiella pneumoniae*; and *fix* genes which are required for symbiotic nitrogen fixation.

(i) Organization and Regulation of Nitrogen Fixation Genes

In *R. meliloti* strains, the *nif* genes are present on a symbiotic megaplasmid (Ausubel et al., 1985). Mutants that map in the *R. meliloti* chromosome or to a second megaplasmid have been shown to result in Fix⁻ phenotypes (Forrai et al., 1983; Finan et al., 1986; Hynes et al., 1986), indicating that some genes required for symbiotic nitrogen fixation may be located on the chromosome or second megaplasmid. Unlike the fast-growing rhizobia, the *nif* and *fix* genes of *B. japonicum* are located on the chromosome. Many of the genes required for symbiotic nitrogen fixation in *B. japonicum* have been mapped into two unlinked gene clusters that are located in a highly specialized region of the chromosome that does not contain genes essential for growth (Adams et al., 1984; Kaluza et al., 1985). At least five other regions of the *B. japonicum* chromosome contain genes with functions that affect symbiotic nitrogen fixation and have been identified by random Tn5 mutagenesis (Regensberger et al., 1986). The organization of the *R. meliloti* and *B. japonicum nif* and *fix* genes is shown in Fig. 8.

A cluster of *nif* structural genes has been identified in *R. meliloti* by homology with the *Klebsiella pneumoniae, nif* HDK genes (Corbin et al., 1982). The structural *nif* genes of *B. japonicum* are located in symbiotic gene cluster I and the organization of these genes in *B. japonicum* differs from that found in *R. meliloti* as the *nif* H and *nif* DK genes are separated by 17 kb of intervening DNA (Kaluza et al., 1983; Fischer and Hennecke, 1984). Through extensive mutagenesis of the DNA between the *nif* DK and *nif* H genes (Noti et al., 1986; Ebeling et al., 1987) and by interspecies homology to *nif* genes from *K. pneumoniae* (Ebeling et al., 1987), the *nif* E, *nif* N, *nif* S and *nif* B genes were identified in cluster I, and mutations in each of these resulted in Nif⁻, Fix⁻ phenotypes, except in the case of *nif* S, which proved to be leaky.

The *nif* E gene in *B. japonicum* is located immediately downstream from the *nif* K gene, and Ebeling et al. (1987) have shown that it is expressed from the *nif* D promoter. However, the detection of mRNA corresponding to *nif* DK (and not including *nif* E) indicates that there may be transcription termination signals between *nif* K and *nif* E that result in less *nif* DKE message than *nif* DK message (Hennecke et al., 1987). The *nif* S gene is located approximately 6 kb 3' to the *nif* N gene, and the *nif* B gene is found about 1 kb downstream from it (Ebeling et al., 1987). The *nif* S promoter contains a perfect *nif* consensus promoter and is preceded by a single putative upstream activator sequence (Ebeling et al., 1987). In *R. meliloti*, *nif* B gene was identified directly downstream of the *nif* A gene by transposon mutagenesis which shared 50% DNA base homology with the *K. pneumoniae nif* B gene.

The *R. meliloti nif* A gene was mapped approximately 5 kb upstream of the *nif* HDK genes cluster (Szeto et al., 1984). In a central region of the protein encompassing 200 aminoacids, it shares 50% homology with both the *K. pneumoniae nif* A and *ntr* C genes, which have been shown to share the same region of homology

R.meliloti

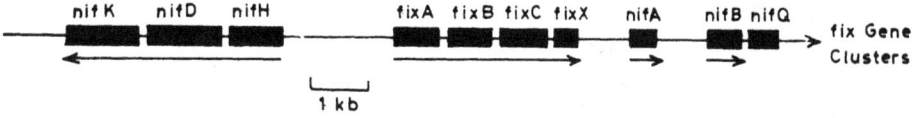

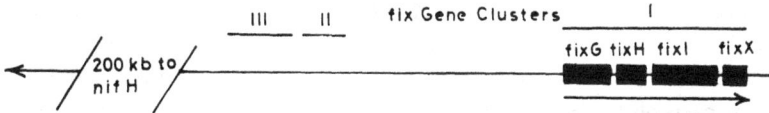

B. japonicum — Cluster I

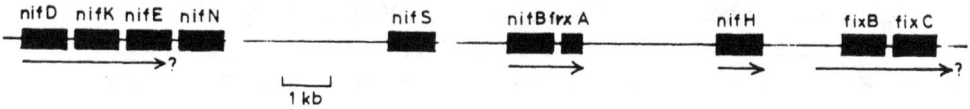

Cluster II

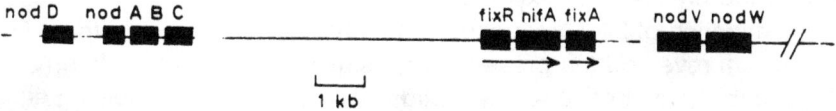

Fig. 8. Organization of genes involved in symbiotic nitrogen fixation in *Rhizobium meliloti* and *Bradyrhizobium japonicum*. The arrows indicate the direction of gene transcription, while the broken vertical lines in *R. meliloti* designate large interruptions in the map. The Fig. is modified from Ligon (1990).

with each other (Buikema et al., 1985). Since the growth of *R. meliloti* strains with mutations in the *nif* A gene region is not impaired on media containing proline, histidine, arginine, or glutamine as sole nitrogen source, the *nif*A gene is likely to be the *R. meliloti* regulatory gene that is specific for the *nif* gene system similar to *K. pneumoniae nif* A. In *B. japonicum* the *nif* A gene is located on the cluster II region that contains the *fix* R *nif* A operon (Fischer et al., 1986; Hennecke et al., 1987). The *fix* R gene encodes a polypeptide consisting of 278 aminoacids, whose function is unknown (Thony et al., 1987). However, deletion of the *fix* R ·gene that permitted transcription of *nif* A, reduced symbiotic fixation by 50%. The *fix* R promoter region contains a consensus *nif* promoter sequence and a region (at-50 to-148) that is required for maximal expression.

Another cluster of genes in *R. meliloti* that is required for symbiotic nitrogen fixation was identified and mapped in the region between the *nif* HDK and *nif* AB gene clusters and is called *fix* ABC (Puhler et al., 1984). Genetic analyses of the *fix* ABC genes indicate that the genes comprise a single operon (Ruvkun et al., 1982; Better et al., 1983; Earl et al., 1987) and that the *fix* A promoter contains a characteristic *nif* consensus promoter sequence (Better et al., 1983) and is activated by *R. meliloti* Nif A (Puhler et al., 1984). The entire *fix* ABC operon has been sequenced and no homology has been detected with DNA from *K. pneumoniae*, an organism that fixes nitrogen only in anaerobic environments (Earl et al., 1987). However, *fix* ABC homology was detected in all *Rhizobium* species examined, including *R. leguminosarum*, *R. trifolii*, *B. japonicum* and *B. sp*, (*Parasponia*) (Earl et al., 1987) and with the free-living nitrogen-fixing bacteria that fix nitrogen aerobically or microaerobically such as *Azospirillum brasilense* (Fogher et al., 1985), *Azotobacter vinelandii* (Gubler and Hennecke, 1986; Earl et al., 1987) and *Azotobacter chroococcum* (Evans et al., 1988). In *B. japonicum*, the *fix* B and *fix* C genes are located adjacent to each other approximatley 2.5 kb downstream from the *nif* H gene on symbiotic gene cluster I (Hennecke et al., 1987). The *fix* A gene is located apart from the *fix* BC genes in cluster II (Henneck et al., 1987), and a consensus *nif* promoter has been found in the *fix* A promoter region (Fuhrmann et al., 1985). Gubler and Hennecke (1986) conducted transcriptional mapping of the *fix* BC and *fix* A operons and demonstrated that the transcription of both the operons is dependent on the presence of the *nif* A gene and on low concentrations of oxygen. Therefore, it is likely that *fix* ABC genes have functions that involve the transport of electrons to nitrogenase which are unique to organisms that fix nitrogen in the presence of oxygen.

Examination of DNA sequence immediately downstream of the *nif* B gene in *B. japonicum* revealed the presence of a small open reading frame (ORF) that has the capability to encode a polypeptide of 74 amino acids (Ebeling et al., 1988). A high degree of homology has been found between this ORF and ferrodoxin genes from several diverse bacteria, and therefore, it has been designated as "*frx* A". A *nif* B- *frx* A-*lac* Z translational fusion containing the *nif* B promoter was shown to be activated in the presence of *K. pneumoniae nif* A in *E. coli*, indicating that *frx* A is cotranscribed with *nif* B from the *nif* B promoter. Mutagenesis of the *frx* A demonstrated that it is not absolutely required for nitrogen fixation in bacteroids, but is required for optimal levels of fixation in free-living cells. Therefore, it appears to have a role in nitrogen fixation, but the nature of its role is unknown. In *R. meliloti*, immediately downstream of the *fix* C gene, another gene *fix* X has been identified by mutation with the indigenous transposable element IS Rm2 (Dusha et al., 1987), and its protein product contains a cluster of cysteine residues that are characteristic of ferrodoxins. A *fix* X gene has now been found in a similar location in other fast-growing rhizobia including *R. leguminosarum* (Gronger et al., 1987) and *R. trifolii* (Lisma and Watson, 1987). It has been suggested by Dusha et al. (1987) that the function of the *fix* X gene product is the transfer of electrons to nitrogenase, similar to the *nif* F gene in *K. pneumoniae*.

Another *fix* gene region containing three *fix* complementation groups was discovered on the *sym* plasmid approximately 200 kb from the *nif* H gene by mutagenesis (Batut et al., 1985 a,b). Kahn et al. (1987) have demonstrated that these *fix* gene clusters are in a 12 kb region and that a 5 kb region of one cluster is duplicated elsewhere in the *R. meliloti* genome and contains functional *fix* genes. The *fix* genes within this duplicated region are transcribed only during symbiosis and independently of *nif* A. These genes were also shown to be highly conserved among fast-growing rhizobia (Kahn et al., 1987). The four *fix* genes in this cluster have been named as *fix* G, H, I and X and all are on a single transcriptional unit (Kahn et al., 1987). The DNA sequencing data suggest that these *fix* genes encode membrane-intergrated proteins that may be involved in a redox process which is specific to symbiotic nitrogen fixation.

Regulation of Nitrogen Fixation

The main component responsible for the reduction of nitrogen to ammonia is nitrogenase. The structure and reactivity of nitrogenase has been reviewed by various workers (Burris, 1984; Bothe et al., 1988; Smith, 1990; Shah et al., 1990). The reaction needs two proteins namely an Fe-protein (dinitrogenase reductase, component II) and a Mo Fe-protein (dinitrogenase, component I), and a Fe Mo cofactor, a strong reductant and MgATP. The basic reaction involves an MgATP activated electron transfer from the Fe-protein to Mo Fe-protein followed by substrate reduction on the Mo Fe-protein. ATP is hydrolysed to ADP and Pi during the protein-protein electron transfer. The reaction needs 2 moles of ATP for every electron transferred and for every mole of nitrogen reduced, a pair of electrons are eliminated in proton reduction to H_2 through an ATP dependent reaction. The currently accepted stoichiometry for nitrogen fixation is:

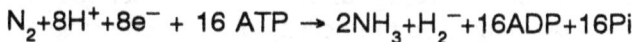

$$N_2 + 8H^+ + 8e^- + 16 ATP \rightarrow 2NH_3 + H_2^- + 16ADP + 16Pi$$

In nitrogenase enzyme, the Mo Fe-protein exists as a tetramer of α_2 and β_2 subunits with a total molecular weight of about 180 (rhizobia) to 220 kD in *Klebsiella* and *Clostridium*, while the Fe-protein exists as a homo dimer of two identical subunits (α_2) with a molecular weight of 32 kD each. The α and β subunits of the dinitrogenase (Mo Fe-protein) are encoded by *nif* D and *nif* K respectively and *nif* H is required for the synthesis of dinitrogenase reductase (Fe-protein). The Mo Fe-protein is known to contain redox centres of two distinct types; two iron-molybdenum cofactors, called FeMoco or M centre and four proposed Fe-S (4 Fe-4S) centres called P clusters. The Fe-protein contains two identical subunits (α_2) with one 4 Fe-4S centre and is responsible for donation of low potential electrons to the Mo

Fe-protein. For each electron transferred to the Mo Fe-protein, the Fe-protein must bind and hydrolyze 2 MgATP to 2 MgADP+2Pi.

In addition to these two components, a third non-protein cofactor called the iron-molybdenum cofactor (FeMoco) has been identified which is composed of Fe, Mo and S (Shah and Brill, 1977; Shah et al., 1984). This cofactor has a Fe: Mo ratio of 6-7:1 and it has also the ability to reduce acetylene to ethylene although at a much lower rate than when associated with Mo Fe-protein. Current evidence suggests that this cofactor may be an important component of the Mo Fe-protein and may actually be the site where nitrogen reduction occurs (Hawkes et al., 1984). In *Klebsiella pneumoniae*, at least six *nif* gene products-Nif Q, Nif B, Nif V, Nif N, Nif E and Nif H (dinitrogen reductase) are involved in the synthesis of active FeMoco.

The regulation of *nif* and *fix* genes in *Rhizobium* and *Bradyrhizobium* is broadly similar to the *nif* gene regulation of *K. pneumoniae*, but there are some important differences (de Bruijn et al., 1990a). Free-living *K. pneumoniae* regulates transcription of *nif* genes via the *nif* A gene product and the *ntr* A-encoded sigma factor. *nif* A gene expression itself is activated in response to low levels of combined nitrogen via the *ntr* C gene product (Gussin et al., 1986; de Bruijn et al., 1990a). The physiological effect of high oxygen concentration or excess nitrogen on *nif* A expression is controlled by the regulatory gene *nif* L. In *Rhizobium* and *Bradyrhizobium* also, the symbiotic activation of *nif* genes is dependent on the *nif* A gene product but *nif* A is not activated by *ntr* C (Szeto et al., 1987) and there is no counterpart of *nif* L, the oxygen-sensing repressor (Fischer and Hennecke, 1987; Beynon et al., 1988). The expression of *nif* A is rather dependent upon low oxygen concentration (Ditta et al., 1987), the effect of which is mediated by two regulatory genes, *fix* L and *fix* J (David et al., 1988; Gilles-Gonzalez et al., 1991). A *nif* A independent *fix* gene (*fix* N) has also been reported in *R. meliloti* (David et al., 1988) and the activation of *Fix* N by *fix* LJ has been shown to require a third regulatory gene, *fix* K (Batut et al., 1989). In addition, *fix* K negatively regulates expression of the *nif*-specific activator *nif* A as well as its own expression by autoregulation. The current cascade model for the regulation of *nif* and *fix* genes in *R. meliloti* proposed by Batut et al. (1989) has been shown in Fig. 9. The *fix* L product seems to be a transmembrane sensor that modulates the activity of the *fix* J product, a cytoplasmic regulator (David et al., 1988). Fix L and Fix J are homologous to a family of bacterial two-component regulators (Nixon et al., 1986), for which the mode of signal transduction is phosphorylation and this regulatory system is responsible to cell environment (in this case low oxygen) (David et al., 1988; Gross et al., 1989) and *fix* K shows a high degree of homology to *E. coli fnr* and *crp* gene products and appears to act as a transcriptional regulator (Batut et al., 1989).

Both *nif* and *fix* gene mutants of *Rhizobium* are still able to cause nodule development, but the nodules do not fix nitrogen (Nod$^+$ Fix$^-$). In addition, the mutations in the dicarboxylate transport (dct) genes have also been found to result in Fix$^-$ nodules (Ronson and Primrose, 1979; Ronson et al., 1981; Finan et al., 1983).

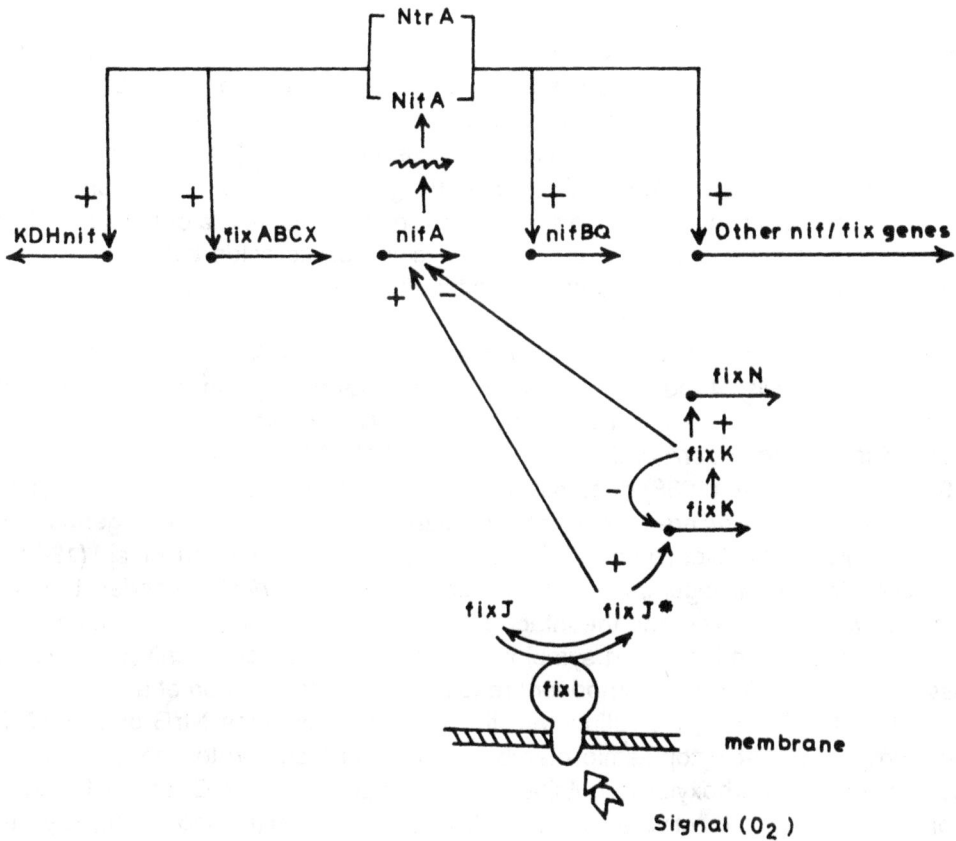

Fig. 9. Regulation of *nif* and *fix* gene expression in *Rhizobium meliloti*. The arrows indicate the direction of gene transcription and filled circles denote promoters for operons, represented by straight lines. Arrows indicate the direction of transcription and/or transcriptional activation exerted by the specified gene product at target promoters. The *fix* LJ products belong to the family of two-component regulatory systems and control the *nif/fix* gene expression in response to oxygen. Modified from Batut et al. (1989).

(ii) Dicarboxylic Acid Transport to Support Symbiotic Nitrogen Fixation

Nitrogen fixation by *Rhizobium*-legume symbiosis is a highly energy-intensive process. Due to microaerobic or nearly anaerobic (dissolved O_2, 10 to 40 nM) conditions in the interior of nodule, carbohydrates from photosynthates are broken down to malate through glycolysis in plants which subsequently may be reduced to fumarate and succinate (Davis, 1980; Vance and Gantt, 1992). These C_4-dicarboxylic acids provide the source of carbon and energy to bacteroid nitrogenase for the conversion of N_2 to ammonia (Guerinot and Chelm, 1987; Mc Rae et al., 1989). Engelke et al. (1987) demonstrated by inhibition studies that in free-living *R. meliloti,* succinate, fumarate and malate are transported by a common active transport, the dicarboxylate transport system (Dct-system). This common Dct-system has also been detected in the fast-growing species *R. leguminosarum* (Glenn et al., 1980; Finan et al., 1983) and *R. leguminosarum* bv *trifolii* (Ronson et al., 1981) as well as in the slow-growing *Bradyrhizobium japonicum* (Mc Allister and Lepo, 1983). Mutants of *R. leguminosarum* bv *viciae* (Arwas et al., 1985), *R. leguminosarum* bv *trifolii* (Ronson et al., 1981), *R. meliloti* (Engelke et al., 1987; Watson, 1990) and *B. japonicum* (San Francisco and Jacobson, 1985) having a defective C_4-Dct system (Dct⁻) were unable to transport C_4-dicarboxylic acids but utilized other carbon sources and these mutants formed ineffective nodules on their respective host plants.

The complementation of *R. leguminosarum* and *R. meliloti* Dct⁻ mutants and sequence analysis of the complementing fragments resulted in the identification of a cluster of three genes: *dctA, dctB* and *dctD*, organised in two divergently transcribed operons *dctA* and *dctBD* (Ronson et al., 1984, 1987a; Engelke et al., 1989). Watson et al. (1988) demonstrated that genes involved in C_4-dicarboxylate transport in *R. meliloti* are located on a megaplasmid that also carries genes that have a role in the biosynthesis of exopolysaccharides. Ronson et al. (1987a) proposed that in *R. leguminosarum*, the structural gene, *dctA*, encoded the C_4-dicarboxylate permease and the intact *dctA* gene was required for transport activity both in the free-living and symbiotic states. The *dctB* and *dctD* genes were assumed to encode a two-component regulatory system (Ronson et al., 1987a,c). Based on the homology existing with the regulatory proteins NtrB and NtrC, it was proposed that cytoplasmic membrane located DctB protein may act as a sensor for C_4-dicarboxylates outside the cytoplasm and the C-terminal kinase domain of DctB could activate the intracellular DctD protein (Ronson et al., 1987a). The activated DctD protein together with the sigma factor σ^{54} (RpoN), initiate the transcription of the *dctA* gene in the free-living state (Ronson et al., 1987b). Mutations in the genes *dctA, dctB, dctD* and *rpoN* in the free-living state, always resulted in a Dct⁻ phenotype (Bolton et al., 1986; Ronson et al., 1984, 1987b; Engelke et al., 1987, 1989).

᠂ The activation of *dctA* in the symbiotic state is regulated in a different way. *R. leguminosarum* and *R. meliloti* strains mutated in the *dctB* and *dctD* genes are able to form effective nodules and to transport C_4-dicarboxylates through the bacteroid membrane (Finan et al., 1983; Watson et al., 1988; Yarosh et al., 1989). However, the acetylene reduction rate of the infected plants and the transport rate of the isolated bacteroids are reduced (Engelke et al., 1987). Therefore, it is assumed that *dctB* and *dctD* are not essential for the Dct-system under symbiotic conditions and an alternative symbiotic activator which complements the function of DctB/DctD system has been postulated for symbiotic *dctA* expression. A consensus sequence with homology to the consensus Nif A-binding sequence upstream of the *dctA* promoter lead to the conclusion that Nif A could be the symbiotic regulator (Ronson, 1988; Engelke et al., 1989). However, Jordig et al. (1992) demonstrated that neither nifA nor the FixL/FixJ system which is responsible for *Nif*A activation in the absence of oxygen (de Philip et al., 1990) have any direct influence on symbiotic *dct*A regulation. Jording et al. (1992) also showed that in *R. meliloti* induced nodules, the *dct*A expression rate was approximately equal in effective and ineffective bacteroids whereas the C_4-dicarboxylate transport rate was reduced to a level of 50% in all Fix⁻ ineffective bacteroids. These results suggested that at least 50% of the C_4-dicarboxylates transported in wild-type bacteroids, is required to maintain the nitrogen fixation process, whereas the remaining portion is necessary for the basic metabolism of the bacteroids.

Birkenhead et al. (1988) transferred *R. meliloti dct* genes to a *B. japonicum* strain. The transconjugant strain of *B. japonicum* containing the *R. meliloti dct* genes showed increased growth rates in media containing dicarboxylic acids as the sole carbon source and increased succinate uptake under aerobic conditions. Moreover, the transconjugant strain demonstrated increased levels of nitrogenase activity over the parent *B. japonicum* strain under free-living nitrogen-fixing conditions. However, it is unclear from this work whether the increase in nitrogenase activity is due to the introduction of a more effective Dct-system in *B. japonicum* or to an increase in *dct* expression through increased copy numbers.

Plant cytosol of nodules has high activity and proteins for phosphoenolpyruvate carboxylase (PEPC=provides a portion of the C-skeletons for malate and aspartate), malate dehydrogenase (MDH) and aspartate aminotransferase (AAT=involved in assimilation of reduced N into aspartate and asparagine). Activities of these enzymes increase coincidently with nitrogenase during effective nodule development and their activities are much higher in nodules than in roots (Reynolds et al., 1982; Egli et al., 1989) Like organic acid concentrations, PEPC, MDH and AAT activities are much reduced in ineffective nodules.

(iii)   Role of Uptake Hydrogenase in Nitrogen Fixation

Among the various factors influencing the efficiency of biological nitrogen fixation, is the hydrogen production during the enzymatic reduction of dinitrogen to ammonia (Eisbrenner and Evans, 1983; Maier, 1986). In the absence of an active uptake hydrogenase, this reduction of protons in legume root nodules results in 25-30% loss of electrons flux through nitrogenase by way of H_2 evolution (Schubert and Evans, 1976; Evans et al., 1981). Certain strains of *R. leguminosarum* (Ruiz Argueso et al., 1978; Dejong et al., 1982), *B. japonicum* (Schubert and Evans, 1976; Ruiz-Argueso et al., 1981) and *Rhizobium* sp. Cowpea miscellany (Gibson et al., 1981; Dadarwal et al., 1985; Sindhu and Dadarwal, 1986) have been shown to recycle the H_2 evolved during nitrogen fixation by inducing an H_2-uptake system (Eisbrenner and Evans, 1983; Dadarwal and Sindhu, 1988). This character has not been demonstrated in native strains of *R. meliloti*, *R. trifolii*, and fast-growing strains of chickpea and cowpea. Such Hup⁺ strains could be easily identified *in vivo* in nodule either by H_2 uptake (Schubert and Evans, 1976) or by $^3H_2$ exchange (Dixon, 1972).

However, *in vitro* H_2 uptake by free-living cells of *Rhizobium* is extremely low compared with bacteroids. Reductions of triphenyl tetrazolium chloride (Maier et al., 1978) and methylene blue (Haugland et al., 1983) in plates containing specific medium or the *ex planta* induction of nitrogenase (Sindhu and Dadarwal, 1986) have been used as indirect tests for identifying Hup[+] *Rhizobium* strains of the soybean and cowpea cross-inoculation groups.

The potential advantages of the *Hup* system in *Rhizobium*-legume symbiosis suggested by Dixon (1972) were: (i) The process utilizes O_2 and thus could protect the O_2-sensitive nitrogenase, (ii) the removal of H_2 could prevent inhibition of nitrogenase by this gas and (iii) oxidation of H_2 could recover some ATP used by nitrogenase in the formation of H_2. Also, the *Hup* system has been demonstrated to provide respiratory protection to allow *ex planta* expression of nitrogenase in cowpea miscellany rhizobia under free-living conditions (Garg and Dadarwal, 1991), the ability to grow chemoautrophically as demonstrated in some of the Hup[+] *B. japonicum* strains (Lepo et al., 1980; Hanus et al., 1981), and to provide survival and competitive advantage to Hup[+] strains over the Hup[−] strains (Dadarwal et al., 1985; Lambert et al., 1985; El Hasan et al., 1986). Agronomic importance of *Hup* character for improving nitrogen fixation and productivity of legume plants by comparison of Hup[+] and Hup[−] strains has been summarized by Eisbrennerr and Evans (1983). In the case of *B. japonicum* (Albrecht et al., 1979; Hanus et al., 1981) and mungbean rhizobia (Pahwa and Dogra, 1981), Hup[+] isolates averaged more N_2 fixation than did Hup[−] strains, but analogous comparisons with *R. leguminosarum* showed little symbiotic advantage for Hup[+] phenotypes relative to Hup[−] strains (Truelsen and Wyndaele, 1984). Use of Hup[+] parent strains with isogenic Hup[−] mutants of *B. japonicum* (Lepo et al., 1981; Evans et al., 1984) and Tn5-induced Hup[−] mutants of *R. leguminosarum* (Cunningham et al., 1985) have shown a significant increase in dry matter yield and total nitrogen content by Hup[+] strains in soybean and pea, respectively.

Genes involved in H_2^- uptake (*hup* genes) are located on the chromosome in *B. japonicum* and have been isolated and characterized (Cantrell et al., 1983; Haugland et al., 1984). In *R. leguminosarum, hup* genes are linked to symbiotic determinants on a plasmid (Brewin et al., 1980; Leyva et al., 1987; Ruiz Argueso et al., 1991). By complementation of Hup[−] *B. japonicum* mutant PJ17 with gene bank library of strain USDA 122 DES, one cosmid p[HU1] was isolated (Cantrell et al., 1983) in which *hup*-specific sequences were found to span in about 16 kb of DNA (Haugland et al., 1984). However, acquisition of p[HU1] by Hup[−] wild type *B. japonicum* and *R. meliloti* strains did not confer *Hup* activity in the free-living state. Lambert et al. (1987) isolated an additional *hup* cosmid p[HU52] which complemented all but one of the Hup[−] mutants and is therefore a promising cosmid for intraspecies transfer of hydrogenase activity. Subsequently, Lambert et al. (1985, 1987) showed that Hup activity encoded by p[HU52] could be expressed in nodules of soybean, alfalfa and clover formed by transconjugant strains of *B. japonicum, R. meliloti* and *R. trifolii*. Recently, Garg and Dadarwal (1993) constructed Hup[+] recombinants by transferring p[HU1] consmid into Hup[−] strain M11 of cowpea miscellany (*Vigna*) and

7.49 to 8.75% increase in symbiotic effectivity in terms of gain in total plant nitrogen over Hup⁻ parent strain was observed.

IV. Host Plant Genes Involved in Symbiosis

Plants represent a diverse group of organisms with relatively simple but unique reproductive, developmental and physiological processes (Goldberg, 1988). Plant genomes are large and complex and physical and functional analysis of plant genes in transformed plants indicate that gene expression is highly regulated and controlled much as it is in the animal kingdon (Klee et al., 1987). Since plants depend greatly on the environment to trigger developmental events, much of that genetic complexity may be used for an array of signals that activate and repress specific genes involved in unique adaptive physiological phenomena. For instance, a complex group of signal transduction mechanisms are used to activate plant defenses against microbial attack (Lamb et al., 1989). Most of the known nitrogen-fixing symbiotic associations also occur between *Rhizobium* and legumes through a complex series of developmental steps involving coordinate regulation between both the organisms. During this process, the plant regulates many functions required for nodule formation and nitrogen fixation including cell division, gene expression, primary and secondary metabolism, morphogenesis, signal transduction and membrane biogenesis (Kondorosi and Kondorosi, 1986; Verma and Fortin, 1989; Nap and Bisseling, 1990a,b; Verma et al., 1992). Many plant mutants have been obtained either by treatment with ethyl methane sulfonate (EMS) or γ-irradiation or by spontaneous mutation in soybean, pea, french bean, chickpea, alfalfa and peanut (reviewed by Caetano-Anolles and Gresshoff, 1991). The phenotype of these mutants range from failure to nodulate (blocked at different stages of nodule development; i.e. cortical cell division or root hair curling) to "supernodulation". Genetic analysis of these mutants has shown that several plant loci are involved in controlling nodulation (Caetano-Anolles and Gresshoff, 1991).

(i) Nodulins

A number of plant genes have been identified, by using classical genetic and molecular approaches, which are expressed exclusively during nodule morphogenesis. These include plant proteins, specifically synthesized by plants during the formation and functioning of a root nodule called nodulins (Delauney and Verma, 1988). Nodulins were first identified in soybean root nodules (Legocki and Verma, 1979, 1980; Fuller et al., 1983). Since then, a large number of nodulins have been identified in eight different legume species either by comparison of *in vitro* translation products from nodule and root mRNA or by the isolation of nodule-specific cDNA clones (Govers et al., 1987; Delauney and Verma, 1988). However, none of these genes associated with any known plant mutation affecting nodule development has been isolated.

Plant nodulins are characterized as "early" and "late" nodulins, based upon the temporal sequence of induction. The proteins from the early nodulin genes are specifically expressed at a distinctive stage of infection process and root nodule morphogenesis (Nap and Bisseling, 1990b; van de Weil et al., 1990b). Most early nodulins are highly proline rich proteins and may therefore be the cell wall components (Nap and Bisseling, 1990b). The late nodulin genes are expressed at the later stages of nodule formation concomitantly with or following the release of bacteria from the infection thread, but prior to the induction of nitrogenase and commencement of nitrogen fixation (Fuller and Verma, 1984; Verma et al., 1988; Verma and Dalauney, 1988). Late nodulins aid to the function of a root nodule by creating the physiological conditions required within the nodule for nitrogen fixation, ammonia assimilation and transport. Among the identified late nodulins are the leghemoglobins (Appleby, 1984), Uricase (Bergmann et al., 1983), subunits of sucrose synthase (Kahn et al., 1985) and glutamine synthetase (Cullimore et al., 1984).

A. Early Nodulins

One of the first early nodulins described is ENOD2 (nodulin-75) from soybean (Franssen et al., 1987). Transcripts of the ENOD2 gene are present in the nodule inner cortex (nodule parenchyma) of pea and soybean (van de Wiel et al., 1990a,b) and of alfalfa (Allen et al., 1991). Amino acid sequence analysis of ENOD2 from different plants (Franssen et al., 1987; Dickstein et al., 1988) revealed that this protein resembles plant cell wall proteins (Averyhart-Fullard et al., 1988; Cassab and Varner, 1988) in that they contain two repeating pentapeptides, Pro-Pro-Glu-Tyr-Glu and Pro-Pro-His-Glu-Lys. It seems likely that the prolines become hydroxylated *in vivo* to yield a hydroxy proline-rich protein and that this protein is further glycosylated and thus may be a part of a wide family of hydroxy-proline rich glyco-proteins (HGRPs) expressed in different plant tissues (Nap and Bisseling, 1990a). Since the nodule inner cortex has been suggested to be an oxygen diffusion barrier (Witty et al., 1986), the presence of ENOD2 in this tissue might indicate the role of nodule specific cell wall proteins in establishing and maintaining the oxygen diffusion barrier in nodules (van de Wiel et al., 1990a). Expression of ENOD2 has also been documented in empty nodules produced by *exo* mutants (van de Weil et al., 1990a), pseudonodules induced by auxin transport inhibitors (Hirsch et al., 1989) and nodules formed in the absence of *Rhizobium* sp. (Truchet et al., 1989a). ENOD2 gene expression has been shown to be induced by high levels of cytokinin which regulates the activity of the ENOD2 gene promoter (Dehlo and de Bruijn, 1992). An extracellular glycoprotein of 95 kD has recently been identified in cell walls and intercellular spaces of nodule parenchyma cells in pea nodules (Rae et al., 1991). Its location suggests that it may be related to ENOD2. However, the 95 kD glycoprotein is also found within the infection thread matrix (Vanden Bosch et al., 1989), a site where ENOD2 transcripts have not been detected.

The expression of two early nodulin genes has been correlated with the infection process in pea. The early nodulin cDNA clones pPsENOD5 (Scheres et al., 1990b) and pPsENOD12 (Scheres et al., 1990a) have been isolated from a pea nodule cDNA library by differential screening. mRNA for PsENOD12 is observed in root hairs and in cortical cells that contain a growing infection thread. The PsENOD12 is also expressed in cells that occur several layers in front of the growing infection threads (Nap and Bisseling, 1990b) and are undergoing the morphological changes that precede penetration by an infection thread (Scheres et al., 1990a). It has been suggested that the putative cell wall protein PsENOD12 may thus be a part of the additional cell wall formed in the cortex cells that prepare for infection thread passage. In addition, the PsENOD12 protein may be a component of the infection thread itself.

In situ hybridization studies with the use of antisense RNA probes show that PsENOD5 gene is only expressed in cells containing growing infection threads (Scheres et al., 1990b). It may also be a part of the plasma membrane of the infection thread as well as a part of the peribacteroid membrane (Franssen et al., 1992). Sequence analysis of PsENOD5 and PsENOD12 shows both early nodulins to be proline-rich proteins. The PsENOD12 early nodulin is for the major part composed of two repeating pentapeptides, each of which contains two prolines and this nodulin is a hydroxy-proline-rich cell wall protein (Scheres et al., 1990a). The PsENOD5 protein besides being rich in proline, is relatively rich in alanine, glycine and serine, suggesting that it is related to arabinogalactan proteins (Scheres et al., 1990b). Both the early nodulins have putative signal peptide at the amino terminal end of their sequence and therefore are probably transported across the membrane. Another early nodulin gene GmENOD40 has been detected in the pericycle of the vascular tissue of soybean nodules (Franssen et al., 1992).

Application of sterile culture filtrates of *R. leguminosarum* bv *viciae* or purified extracellular Nod factor (Nod Rlv) to pea seedlings elicit PsENOD5 and PsENOD12 gene expression in root hairs (Nap and Bisseling, 1990b; Stacey et al., 1992), suggesting that Nod factors can also induce responses of infection. Recently, transcripts of PsENOD12 and ENOD2 have been detected in naphthyl pthalamic acid auxin transport inhibitor) induced pseudonodules formed on roots of pea cv. Afghanistan (Hirsch, 1992), suggesting that a change in the hormone level is an early event during the nodulation process. The expression of several early plant nodulins (ENOD5, ENOD12 and the meristematically associated ENOD40) in tissues after treatment with the nodulation signal, Nod factors, indicated that bacterial infection is not required for the expression of some of these early nodulin genes.

In the early symbiotic zone, two other early nodulin genes PsENOD3 and PsENOD14, are exclusively expressed in the infected cells of the central tissue (Scheres et al., 1990b). Sequence analyses of PsPNOD3 and PsENOD14 clones show that the mRNAs encode 6 kD polypeptides that are 55% homologous. Due to the presence of a putative signal peptide at the amino terminal end and four cysteine residues with a spatial distribution resembling to that of metal-binding proteins (Scheres et al., 1990b), it has been suggested that PsENOD3 and PsENOD14 nodulins may be metal-binding proteins involved in, for instance, transport of Fe or Mo to the bacteroids for a functional nitrogenase/hydrogenase. Similar to PsENOD12 and PsENOD5, the PsENOD3 and PsENOD14 genes are transiently expressed during the development of nodule. In the late symbiotic zone, the amounts of PsENOD3 and PsENOD14 mRNAs decrease and late nodulin mRNAs such as for leghemoglobin, become detectable.

B.   Late Nodulins

Among the large groups of late nodulins, leghemoglobin (Lb) is the most abundant and also the best characterized nodulin (Appleby, 1984). The biosynthesis and its role in oxygen regulation in nodules of different legumes has been discussed separately. Other known late nodulins are involved in the assimilation of reduced nitrogen, for example N-35 (Ngm-35) which is the second most abundant nodulin in soybean nodules (Legocki and Verma, 1979), is the 33 kD subunit of n-uricase (or Uricase II) (Verma et al., 1986), a key enzyme in the ureide biosynthetic pathway in soybean involved in ammonia assimilation. Immunofluorescent localization and immunogold labelling of N-35 in soybean nodules indicates that N-35 is exclusively found in the peroxisomes of uninfected nodule cells (Bergmann et al., 1983; Nguyen et al., 1985). Xue et al. (1991) have shown that exposure of soybean stem callus cells to low oxygen concentrations induces a 4- to 6-fold increase in the synthesis of nodule specific uricase (Nodulin-35). Another enzyme glutamine synthetase (GS) plays a central role in assimilation of ammonia in nodules. GS is encoded by a small family of genes, the members of which are expressed in different organs of plants, giving rise to many functional forms of this enzyme (Tingey et al., 1987; Verma, 1989). Nodule-specific GS isoforms have been identified in *Phaseolus vulgaris* (Lara et al., 1983; Forde et al., 1989), alfalfa (Dunn et al., 1988) and lupin (Konieczny et al., 1988). In soybean, enhanced GS activity in nodules has been reported to be· due to the synthesis of nodule-specific GS isoforms (Sengupta-Gopalan and Pitas, 1986). However, recent studies have shown that the increase in soybean nodule GS is due to the NH_4^+-stimulated expression of GS isoforms in the root (Hirel et al., 1987; Miao et al., 1991). All GS sequences so far isolated, either through screening the soybean nodule cDNA library (Hirel et al., 1987) or by direct complementation of *E. coli glnA* mutant (Miao et al., 1991), seem to be expressed in both roots and nodules.

Sucrose is the main sugar translocated from the host plant leaves to the nodules. Sucrose synthetase in the nodules appears to be involved in the breakdown of sucrose to support the carbon requirements of the nodules (Reibach and Streeter, 1983). Nodulin-100 has been shown to encode a subunit of sucrose synthetase (Thummler and Verma, 1987). The expression of maize sucrose synthetase in transgenic plants showed that it is anaerobically induced in roots and is specifically expressed in the phloem cells (Yang and Russell, 1990). It is possible that the expression of the soybean sucrose synthetase in nodules is also controlled by the low level of oxygen. Also, the activity of sucrose synthetase seems to be regulated by the binding of free heme, which dissociates this enzyme into subunits (Thummler and Verma, 1987) and thereby the availability of free heme during nodule senescence may control the supply of carbon to the bacteroids and thus may prevent bacteria from becoming pathogenic to the plant cell (Verma, 1989). Moreover, several nodule enzymes have been shown to differ in physical, kinetic, or immunological characteristics from their root counterparts. Examples include phosphoenolpyruvate carboxylase (Deroche et al., 1983), Choline kinase (Mellor et al., 1986). Xanthine dehydrogenase (Verma et al., 1986), purine nucleosidase and malate dehydrogenase (Waters et al., 1985; Egli et al., 1989). It is unknown whether the nodule-specific forms of these enzymes originate from the expression of nodulin genes or result from nodule-specific, post-translational modifications of gene products also found elsewhere in the plant.

During the formation of the peribacteroid membrane (PBM) a number of nodulins are induced and are specifically targetted to this membrane (Fortin et al., 1985). Many of the PBM nodulins are glycosylated in the endoplasmic reticulum (ER) and further modified before they are integrated into the membrane (Verma and Fortin, 1989). The best studied is the soybean nodulin-26 (N-26), which is a major intrinsic membrane protein (Fortin et al., 1987). N-26 shares significant sequence homology with several proteins identified in other plants and other species (see review Verma et al., 1992) and with the *Escherichia coli* glycerol facilitator, GlpF (Sweet et al., 1990). It has been shown that the GlpF is a pore-type ion channel in the cytoplasmic membrane of *E. coli* (Sweet et al., 1990). N-26 contains six membrane-spanning domains (Miao et al., 1990) and lacks any signal sequence on its amino terminus, but it may have an internal signal for membrane translocation. The carboxy end of N-26 contains three phosphorylation sites which may interact with the acidic lipids present in the PBM and a nodule kinase has been shown to phosphorylate this protein (Weaver et al., 1991). Based on its secondary structure and sequence homology with GlpF, it is very likely that N-26 functions as an ion channel in the PBM. Since the expression of this gene occurs prior to and independent of nitrogen fixation, N-26 may be involved in translocation of specific metabolites, like the carbon source succinate, across the PBM to support the bacteroids.

Nodulin-24 (N-24) represents a surface PBM protein, facing the peribacteroid space. N-24 is synthesized on the ER, where its signal sequence is removed (co-translationally) and the protein is finally translocated to the PBM via Golgi vesicles (Katinakis and Verma, 1985). Western blot analysis of N-24 in the PBM revealed a cross-reacting band of 32 kD (Fortin et al., 1987), which suggests that extensive post-translational processing through Golgi increases its size. Recently, an N-24 homologous sequence, nodulin-16, has been isolated from soybean nodules (Nirun-suksiri and Sengupta-Gopalan, 1990).

Another nodulin, N-23 (Ngm-23) belongs to a group consisting of five related nodulins, Ngm-20, Ngm-23, Ngm-26b, Ngm-27 and Ngm-44 (Jacobs et al., 1987). These nodulins have regions in common and regions unique to each particular nodulin.

The spatial distribution of cysteine residues and possession of conserved hydrophobic signal sequence (except for Ngm-26b) suggests that most members of this family consist of metal-binding polypeptides, that either cross a membrane or are associated with it. It has been shown that Ngm-23 is located in the peribacteroid membrane, but the location of the other members of this family has not been established yet. Future research will involve the characterization of the function of these nodulins and elucidation of the bacterial-plant signals triggering their induction.

(ii) Leghemoglobin Biosynthesis and its Function in Nodules

Leghemoglobin (Lb) is an essential component for N_2 fixation in legume nodules and constitutes upto 25% of the total soluble protein in a mature nodule (Appleby, 1984). It is a 16kD myoglobin-like hemoprotein with high affinity for oxygen and controlls the concentration of free oxygen supply to the bacteroid in such a way that balances protection of nitrogenase against oxygen damage with support of respiration (Appleby, 1984). *Rhizobium* and *Bradyrhizobium* species synthesize varying levels of Lb and a positive correlation has been observed of the Lb quantity with the symbiotic effectivity of *Rhizobium* strains in soybean, clover and alfalfa host (Jordan and Gerrad, 1951; Graham and Parker,1961; Damrey and Alexander, 1979). Lb is localized in the cytoplasm of the infected cells of the nodule (Robertson et al., 1984; Nguyen et al., 1985) and no Lb exists inside the peribacteroid compartment. A small amount of Lb has also been detected in the uninfected cells of soybean (Vanden Bosch and Newcomb, 1988). The high demand for oxygen by the bacteroids in infected cells of the nodule may cause hypoxic conditions and Lb has been proposed as a 'defence' molecule against hypoxia (Verma et al., 1990).

Leghemoglobin is apparently a true product of symbiosis: the globin proteins are encoded by the host plant while the heme moiety is primarily made by the bacteria and transported into the host cytoplasm (Lee and Verma, 1984; Nap and Bisseling, 1990b). Evidence that the apoprotein of Lb is plant-encoded came from the observations that a common globin is formed in nodules of *Vigna sesquipedalis* induced by *Rhizobium* strains of divergent origin and host infective groups (Broughton and Dilworth, 1971). Tracer experiments indicate that the heme prosthetic group of Lb is synthesized by bacteroids (Cutting and Schulman, 1972) probably via the bacterial C-4 haem-biosynthetic pathway in which the first step is catalysed by δ-aminolaevulinic acid synthase (ALAS) (Nadler and Avissar, 1977). Recent studies have shown that heme biosynthetic pathway may be spatially separated between the two organisms. The heme precursor, δ-aminolaevulinic acid, is synthesized in the plant cell while the subsequent heme biosynthesis steps are completed by bacteria (Sangwan and O'Brian, 1991). Molecular genetic analysis of the *hem* A gene coding ALAS shows that site-directed *hem* A mutants in *R. meliloti* lack ALAS and are Fix⁻ on alfalfa (Leong et al., 1982). However, a similar site directed *hem* A mutant of *B. japonicum* has been observed to be Fix⁺ soybean (Guerinot and Chelm, 1986). Stanley et al. (1988) cloned the NGR234 *hem* A gene and constructed a site directed *hem* A mutant, which produced Fix⁻ nodules on either determinat

(*Lablab, Vigna* and *Macroptilium*) or indeterminate (*Leucaena*) nodule type plants. The result supports the general concept that bacteroid ALAS is required for synthesis of the prosthetic group of nodule Lb. The *B. japonicum* results might be explained by the presence of tRNA-glu dependent C-5 haem synthesis in this bacterium. This pathway, which exists in higher plant plastids, has been shown to exist in *E. coli* (Li et al., 1989).

In soybean, there are four major forms of Lb species i.e. Lba, Lbcl, Lbc2 and Lbc3 (Lee et al., 1983), differing only slightly in their amino acid composition plus four minor forms probably arising from post-transcriptional modifications. Leghemoglobin genes are arragned in two multigene clusters and bear a strong similarity to hemoglobin genes found in animals and insects (Appleby, 1984). The hemoglobin genes equivalent to leghemoglobin are also induced in non-leguminuous woody shrubs during the development of N_2-fixing root nodules containing *Frankia*, a Gram positive actinorhizal microorganism. Moreover, when the hemoglobin gene promoter derived from *Trema*, a non-nodulating non-legume, was introduced into *Lotus*, it showed exactly the same pattern of nodule specific expression as the endogenous leghemoglobin gene (Bogusz et al., 1990). The regulation of expression of one of the soybean genes, lbc3, has been studied in *Lotus*, following transformation using *Agrobacterium rhizogenes* as vector, by linking various sequences of the promoter region of the gene to the coding sequences of the reporter gene *Cat* (Chloramphenicol acetyl transferase) (de Bruijn et al., 1990b; Stougaard et al., 1990). Two regions of the gene, at least, appear to be essential for nodule-specific gene expression. Corresponding studies using the *Sesbania* gene *Srg lb3* have indicated both similarities and differences with respect to the soybean gene *lbc3* promoters (Szabados et al., 1990). Apparently, the Lb promoter region not only carries all required *cis*-regulatory sequences, but all relevant *trans* acting factors are also conserved among different leguminous species. Deletion analysis of the soybean Lb promoter defines a relatively small region as responsible for the expression pattern (Stougaard et al., 1986). In soybean, a nodule specific *trans*-acting factor interacting with Lb promoter has been identified (Jacobson et al., 1990). The factors responsible for the expression of *lb* genes are to the present time, not known. They may be bacterial signals or host plant-derived factors which are a consequences of nutritional and/or development changes occasioned by the presence of bacteria.

Dadarwal et al. (1982) showed that Hup⁺ strains possess low levels of Lb compared to Hup⁻ native strains and the nodules formed by Hup⁻ strain M11 contained 2-3 times more of the Ferric-oxy Lb fraction than the nodules formed by Hup⁺ strain S24 in six species of cowpea miscellany hosts. The chemical nature of Lb and conditions inside nodules, such as slightly acid pH, microaerobic conditions and the presence of metal ions, chelators and toxic metabolites (nitrite, superoxide radical, peroxides), are conducive for oxidation of Ferrous Lb (Lb^{2+}) or its oxygenated form (LbO_2) to non-functional ferric Lb (Lb^{3+}) and ferryl Lb (Becana and Klucas, 1992). It has been suggested that some mechanism operates in nodules to regulate functional form of Lb for effective symbiosis. Kretovich et al. (1982) demonstrated that a Lb^{3+} reductase enzyme with FLbR activity may be involved in Lb^{3+} reduction *in vivo* and subsequently a protein with FlbR activity has been purified to homogenity from soybean nodules (Puppo et al., 1980; Saari and Klucas, 1984). Moreover, nodules contain many potential reductants of Lb^{3+} including NAD(P)H, ascorbate, reduced glutathione, and cysteine. Flavins are intermediate electron carriers between NAD(P)H and Lb^{3+} (Becana and Klucas, 1990). In the presence of NAD(P)H, free flavins efficiently reduce Lb^{3+} without the formation of superoxide or peroxide. The abundance of free flavins, especially riboflavin, in nodules and stimulation of Lb^{3+} reduction reaction by flavins in nearly anaerobic conditions suggest that flavins may play an important role in reducing Lb^{3+} in the microaerobic conditions inside the nodules (Becana and Klucas, 1990).

V. Genetic Engineering for Increasing Nodulation and Nitrogen Fixation

Symbiotic nitrogen fixation is widely recognized as a trait that is beneficial to the plant host, and many of the plants capable of forming this type of symbiosis have agronomic and/or ecological importance. It is therefore only natural that mechanisms for increasing nodulation and nitrogen fixation within the symbiotic relationship, or extending the ability to fix nitrogen in plants outside they symbiotic relationship, are of tremendous interest. With the introduction of molecular biology techniques to the study of symbiotic nitrogen fixation in the mid-70s, the research in this area has advanced greatly. However, due to the unexpected complexity of the interrelationships, molecular approaches have not improved the symbiotic performance of existing *Rhizobium*-legume associations as once expected. Even if improvement is achieved in the laboratory, the inability to ensure nodulation by engineered strains, in competition with naturally-occurring strains will remain a major obstacle.

One novel approach has been to use inhibitors of *nod* gene induction to regulate nodule occupancy. It has been shown that flavonoids, isoflavones and chalcones induce expression of *nod* genes in *Rhizobium* and *Bradyrhizobium* spp. (Firmin et al., 1986; Djordjevic et al., 1987; Kosslak et al., 1987; Denarie et al., 1992). Inhibitors of *nod* gene induction by flavonoids have also been identified (Firmin et al., 1986; Djordjevic et al., 1987; Cunningham et al., 1989, 1991; Kosslak et al., 1990). By mutation or recombination, the structure of *nod* D can be altered such that either the expression of the *nod* genes is flavonoid-independent or the *nod* genes are induced by a broader spectrum of inducer molecules (Burn et al., 1987, 1989; Johansen et al., 1988; Spaink et al., 1989a). Cunningham et al. (1989) showed that inhibitors of *nod* gene induction were also capable of inhibiting nodulation in soybean by *B. japonicum*. Thus, a flavonoid-independent strain should be able to nodulate soybean exclusively in the presence of indigenous bradyrhizobia and the appropriate inhibitor of *nod* gene induction. Spaink et al. (1989a) have constructed a hybrid *nod* D gene that does not require a flavonoid for *nod* gene expression. A *Rhizobium* strain with this hybrid *nod* D gene expresses increased nitrogenase activity in the nodule. A flavonoid-independent *nod* D-containing strain would constitutively express *nod* genes and not be affected by inhibitors of *nod* gene expression. However, the commercial utility of such inhibitors is limited by the observation that strains of *Bradyrhizobium* differ in the types of inhibitor necessary to prevent *nod* gene induction (Cunningham et al., 1989, 1991) and secondly, the inhibitors of *nod* gene induction are ustable in soil (Kosslak et al., 1990). Although halogenation of these inducers may enhance their stability in soil a minimum of four halogenated inhibitors would be required to obtain effective inhibition of soybean nodulation in the field (Cunningham et al., 1989).

Attempts to impove nitrogen fixation and legume productivity by inoculation with superior genetically manipulated rhizobia often fail under field conditions because the indigenous strains occupied the root nodules rather than the inoculant strains. Therefore, it is essential that constructed superior strains should compete successfully for the establishment in nodules. Several research groups have isolated genetic determinants thought to confer increased nodulation competitiveness (Triplett and Sadowsky, 1992). Some strains of *Rhizobium* produce bacteriocin, antibiotics or possess phages (Schwinghamer and Belkengren, 1968; Joseph et al., 1983; Hodgson et al., 1985) which specifically inhibit other *Rhizobium* strains. The most highly characterized anti-rhizobial bacteriocin is trifolitoxin a small peptide produced by *R. leguminosarum* bv *trifolii* strain T 24 (Schwinghamer and Belkengren, 1968; Triplett and Barta, 1987). This strain T 24 was found highly competitive for nodulation against a bacteriocin-sensitive strain in a co-inoculation experiment. Triplett (1988) has isolated the trifolitoxin production and resistance genes (*tfx*) and stably integrated into the genome of *R. leguminosarum* bv *trifolii* TA1. The recombinant strain showed increased nodulation competitiveness compared with the wild-type strain (Triplett, 1990). The genes for competitiveness have also been isolated in *R. leguminosarum* bv *viciae* strain PF 2 (Dowling et al., 1987). This strain showed competitive nodulation blocking of the TOM strain, the only strain capable of nodulating the Afghanistan cultivar of peas. The genes which code for this blocking (Cnb[+]) are located on the *sym* plasmid of PF2 (Dowling et al., 1987, 1989) and mutations in *nod* C, *nod* D and *nod* E eliminated the competitive blocking in PF2.

Caetano-Anolles et al (1988a) found that nonmotile spontaneous mutants of *R. meliloti* had lower nodulation competitiveness and reduced efficiency of nodule initiation compared with wild-type parent strain. Similarly Tn5 mutants of *R. leguminosarum* bv *phaseoli*, *R. tropici* and *B. japonicum* deficient in exopolysaccharide (EPS) production were found less competitive for nodule occupancy than the corresponding wild-type strains (Araujo and Handelsman, 1990; Bhagwat et al., 1991). Recently, Legares et al. (1992) have shown that the lipopolysaccharide in *R. meliloti* also play a role in nodulation competitiveness. Sanjuan and Olivares (1989) identified a region, referred to as *nfe* on a cryptic plasmid of a *R. meliloti* GR4 that delay nodule formation by one or two days. Insertions of Tn3-HoHo1 in this *nfe* region reduced nodulation competitiveness of strain GR4. Moreover, Nif A[−] mutant of *R. meliloti* was found to have reduced nodulation competitiveness and addition of multiple copies of *nif* A from *Klebsiella pneumoniae* also resulted in increased nodulation competitiveness of *R. meliloti* strains (Sanjuan and Olivares, 1991). Thus, strains with an additional copy of *nif* A stably integrated into the genome may be useful in improving an inoculants' ability to occupy nodules to the exclusion of indigenous rhizobia.

Nitrate and ammonia in the soil can inhibit nodulation and nitrogen fixation by *Rhizobium*. When nitrate is applied to a nodulated plant, nitrogen fixation is repressed and the nodules senesce prematurely. Therefore, the presence of nitrogen fertilizer in the soil can be deterimental to the establishment of an effective N_2-fixing symbiosis. Such a situation can occur when leguminous crops are intercropped with other crops, such as corn, which require heavy use of nitrogen fertilizers. Exogenous application of nitrate has been shown to suppress nodulation and nitrogen fixation through multiple and complex effects (see review Streeter, 1988; Carroll and Mathews, 1990). Nitrate has its greatest inhibiting effects on infection events that are completed during the first 18 h after inoculation of soybean (Malik et al., 1987). In addition to a wide range of effects of nitrate on plant metabolism and nodule morphogenesis (Carroll and Mathews, 1990), some aspects of bacterial *nod* gene activity are also regulated by nitrate or ammonia (Wang and Stacey, 1990). In the case of *R. meliloti*, transcription of the *nod* ABC genes is nitrogen-regulated (Dusha et al., 1989), and the *syr* M gene seems to be involved in these effects (Sharma and Signer, 1990). Experiments indicate that the nitrogenase proteins continue to be synthesized in the nodule even in the presence of added nitrate (Noel et al., 1982). The product of the enzyme nitrate reductase, nitrite and also nitric oxide, inhibit and possibly denature preformed nitrogenase (Trinchant and Rigaud, 1982). Other studies suggest that competition for reductant by nitrate (Shanmugam et al., 1981), and de-enerization of the *Rhizobium* membrane system by nitrate (Veeger et al., 1981) as partial explanations for decreased nitrogenase activity in the presence of nitrate.

Mutants of *B. japonicum* selected as resistant to chlorate were analyzed for acetylene reduction in the presence of nitrate (Hua et al., 1981). One mutant lacked nitrate reductase (NR) activity and had 2-fold higher nitrogenase activity in the presence of nitrate than the parental type in the absence of nitrate. Gordon and Brill (1972) showed the derepression of nitrogenase in the presence of ammonium ions in azide resistant mutants of *Azotobacter chroococcum*. The azide resistant mutants of cowpea miscellany rhizobia showed less decrease in *ex planta* nitrogenase activity in the presence of 15 mM ammonium chloride and the competitive ability and symbiotic effectivity was increased in comparison with native rhizobia (Yadav et al., 1992). Several soybean mutants with the ability to form large number of nodules even in the presence of nitrate have been isolated (Carroll et al., 1985a, b). These nitrate-tolerant symbiotic (*nts*) mutants usually form 3 to 40 times as many nodules as the parent, have increased nitrogen fixation ability in the presence of nitrate (Hansen et al., 1989), and are partially tolerant to the inhibitory effects of soil acidity (Alva et al., 1988). Twelve independent *nts* mutants were characterized using complementation analysis to define a single supernodulation locus (Delves et al., 1988).

Ecological factors such pH, temperature or moisture also affect the successful establishment of an introduced inoculant. With *R. trifolii*, overall numbers of nodule bacteria are normally very low in acid soils and of those present, many are very poor nitrogen fixers (Holding and King, 1963). In order to survive, multiply and be effective in such soils, the introduced inoculum must be tolerant of chemical factors such as acidity and physical factors such as temperature. Recently, Chen et al. (1991) constructed an acid-tolerant strain of *R. leguminosarum* bv *trifolii* that expressed increased nitrogen fixation compared with the acid-sensitive parent strain. This acid-tolerant strain was developed by curing the acid-sensitive strain of its *sym* plasmid and replacing it with pBRIAN, a self mobilizable *sym* plasmid from *R. leguminosarum* bv *trifolii*. An interesting observation of practical significance to the growth of legumes in acid soils is the observation of a net reduction of *nod* gene inducing factors in the exudates of clover seedlings grown at low pH. Also, the induction of *nod* gene expression by flavonoids in clover rhizobia, even those able to grow under low pH conditions, is affected adversely by low pH (Richardson et al., 1988). Consequently, modification of the host to produce higher levels of flavonoids at low pH and/or the manipulation of *nod* genes to be readily induced at low pH, will be essential to overcome nodulation problems in acid soils.

The most promising avenues for increasing symbiotic nitrogen fixation is by increasing the number of nitrogen fixation (*nif*) genes in the endosymbiont with a concomitant increase in the number of active nitrogenase molecules. Since the rate-limiting step in nitrogen fixation appears to be the cycle of binding reduced dinitrogenase reductase, the *nif*H gene product, to dinitrogenase followed by one electron transfer, it has been proposed that increased copies of the *nif* H gene and its protein product may result in increasing the turnover rate of nitrogenase. An excess of the *nif* H gene product has been shown to be required for maximal nitrogen fixation activity in *K. pneumoniae* (Thorneley and Lowe, 1984). This may be the basis for the presence of more than one copy of the *nif*H gene in some diazotrophs such as *Azotobacter vinelandii* (Jacobson et al., 1986), *Rhizobium phaseoli* (Quinto et al., 1985) and *Azorhizobium sesbaniae* (Norel and Elmerich, 1987). Another gene *nif*A which serves as the positive regulator of other *nif* genes, when present in multiple copies and over-expressed, is thought by Cannon et al. (1988) and Ronson et al. (1990) to result in increased symbiotic nitrogen fixation followed by increased grain yield. Critical experiments to test this hypothesis, however, remains to be done. It has also been demonstrated that purified nitrogenase from *A. vinelandii* has a specific activity that is approximately threefold higher than that of *B. japonicum* (Burgess et al., 1980). Therefore, it may be possible to attain increased symbiotic nitrogen fixation by introducing a more active nitrogenase from another diazotroph, or similarly, by replacing the indigenous nitrogenase system with a more active nitrogenase (reviewed by Robson et al., 1983).

Nitrogen fixation in the *Rhizobium*-legume symbiosis is thought to be limited by the amount of plant-derived photosynthate available to bacteroids (Hardy and Havelka, 1975; Ryle et al., 1979). Birkenhead et al. (1988) suggested that increasing the ability of the endosymbiont to utilize photosynthate in the nodule may lead to increased N_2-fixation rates. These authors transferred *R. meliloti dct* genes to *B. japonicum* and demonstrated a 50% increase in the uptake of C_4-dicarboxylic acids, increased gorwth rates on succinate as the sole carbon source, and a corresponding increase in the levels of symbiotic nitrogen fixation. Ronson et al. (1990) have also proposed that increased expression of *dct*A (structural gene for dicarboxylate transport) can increase symbiotic nitrogen fixation in alfalfa and soybean.

Uptake hydrogenase activity in root nodulate bacteroids has been shown to improve yield of soybean using near-isogenic strains of *B. japonicum* (Hanus et al., 1981; Evans et al., 1987; Hungria et al., 1989). The evidence for uptake hydrogenase-mediated yield enhancement is less clear in other legumes, due to lack of comparisons by using near-isogenic strains. Increase in C_2H_2 reduction rates have been observed in H_2 atmosphere in Hup[+] *Rhizobium* strains of cowpea miscellany which exhibit *ex planta* nitrogenase activity (Dadarwal et al., 1985). Similarly, enhanced C_2H_2 reduction rates have been reported in nodules and bacteroids of soybean, pea and *Vigna* group of hosts, formed by inoculation of Hup[+] strains (Emerich et al., 1979; Nelson and Salminen, 1982; Dadarwal et al., 1985). Thus another approach to improve yields could be to increase the activity of the uptake hydrogenase in strains that already possess it. Merberg and Maier (1983) have succeeded in doing this by mutagenizing a Hup[+] *B. japonicum* strain and selecting for growth on higher than normal levels of oxygen. Similarly, mutants of cowpea miscellany *Rhizobium* strains S24 and GR4 were isolated for increased hydrogenase activity by nitrosoguanidine mutagenesis and these mutants showed increase in dry matter yield when inoculated onto greengram and blackgram (Sindhu and Dadarwal, 1992).

The possibility of transferring bacterial nitrogen-fixing genes to plants has been discussed for the last two decades (Sprent, 1986). One serious obstacle to this approach is the necessity to provide protection to the oxygen-labile nitrogenase enzyme from oxygen in an aerobic organism. In this respect, the nitrogen-fixing cyanobacteria provide the closest analogy to a nitrogen-fixing plant cell as they have the additional problem of protecting nitrogenase from their own, photosynthetically produced oxygen (Gallon and Chaplin, 1987). Of those cyanobacteria which fix nitrogen aerobically, most possess heterocysts which lack photosystem II of photosynthesis and therefore, provide an environment in which nitrogen fixation is spatially separated from photosynthetic oxygen evolution. It has been proposed to introduce the nitrogen fixation apparatus into chloroplasts in which the oxygen generating activity of photosystem II has been inactivated. In the future it should be possible to identify promoters in the plastid which are subject to different developmental controls. Consequently, the choice of appropriate promoters for the *nif* regulatory genes should allow control of the conditions under which nitrogenase enzymes is expressed in a plant. Alternatively, the bundle sheath cells of C_4 plants such as maize in which oxygen evolution does not occur, could provide a relatively 'low oxygen' environment, comparable to that found in the cyanobacterial heterocysts. The transcripts of large subunits of ribulose biphosphate carboxylase (*rbc* L) gene have also been found to be expressed in bundle sheath cells and are not present in photosynthetic mesohpyll cells (Jolly et al., 1981). By linking the *nif* regulatory genes to the *rbc* L promoter, the tissue-specific expression of *nif* genes might be achieved in the bundle sheath cells of C_4 plants. Finally, it may be possible to regulate the genes temporally so that they are expressed only on the dark; a situation analogous to that found in non-heterocystous cyanobacteria, *Gleocapsa.*

Another approach to extend the range of nodulated plants involves the transfer of symbiotic plasmids or defined nodulation genes in heterologous rhizobia or agro-bacteria (Hooykaas et al., 1982; Kondorosi et al., 1982; Truchet et al., 1984; Hirsch et al., 1985; Sindhu and Dadarwal, 1993b). However, such recombinant strains have been found to form white pseudonodules on the host plants and none of the nodules developed to the nitrogen-fixing stage. There is also a possibility for modelling *Rhizobium* to nodulate non-leguminous genera, since it happens naturally in the genus *Parasponia.* Its nodules are produced following epidermal infections and active nitrogen-fixing cells contain bacteria which are retained with in the infection threads bounded by wall-like material secreted by the host cell (Sprent, 1986). This characteristic has now also been found in primitive woody legumes (Faria et al., 1988). Studies of such nodules might help to extend the range of nodulation even to non-leguminous crop plants.

Recently, Cocking et al. (1990) have reported the formation of nodule-like structures on the roots of rice and wheat following inoculation with *R. loti* and *B. japonicum* after treatment of roots with a cellulase-pectolyase enzyme solution. However, nitrogenase activity was very low and the bacteria were confined to intercellular spaces. They also reported the induction of nodule-like structures on oilseed rape (*Brassica napus*) following inoculation with rhizobia isolated from nodules of the non-legume, *Parasponia.* Jing et al. (1990) reported pseudonodule formation on barley roots induced by *Rhizobium astragali.* Tchan and Kennedy (1989) have shown the induction of 'para nodules' on wheat following treatment with the herbicide 2,4-D or with the auxins IAA (indole-3-acetic acid) and NAA (Naphthyl-acetic acid), and by inoculation with rhizobia or *Azospirillum.* Low levels of nitrogenase activity were found only after inoculation with *Azospirillum* and there was little evidence for bacterial infection. Rolfe and Bender (1991) have reported the formation of similar structures on rice roots following inoculation with a strain of *Rhizobium* containing a *nod* D allele whose gene product interacts with rice root exudate. True infection was apparent in the few nodules that were formed, but nitrogenase activity could not be detected. The development of symbiotic capability in agriculturally-important non-legumes requires an understanding of the various physiological and genetic processes within the plants and the bacteria and to identify the essential characters that are present in legumes which must be introduced (Kennedy and Tchan, 1992). Furthermore, these modified lateral roots of cereals having nodule-like structures must contain some sort of a micro-aerobic environment for protection of the O_2-sensitive nitrogenase. For this requirement the plant could be engineered so that the intercellular space becomes filled with polysaccharides or other O_2-excluding material upon infection. Much efforts and coordination are required in genetics, molecular biology and developmental biology to achieve a complete understanding of the *Rhizobium*-legume symbiosis which could then be translated in non-legume-rhizobial associations.

Literature Cited

Adams, T.H., McClung, C.R., and Chelm, B.K. 1984. Physical organization of the *Bradyrhizobium japonicum* nitrogenase gene region. *J. Bacteriol.* **159**: 857-862.

Albrecht, S.L., Maier, R.J., Hanus, F.J., Russell, S.A., Emerich, D.W., and Evans, H.J. 1979. Hydrogenase in *Rhizobium japonicum* increases nitrogen fixation by nodulated soybeans. *Science.* **203**: 1255-1257.

Allen, O.N., and Allen, E.K. 1981. The Leguminosae. A Source Book of Characteristics, Uses and Nodulation. Wisconsin University Press, Madison. pp. 604-607.

Allen, T., Raja, S., and Dunn, K. 1991. Cells expressing ENOD2 show differential organization during the development of alfalfa root nodules. *Mol. Plant-Microbe Interact.* **4**: 139-146.

Alva, A.K., Edwards, D.G., Carroll, B.J., Asher, C.J., and Gresshoff, P.M. 1988. Effects of soil infertility factors on nodulation and growth of soybean mutants with increased nodulation capacity. *Agron. J.* **80**: 836-841.

Appelbaum, E.R., Thompson, D.V., Idler, K., and Chartrain, N. 1988. *Bradyrhizobium japonicum* USDA 1&1 has two *nod* D genes that differ in primary structure and function. *J. Bacteriol.* **170**: 12-20.

Appleby, C.A. 1984. Leghemoglobin and *Rhizobium* respiration. *Annu. Rev. Plant Physiol.* **35**: 443-478.

Araujo, R.S., and Handelsman, J. 1990. Characteristics of exopolysaccharide-deficient mutants of *Rhizobium* spp. with altered nodulation competitiveness. *In Nitrogen Fixation: Achievements and Objectives.* (Gresshoff, P.M., Roth, L.E., Stacey, G. and Newton, W.E. ed.), Chapman and Hall, New York. pp. 247.

Arora, N., Skoog, F., and Allen, O.N. 1959. Kinetin induced pseudonodules on tobacco roots. *Am. J. Bot.* **46**: 610-613.

Arwas, R., McKay, I.A., Rowney, F.R.P., Dilworth, M.J., and Glenn, A.R. 1985. Properties of organic acid utilization mutants of *Rhizobium leguminosarum* strain 300. *J. Gen. Microbiol.* **131**; 2059-2066.

Ausubel, F.M., Buikema, W.J., Earl, C.D., Klingensmith, J.A., Nixon, B.T., and Szeto, W.W. 1985. Organization and regulation of *Rhizobium meliloti* and *Parasponia Bradyrhizobium* nitrogen fixation genes. *In Nitrogen Fixation Research Progress, (Evans, H.J., Bottomley, P.J. and Newton, W.E. eds.), Martinus Nijhoff, Dordrecht.* pp. 165-171.

Averyhart-Fullard, V., Datta, K., and Marcus, A. 1988. A hydroxy proline rich protein in the soybean cell wall. *Proc. Natl. Acad. Sci., USA.* **85**: 1082-1085.

Baev, N., Endre, G., Petrovics, G., Banfalvi, Z., and Kondorosi, A. 1991. Six nodulation genes of *nod* box locus-4 in *Rhizobium meliloti* are involved in nodulation signal production-*nod* M codes for D-glucosamine synthetase. *Mol. Gen. Genet.* **228**: 113-124.

Baev, N., Schultze, M., Barlier, I., Ha, D.C., Virelizier, H., Kondorosi, E., and Kondorosi, A. 1992. *Rhizobium nod* M and *nod* N genes are common *nod* genes: *nod* M encodes functions for efficiency of Nod signal production and bacteroid maturation. *J. Bacteriol.* **174**: 7555-7565.

Bakhuzien, R. 1988. The plant cytoskeleton in the *Rhizobium* - legume symbiosis. *Ph.D. Thesis. University of Leiden, The Netherlands.*

Banfalvi, Z., Nieuwkoop, A., Schell, M., Besi, L., and Stacey, G. 1988. Regulation of *nod* gene expression in *Bradyrhizobium japonicum*. *Mol. Gen. Genet.* **214**: 420-424.

Barbour, W.M., Wang, S.P., and Stacey, G. 1992. Molecular genetics of *Bradyrhizobium* symbiosis. *In Biological Nitrogen Fixation. (Stacey, G., Evans, H.J. and Burris, R.H. eds.Chapman and Hall, New York.* pp. 645-681.

Battisti, L., Lara, J.C., and Leigh, J.A. 1992. A specific oligo-saccharide form of the *Rhizobium meliloti* exopolysaccharide promotes nodule invasion in alfalfa. *Proc. Natl. Acad. Sci. USA.* **89**: 5625-5629.

Batut, J., Boistard, P., Debelle, F., Denarie, J., Ghai, J., Huguet, T., Infante, D., Martinez, E., Rosenberg, C., Vasse, J.,and Truchet, G. 1985b. Developmental biology of the *Rhizobium meliloti* - alfalfa symbiosis: a joint genetic and cytological approach. *In Nitrogen Fixation Research Progress,* (Evans, H.J., Bottomley, P.J. and Newton, W.E. eds.), Martinus Nijhoff, Dordrecht. The Netherlands. pp. 109-115.

Batut, J., Daveran-Mingot, M.L., David, M., Jacobs, J, Garnerone, A.M.,and Kahn, D. 1989. *fix K,* a gene homologous with *fnr* and *crp* from *Escherichia coli,* regulates nitrogen fixation genes both positively and negatively in *Rhizobium meliloti. EMBO J.* **8**: 1279-1286.

Batut, J., Terzaghi, B., Gherardi, M., Huguet, M., Tezaghi, E., Garnerone, A.M., Boistard, P., and Huguet, T. 1985a. Localization of a symbiotic *fix* region more than 200 Kilobases from the *nod-nif* region. *Mol. Gen. Genet.* **199**: 232-239.

Bauer, W.D. 1981. Infection of legumes by rhizobia. *Annu. Rev. Plant Physiol.* **32**: 407-449.

Becana, M.,and Klucas, R.V. 1990. Enzymatic and nonenzymatic mechanisms for ferric leghemoglobin reduction in legume root nodules. *Proc. Natl. Acad. Sci., USA.* **87**: 7295-7299.

Becana, M.,and Klucas, R.V. 1992. Oxidation and reduction of leghemoglobin in root nodules of leguminous plants. *Plant Physiol.* **98**: 1217-1221.

Bergersen, F.J. 1982. Root nodules of legumes: structure and functions. pp. 164. Wiley, Chichester.

Bergmann, J., Preddie, E.,and Verma, D.P.S. 1983. Nodulin-35: A subunit of specific uricase (Uricase II) induced and localized in the uninfected cells of soybean nodules. *EMBO J.* **2**: 2333-2339.

Better, M., Lewis, B., Corbin, D., Ditta, G., and Helinski, D.R. 1983. Structural relationships among *Rhizobium meliloti* Symbiotic promoters. *Cell.* **35**: 479-485.

Beynon, J.L., Beringer, J.E., and Johnston, A.W.B. 1980. Plasmids and host range in *Rhizobium leguminosarum* and *Rhizobium phaseoli. J. Gen. Microbiol.* **120**: 421-429.

Beynon, J.L., Williams, M.K., and Cannon, F.C. 1988. Expression and functional analysis of the *Rhizobium meliloti nif* A gene. *EMBO J.* **7**: 7-14.

Bhagwat, A.A., Tully, R.E., and Keister, D.L. 1991. Isolation and characterization of a competition-defective *Bradyrhizobium japonicum* mutant. *Appl. Environ. Microbiol.* **57**: 3496-3501.

Bhat, U.R., Mayer, H., Yokota, A., Hollingsworth, R.I., and Carlson, R.W. 1991. Occurrence of lipid A variants with 27-hydroxyoctacosanoic acid in lipopolysaccharides from members of the family *Rhizobeaceae. J. Bacteriol.* **173**: 2155-2159.

Bhuvaneswari, T.V., Bhagwat, A.A., and Bauer, W.D. 1981. Transient susceptibility of root cells in four common legumes to nodulation by rhizobia. *Plant Physiol.* **68**: 1144-1149.

Birkenhead, K., Manian, S.S., and O'Gara, F. 1988. Dicarboxylic acid transport in *Bradyrhizobium japonicum:* Use of *Rhizobium meliloti dct* gene(s) to enhance nitrogen fixation. *J. Bacteriol.* **170**: 184-189.

Bogusz, D., Llewellyn, D.J., Craig, S., Dennis, E.S., Appleby, C.A., and Peacock, W.J. 1990. Non-legume hemoglobin genes retain organ-specific expression in heterologous transgenic plants. *Plant Cell.* **2**: 633-641.

Bolton, E., Higgisson, B., Harrington, A., and O'Gara, F. 1986. Dicarboxylic acid transport in *Rhizobium mililoti:* isolation of mutants and cloning of dicarboxylic acid transport genes. Arch Microbiol. **144**: 142-146.

Bond, L. 1948. Origin and developmental morphology of root nodules of *Pisum sativum*. Botanical *Gazette*. **109**: 411-434.

Borthakur, D., Barber, C.E., Lamb, J.E., Daniels, M.J., Downie, J.A., and Johnston, A.W.B. 1986. A mutation that blocks exopolysaccharide synthesis prevents nodulation of peas by *Rhizobium leguminosarum* but not of beans by *R. phaseoli* and is corrected by cloned DNA from *Rhizobium* or the phytopathogen *Xanthomonas*. *Mol. Gen. Genet*. **203**: 320-323.

Bothe, H., de Bruijn, F.J.,and Newton, W.E. 1988. *Nitrogen Fixation: Hundred years after*. Gustav Fischer, Stuttgart.

Bradley, D.J., Butcher, G.W., Galfre, G., Wood, E.A., and Brewin, N.J. 1986. Physical association between the peribacteroid membrane and lipopolysaccharide from the bacteroid outer membrane in *Rhizobium* - infected pea root nodule cells. *J. Cell Sci*. **85**: 47-61.

Brewin, N.J. 1990. The role of plant plasma membrane in symbiosis. *In The Plant Plasma Membrane*. (Larsson, C. and Moller, I.M. eds.) Springer Verlag, Berlin, Heidelberg. pp. 351-375.

Brewin, N.J. 1991. Development of the legume root nodule. *Annu. Rev. Cell Biol*. **7**: 191-226.

Brewin, N.J., De Jong, T.M., Phillips, D.A.,and Johnston, A.W.B. 1980. Co-transfer of determinants for hydrogenase activity and nodulation ability in *Rhizobium leguminosarum*. *Nature*. **288**: 77-79.

Brewin, N., Rae, A.L., Perotto, S., Knox, J.P., Roberts, K., LeGal, M.F., Sindhu, S.S., Wood, E.A., and Kannenberg, E.L. 1990. Immunological dissection of the plant-microbe interface in pea nodules. *In Nitrogen fixation: Achievements and Objectives*. (Gresshoff, P.M., Roth, L.E., Stacey, G. and Newton, W.E. eds.), Chapman and Hall, New York. pp. 227-234.

Brewin, N.J., Robertson, J.G., Wood, E.A., Wells, B., Larkins, A.P., Galfre, G., and Butcher, G.W. 1985. Monoclonal antibodies to antigens in the peribacteroid membrane from *Rhizobium* -induced root nodules of pea cross-react with plasma membranes and Golgi bodies. *EMBO J*. **4**: 605-611.

Brewin, N.J.,and Vaneden Bosch, K.A. 1988. Inside the legume root nodule. Plants Today. **1**: 114-121.

Brink, B.A., Miller, J., Carlson, R.W.,and Noel, K.D. 1990. Expression of *Rhizobium leguminosarum* CFN 42 genes for lipopolysaccharide in strains derived from different *Rhizobium leguminosarum* soil isolates. *J. Bacteriol*. **172**: 548-555.

Broughton, W.J.,and Dilworth, M.J. 1971. Control of leghemoglobin synthesis in snake beans. *Biochem. J*. **125**: 1075-1080.

Buchmeier, N.A.,and Heffron, F. 1990. Induction of *Salmonella* stress proteins upon infection of macrophages. *Science*. **248**: 730-732.

Buikema, W.J., Szeto, W.W., Limley, P.V., Orme-Johnson, W.H.,and Ausubel, F.M. 1985. Nitrogen fixation specific regulatory genes of *Klebsiella pneumoniae* and *Rhizobium meliloti* share homology with the general nitrogen regulatory gene *ntr* C of *K. pneumoniae*. *Nucleic Acids Res*. **13**: 4539-4555.

Burgess, B.K., Jacobs, D.B., and Stiefel, E.I. 1980. Large-scale purification of high activity *Azotobacter vinelandii* nitrogenase. *Biochim. Biophys. Acta*. **614**: 196-209.

Burn, J.W., Hamilton, W.D., Wootton, J.C.,and Johnston, A.W.B. 1989. Single and multiple mutations affecting properties of the regulatory gene *nod* D of *Rhizobium. Mol. Microbiol*. **3**: 1567-1577.

Burn, J.E., Rossen, L., and Johnston, A.W.B. 1987. Four classes of mutations in the *nod* D gene of *Rhizobium leguminosarum* biovar *viciae* that effect its ability to autoregulate and/or activate other *nod* genes in the presence of flavonoid inducers. *Genes Dev*. **1**: 456-464.

Burris, R.H. 1980. The global nitrogen budget: Science of Seance? *In Free Living Systems and Chemical Models. Nitrogen Fixation*. Vol. I. (Newton, W.E. and Orme-Johnson, W.H. eds.), University Park Press, Baltimore. pp. 7-16.

Burris, R.H. 1984. Enzymology of nitrogenase. *In Advance in Nitrogen Fixation Research.* (Veeger, C. and Newton, W.E. eds.), Martinus Nijhoff, The Hague. pp. 143-144.

Caetano-Anolles, G., and Gresshoff, P.M. 1991. Plant genetic control of nodulation. *Annu. Rev. Microbiol.* **45**: 345-382.

Caetano-Anolles, G., Christ-Estes, D.K., and Bauer, W.D. 1988b. Chemotaxis of *Rhizobium mililoti* to the plant flavone luteolin requires functional nodulation genes. *J. Bacteriol.* **170**: 3164-3169.

Caetano-Anolles, G., Wall, L.G., De Micheli, A.T., Macchi, E.M., Bauer, W.D., and Favelukes, G. 1988a. Role of motility and chemotaxis in efficiency of nodulation by *Rhizobium meliloti. Plant Physiol.* **86**: 1228-1235.

Callaham, D.A., and Torrey, J.G. 1981. The structural basis for infection of root hairs by *Trifolium repens* by *Rhizobium. Can. J. Bot.* **59**: 1647-1664.

Calvert, H.E., Pence, M.K., Pierce, M., Malik, N.S.A.,and Bauer, W.D. 1984. Anatomical analysis of the development and distribution of *Rhizobium* infections in soybean roots. *Can. J. Bot.* **62**: 2375-2384.

Cannon F.C., Beynon, J., Hankinson, T., Kwiatkowski, R., and Legocki, R.P. 1988. Increasing biological nitrogen fixation by genetic manipulation. In *Nitrogen Fixation : Hundred Years After.* (Bothe, H., de Bruijn, F.J. and Newton, W.E. eds.), Gustav Fischer, Stuttgart. pp. 735-740.

Cantrell, M.A., Haugland, R.A., and Evans, H.J. 1983. Construction of a *Rhizobium japonicum* gene bank and use in the isolation of a hydrogen uptake gene. *Proc. Natl. Acad. Sci. USA.* **80**: 181-185.

Carlson, R.W. 1984. The heterogeneity of *Rhizobium* lippolysaccharide. *J. Bacteriol.* **158**: 1012-1017.

Carroll, B.J.,and Mathews, A. 1990. Nitrate Inhibition of nodulation in legumes. In *Molecular Biology of Symbiotic Nitrogen Fixation.* (Gresshoff, P.M. ed.), CRC Press, Boca Raton, Fla. pp. 159-180.

Carroll, B.J., McNeil, D.L., and Gresshoff, P.M. 1985a. Isolation and properties of soybean (*Glycine max*) mutants that nodulate in the presence of high nitrate concentrations. *Proc. Natl. Acad. Sci., USA.* **82**: 4164-4166.

Carroll, B.J., McNeil, D.L., and Gresshoff, P.M. 1985b. A supernodulation and nitrate tolerant symbiotic (*nts*) soybean mutant. *Plant Physiol.* **78**: 34-40.

Carroll, B.J., McNeil, D.L., and Gresshoff, P.M. 1986. Mutagenesis of soybean (*Glycine max* (L.). Merr.) and the isolation of non-nodulating mutants. *Plant Sci.* **47**: 109-114.

Cassab, G.I., and Varner, J.E. 1988. Cell wall proteins. *Annu. Rev. Plant Physiol. Plant Mol-Biol.* **39**: 321-353.

Cava, J.R., Elias, P.M., Turowski, D.A., and Noel, K.D. 1989. *Rhizobium leguminosarum* CFN 42 genetic regions encoding lipopolysaccharide structures essential for complete nodule development on bean plants. *J. Bacteriol.* **171**: 8-15.

Chen, H., Batley, M., Redmond, J., and Rolfe, B.G. 1985. Alternation of the effective nodulation properties of a fast-growing broad host range *Rhizobium* due to changes in exopolysaccharide synthesis. *J. Plant Physiol.* **120**: 331-349.

Chen, H., Richardson, A.E., Gartner, E., Djordjevic, M.A., Roughley, R.J., and Rolfe, B.G. 1991. Construction of an acid-tolerant *Rhizobium leguminosarum* bv. *trifolii* strain with enhanced capacity for nitrogen fixation. *Appl. Environ. Microbiol.* **57**: 2005-2011.

Clarke, H.R.G., Leigh, J.A., and Douglas, C.J. 1992. Molecular signals in the interactions between plants and microbes. *Cell.* **71**: 191-199.

Clover, R.J., Kieber, J., and Signer, E.R. 1989. Lipopolysaccharide mutants of *Rhizobium meliloti* are not defective in symbiosis. *J. Bacteriol.* **171**: 3961-3967.

Cocking, E.C., Al-Mallah, M.K., Bensen, E., and Davey, M.R. 1990. Nodulation of non-legumes by rhizobia. *In Nitrogen Fixation: Achievements and Objectives. (Gresshoff, P.M., Roth, L.E., Stacey, G. and Newton, W.E. eds.), Chapman and Hall,* New York. pp. 813-823.

Corbin, D., Ditta, G., and Helinski, D.R.,1982. Clustering of nitrogen fixation (*nif*) genes in *Rhizobium meliloti. J. Bacteriol.* **149**: 221-228.

Corby, H.D.L. 1988. Types of rhizobial nodules and their distribution among the *Leguminosae. Kirkia.* **13**: 53-123.

Cullimore, J.V., Gebhardt, C., Saarelainen, R., Miflin, B.J., Idler, K.B., and Barker, R.F. 1984. Glutamine synthetase of *Phaseolus vulgaris* L.: organ specific expression of a multigene family. *J. Mol. Appl. Genet.* **2**: 589-599.

Cunningham, S.D., Kapulnik, Y., Brewin, N.J., and Phillips, D.A. 1985. Uptake hydrogenase activity determined by plasmid pRL6J1 in *Rhizobium leguminosarum* does not increase symbiotic nitrogen fixation. *Appl. Environ. Microbiol.* **50**: 791-794.

Cunningham, S.D., Kollmeyer, W.D., and Stacey, G. 1989. The selection and use of exogenously applied chemicals to control competition for nodule formation between *Bradyrhizobium japonicum* strains. In *Proceedings of the 12th North American Symbiotic Nitrogen Fixation Conference.* pp. 26.

Cunningham, S.D., Kollmeyer, W.D., and Stacey, G. 1991. Chemical control of interstrain competition for soybean nodulation by *Bradyrhizobium japonicum. Appl. & Environ. Microbiol.* **57**: 1886-1892.

Cutting, J.A., and Schulman, J.A. 1972. The control of heme synthesis in soybean root nodules. *Biochem. Biophys. Acta.* **261**: 432-441.

Dadarwal, K.R., Grover, R., and Tauro, P. 1982. Uptake hydrogenase in *Rhizobium* and nodule leghemoglobin in cowpea miscellany hosts. *Arch. Microbiol.* **133**: 303-306.

Dadarwal, K.R., and Sindhu, S.S. 1988. Significance of uptake hydrogenase in legume - *Rhizobium* symbiosis. In *Advances in Frontier Areas of Plant Biochemistry.* (Singh, R. and Sawhney, S.K. eds.), Prentice Hall of India, New Delhi. pp. 317-333.

Dadarwal, K.R., Sindhu, S.S., and Batra, R. 1985. Ecology of Hup⁺ *Rhizobium*, strains of cowpea miscellany : native frequency and competence. *Arch. Microbiol.* **141**: 255-259.

Damrey, J.T., and Alexander, M. 1979. Physiological differences between effective and ineffective strains of *Rhizobium. Soil Sci.* **108**: 209-216.

Dart, P.J. 1977. Infection and development of leguminuous nodules. In A Treatise on Dinitrogen Fixation. (Hardy, R.W.F. ed.), Wiley, New York. pp. 367-472.

Darvil, A.G., Albersheim, P.A., Buchelt, P., Doares, S., Doubrava, N., Eberhard, S., Gollin, D.J., Hahn, M.G., Marfa-Riera, V., York, W.S., and Mohnen, D. 1989. Oligosaccharins-plant regulatory molecules. In *Signal Molecules in Plants and Plant-Microbe Interactions*, NATO. ASI Series. Vol. H36 (Lugtenberg, B.J.J. ed.), Springer-Verlag, Berlin, Heidelberg. pp. 41-48.

David, M., Daveran-Mingot, M.L., Batut, J., Dedieu, A., Domergue, O., Ghai, J., Hertig, C., Boistard, P., and Kahn, D. 1988. Cascade reglation of *nif* gene expression in *Rhizobium meliloti. Cell.* **54**: 671-683.

Davis, D.D. 1980. Anaerobic metabolism and the production of organic acids. In *The Biochemistry of Plants: an Advanced Treatise, Vol. 2* (Stumpf. P.K. and Conn, E.E. eds.), Academic Press, New York. pp. 581-611.

Davis, E.O., Evans, I.J., and Johnston, A.W.B. 1988. Identification of *nod* X, a gene that allows *Rhizobium leguminosarum* biovar *viciae* strain TOM to nodule Afghanistan peas. *Mol. Gen. Genet.* **212**: 531-535.

Davis, E.O., and Johnston, A.W.B. 1990. Analysis of 3 *nod* D genes in *Rhizobium leguminosarum* biovar *pheseoli-* nod DI is preceded by *nol* E, a gene whose product is secreted from the cytoplasm. *Mol. Microbiol.* **4**: 921-932.

Day, D.A., Yang, L.J.O., and Udvardi, M.K. 1990. Nutrient exchange across the peribacteroid membrane of isolated symbiosomes. In *Nitrogen fixation : Achievements and Objectives.* (Gresshoff,

P.M., Roth, L.E., Stacey, G., and Newton, W.E. eds.), Chapman and Hall, New York. pp. 219-226.

Dazzo, F.B., Truchet, G.L., Sherwood, J.E., Hrabak, E.M., Abe, M., and Pankratz, S.H. 1984. Specific phases of root hair attachment in the *Rhizobium trifolii* - clover symbiosis. *Appl. Environ. Microbiol.* **48**: 1140-1150.

Debelle, F., Rosenburg, C., Vasse, J., Maillet, F., Martinez, E., Denarie, J., and Truchet, G. 1986. Assignment of symbiotic developmental phenotypes to common and specific nodulation (*nod*) genetic loci of *Rhizobium meliloti. J. Bacteriol.* **168**: 1075-1086.

de Bruijn, F.J., Hilgert, U., Stigter, J., Schneider, M., Meyer, H.A., Klosse, U., and Pawlowski, K. 1990a. Regulation of nitrogen fixation and assimilation genes in the free-living versus symbiotic state. *In Nitrogen Fixation : Achievements and Objectives* (Gresshoff, P.M., Roth, L.E., Stacey, G. and Newton, W.E. eds.), Chapman and Hall, New York. pp. 33-44.

de Bruijn, F.J., Szabados, L., and Schell, J. 1990b. Chimeric genes and transgenic plants are used to study the regulation of genes involved in symbiotic plant-microbe interactions (nodulin genes). *Dev. Genet.* **11**: 182-196.

deFaria, S.M., Lewis, G.P., Sprent, J.I., and Sutherland, J.M. 1989. Occurrence of nodulation in the *Leguminosae. New Phytol.* **111**: 607-619.

Dehlo, C., and de Bruijn. F.J. 1992. The early nodulin gene SrENOD2 from *Sesbania rostrata* is inducbile by cytokinin. *Plant J.* **2**: 117-128.

Dejong, T.M., Brewin, N.J., Johnston, A.W.B., and Phillips, D.A. 1982. Improvement of symbiotic properties in *Rhizobium leguminosarum* by plasmid transfer. *J. Gen. Microbiol.* **128**: 1829-1838.

Delauney, A.J., and Verma, D.P.S. 1988. Cloned nodulin genes for symbiotic nitrogen fixation. *Plant Mol. Biol. Rep.* **6**: 279-285.

Delves, A.C., Carroll, B.J., and Gresshoff, P.M. 1988. Genetic analysis and complementation studies on a number of mutant supernodulating soybeans. *J. Genet.* **67**: 1-8.

De Maagd, R.A., de Rijk, R., Mulder, I.H.M., and Lugtenberg, B.J.J. 1989a. Immunological characterization of *Rhizobium leguminosarum* outermembrane antigens by use of polyclonal and monoclonal antibodies. *J. Bacteriol.* **171**: 1136-1142.

De Maagd, R.A., Rao, A.S., Mulders, I.H.M., Goosen-de Roo, L., van Loosdrecht, M.C.M., Wijffelman, C.A., and Lugtenberg, B.J.J. 1989b. Isolation and characterization of mutants of *Rhizobium leguminosarum* bv. viciae 248 with altered lipopolysaccharides : possible role of surface charge or hydrophobicity in bacterial release from the infection thread. J. Bacteriol. **171**: 1143-1150.

Denarie, J., Debelle, F., and Rosenberg, C. 1992. Signalling and host range variation in nodulation. *Annu. Rev. Microbiol.* **46**: 497-531.

De Philip, P., Batut, J., and Boistard, P. 1990. *Rhizobium meliloti* fix L is an oxygen sensor and regulates *R. meliloti* nif A and fix K genes differently in *Escherichia coli J. Bacteriol.* **172**: 4255-4262.

Deroche, M.E., Carrayol, E., and Jolivet, E. 1983. Phosphoenol-pyrvate carboxylase in legume nodules. *Physiol. Veg.* **21**: 1075-1081.

Deshmane, N., and Stacey, G. 1989. Identification of *Bradyrhizobium nod* genes involved in host-specific nodulation. *J. Bacteriol.* **171**: 3324-3330.

Dickstein, R., Bisseling, T., Reinhold, V.N., and Ausubel, F.M. 1988. Expression of nodule-specific genes in alfalfa root nodules blocked at an early stage of development. *Genes Dev.* **2**: 677-687.

Dilworth, M., and Glenn, a. 1984. How does a legume nodule work. *Trends in Biochem. Sciences.* **9**: 519-523.

Ditta, G., Virts, E., Palomares, A., and Kim, C.-H. 1987. The *nif* A gene of *Rhizobium meliloti* is oxygen regulated. *J. Bacteriol.* **169**: 3217-3223.

Dixon, R.O.D. 1972. Hydrogenase in legume root nodule bacteroids: Occurrence and properties. *Arch. Microbiol.* **85**: 193-201.

Djordjevic, M.A., Innes, R.W., Wijffelman, C.A., Schofield, P.R., and Rolfe, B.G. 1987. Clovers secrete specific phenolic compounds which either stimulate or repress *nod* gene expression in *Rhizobium trifolii*. *EMBO J.* **6**: 1173-1179.

Djordjevic, M.A., Schofield, P.R., and Rolfe, B.G. 1985. Tn5 mutagenesis of *Rhizobium trifolii* host specific nodulation genes results in mutants with altered host range ability. *Mol. Gen. Genet.* **200**: 463-471.

Dowling, D.N., Samrey, U., Stanley, J., and Broughton, W.J. 1987. Cloning of *Rhizobium leguminosarum* genes for competitive nodulation blocking on peas. *J. Bacteriol.* **169**: 1345-1348.

Dowling, D.N., Stanley, J.,and Broughton, W.J. 1989. Competitive nodulation blocking of Afghanistan pea is determined by *nod* DABC and *nod* Fe alleles in *Rhizobium leguminosarum*. *Mol. Gen. Genet.* **216**: 170-174.

Downie, J.A. 1989. The *nod* L gene from *Rhizobium leguminosarum* is homologous to the acetyl transferase encoded by *lac* A and *cys* E. *Mol. Microbiol.* **3**: 1649-1651.

Downie, J.A., Knight, C.D., Johnston, A.W.B., and Rossen, L. 1985. Identification of genes and gene products involved in nodulation of peas by *Rhizobium leguminosarum*. *Mol. Gen. Genet.* **198**: 255-262.

Dreyfus, B., Garcia, J.L., and Gillis, M. 1988. Characterization of *Azorhizobium caulinodans* gen. nov., sp. nov., a stem-nodulating nitrogen-fixing bacterium isolated from *Sesbania rostrata*. *Int. J. Syst. Bacteriol.* **38**: 89-98.

Dudman, W.F. 1984. The polysaccharide and oligosaccharides of *Rhizobium* and their role in the infection process. In *Advances in Nitrogen Fixation Research*. (Veeger, C. and Newton, W.E. eds.), Nijhoff, The Hague. pp. 397-404.

Dunn, K., Dickstein, R., Feinbaum, R., Burnett, B.K., Peterman, T.K., Thoidis, G., Goodman, H.M., and Ausubel, F.M. 1988. Developmental regulation of nodule specific genes in alfalfa root nodules. *Mol. Plant-Microbe Interact.* **1**: 66-74.

Dusha, I., Bakos, A., Kondorosi, A., de Brujn, F.J.,and Schell, J. 1989. The *Rhizobium meliloti* early nodulation genes (*nod* ABC) are nitrogen regulated : isolation of a mutant strain with efficient nodulation capacity on alfalfa in the presence of ammonium. *Mol. Gen. Genet.* **219**: 89-96.

Dusha, I., Kovatenko, S., Banfalvi, Z.,and Kondorosi, A. 1987. *Rhizobium meliloti* insertion element ISRm2 and its use for identification of the *fix* X gene. *J. Bacteriol.* **169**: 1403-1409.

Dylan, T., Ielpi, L., Stanfield, S., Kashyap, L., Douglas, C., Yanofsky, M., Nester, E., Helinski, D.R., and Ditta, G. 1986. *Rhizobium meliloti* genes required for nodule development are related to chromosomal virulence genes in *Agrobacterium tumefaciens*. *Proc. Natl. Acad. Sci., USA.* **83**: 4403-4407.

Dylan, T., Nagpal, P., Helinski, D.R., and Ditta, G.S. 1990. Symbiotic pseudorevertants of *Rhizobium meliloti* ndv mutants. *J. Bacteriol.* **172**: 1409-1416.

Earl, C.D., Ronson, C.W., and Ausubel, F.M. 1987. Genetic and structural analysis of the *Rhizobium meliloti* fix A, fix B, fix C and fix X genes. *J. Bacteriol.* **169**: 1127-1136.

Ebeling, S., Hahn, M., Fischer, H.M., and Hennecke, H. 1987. Identification of *nif* E-, *nif* N- and *nif* S- like genes in *Bradyrhizobium japonicum*. *Mol. Gen. Genet.* **207**: 503-508.

Ebeling, S., Noti, J.D., and Hennecke, H. 1988. Identification of new *Bradyrhizobium japonicum* (frx A) encoding a ferrodoxin-like protein. *J. Bacteriol.* **170**: 1999-2001.

Economou, A., Hamilton, W.D.O., Johnston, A.W.B., and Downie, J.A. 1990. The *Rhizobium* nodulation gene *nod* O encodes a Ca^{2+}-binding protein that is exported without N-terminal cleavage and is homologous to hemolysin and related proteins. *EMBO J.* **9**: 349-354.

Egelhoff, T.T., and Long, S.R. 1985. *Rhizobium meliloti* nodulation genes : identification of *nod* ABC gene products, purification of *nod* A protein and expression of *nod* A in *Rhizobium. J. Bacteriol.* **164**: 591-599.

Egli, M.A., Griffith, S.M., Miller, S.S., Anderson, M.P., and Vance, C.P. 1989. Nitrogen assimilating enzyme activity and enzyme protein during development and senescence of effective and plant - controlled ineffective alfalfa nodules. *Plant Physiol.* **91**: 898-904.

Ehrhardt, D.W., Atkinson, E.M., and Long, S.R. 1992. Depolarization of alfalfa root hair membrane potential by *Rhizobium meliloti* Nod factors. *Science.* **256**: 998-1000.

Eisbrenner, G., and Evans, H.J. 1983. Aspects of hydrogen metabolism in nitrogen-fixing legumes and other microassociations. *Annu. Rev. Plant Physiol.* **34**: 105-136.

El Hasan, G.A., Hernandez, B.S., and Focht, D.D. 1986. Comparison of Hup trait and intrinsic antibiotic resistance for assessing rhizobial competitiveness axenically and in soil. *Appl. Environ. Microbiol.* **51**: 546-551.

Emerich, D.W., Ruiz-Argueso, T., Ching, T.M., and Evans, H.J. 1979. Hydrogen dependent nitrogenase activity and ATP formation in *Rhizobium japonicum* bacteroids. *J. Bacteriol.* **137**: 153-160.

Engelke, T., Jagdish, M.N., and Puhler, A. 1987. Biochemical and genetic analysis of *Rhizobium meliloti* mutants defective in C$_4$-dicarboxylate transport. *J. Gen. Microbiol.* **133**: 3019-3029.

Engelke, T., Jording, D., Kapp, D., and Puhler, A. 1989. Identification and sequence analysis of the *Rhizobium meliloti dct* A gene encoding the C$_4$-dicarboxylate carrier. *J. Bacteriol.* **171**: 5551-5560.

Estabrook, E.M., and Sengupta-Gopalan, C. 1990. Differential expression of phenylalanine ammonia-lyase and chalcone synthase during soybean nodule development. *Plant Cell.* **3**: 299-308.

Evans, D., Jones, R., Woodley, P., and Robson, R. 1988. Further analysis of nitrogen fixation (*nif*) genes in *Azotobacter chroococcum*: identification and expression in *Klebsiella pneumoniae* of *nif* S, *nif* V, *nif* M and *nif* B genes and localization of *nif* E/N-, *nif* U-, *nif* A- and *fix* ABC genes. *J. Gen. Microbiol.* **134**: 931-942.

Evans, H.J., Hanus, F.J., Haugland, R.A., Cantrell, M.A., Xu, L.S., Russell, S.A., Lambert, G.R., and Harker, A.R. 1984. Hydrogen recycling in nodules affects nitrogen fixation and growth of soybean. *In Proceedings of World Soybean Research Conference III.* (Shibles, R. ed.), Westview Press, Bonlder, London. pp. 935-942.

Evans, H.J., Harker, A.R., Papen, H., Russell, S.A., Hanus, F.J., and Zuber, M. 1987. Physiology, biochemistry and genetics of the uptake hydrogenase in rhizobia. *Annu. Rev. Microbiol.* **41**: 335-361.

Evans, H.J., Purohit, K., Cantrell, M.A., Eisbrenner, G., Russell, S.A., Hanus, F.J., and Lepo, J.E. 1981. Hydrogen losses and hydrogenase in nitrogen-fixing organisms. *In Current Perspectives in Nitrogen Fixation.* (Gibson, A.H. and Newton, W.E. eds.), Griffith Press, Netley, South Australia. pp. 84-96.

Evans, I.J., and Downie, J.A. 1986. The *nod* I product of *Rhizobium leguminosarum* is closely related to ATP -binding bacterial transport proteins: nucleotide sequence of the *nod* I and *nod* J genes. *Gene.* **43**: 95-101.

Faria, S.M., de McInroy, S.G., Rowell, P., and Sprent, J.I. 1988. Some properties of 'primitive' legume nodules. *In Nitrogen Fixation: Hundred Years After* (Bothe, H., de Bruijn, F.J. and Newton, W.E. eds.), Gustav Fischer, Stuttgart. pp. 524.

Faucher, C., Camut, S., Denarie, J., and Truchet, G. 1989. The *nod* H and *nod* Q host range genes of *Rhizobium meliloti* behave as a virulence genes in *R. leguminosarum* bv. *viciae* and determine changes in the production of plant-specific extracellular signals. *Mol. Plant-Microbe Interact.* **2**: 291-300.

Finan, T.M., Hirsch, A.M., Leigh, J.A., Johannsen, E., Kaldau, G.A., Deegan, S., Walker, G.C., and

Signer, E.R. 1985. Symbiotic mutants of *Rhizobium meliloti* that uncouple plant from bacterial differentiation. *Cell.* **40**: 869-877.

Finan, T.M., Kunkel, B., De Vos, F., and Signer, E.R. 1986. Second symbiotic megaplasmid in *Rhizobium meliloti* carrying exopolysaccharide and thiamine synthesis genes. *J. Bacteriol.* **167**: 66-72.

Finan, T.M., Oresnik, I.,and Bottacin, A. 1988. Mutants of *Rhizobium meliloti* defective in succinate metabolism. *J. Bacteriol.* **170**: 3396-3403.

Finan, T.M., Wood, J.M.,and Jordan, D.C. 1983. Symbiotic properties of C_4-dicarboxylic acid transport mutants of *Rhizobium leguminosarum. J. Bacteriol.* **154**: 1403-1413.

Firmin, J.L., Wilson, K.E., Rossen, L., and Johnston, A.W.B. 1986. Flavonoid activation of nodulation genes in *Rhizobium* reversed by other compounds present in plants. *Nature.* **324**: 90-92.

Fischer, H.-M., Alvarez-Morales, A., and Hennecke, H. 1986. The pleiotropic nature of symbiotic regulatory mutants: *Bradyrhizobium japonicum nif* A gene is involved in control of *nif* gene expression and formation of determinate symbiosis. *EMBO J.,* **5**: 1165-1173.

Fischer, H.-M.,and Hennecke, H. 1984. Linkage map of the *Rhizobium japonicum nif* H and *nif* DK operons encoding the polypeptides of the nitrogenase enzyme complex. *Mol. Gen. Genet.* **196**: 537-540.

Fischer, H.-M.,and Hennecke, H. 1987. Direct response of *Bradyrhizobium japonicium nif* A-mediated *nif* gene regulation to cellular oxygen status. *Mol. Gen. Genet.* **209**: 621-626.

Fisher, R.F., Egelhoff, T.T., Mulligan, J.T., and Long, S.R. 1988. Specific binding of proteins from *Rhizobium meliloti* cell-free extracts containing *nod* D to DNA sequences upstream of inducible nodulation genes. *Genes Dev.* **2**: 282-293.

Fisher,R.F.,and Long, S.R. 1992. *Rhizobium* - plant signal exchange. *Nature.* **357**: 655-660.

Fogher, L., Dusha, I., Barbot, P., and Elmerich, C. 1985. Heterologous hybridization of *Azospirillum* DNA to *Rhizobium nod* and *fix* genes. *FEMS Microbiol. Lett.* **30**: 245-249.

Forde, B.G., and Cullimore, J.V. 1989. The molecular biology of glutamine synthetase in higher plants. *Oxford Surv. Plant Mol. Cell. Biol.* **6**: 247-296.

Forde, B.G., Day, H.M., Turton, J.F., Shen, W.J. Cullimore, J.V., and Oliver, J.E. 1989. Two glutamine synthetase genes from *Phaseolus vulgaris* L. display contrasting developmental and spatial patterns of expression in transgenic *Lotus corniculatus* plants. *Plant Cell.* **I**: 391-401.

Forrai, T., Vincze, E., Vanfalvi, S., Kiss, G.B., Randhawa, G.S., and Kondorosi, A. 1983. Localization of symbiotic mutations in *Rhizobium meliloti. J. Bacteriol.* **153**: 635-643.

Fortin, M.G., Morrison, N.A., and Verma, D.P.S. 1987. Nodulin-26, a peribacteroid membrane nodulin is expressed independently of the development of the peribacteroid compartment. *Nucleic Acids Res.* **15**: 813-824.

Fortin, M.G., Zelechowa, M., and Verma, D.P.S. 1985. Specific targeting of the membrane nodulins to the bacteroid enclosing compartment in soybean nodules. *EMBO J.* **4**: 3041-3046.

Franssen, H.J., Nap, J.P., Gloudemans, T., Stiekema, W., van Dam, H., Govers, F., Louwerse, J., van Kammen, A.,and Bisseling, T. 1987. Characterization of cDNA for nodulin-75 of soybean: A gene product involved in early stages of root nodule development. *Proc. Natl. Acad. of Sci.,* USA. **84**: 4495-4499.

Franssen, H.J., Vijn, I., Yang, W.C., and Bisseling, T. 1992. Developmental aspects of the *Rhizobium* - legume symbiosis. *Plant Mol. Biol.* **19**: 89-107.

Fuhrmann, M., Fischer, H.-M.,and Hennecke, H. 1985. Mapping of *Rhizobium japonicum nif* B-, *fix* BC- and *fix* A - like genes and identification of the *fix* A promoter. *Mol. Gen. Genet.* **199**: 315-322.

Fuller, F., Kunstner, P.W., Nguyen, T., and Verma, D.P.S. 1983. Soybean nodulin genes: Analysis of cDNA clones reveals several major tissue-specific sequences in nitrogen-fixing root nodules. *Proc. Natl. Acad. Sci.,* USA. **80**: 2594-2598.

Fuller, F., and Verma, D.P.S. 1984. Appearance and accumulation of nodulin mRNAs and their relationship to the effectiveness of root nodules. *Plant Mol. Biol.* **3**: 21-28.

Gallon, J.R., and Chaplin. A.E. 1987. *An Introduction to Nitrogen Fixation.* Cassell.

Garg, R.P., and Dadarwal, K.R. 1991. *Ex planta* nitrogenase expression associated with *Hup* character in cowpea miscellany *Rhizobium*. *Indian J. Exp. Biol.* **29**: 371-374.

Garg, R.P., and Dadarwal, K.R. 1993. Symbiotic effectivity of Hup$^+$ and Hup$^-$ isogenic derivatives of cowpea miscellany *Rhizobium* strains. *Indian J. Microbiol.* **33**: 35-41.

Gibson, A.H., Drefus, B.L., Lawn, R.J.,and Sprent, J.I. 1981. Host and environmental factors affecting hydrogen evolution and uptake. *In Current perspectives in nitrogen fixation.* (Gibson, A.H. and Newton, W.E. eds.), Griffith, Press, Netley, South Australia. pp. 373.

Gilles-Gonzalez, M.A., Ditta, G.S., and Helinski, D.R. 1991. A hemoprotein with kinase activity encoded by the oxygen sensor of *Rhizobium meliloti*. *Nature* **350**: 170-172.

Glazebrook, J.,and Walker, G.C. 1989. A novel exopolysaccharide can function in place of the calcofluor-binding exopolysaccharide in nodulation of alfalfa by *Rhizobium meliloti*. *Cell.* **55**: 661-672.

Glenn, A.R., Poole, P.S.,and Hudman, J.F. 1980. Succinate uptake by free-living and bacteroid forms of *Rhizobium leguminosarum*. *J. Gen. Microbiol.* **119**: 267-271.

Goethals, K., Gao, M., Tomekpe, K., Van Montagu, M., and Holsters, M. 1989. Common *nod* ABC genes in *nod* locus 1 of *Azorhizobium caulinodans*: nucleotide sequence and plant-inducible expression. *Mol. Gen. Genet.* **219**: 289-298.

Goethals, K., Van Montagu, M., and Holsters, M. 1992. Conserved motifs in a divergent *nod* box of *Azorhizobium caulinodans* ORS 571 reveal a common structure in promoters regulated by Lys R-type proteins. *Proc. Natl. Acad. Sci., USA.* **89**: 1646-1650.

Goldberg, R.B. 1988. Plants : novel developmental processes. *Science.* **240**: 1460-1467.

Gordon, J.K., and Brill, W.J. 1972. Mutants that produce nitrogenase in presence of ammonia. *Pro. Natl. Acad. Sci., USA.* **69**: 3501-3503.

Göttfert, M., Grob, P., and Hennecke, H. 1990a. Proposed regulatory pathway encoded by the *nod* V and *nod* W genes, determinants of host specificity in *Bradyrhizobium japonicum*. *Proc. Natl. Acad. Sci., USA.* **87**: 2680-2684.

Göttfert, M., Hitz, S., and Hennecke, H. 1990b. Identification of *nod* S and *nod* U, two inducible genes inserted between the *Bradyrhizobium japonicum nod* YABC and *nod* IJ genes. *Mol. Plant-Microbe Interact.* **3**: 308-316.

Göttfert, M., Horvath, B., Kondorosi, E., Putnoky, P., Rodriguez-Quinones, F., and Kondorosi, A. 1986. At least two *nod* D genes are necessary for efficient nodulation of alfalfa by *Rhizobium meliloti*. *J. Mol. Biol.* **191**: 411-420.

Göttfert, M., Lamb, J.W., Gasser, R., Semenza, J.,and Hennecke, H. 1989. Mutational analysis of the *Bradyrhizobium japonicum* common *nod* genes and further *nod* box-linked genomic DNA regions. *Mol. Gen. Genet.* **215**: 407-415.

Govers, F., Nap, J.P., Moerman, M., Franssen, H., van Kammen, A.,and Bisseling, T. 1987. cDNA cloning and developmental expression of pea nodulin genes. *Plant Mol. Biol.* **8**: 425-435.

Graham, P.H.,and Parker, C.A. 1961. Leghemoglobin and symbiotic nitrogen fixation. *Aust. J. Sci.* **23**: 231-232.

Graham, T.L. 1991. Flavonoid - and isoflavonoid distribution in developing soybean seeding tissues and in seed and root exudates. *Plant Physiol.* **95**: 594-603.

Gray, J.X.,and Rolfe, B.G. 1990. Exopolysaccharide production in *Rhizobium* and its role in invasion. *Mol. Microbiol.* **4**: 1425-1431.

Gronger, P., Manian, S.S., Reiländer, H., O'Connell, M., Priefer, U.B., and Pühler, A. 1987. Organization and partial sequence of a DNA region of the *Rhizobium leguminosarum* symbiotic plasmid pRL6JI containing the genes *fix* ABC, *nif* A, *nif* B and a novel open reading frame. *Nucleic Acids Res.* **15**: 31-49.

Gross, R., Arico, B., and Rappuoli, R. 1989. Families of bacterial signal-transducing proteins. *Mol. Microbiol.* **3**: 1661-1667.

Gubler, M., and Hennecke, H. 1986. *fix* A, B and C genes are essential for symbiotic and free-living, microaerobic nitrogen fixation genes. *FEBS Lett.* **200**: 186-192.

Guerinot, M.L., and Chelm, B.K. 1986. Bacterial δ-aminolaevulinic acid synthase activity is not essential for leghemoglobin formation in the soybean - *Bradyrhizobium japonicum* symbiosis. *Proc. Natl. Acad. Sci., USA.* **83**: 1837-1841.

Guerinot, M.L., and Chelm, B.K. 1987. Molecular aspects of the physiology of symbiotic nitrogen fixation in legumes. In *Plant-Microbe Interactions, Molecular and Genetic Perspectives* Vol. 2. (Kosuge, T. and Nester, E.W. eds.), Macmillan Press, New York. pp. 103-146.

Gulash, M., Ames, P., Larosiliere, R.C., and Bergman, K. 1984. *Rhizobium* are attracted to localized sites on legume roots. *Appl. Environ. Microbiol.* **48**: 149-152.

Gussin, G.N., Ronson, C.W., and Ausubel, F.M. 1986. Regulation of nitrogen fixation genes. *Annu. Rev. Genet.* **20**: 567-591.

Györgypal, Z., Iver, N., and Kondorosi, A. 1988. Three regulatory *nod* D alleles of diverged flavonoid-specificity are involved in host-dependent nodulation by *Rhizobium meliloti. Mol. Gen. Genet.* **212**: 85-92.

Györgypal, Z., Kondorosi, E., and Kondorosi, A. 1991. Diverse signal sensitivity of *nod* D protein homologs from narrow and broad host range rhizobia. *Mol. Plant-Microbe Interact.* **4**: 356-364.

Halverson, L.J., and Stacey, G. 1986. Signal exchange in plant-microbe interactions. *Microbiol. Rev.* **50**: 193-211.

Hansen, A.P., Peoples, M.B., Gresshoff, P.M., Atkins, C.A., Pate, J.S., and Carroll, B.J. 1989. Symbiotic performance of supernodulating soybean (*Glycine max* (L.) Merr.) mutants during development on different nitrogen regimes. *J. Expt. Bot.* **40**: 715-724.

Hanus, F.J., Albrecht, S.L., Zablotowicz, R.M., Emerich, D.W., Russell, S.A., and Evans, H.J. 1981. The effect of the hydrogenese system in *Rhizobium japonicum* inocula on the nitrogen content and yield of soybean seed in field experiments. *Agron. J.* **73**: 368-372.

Hardy, R.W.F., and Havelka, U.D. 1975. Nitrogen fixation research : a key to world food? *Science.* **188**: 633-643.

Hartwig, U.A., Maxwell, C.A., Joseph, C.M., and Phillips, D.A. 1989. Interactions among flavonoid *nod* gene inducers released from alfalfa seeds and roots. *Plant Physiol.* **91**: 1138-1142.

Hata, S., Kouchi, H., Suzuka, I., and Ishii, T. 1991. Isolation and characterization of cDNA clones for plant cyclins. *EMBO J.* **10**: 2681-2688.

Haugland, R.A., Cantrell, M.A., Beaty, J.S., Hanus, F.J., Russell, S.A., and Evans, H.J. 1984. Characterization of *Rhizobium japonicum* hydrogen uptake genes. *J. Bacteriol.* **159**: 1006-1012.

Haugland, R.A., Hanus, F.J., Cantrell, M.A., and Evans, H.J. 1983. Rapid colony screening method for identifying hydrogenase activity in *Rhizobium japonicum. Appl. Environ. Microbiol.* **45**: 892-897.

Hawkes, T.R., McLean, P.A., and Smith, B.E. 1984. Nitrogenase from *nif* V mutants of *Klebsiella pneumoniae* contains an altered form of the iron-molybdenum cofactor. *Biochem. J.* **217**: 317-321.

Henikoff, S., Haughn, G.W., Calvo, J.M., and Wallace, J.C. 1988. A large family of bacterial activator proteins. *Proc. Natl. Acad. Sci., USA.* **85**: 6602-6606.

Hennecke, H. 1990. Nitrogen fixation genes involved in the *Bradyrhizobium japonicum*-soybean symbiosis. *FEBS Lett.* **268**: 422-426.

Hennecke, H., Fischer, H.-M., Ebeling, S., Gubler, M., Thony, B., Gottfert, M., Lamb, J., Hahn, M.,

Ramseier, T., Regensberger, B., Alvarez-Morales, A., and Studer, D. 1987. *nif, fix* and *nod* gene clusters in *Bradyrhizobium japonicum*, and *nif* A - mediated control of symbiotic nitrogen fixation. In *Molecular Genetics of Plant-Microbe Interactions* (Verma, D.P.S. and Brisson, N. eds.), Martinus Nijhoff, Dordrecht. pp. 191-196.

Hennecke, H., Kaluza, K., Thony, B., Fuhrmann, M., Ludwig, W., and Stackebrandt, E. 1985. Concurrent evolution of nitrogenase genes and 16S rRNA in *Rhizobium* species and other nitrogen-fixing bacteria. *Arch. Microbiol.* **142**: 342-348.

Hirel, B., Bouet, C., King, B., Layzell, D., Jacobs, F., and Verma, D.P.S. 1987. Glutamine synthetase genes are regulated by ammonia provided externally or by symbiotic nitrogen fixation. *EMBO J.* **5**: 1167-1171.

Hirsch, A.M. 1992. Development biology of legume nodulation. New Phytol. **122**: 211-237.

Hirsch, A.M., Bhuvaneswari, T.V., Torrey, J.G., and Bisseling, T. 1989. Early nodulin genes are induced in alfalfa root outgrowths elicited by auxin transport inhibitors. *Proc. Natl. Acad. Sci.,* USA. **86**: 1244-1248.

Hirsch, A.M., Drake, D., Jacobs, T.W., and Long, S.R. 1985. Nodules are induced on alfalfa roots by *Agrobacterium tumefaciens* and *Rhizobium trifolii* containing small segments of the *Rhizobium meliloti* nodulation region. *J. Bacteriol.* **161**: 223-230.

Hirsch, A.M., and Smith, C.A. 1987. Effects of *Rhizobium meliloti nif* and *fix* mutants on alfalfa root nodule development. *J. Bacteriol.* **169**: 1137-1146.

Hodgson, A.L.M., Roberts, W.P., and Waid, J.S. 1985. Regulated nodulation of *Trifolium subterraneum* inoculated with bacteriocin-producing strains of *Rhizobium trifolii. Soil Biol. Biochem.* **17**: 475-478.

Holding, A.J.,and King, J. 1963. The effectiveness of indigenous populations of *Rhizobium trifolii* in relation to soil factors. *Plant Soil.* **18**: 191-198.

Hollingsworth, R.I., Philip-Hollingsworth, S.,and Dazzo, F.B. 1990. Isolation, characterization and structural elucidation of a '*nod* signal' excreted by *Rhizobium trifolii* ANU 843 which induces root hair branching and nodule-like primordia in axenic white clover seedlings. *In Nitrogen Fixation: Achievements and Objectives* (Gresshoff, P.M., Roth, L.E., Stacey, G. and Newton, W.E. eds.), Chapman and Hall, New York. pp. 193-198.

Hong, G.F., Burn, J.E., and Johnston, A.W.B. 1987. Evidence that DNA involved in the expression of nodulation (*nod*) genes in *Rhizobium* binds to the product of the regulatory gene *nod* D. *Nucleic Acids Res.* **15**: 9677-9690.

Honma, M.A., Asomaning, M., and Ausubel, F.M. 1990. *Rhizobium meliloti nod* D genes mediate host-specific activation of *nod*ABC. *J. Bacteriol.* **172**: 901-911.

Honma, M.A., and Ausubel, F.M. 1987. *Rhizobium meliloti* has three functional copies of the *nod* D symbiotic regulatory gene. *Proc. Natl. Acad. Sci.,* USA. **84**: 8558-8562.

Hooykaas, P.J.J., Snijdewint, F.G.M., and Schilperoort, R.A. 1982. Identification of the *sym* plasmid of *Rhizobium leguminosarum* strain 1001 and its transfer to and expression in other rhizobia and *Agrobacterium tumerfaciens. Plasmid.* **8**: 73-82.

Horvath, B., Bachem, C.W.B., Schell, J., and Kondorosi, A. 1987. Host-specific regulation of nodulation genes in *Rhizobium* is mediated by a plant signal, interacting with the *nod* D gene product. *EMBO J.* **6**: 841-848.

Horvath, B., Kondorosi, E., John, M., Schmidt, J., Torok, I., Györgypal, Z., Barabas, I., Wieneke, U., Schell, J., and Kondorosi, A. 1986. Organization, structure and symbiotic function of *Rhizobium meliloti* nodulation genes determining host specificity for alfalfa. *Cell.* **46**: 335-344.

Hotter, G.S., and Scott, D.B. 1991. Exopolysaccharide mutants of *Rhizobium loti* are fully effective on a determinate nodulating host but are ineffective on an indeterminate nodulating host. *J. Bacteriol.* **173**: 851-859.

Hua, S.S.T., Scot, D.B.,and Lim, S.T. 1981. A mutant of *Rhizobium japonicum* 110 with elevated *nif* activity in free-living culture. In *Genetic Engineering of Symbiotic Nitrogen Fixation* (Lyons, J.M., Valentine, R.C., Philips, D.A., Rains, D.W. and Huffaker, R.C. eds.), Plenum Press, New York. pp. 95-105.

Hungria, M., Joseph, C.M., and Phillips, D.A. 1991a. Anthocyanidins and flavonols, major *nod* gene inducers from seeds of a black-seeded common bean (*Phaseolus vulgaris* L.). *Plant Physiol.* 97: 751-758.

Hungria, M., Joseph, C.M., and Phillips, D.A. 1991b. *Rhizobium nod* gene inducers exuded naturally from roots of common bean (*Phaseolus vulgarus* L.). *Plant Physiol.* 97: 759-764.

Hungria, M., Neves M.C.P., and Dobereiner, J. 1989. Relative efficiency, ureide transport and harvest index in soybeans inoculated with isogenic *Hup* mutants of *Bradyrhizobium japonicum*. *Biol. Fert. Soils.* 7: 325-329.

Hynes, M.F., Simon, R., Muller, P.R., Niehaus, K., Labes, M., and Pühler, A. 1986. The two megaplasmids of *Rhizobium meliloti* are involved in the effective nodulation of alfalfa. *Mol. Gen. Genet.* 202: 355-362.

Iismaa, S.E., and Watson, J.M. 1987. A gene upstream of the *Rhizobium trifolii nif* A gene encodes a ferrodoxin-like protein. *Nucleic Acids Res.* 15: 3180-3184.

Jacobs, F.A., Zhang, M., Fortin, M.G., and Verma, D.P.S. 1987. Several nodulins of soybean share structural domains but differ in their subcellular location. *Nucleic Acids Res.* 15: 1271-1280.

Jacobs, T.W., Egelhoff, T.T.,and Long, S.R. 1985. Physical and genetic map of a *Rhizobium meliloti* gene region and nucleotide sequence of *nod* C. *J. Bacteriol.* 162: 469-476.

Jacobsen, K., Laursen, N.B., Jensen, E.O., Marcker, A., Poulson, C.,and Marcker, K.A. 1990. HMG-like proteins from leaf and nodule nuclei interact with different AT-motifs in soybean nodule promoters. *Plant Cell.* 2: 85-94.

Jacobson, M.R., Premakumar, R.,and Bishop, P.E. 1986. Transcriptional regulation of nitrogen fixation by molybdenum in *Azotobacter vinelandii*. *J. Bacteriol.* 167: 480-486.

Jarvis, B.D.W., Gills, M.,and De Ley, J. 1986. Intra- and intergeneric similarities between the ribosomal ribonucleic acid cistrons of *Rhizobium* and *Bradyrhizobium* species and some related bacteria. *Int. J. Syst. Bacteriol.* 36: 129-138.

Jing, Y., Zhang, B.T., and Shan, X.Q. 1990. Pseudonodule formation on barley roots induced by *Rhizobium astragali*. *FEMS Microbiol. Lett.* 69: 123-128.

Johansen, E., Stencel, L., and Applebaum, E. 1988. Isolation of *Bradyrhizobium japonicum* mutants with hyperinducible common *nod* genes. In *Nitrogen Fixation : Hundred Years After* (Bothe, H., de Bruijn, F.J. and Newton, W.E. eds.), Gustav Fischer, Stuttgart. pp. 464.

Jolly, S.O., McIntosh, L. Link, G., and Bogorad, L. 1981. Differential transcription *in vivo* and *in vitro* of two adjacent maize chloroplast genes : The large subunit of ribulose-biphosphate carboxylase and the 2.2-kilobase gene. *Proc. Natl. Acad. Sci., USA.* 78: 6821-6825.

Jordan,D.C.,and Garrad, E.H. 1951. Studies on the legume root nodule bacteria-1. Detection of effective and ineffective strains. *Can. J. Bot.* 29: 360-372.

Jording, D., Sharma, P.K., Schmidt, R., Engelke, T., Uhde, C., and Puhler, A. 1992. Regulatory aspects of the C_4-dicarboxylate transport in *Rhizobium meliloti*: transcriptional activation and dependence on effective symbiosis. *J. Plant Physiol.* 141: 18-27.

Joseph, M.V., Desai, J.D., and Desai, A.J. 1983. Production of antimicrobial and bacteriocin - like substances by *Rhizobium trifolii*. *Appl. Environ. Mirobiol.* 45: 532-535.

Joshi, P.A., Caetano-Anolles, G., Graham, E.T.,and Gresshoff, P.M. 1991. Ontogeny and ultrastructure of spontaneous nodules in alfalfa (*Medicago sativa*). *Protoplasma.* 162: 1-11.

Kahn, D., Batut, J., Boistard, P., Daveran, M.L., David, M., Domergue, O., Garnerone, A.M., Ghai, J., Hertig, C., Infante, D., and Renalier, M.H. 1987. Molecular analysis of a *fix* cluster from

Rhizobium meliloti. In *Molecular Genetics of Plant-Microbe Interactions* (Verma, D.P.S. and Brisson, N. eds.), Martinus Nijhoff, Dordrecht. pp. 258-268.

Kahn, M., Kraus, J., and Somerville, J.E. 1985. A model for nutrient exchange in the *Rhizobium*-legume symbiosis. In Nitrogen Fixation Research Progress (Evans, H.J., Bottomley, P.J. and Newton, W.E. eds.), Martinus Nijhoff Dordrecht. pp. 193-199.

Kaluza, K., Fuhrmann, M., Hahn, M., Regensburger, B., and Hennecke, H. 1983. In *Rhizobium japonicum* the dinitrogenase genes *nif* H and *nif* DK are separated. *J. Bacteriol.* **155**: 915-918.

Kaluza, K., Hahn, M., and Hennecke, H. 1985. Repeated sequences similar to insertion elements clustered around the *nif* region of the *Rhizobium japonicum* genome. *J. Bacteriol.* **162**: 535-542.

Kaneko, Y., and Newcomb, E.H. 1990. Specialization for ureide biogenesis in the root nodules of black locust (*Robina pseudoacacia* L.), an amide exporter. *Protoplasma.* **157**: 102-111.

Kannenberg, E.L., and Brewin, N.J. 1989. Expression of a cell surface antigen from *Rhizobium leguminosarum* 3841 is regulated by oxygen and pH. *J. Bacteriol.* **171**: 4543-4548.

Kannenberg, E.L., Rathbun, E.A., and Brewin, N.J. 1992. Molecular dissection of structure and function in the lipopolysaccharide of *Rhizobium leguminosarum* strain 3841 using monoclonal antibodies and genetic analysis. *Mol. Microbiol.* **6**: 2477-2487.

Kape, R., Parniske, M., and Werner, D. 1991. Chemotaxis and *nod* gene activity of *Bradyrhizobium japonicum* in response to hydroxycinamic acids and isoflavonoids. *Appl. Environ. Microbiol.* **57**: 316-319.

Katinakis, P., Klein-Lankhorst, R.M., Louwerse, H., van Kammen, A., and Vanden Bos, R.C. 1988a. Bacteriod-encoded proteins are secreted into the peribacteroid space by *Rhizobium leguminosarum*. *Plant Mol. Biol.* **11**: 183-190.

Katinakis, P., van Kammen, A., and Vanden, Bos, R.C. 1988b. Protein composition of the peribacteroid space in root nodules of *Pisum sativum* and *Vicia faba* induced by *Rhizobium leguminosarum* biovar *viciae*. *Lett. Appl. Microbiol.* **7**: 115-118.

Katinakis, P., and Verma, D.P.S. 1985. Nodulin-24 gene of soybean codes for a peptide of the peribacteroid membrane and was generated by tandem duplication of an insertion element. *Proc. Natl. Acad. Sci., USA.* **82**: 4157-4161.

Keen, N.T., and Staskawicz, B.J. 1988. Host range determinants in plant pathogens and symbionts. *Annu. Rev. Phytopathol.* **42**: 421-440.

Kennedy, I.R., and Tchan, Y.T. 1992. Biological nitrogen fixation in non-leguminous field crops: Recent advances. *Plant Soil.* **141**: 93-118.

Kijne, J.W. 1992. The *Rhizobium* infection process. In *Biological Nitrogen Fixation* (Stacey, G., Burris, R.H. and Evans, H.J. eds.), Chapman and Hall, New York. pp. 349-398.

Kim, C.H., Tully, R.E., and Keister, D.L. 1989. Exopolysaccharide-deficient mutants of *Rhizobium fredii* HH303 which are symbiotically effective. *Appl. Environ. Microbiol.* **55**: 1852-1854.

Klee, H., Horsch, R., and Rogers, S. 1987. *Agrobacterium* - mediated plant transformation and its further applications to plant biology. *Annu. Rev. Plant Physiol.* **38**: 467-486.

Kneen, B.F., La Rue, T.A., Hirsch, A.M., Smith, C.A., and Weeden, N.F. 1990. *sym* 13-a gene conditioning ineffective nodulation in *Pisum sativum*. *Plant Physiol.* **94**: 899-905.

Kondorosi, A. 1992. Regulation of nodulation genes in rhizobia. In *Molecular Signals in Plant-Microbe Communications* (Verma, D.P.S. ed.), CRC Press, Boca Raton pp. 325-340.

Kondorosi, E., Banfalvi, Z., and Kondorosi, A. 1984. Physical and genetic analysis of a symbiotic region of *Rhizobium meliloti*: identification of nodulation genes. *Mol. Gen. Genet.* **193**: 445-452.

Kondorosi, E., Burie, M., Cren, M., Iyer, N., Hoffman, B., and Kondorosi, A. 1991. Involvement of the *syr* M and *nod* D3 genes of *Rhizobium meliloti* in *nod* gene activation and in optimal nodulation of the plant host. *Mol. Microbiol.* **5**: 3035-3048.

Kondorosi, E., Gyuris, J., Schmidt, J., John, M., and Duda, E. 1989. Positive and negative control of *nod* gene expression in *Rhizobium meliloti* is required optimal nodulation. *EMBO J.* **8**: 1331-1340.

Kondorosi, E., and Kondorosi, A. 1986. Nodule induction on plant roots by *Rhizobium*. *Trends Biochem. Sci.* **11**: 296-299.

Kondorosi, A., Kondorosi, E., Pankhurst, C.E., Broughton, W.J., and Banfalvi, Z. 1982. Mobilization of *Rhizobium meliloti* megaplasmid carrying nodulation and nitrogen fixation genes in other rhizobia and *Agrobacterium*. *Mol. Gen. Genet.* **188**: 433-439.

Konieczny, A., Szezyglowski, K., Boron, L., Przybylska, M., and Legocki, A.B. 1988. Expression of lupin nodulin genes during root nodule development. *Plant Sci.* **55**: 145-149.

Kosslak, R.M., Bookland, R., Barkei, J., Paaren, H.E., and Appelbaum, E.R. 1987. Induction of *Bradyrhizobium japonicum* common *nod* genes by isoflavones isolated from *Glycine max*. *Proc. Natl. Acad. Sci., USA*. **84**: 7428-7432.

Kosslak, R.M., Joshi, R.S., Bowen, B.A., Paaren, H.E., and Appelbaum, E.R. 1990. Strain-specific inhibition of *nod* gene induction in *Bradyrhizonium japonicum* by flavonoid compounds. *Appl. Environ. Microbiol.* **5**: 1333-1341.

Kretovich, V.L., Melik-Sarkissyan, S.S., Bashirova, N.F., and Topunov, A.F. 1982. Enzymatic reduction of leghemoglobin in lupin nodules. *J. Appl. Biochem.* **4**: 209-217.

Lamb, C.J., Lawton, M.A., Dron, M., and Dixon, R.A. 1989. Signals and transduction mechanisms for activation of plant defenses against microbial attack. *Cell.* **56**: 215-224.

Lambert, G.R., Cantrell, M.A., Hanus, F.J., Russell, S.A., Hadded, K.R., and Evans, H.J. 1985. Intra and interspecies transfer and expression of *Rhizobium japonicum* hydrogen uptake genes and autotrophic growth capability. *Proc. Natl. Acad. Sci., USA*. **82**: 3232-3236.

Lambert, G.R., Harker, A.R., Cantrell, M.A., Hanus, F.J., Russell, S.A., Haugland, R.A., and Evans, H.J. 1987. Symbiotic expression of cosmid-borne *Bradyrhizobium japonicum* hydrogenase genes. *Appl. Environ. Microbiol.* **53**: 422-428.

Lara, M., Cullimore, J.V., Lea, P.J., Miflin, B.J., Johnston, A.W.B., and Lamb, J.W. 1983. Appearance of a novel form of plant glutamine synthetase during nodule development in *Phaseolus vulgaris*. *Planta* **157**: 254-258.

Larsen, K., and Jochimsen, B.U. 1986. Expression of nodule-specific uricase in soybean callus tissue is regulated by oxygen. *EMBO J.* **5**: 1519-1524.

Lee, J.S., Brown, G.G., and Verma, D.P.S. 1983. Chromosomal arrangement of leghemoglobin genes in soybean. *Nucleic Acids Res.* **11**: 5541-5553.

Lee, J.S., and Verma, D.P.S. 1984. Structure and chromosomal arrangement of leghemoglobin genes in kidney bean suggest divergence in soybean leghemoglobin gene loci following tetraploidization. EMBO J. **3**: 2745-2752.

Legares, A., Caetano-Anolles, G., Niehaus, K., Lorenzen, J., Ljunggren, H.D., Puhler, A., and Favelukes, G. 1992. A *Rhizobium meliloti* lipopolysaccharide mutant altered in competitiveness for nodulation of alfalfa. *J. Bacteriol.* **174**: 5941-5952.

Legocki, R.P., and Verma, D.P.S. 1979. A nodule-specific plant protein (nodulin-35) from soybean. *Science.* **205**: 190-192.

Legocki, R.P., and Verma, D.P.S. 1980. Identification of nodule-specific host proteins (nodulins) involved in the development of *Rhizobium*-legume symbiosis. *Cell.* **20**: 153-163.

Leigh, J.A., and Coplin, D.L. 1992. Exopolysaccharides in plant-bacterial interactions. *Annu. Rev. Microbiol.* **46**: 307-346.

Leigh, J.A., Reed, J.W., Hanks, J.F., Hirsch, A.M.,and Walker, G.C. 1987. *Rhizobium meliloti* mutants that fail to succinylate their calcofluor-binding exopolysaccharide are defective in nodule invasion. *Cell.* **51**: 579-587.

Leigh, J.A., Signer, E.R.,and Walker, G.C. 1985. Exopolysaccharide deficient mutants of *Rhizobium meliloti* that form ineffective nodules. *Proc. Natl. Acad. Sci., USA*. **82**: 6231-6235.

Leong, S.A., Ditta, G.S.,and Helinski, D.R. 1982. Heme biosynthesis in *Rhizobium*. *J. Biol. Chem.* **257**: 8724-8730.

Lepo, J.E., Hanus, F.J., and Evans, H.J. 1980. Chemoautotropic growth of hydrogen uptake positive strains of *Rhizobium japonicum*. *J. Bacterial.* **141**: 664-670.

Lepo, J.E., Hickok, R.E., Cantrell, M.A., Russell, S.A., and Evans, H.J. 1981. Revertible hydrogen uptake deficient mutants of *Rhizobium japonicum*. *J. Bacterial.* **146**: 614-620.

Lerouge, P., Roche, P., Faucher, C., Maillet, F., Truchet, G., Prome, J.C., and Denarie, J. 1990. Symbiotic host-specificity of *Rhizobium meliloti* is determined by a sulphated and acylated glucosamine oligosaccharide. *Nature.* **344**: 781-784.

Lewin, A., Cervantes, E., Wong, C.H., and Broughton, W.J. 1990. *nod* SU, two new *nod* genes of the broad host-range *Rhizobium* strain NGR 234 encode host specific nodulation of the tropical tree *Leucaena leucocephala. Mol. Plant-Microbe Interact.* **3**: 317-326.

Leyva, A., Palacios, J.M., Mozo, T., and Ruiz-Argueso, T. 1987. Cloning and characterization of hydrogen uptake genes from *Rhizobium leguminosarum. J. Bacteriol.* **169**: 4929-4934.

Li, J.M., Brathwaite, O., Cosloy, S.D., and Russell, C.S. 1989. 5-aminolevulinic acid synthesis in *Escherichia coli. J. Bacteriol.* **171**: 2547-2552.

Libbenga, K.R., and Harkes, P.A.A. 1973. Initial proliferation of cortical cells in the formation of root nodules in *Pisum sativum* L. *Planta.* **114**: 17-28.

Libbenga, K.R., van Iren, F., Bogers, R.J., and Schraag-Lammers, M.F. 1973. The role of hormones and gradients in the initiation of cortex proliferation and nodule formation in *Pisum sativum* L. *Planta.* **114**: 29-39.

Ligon, J.M. 1990. Molecular genetics of nitrogen fixation in plant-bacteria symbioses. In *Biotechnology of Plant-Microbe Interactions.* (Nakas, J.P. and Hagedorn, C. eds.), Mc Graw-Hill Publishing Company, New York. pp. 145-187.

Lloyd, C.W , Pearce, K.J., Rawlins, D.J., Ridge, R.W., and Shaw, P.J. 1987. Endoplasmic microtubules connect the advancing nucleus to the tip of legume root hairs, but F-actin is involved in basipetal migration. *Cell Motility Cytoskeleton.* **8**: 27-36.

Long, S.R. 1989a. *Rhizobium* genetics. *Annu. Rev. Genet.* **23**: 483-506.

Long, S.R. 1989b. *Rhizobium*-legume nodulation: Life together in the underground. *Cell.* **56**: 203-214.

Long, S.R., Fisher, R.F., Ogawa, J., Swanson, J., Ehrhardt, D.S., Atkinson, E.M., and Schwedock, J.S. 1991. *Rhizobium* meliloti nodulation gene regulation and molecular signals. In *Advances in Molecular genetics of Plant-Microbe Interactions* (Hennecke, H. and Verma, D.P.S. ed.), Kluwer Academic Press, Dordrecht. pp. 127-133.

Maier, R.J. 1986. Biochemistry, regulation and genetics of hydrogen oxidation in *Rhizobium. Critical Rev. Biotechnol.* **3**: 17-38.

Maier, R.J., Campbell, N.E.R., Hanus, F.J., Simpson, F.B., Russell, S.A., and Evans, H.J. 1978. Expression of hydrogenase activity in free-living *Rhizobium japonicum.Proc. Natl. Acad. Sci., USA.* **75**: 3258-3262.

Maillet, F., Debelle, F., and Denarie, J. 1990. Role of the *nod* D and *syr* M genes in the activation of the regulatory gene *nod* D3, and of the common and host-specific *nod* genes of *Rhizobium meliloti. Mol. Microbiol.* **4**: 1975-1984.

Maiti, T.K., and Podder, S.K. 1989. Differential binding of peanut agglutinin with lipopolysaccharide of homologous and heterologous *Rhizobium. FEMS Microbiol. Lett.* **65**: 279-284.

Malik, N.S.A., Calvert, H.E., and Bauer, W.D. 1987. Nitrate induced regulation of nodule formation in soybean. *Plant Physiol.* **84**: 266-271.

Martinez, E., Romero, D., and Palacois, R. 1990. The *Rhizobium* genome. *Critical Rev. Plant Sci.* **9**: 59-93.

Mathews, A., Carroll, B.J., and Gresshoff, P.M. 1987. Characterization of non-nodulation mutants of soybean (*Glycine max* (L.) Merr.): *Bradyrhizobium* effects and absence of root hair curling. *J. Plant Physiol.* **131**: 349-361.

Mathews, A., Carroll, B.J., and Gresshoff, P. M. 1989. Development of *Bradyrhizobium* infections in a super nodulating and non-nodulating mutant of soybean (*Glycine max.* (L.) Merr.). *Protoplasma.* **150**: 40-47.

Mc Allister, C.F., and Lepo, J.E. 1983. Succinate transport by free-living forms of *Rhizobium japonicum. J. Bacteriol.* **153**: 1155-1162.

Mc Iver, J., Djordjevic, M.A., Weinman, J.J., Bender, G.L., and Rolfe, B.G. 1989. Extension of host range of *Rhizobium leguminosarum* bv. *trifolii* caused by point mutations in *nod* D that result in alterations in regulatory function and recognition of inducer molecules. *Mol. Plant-Microbe Interact.* **2**: 97-106.

McRae, D.G., Miller, R.W., Berndt, W.B., and Joy, K. 1989. Transport of C_4-dicarboxylates and aminoacids by *Rhizobium meliloti* bacteriods. *Mol. Plant-Microbe Interact.* **2**: 273-278.

Mellor, R.B. 1989. Bacteroids in the *Rhizobium*-legume symbiosis inhabit a plant internal lytic compartment: implications for other microbial endosymbioses. *J. Expt. Bot.* **40**: 831-839.

Mellor, R.B., Thierfelder, H., Pausch, G., and Werner, D. 1986. The occurrence of choline kinase II in the cytoplasm of soybean root nodules infected with various strains of *Bradyrhizobium japonicum. J. Plant Physiol.* **128**: 169-172.

Merberg, D., and Maier, R.J. 1983. Mutants of *Rhizobium japonicum* with increased hydrogenase activity. *Science.* **220**: 1064-1065.

Miao, G.H., Hirel, B., Marsolier, M.C., Ridge, R.W., and Verma, D.P.S. 1991. Ammonia-regulated expression of a soybean gene encoding cytosolic glutamine synthetase in transgenic *Lotus corniculatus. Plant Cell.* **3**: 11-22.

Miao, G.H., Joshi, C.P., and Verma, D.P.S. 1990. Nodulin-26: Topology and function in root nodules. *In Nitrogen Fixation: Achievements and Objectives* (Gresshoff, P.M., Roth, L.E., Stacey, G., and Newton, W.E. eds.), Chapman and Hall, New York. pp. 755.

Miller, J.E., Viands, D.R., and La Rue, T.A. 1991. Inheritance of non-nodulating mutants of sweetclover. *Crop Sci.* **31**: 948-952.

Mulligan, J.T., and Long, S.R. 1985. Induction of *Rhizobium meliloti nod* C expression by plant exudate requires *nod* D. *Proc. Natl. Acad. Sci., USA.* **82**: 6609-6613.

Mulligan, J.T., and Long, S.R. 1989. A family of activator genes regulates expression of *Rhizobium meliloti* nodulation genes. *Genet.* **122**: 7-18.

Nadler, K.D., and Avissar, Y.J. 1977. Heme synthesis in soybean root nodules. *Plant Physiol.* **60**: 433-436.

Nap, J.P., and Bisseling, T. 1990a. Nodulin function and nodulin gene regulation in root nodule development. In *Molecular Biology of Symbiotic Nitrogen Fixation* (Gresshoff, P.M. ed.), CRC Press, Boca Raton. pp. 181-229.

Nap, J.P., and Bisseling, T. 1990b. Developmental biology of a plant-prokaryote symbiosis: the legume root nodule. *Science.* **250**: 948-954.

Nelson, L.M., and Salminen, S.O. 1982. Uptake hydrogenase activity and ATP formation in *Rhizobium leguminosarum* bacteriods. *J. Bacteriol.* **151**: 989-995.

Newcomb, W. 1981. Nodule morphogenesis and differentiation. In *Biology of Rhizobiaceae, supplement 13, International Review of Cytology* (Giles, K.L. and Atherly, A.G. eds.), Academic Press, New York. pp. 247-298

Newcomb, W., and Mc Intrye, L. 1981. Development of root nodules of mungbean (*Vigna radiata*): a reinvestigation of endocytosis. *Can. J. Bot.* **59**: 2478-2499.

Newcomb, W., Sippel, D., and Peterson, R.L. 1979. The early morphogenesis of *Glycine max* and *Pisum sativum* root nodules. *Can. J. Bot.* **57**: 2603-2616.

Nguyen, T., Zelechowska, M., Foster, V., Bergmann, H., and Verma, D.P.S. 1985. Primary structure of the soybean nodulin-35 gene encoding uricase II localized in the peroxisomes of uninfected cells of nodules. *Proc. Natl. Acad. Sci., USA.* **82**: 5040-5044.

Nirunsuksiri, W., and Sengupta-Gopalan, C. 1990. Characterization of a novel nodulin gene in soybean that shares sequence similarity to the gene for nodulin-24. *Plant Mol. Biol.* **15**: 835-849.

Noel, K.D. 1991. Rhizobial polysaccharides required in symbiosis with legumes. In *Molecular Signals in Plant-Microbe Communications* (Verma, D.P.S. ed.), CRC Press, Boca Raton. pp. 341-357.

Noel, D.K., Carneol, M., and Brill, W.J. 1982. Nodule protein synthesis and nitrogenase activity of soybeans exposed to fixed nitrogen. *Plant Physiol.* **70**: 1236-1241.

Noel, K.D., Vanden Bosch, K.A., and Kulpaca, B. 1986. Mutations in *Rhizobium phaseoli* that lead to arrested development of infection threads. *J. Bacteriol.* **168**: 1392-1401.

Norel, F., and Elmerich, C. 1987. Nucleotide sequence and functional analysis of the two *nif*H copies of *Rhizobium* ORS 571. *J. Gen. Microbiol.* **133**: 1563-1576.

Noti, J.D., Folkerts, O., Turken, A.N., and Szalay, A.A. 1986. Nitrogenase promoter-*lac* z fusion studies of essential nitrogen fixation genes in *Bradyrhizobium japonicum* T-110. *J. Bacteriol.* **167**: 784-791.

Nixon B.T., Ronson, C.W., and Ausubel, F.M. 1986. Two-component regulatory systems responsive to environmental stimuli share strongly conserved domains with the nitrogen assimilation genes *ntr* A and *ntr* C. *Proc. Natl. Acad. Sci., USA.* **83**: 7850-7854.

Pahwa, K., and Dogra, R.C. 1981. H_2-recycling system in mungbean *Rhizobium* in relation to N_2-fixation. *Arch. Microbiol.* **129**: 380-383.

Peters, N.K., Frost, J.W., and Long, S.R. 1986. A plant flavone, luteolin induces expression of *Rhizobium meliloti* nodulation genes. *Science.* **233**: 977-980.

Peters, N.K., and Long, S.R. 1988. Alfalfa root exudates and compounds which promote or inhibit induction of *Rhizobium meliloti* nodulation genes. *Plant Physiol.* **88**: 396-400.

Philip-Hollingsworth, S., Hollingsworth, R.I., and Dazzo, F.B. 1991. N-Acetylglutamic acid: an extracellular *nod* signal of *Rhizobium trifolii* ANU843 that induces root hair branching and nodule-like primordia in white clover roots. *J. Biol. Chem.* **266**: 16854-16858.

Phillips, D.A., Joseph, C.M.,and Maxwell, C.A. 1992. Trigonelline and stachydrine released from alfalfa seeds activate *nod* D2 protein in *Rhizobium meliloti*. *Plant Physiol.* **99**: 1526-1531.

Polhill, R.M., and Raven, P.H. 1981. Advances in Legume Systematics. Royal Botanic Gardens, Kew, England.

Priefer, U.B. 1989. Genes involved in lipopolysaccharide production and symbiosis are clustered on the chromosome of *Rhizobium leguminosarum* biovar *viciae* VF39. *J. Bacteriol.* **171**: 6161-6168.

Pühler, A., Aguilar, M.O., Hynes, M., Müller, P., Klipp, W., Priefer, U., Simon R., and Weber, G. 1984. Advances in the genetics of free-living and symbiotic nitrogen fixing bacteria. In *Advances in Nitrogen Fixation Research* (Veeger, C. and Newton, W.E. eds.), Nijhoff/Junk, The Hague. pp. 609-619.

Pühler, A., Arnold, W., Buendia-Claveria, A., Kapp, D., Keller, M., Niehaus, K., Quant, J., Roxlau, A., and Weng, W.M. 1991. The role of *Rhizobium meliloti* exopolysaccharides EPSI and EPSII in the infection process of alfalfa nodules. In *Advances in Molecular Genetics of Plant-Microbe Interactions. Vol. I.* (Hennecke, H. and Verma, D.P.S. eds.), Kluwer Academic Publishers, Dordrecht. pp. 189-194.

Puppo, A., Rigaud, J., and Job, D. 1980. Leghemoglobin reduction by a nodule reductase. *Plant Sci. Lett.* **20**: 1-6.

Puvanesarajah, V., Schell, F.M., Gerhold, D., and Stacey, G. 1987. Cell surface polysaccharides from *Bradyrhizobium japonicum* and a non-nodulating mutant. *J. Bacteriol.* **169**: 137-141.

Quinto, C., de la Vega, H., Flores, M., Leemans, J., Cevallos, M.A., Pardo, M.A., Azpiroz, R., de Lourdes, G.M., Calva, E., and Palacois, R. 1985. Nitrogenase reductase : a functional multigene family in *Rhizobium phaseoli*. *Proc. Natl. Acad. Sci., USA.* **82**: 1170-1174.

Rae, A.L., Perotto, S., Knox, J.P., Kannenberg, E.L.,and Brewin, N.J. 1991. Expression of extra-cellular glycoproteins in the uninfected cells of developing pea nodule tissue. *Mol. Plant Microbe Interact.* **4**: 563-570.

Regensburger, B., Meyer, L., Filser, M., Weber, J., Studer, D., Lamb, J.W., Fischer, H.M., Hahn, M., and Hennecke, H. 1986. *Bradyrhizobium japonicum* mutants defective in root-nodule bac-teroid development and nitrogen fixation. *Arch. Microbiol.* **144**: 355-366.

Reibach, P.H., and Streeter, J.G. 1983. Metabolism of ^{14}C-labelled photosynthate and distribution of enzymes of glucose metabolism in soybean nodules. *Plant Physiol.* **72**: 634-640.

Reuber, T.L., Reed, J.W., Glazebrook, J., Urzainqui, A., and Walker, G.C. 1991. Analysis of the roles of *Rhizobium meliloti* exopolysaccharides in nodulation. In *Advances in Molecular Gen-etics of Plant-Microbe Interactions.* Vol. I. (Hennecke, H. and Verma, D.P.S. eds.), Kluwer Academic Press, Dordrecht. pp. 182-188

Reynolds, P.H.S., Boland, M.J., Blevins, D.G., Schubert, K.R., and Randall, D.D. 1982. Enzymes of amide and ureide biogenesis in developing soybean nodules. *Plant Physiol.* **69**: 1334-1338.

Richardson A.E., Simpson, R.J., Djordjevic, M.A., and Rolfe, B.G. 1988. Expression of nodulation genes in *Rhizobium leguminosarum* bv. *trifolii* is affected by low pH, Ca and Al. *Appl. Envi-ron. Microbiol.* **54**: 2541-2548.

Robertson, J.G., and Lyttleton, P. 1982. Coated and smooth vesicles in the biogenesis of cell walls, plasma membranes, infection threads and peribacteroid membranes in root hairs and nodules of white clover. *J. Cell Sci.* **58**: 63-78.

Robertson, J.G., Lyttleton, P., Bullivant, S., and Grayston, G.F. 1978. Membranes in lupin root nodules 1. The role of Golgi bodies in the biogenesis of infection threads and peribacteroid mem-branes. *J. Cell Science.* **30**: 129-149.

Robertson, J.G., Wells, B., Bisseling, T., Farnden, K.J.F., and Johnson, A.W.B. 1984. Immunogold localization of leghemoglobin in cytoplasm in nitrogen-fixing root nodules of pea (*Pisum sat-ivum*). *Nature.* **311**: 254-256.

Robson, R., Kennedy, C., and Postgate, J. 1983. Progress in comparative genetics of nitrogen fix-ation. *Can. J. Microbiol.* **29**: 954-967.

Roche, P., Debelle, F., Maillet, F., Lerouge, P., Faucher, C., Truchet, G., Denarie, J., and Prome, J.C. 1991. Molecular basis of symbiotic host specificity in *Rhizobium meliloti*: nod H and nod PQ genes encode the sulfation of lipooligosaccharide signals. *Cell.* **67**: 1-20.

Rodrigues-Quinones, F., Banfalvi, Z., Murphy, P., and Kondorosi, A. 1987. Interspecies homology of nodulation genes in *Rhizobium*. *Plant Mol. Biol.* **8**: 61-76.

Rolfe, B.G.,and Bender, G.L. 1991. Evolving a *Rhizobium* for non-legume nodulation. In *Nitrogen Fixation : Achievements and Objectives* (Gresshoff, P.M., Roth, L.E., Stacey, G. and Newton, W.E. eds.), Chapman and Hall, New York. pp. 779-780.

Rolfe, B.G.,and Gresshoff, P.M. 1988. Genetic analysis of legume nodule initiation. *Annu. Rev. Plant Physiol. Plant Mol. Biol.* **39**: 297-319.

Ronson, C.W. 1988. Genetic regulation of C_4-dicarboxylate transport in rhizobia. In *Nitrogen Fix-ation : Hundred Years After* (Bothe, H., de Bruijn, F.J. and Newton, W.E. eds.), Gustav Fischer-Verlag, Stuttgart. pp. 547-551

Ronson, C.W., Astwood, P.M.,and Downie, J.A. 1984. Molecular cloning and genetic organization of C_4-dicarboxylate transport genes from *Rhizobium leguminosarum*. *J. Bacteriol.* **160**: 903-909.

Ronson, C.W., Astwood, P.M., Nixon, B.T.,and Ausubel, F.M. 1987a. Deduced products of C_4-dicarboxylate transport regulatory genes of *Rhizobium leguminosarum* are homologous to nitrogen regulatory gene products. *Nucleic Acid Res.* **15**: 7921-7934.

Ronson, C.W., Bosworth, A., Genova, M., Gudbrandsen, S., Hankinson, T., Kwaitkowski, R., Rat-cliffe, H., Robie, C., Sweeney, P., Szeto, W., Williams, M., and Zablotowicz, R. 1990. Field release of genetically engineered *Rhizobium meliloti* and *Bradyrhizobium japonicum* strains. In Nitrogen Fixation : Achievements and Objectives (Gresshoff, P.M., Roth, L.E., Stacey, G. and Newton, W.E. eds.), Chapman and Hall, New York. pp. 397-403.

Ronson, C.W., Lyttleton, P., and Robertson, J.G. 1981. C$_4$-dicarboxylate transport mutants of *Rhizobium trifolii* form ineffective nodules on *Trifolium repens*. *Proc. Natl. Acad. Sci., USA*. **78**. 4284-4288.

Ronson, C.W., Nixon, B.T., Albright, L.M., and Ausubel, F.M. 1987b. *Rhizobium meliloti ntr* A (*rpo* N) gene is required for diverse metabolic functions. *J. Bacteriol. 169*: 2424-2431.

Ronson, C.W., Nixon, B.T., and Ausubel, F.M. 1987c. Conserved domains in bacterial regulatory proteins that respond to environmental stimuli. *Cell*. **49**: 579-581.

Ronson, C.W., and Primrose, R.B. 1979. Carbohydrate metabolism in *Rhizobium trifolii* : identification and symbiotic properties of mutants. *J. Gen. Microbiol.* **112**: 77-88.

Rossen, L, Johnston, A.W.B.,and Downie, J.A. 1984. DNA sequence of the *Rhizobium leguminosarum* nodulation genes *nod* AB and C required for root hair curling. *Nucleic Acids Res.* **12**: 9497-9508.

Rossen, L., Shearman, C.A., Johnston, A.W.B., and Downie, J.A. 1985. The *nod* D gene of *Rhizobium leguminosarum* is autoregulatory and in the presence of plant exudate induces the *nod* A,B,C genes. *EMBO J.* **4**: 3369-3373.

Rostas, K., Kondorosi, E., Horvath, B., Simoncsits, A.,and Kondorosi, A. 1986. Conservation and extended promoter regions of nodulation genes in *Rhizobium*. *Proc. Natl. Acad. Sci., USA*. **83**: 1757-1761.

Roth, L.E., and Stacey, G. 1989a. Bacterium release into host cells of nitrogen-fixing soybean nodules: the symbiosome membrane comes from three sources. *Eur. J. Cell Biol.* **49**: 13-23.

Roth, L.E., and Stacey, G. 1989b. Cytoplasmic membrane systems involved in bacterium release into soybean nodule cells as studied with two *Bradyrhizobium japonicum* mutant strains. Eur. *J. Cell Biol.* **49**: 24-32.

Ruiz-Argueso, T., Cabrera, E., and de Bertalmio, M.B. 1981. Selection of symbiotically energy efficient strains of *Rhizobium japonicum* by their ability to induce a hydrogen uptake hydrogenase in the free-living state. *Arch. Microbiol.* **128**: 275-279.

Ruiz-Argueso, T., Hanus, F.J., and Evans, H.J. 1978. Hydrogen production and uptake by pea nodules as affected by strains of *Rhizobium leguminosarum*. *Arch. Microbiol.* **116**: 113-118.

Ruiz-Argueso, T., Hidalgo, E., Murillo, J., Rey, L., and Palacios, J.M. 1991. Molecular genetics of the hydrogen uptake system of *Rhizobium leguminosarum*. In *Advances in Molecular Genetics of Plant-Microbe Interactions*. *Vol. I*. (Hennecke, H. and Verma, D.P.S. eds.), Kluwer Academic Press, Dordrecht. pp. 222-225.

Rushing, B.G., Yelton, M.M., and Long, S.R. 1991. Genetic and Physical analysis of the *nod* D3 region of *Rhizobium meliloti*. *Nucleic acids Res.* **19**: 921-927.

Ruvkun, G.B., Sundaresan, V., and Ausubel, F.M. 1982. Directed transposon Tn5 mutagenesis and complementation analysis of *Rhizobium meliloti* symbiotic nitrogen fixation genes. *Cell*. **29**: 551-559.

Ryle, G.J.A., Powell, C.E., and Gordon, J.A. 1979. The respiratory costs of nitrogen fixation in soybean, cowpea and white clover. 1. Nitrogen fixation and the respiration of the nodulated root. *J. Exp. Bot.* **30**: 135-144.

Saari, L.L.,and Klucas, R.V. 1984. Ferric leghemoglobin reductase from soybean root nodules. *Arch. Biochem. Biophys.* **231**: 102-113.

Sadowsky, M.J., Cregan, P.B., Gottfert, M., Sharma, A., Gorhold, D., Rodriguez-Quinones, F., Keyser, H.H., Hennecke, H.,and Stacey, G. 1991. The *Bradyrhizobium japonicum nol* A gene and its involvement in the genotype-specific nodulation of soybean. Proc. Natl. Acad. Sci., USA. **88**: 637-641.

San Francisco, M.J.D.,and Jacobson, G.R. 1985. Uptake of succinate and malate in cultured cells and bacteroids of two slow-growing species of *Rhizobium. J. Gen. Microbiol.* **131**: 765-773.

Sangwan, I.,and O'Brian, M.R. 1991. Evidence for an interorganismic heme biosynthetic pathway in symbiotic soybean root nodules. *Science.* **251**: 1220-1222.

Sanjuan, J.,and Olivares, J. 1989. Implication of *nif* A in regulation of genes located on a *Rhizobium meliloti* cryptic plasmid that affect nodulation efficiency. *J. Bacteriol.* **171**: 4154-4161.

Sanjuan, J., and Olivares, J. 1991. Multicopy plasmids carrying the *Klebsiella pneumoniae nif* A gene enhance *Rhizobium meliloti* nodulation competitiveness on alfalfa. *Mol. Plant-Microbe Interact.* **4**: 365-369.

Scheres, B., van de Weil, C., Zalensky, A., Horvath, B., Spaink, H., van Eck, H., Zwartkruis, F., Wolters, A.M., Gloudemans, T., van Kammen. A., and Bisseling, T. 1990a. The ENOD12 gene product is involved in the infection process during the pea-*Rhizobium* interaction. *Cell.* **60**: 281-294.

Scheres, B., van Engelen, F., van der Knaap, E., van de Weil, C., van Kammen, A., and Bisseling, T. 1990b. Sequential induction of nodulin gene expression in the developing pea nodule. *Plant Cell.* **2**: 687-700.

Schofield, P.R.,and Watson, J.M. 1986. DNA sequence of *Rhizobium trifolii* nodulation genes reveals a reiterated and potentially regulatory sequence preceding *nod* ABC and *nod* FE. *Nucleic Acids Res.* **14**: 2891-2903.

Schubert, K.R.,and Evans, H.J. 1976. Hydrogen evolution : a major factor affecting the efficiency of nitrogen-fixation in nodulated symbionts. *Proc. Natl. Acad. Sci., USA.* **73**: 1207-1211.

Schultze, M., Quiclet-Sire, G., Kondorosi, E., Virelizier, H., Glushka, J.N., Endre, G., Gero, S.D., and Kondorosi, A. 1992. *Rhizobium meliloti* produces a family of sulfated lipooligosaccharides exhibiting different degrees of plant host specificity. *Proc. Natl, Acad. Sci., USA.* **89**: 192-196.

Schwedock, J., and Long, S.R. 1990. ATP sulfurylase activity of the *nod* P and *nod* Q gene products of *Rhizobium meliloti. Nature.* **348**: 644-647.

Schwedock, J., and Long, S.R. 1992. *Rhizobium meliloti* genes involved in sulfate activation - The two copies of *nod* PQ and a new locus, saa. *Genetics* **132**: 899-909.

Schwinghamer, E.A., and Belkengren, R.P. 1968. Inhibition of rhizobia by a strain of *Rhizobium trifolii*: some properties of the antibiotic and of the strain. *Arch. Microbiol.* **64**: 130-145.

Sengupta-Gopalan, C., and Pitas, J.W. 1986. Expression of nodule-specific glutamine synthetase genes during nodule development in soybeans. *Plant Mol. Biol.* **7**: 189-199.

Shah, V.K., and Brill, W.J. 1977. Isolation of an iron-molybdenum cofactor from nitrogenase. Proc. Natl. Acad. Sci., USA. **74**: 3249-3253.

Shah, V.K., Madden, M.S., and Ludden, P.W. 1990. *in vitro* synthesis of theiron-molybdenum cofactor and its analogs : requirement of a non-*nif* gene product for the synthesis, and altered properties of dinitrogenase. In *Nitrogen Fixation: Achievements and Objectives*, (Gresshoff, P.M., Roth, L.E., Stacey, G. and Newton, W.E. eds.), Chapman and Hall, New York. pp. 87-94.

Shah, V.K., Vgalde, R.H., Imperial, J.,and Brill, W.J. 1984. Mo in nitrogenase. *Annu. Rev. Biochem.* **53**: 231-257.

Shanmugam, K.T., Lim, S.T., Hom, S.S.M., Scott, D.B., and Hennecke, H. 1981. "Redox control" of nitrogen fixation : an overview. In *Genetic Engineering of Symbiotic Nitrogen Fixation* (Lyons, J.M., Valentine, R.C., Phillips, D.A., Rains, D.W. and Huffaker, R.C. eds.), Plenum Press, New York. pp. 79-93.

Sharma, S.B., and Signer, E.R. 1990. Temporal and spatial regulation of the symbiotic genes of *Rhizobium meliloti in planta* revealed by transposon Tn5-*gus* A. *Genes Dev.* **4**: 344-356.

Shearman, C.A., Rossen, L., Johnston, A.W.B., and Downie, J.A. 1986. The *Rhizobium* gene *nod*

F encodes a protein similar to acyl carrier protein and is regulated by *nod* D plus a factor in pea root exudate. *EMBO J.* **5**: 647-652.

Sindhu, S.S., Brewin, N.J., and Kannenberg, E.L. 1990. Immunochemical analysis of lipopolysaccharides from free-living and endosymbiotic forms of *Rhizobium leguminosarum. J. Bacteriol.* **172**: 1804-1813.

Sindhu, S.S., and Dadarwal, K.R. 1986. *Ex planta* nitrogenase induction and uptake hydrogenase in *Rhizobium* sp. (cowpea miscellany). *Soil Biol. Biochem.* **18**: 291-295.

Sindhu S.S.,and Dadarwal, K.R. 1992. Symbiotic effectivity of cowpea miscellany *Rhizobium* mutants having increased hydrogenase activity. *Indian J. Microbiol.* **32**: 411-416.

Sindhu, S.S.,and Dadarwal, K.R. 1993b. Broadening of host-range infectivity in cowpea miscellany *Rhizobium* by protoplast fusion. *Indian J. Exp. Biol.* **31**: 521-528.

Smit, G., Kijne, J.W.,and Lugtenberg, B.J.J. 1987. Involvement of both cellulose fibrils and Ca^{2+}-dependent adhesion in the attachment of *Rhizobium leguminosarum* to pea root hair tips. *J. Bacteriol.* **169**: 4294-4301.

Smith, B.E. 1990. Recent studies on the biochemistry and chemistry of nitrogenases. In *Nitrogen Fixation: Achievements and Objectives.* (Gresshoff, P.M., Roth, L.E., Stacey, G. and Newton, W.E. eds.), Chapman and Hall, New York. pp. 3-13.

Spaink, H.P., Geiger, O., Sheeley, D.A., van Brussel, A.A.N., York, W.S., Reinhold, V.N., Lugtenberg, B.J.J., and Kennedy, C.P. 1991a. The biochemical function of the *Rhizobium leguminosarum* proteins involved in the production of host specific signal molecules. In *Advances in Genetics of Plant-Microbe Interactions. Vol. I.* (Hennecke, H. and Verma, D.P.S. eds.), Kluwer Academic Publishers, Dordrecht. pp. 142-149.

Spaink, H.P., Okker, R.J.H., Wijffelman, C.A., Tak, T., Roo, L.G., Pees, E., van Brussel, A.A.N., and Lugtenberg, B.J.J. 1989a. Symbiotic properties of rhizobia containing a flavonoid-independent hybrid *nod* D product. *J. Bacteriol.* **171**: 4045-4053.

Spaink, H.P., Sheeley, D.M., van Brussel, A.A.N., Glushka, J., York, W.S., Tak, T., Geiger, O., Kennedy, E.P., Reinhold, V.N., and Lugtenberg, B.J.J. 1991b. A novel highly unsaturated fatty acid moiety of lipo-oligosaccharide signals determines host specificity of *Rhizobium. Nature.* **354**: 125-130.

Spaink, H.P., Weinman, J., Djordjevic, M.A., Wijffelman, C.A., Okker, R.J.H., and Lugtenberg, B.J.J. 1989b. Genetic analysis and cellular localization of the *Rhizobium* host specificity determining Nod E protein. *EMBO J.* **8**: 2811-2818.

Spaink, H.P., Wijffelman, C.A., Okker, R.J.H.,and Lugtenberg, B.J.J. 1989c. Localization of functional regions of the *Rhizobium nod* D product using hybrid *nod* D genes. *Plant Mol. Biol.* **12**: 59-73.

Sprent, J.I. 1986. Benefits of *Rhizobium* to agriculture. *Trends in Biotechnol.* **4**: 124-129.

Sprent, J.I. 1989. Which steps are essential for the formation of functional legume nodules? *New Phytol.* **111**: 129-153.

Stacey, G. 1990. Workshop summary: Compilation of the *nod, fix* and *nif* genes of rhizobia and information concerning their function. In *Nitrogen Fixation: Achievements and Objectives.* (Gresshoff, P.M., Roth, L.E., Stacey, G. and Newton, W.E. eds.), Chapman and Hall, New York. pp. 239-244.

Stacey, G., Gresshoff, P.M., and Keen, N.T. 1992. Friends and Foes: New insights into plant-microbe interactions. The highlights of oral and poster presentations of the 6th International Symposium on Molecular Plant-Microbe Interactions. *Plant Cell.* **4**: 1173-1179.

Stacey, G., So, J.S., Roth, L.E., Bhagya Lakshmi, S.K.,and Carlson, R.W. 1991. A lipopolysaccharide mutant of *Bradyrhizobium japonicum* that uncouples plant from bacterial differentiation. *Mol. Plant-Microbe Interact.* **4**: 332-340.

Stanley, J., Dowling, D.N.,and Broughton, W.J. 1988. Cloning of *hem* A from *Rhizobium* sp. NGR

234 and symbiotic phenotype of a site-directed mutant in diverse legume genera. *Mol. Gen. Genet.* **217**: 528-532.

Stougaard, J., Jorgensen, J.E., Christensen, T., Kuhle, A., and Marcker, K.A. 1990. Inter-dependence and nodule specificity of *cis*-acting regulatory elements in the soybean leghemoglobin *1bc* 3 and N23 gene promoters. *Mol. Gen. Gene.* **220**: 353-360.

Stougaard, J., Marcker, K.A., Otten, L., and Schell, J. 1986. Nodule-specific expression of a chimeric soybean leghemoglobin gene in transgenic *Lotus corniculatus. Nature.* **321**: 663-674.

Streeter, J. 1988. Inhibition of legume nodule formation and nitrogen fixation by nitrate. *CRC Critical Rev. Plant Sci.* **7**: 1-23.

Surin, B.P.,and Downie, J.A. 1988. Characterization of the *Rhizobium leguminosarum* genes *nod* LMN involved in efficient host-specific nodulation. *Mol. Microbiol.* **2**: 173-183.

Surin, B.P., Watson, J.M., Hamilton, W.D.O., Economou, A., and Downie, J.A. 1990. Molecular characterization of the nodulation gene, *nod* T, from two biovars of *Rhizobium leguminosarum. Mol. Microbiol.* **4**: 245-252.

Sutton, W.D. 1983. Nodule development and senescence. In *Nitrogen Fixation. Vol. 3. legumes* (Broughton, W.J. ed.), Clarendon Press, Oxford. pp. 144-212.

Sweet, G., Ganor, C., Voeglele, R., Wittekindt, N.A., Beuerle, J., Truniger, V., Lin, E.C.C., and Winfried, B. 1990. Glycerol facilitator of *Escherichia coli*: cloning of *glp* F and identification of glp F product. *J. Bacteriol.* **172**: 424-430.

Szabados, L., Ratet, P.,Grunenberg, B., Schell, J., and de Bruijn, F.J. 1990. Functional analysis of the *Sesbania rostrata* leghemoglobin g1b 3 gene 5' upstream region in transgenic *Lotus corniculatus* and *Nicotiana tobacum* plants. *Plant Cell.* **2**: 973-986.

Szeto, W.W., Nixon, B.T., Ronson, C.W., and Ausubel, F.M. 1987. Identification and characterization of *Rhizobium meliloti ntr* C gene: *R. meliloti* has separate regulatory pathways for activation of nitrogen fixation genes in free-living and symbiotic cells. *J. Bacteriol.* **169**: 1423-1432.

Szeto, W.W., Zimmerman, J.L., Sundaresan, V., and Ausubel, F.M. 1984. A *Rhizobium meliloti* symbiotic regulatory gene. *Cell.* **36**: 534-543.

Tchan, Y.T.,and Kennedy, I.R. 1989. Possible N_2-fixing root nodules induced in non-legumes. *Agricul. Sci.* **2**: 57-59.

Thöny, B., Fischer, H.-M., Anthamatten, D., Bruderel, T., and Hennecke, H. 1987. The symbiotic nitrogen fixation regulatory operon (*fixR nifA*) of *Bradyrhizobium japonicum* is expressed aerobically and is subject to a novel, *nif* A-independent type of activation. *Nucleic Acids Res.* **15**: 8479-8499.

Thorneley, R.N.F.,and Lowe, D.J. 1984. The mechanism of *Klebsiella pneumoniae* nitrogenase action: Stimulation of the dependence of hydrogen evolution rate on component protein concentration and ratio and sodium dithionite concentration. *Biochem. J.* **224**: 903-910.

Thummler, F.,and Verma, D.P.S. 1987. Nodulin-100 of soybean is the subunit of sucrose synthase regulated by the availability of free heme in nodules. *J. Biol. Chem.* **262**: 14730-14736.

Tingey, S.V., Walker, E.L.,and Coruzzi, G.M. 1987. Glutamine synthetase genes of pea encode distinct polypeptides which are differentially expressed in leaves, roots and nodules. *EMBO J.* **6**: 1-9.

Torok, I., Kondorosi, E., Strepkowski, T., Postfai, J.,and Kondorosi, A. 1984. Nucleotide sequence of *Rhizobium meliloti* nodulation genes. *Nucleic Acids Res.* **12**: 9509-9524.

Trinchant, J.C.,and Rigaud, J. 1982. Nitrite and nitric oxide as inhibitors of nitrogenase from soybean bacteroids. *Appl. Environ. Microbiol.* **44**: 1385-1388.

Triplett, E.W. 1988. Isolation of genes involved in nodulation competitiveness form *Rhizobium leguminosarum* bv. *trifolii* T24. *Proc. Natl. Acad. Sci., USA.* **85**: 3810-3814.

Triplett, E.W. 1990. Construction of a symbiotically effective strain of *Rhizobium leguminosarum* bv. *trifolii* with increased nodulation competitiveness. *Appl. Environ. Microbiol.* **56**: 98-103.

Triplett, E.W., and Barta, T.M. 1987. Trifolitoxin production and nodulation are necessary for the expression of superior nodulation competitiveness by *Rhizobium leguminosarum* bv. *trifolii* strain T24 on clover. *Plant Physiol.* **85**: 335-342.

Triplett, E.W., and Sadowsky, M.J. 1992. Genetics of competition for nodulation of legumes. *Annu. Rev. Microbiol.* **46**: 399-428.

Truchet, G., Barker, D.G., Camut, S., de Billy, F., Vasse, J., and Haguet, T. 1989a. Alfalfa nodulation in the absence of *Rhizobium*. *Mol. Gen. Genet.* **219**: 65-68.

Truchet, G., Camut, S., de Billy, F., Odorico, R.,and Vasse, J. 1989b. The *Rhizobium*-legume symbiosis: The methods to discriminate between nodules and other root-derived structures. *Protoplasma.* **149**: 82-88.

Truchet, G., Roche, P., Lerouge, P., Vasse, J., Camut, S., de Billy, F., Prome, J.C. and Denarie, J.. 1991. Sulphated lipooligosaccharide signals of *Rhizobium meliloti* elicit root nodule organogenesis in alfalfa. *Nature.* **351**: 670-673.

Truchet, G., Rosenburg, C., Vasse, J., Julliot, J.S., Camut, S.,and Denarie, J. 1984. Transfer of *Rhizobium meliloti sym* genes into *Agrobacterium tumefaciens*: Host specific nodulation by atypical infection. *J. Bacteriol.* **157**: 134-142.

Truelsen, T.A.,and Wyndaele, R. 1984. Recycling efficiency in hydrogenase uptake positive strains of *Rhizobium leguminosarum*. *Physiol. Plant.* **62**: 45-50.

Udvardi, M.K.,and Day, D.A. 1989. Electrogenic ATPase activity on the peribacteroid membrane of soybean (*Glycine max* L.) root nodules. *Plant Physiol.* **90**: 982-987.

Urzainqui, A.,and Walker, G.C. 1992. Exogenous suppression of the symbiotic deficiencies of *Rhizobium meliloti exo* mutants. *J. Bacteriol.* **174**: 3403-3406.

Van Brussel, A.A.N., Bakhuizen, R., van Spronsen, P.C., Spaink, H.P., Tak, T., Lugtenberg, B.J.J., and Kijne, J.W. 1992. Induction of pre-infection thread structures in the leguminous host plant by mitogenic lipooligosaccharides of *Rhizobium*. *Science.* **257**: 70-72.

Vance, C.P.,and Gantt, J.S. 1992. Control of nitrogen and carbon metabolism in root nodules. *Physiol. Plant.* **85**: 266-274.

Van de Weil, C., Norris, J.H., Bochenek, B., Dickstein, R., Bisseling, T.,and Hirsch, A.M. 1990a. Nodulin gene expression and ENOD2 localization in effective, nitrogen-fixing and ineffective, bacteria-free nodules of alfalfa. *Plant Cell.* **2**: 1009-1017.

Van de Wiel, C., Scheres, B., Franssen, H., van Lierop, M.J., van Lammeren, A., van Kammen, A., and Bisseling, T. 1990b. The early nodulin transcript ENOD2 is located in the nodule parenchyma (inner cortex) of pea and soybean root nodules. *EMBO J.* **9**: 1-7.

Vanden Bosch, K.A., Bradley, D., Knox, J.P., Perotto, S, Butcher, G.W., and Brewin, N.J. 1989. Common components of the infection thread matrix and the intercellular space identified by immunocytochemical analysis of pea nodules and uninfected roots. *EMBO J.* **8**: 335-342.

Vanden Bosch, K.A., and Newcomb, E.H. 1986. Immunogold localization of nodule-specific uricase in developing soybean root nodules. *Planta.* **167**: 425-436.

Vanden Bosch, K.A.,and Newcomb, E.H. 1988. The occurrence of leghemoglobin protein in the uninfected interstitial cells of soybean root nodules. *Planta.* **175**: 442-451.

Vasse, J., de Billy, F., Camut, S., and Truchet, G. 1990. Correlation between ultrastructural differentiation of bacteroids and nitrogen fixation in alfalfa nodules. *J. Bacteriol.* **172**: 4295-4306.

Veeger, C., Haaker, H., and Laane, C. 1981. Energy transduction and nitrogen fixation. In *Current Perspectives in Nitrogen Fixation*. (Gibson, A.H. and Newton, W.E. eds.), Australian Academy of Science, Canberra. pp. 101-104.

Verma, D.P.S. 1989. Plant genes involved in carbon and nitrogen assimilation in root nodules.

In *Plant Nitrogen Metabolism* (Poulton, J.E., Romeo, J.T. and Conn, E.E. eds.), Plenum Press, New York. pp. 43-63.

Verma, D.P.S. 1992. Signals in root nodule organogenesis and endocytosis of *Rhizobium*. *Plant Cell*. **4**: 373-382.

Verma, D.P.S., and Delauney, A.J. 1988. Root nodule symbiosis: Nodulins and nodulin genes. In *Plant Gene Research: Temporal and Spatial Regulation of Plant Genes* (Verma, D.P.S. and Goldberg, R. eds.), Springer Verlag, New York. pp. 169-199.

Verma, D.P.S., Delauney, A.J., Guida, M., Hirel, BB., Schafer, R., and Koh, S. 1988. Control of expression of nodulin genes. In *Molecular Genetics of Plant-Microbe Interactions* (Palacios, R. and Verma, D.P.S. eds.), APS Press, St Paul, MN. pp. 315-320.

Verma, D.P.S., and Fortin, M.G. 1989. Nodule development and formation of the endosymbiotic compartment. In *Cell Culture and Somatic Cell Genetics of Plants, Vol. VI. The Molecular Biology of Plant Nuclear Genes* (Schell, J. and Vasil, I.K. eds.), Academic Press, New York. pp. 329-353.

Verma, D.P.S., Fortin, M.G., Stanley, J., Mauro, V., Purohit, S., and Morrison, N. 1986. Nodulins and nodulin genes of *Glycine max*. *Plant Mol. Biol.* **7**: 51-61.

Verma, D.P.S., Hu, C.A., and Zhang, M. 1992. Root nodule development: origin, function and regulation of nodulin genes. *Physiol. Plant.* **85**: 253-265.

Verma, D.P.S., Kazazian, V., Zogbi, V., and Bal, A.K. 1978. Isolation and characterization of the membrane envelope enclosing the bacteriods in soybean root nodules. *J. Cell Sci.* **78**: 919-938.

Verma, D.P.S., and Long, S. 1983. The molecular biology of *Rhizobium* - legume symbiosis. In *International Review of Cytology, Supplement 14: Intracellular Symbiosis* (Jeon, K. ed.), Academic Press, New York. pp. 211-245.

Verma, D.P.S., Miao, G.H., Cheon, C.I., and Suzuki, H. 1990. Genesis of root nodules and function of nodulins. In *Advances in Molecular Genetics of Plant-Microbe Interactions* (Hennecke, H. and Verma, D.P.S. eds.), Kluwer Academic Press, Dordrecht. pp. 291-299.

Vesper, J.S., and Bauer, W.D. 1986. Role of pili (fimbriae) in attachment of *Bradyrhizobium japonicum* on soybean roots. *Appl. Environ. Microbiol.* **52**: 134-141.

Wang, S.P., and Stacey, G. 1990. Ammonia regulation of *nod* genes in *Bradyrhizobium japonicum*. *Mol. Gen. Genet.* **233**: 329-331.

Wang, S.P., and Stacey, G. 1991. Studies of the *Bradyrhizobium japonicum nod* DI promoter: a repeated structure for the *nod* box. *J. Bacteriol.* **173**: 3356-3365.

Waters, J.K., Karr, D.B., and Emerich, D.W. 1985. Malate dehydrogenase from *Rhizobium japonicum* 311b-143 bacteroids and *Glycine max* root-nodule mitochondria. *Biochemistry*. **24**: 6479-6486.

Watson, R.J. 1990. Analysis of the C_4-dicarboxylate transport genes of *Rhizobium meliloti*: nucleotide sequence and deduced products of *dct* A, *dct* B and *dct* D. *Mol. Plant-Microbe Interact.* **3**: 174-181.

Watson, R.J., Chan, Y.K., Wheatcroft, R., Yang, A.F., and Han, S. 1988. *Rhizobium meliloti* genes required for C_4-dicarboxylate transport and symbiotic nitrogen fixation are located on a megaplasmid. *J. Bacteriol.* **170**: 927-934.

Weaver, C.D., Crombie, B., Stacey, G., and Roberts, D.M. 1991. Calcium-dependent phosphorylation of symbiosome membrane protein from nitrogen-fixing soybean nodules. *Plant Physiol.* **95**: 222-227.

Werner, D., Morschel, E., Kort, R., Mellor, R.B., and Bassarab, S. 1984. Lysis of bacteroids in the vicinity of the host cell nucleus in ineffective (Fix⁻) root nodules of soybean (*Glycine max*). *Planta.* **162**: 8-16.

Williams, M.N.V., Hollingsworth, R.I., Klein, S., and Signer, E.R. 1990. The symbiotic defect of

Rhizobium meliloti exopolysaccharide mutants is suppressed by *lps* Z+, a gene involved in lipo-polysaccharide biosynthesis. *J. Bacteriol.* **172**: 2622-2632.

Witty, J.F., Minchin, F.R., Şkot, L., and Sheehy, J.E. 1986. Nitrogen fixation and oxygen in legume root nodules. *Oxford Surveys in Plant Molecular and Cellular Biol.* **3**: 275-314.

Wood, S.E., and Newcomb, W. 1989. Nodule morphogenesis: the early infection of alfalfa (*Medicago sativa*) root hairs by *Rhizobium meliloti. Can. J. Bot.* **67**: 3108-3122.

Xue, Z.T., Larsen, K., and Jochimsen, B.U. 1991. Oxygen regulation of uricase and sucrose synthase synthesis in soybean callus tissue is excreted at the mRNA level. *Plant Mol. Biol.* **16**: 899-906.

Yadav, K.S., Garg, R.P., Nilum,and Dadarwal, K.R. 1992. Symbiotic effectivity and competitiveness of azide resistant mutants of *Rhizobium* sp. (*Vigna*). *Indian J. Microbiol.* **32**: 423-427.

Yang, C., Signer, E.R., and Hirsch, A.M. 1992. Nodules initiated by *Rhizobium meliloti* exopolysaccharide (*exo*) mutants lack a discrete, persistent nodule meristem. *Plant Physiol.* **98**: 143-151.

Yang, L.J.O., Udvardi, M.K.,and Day, D.A. 1990. Specificity and regulation of the dicarboxylate carrier on the peribacteroid membrane of soybean nodules. *Planta.* **182**: 437-444.

Yang, N.S.,and Russell, D. 1990. Maize sucrose synthase-I promoter directs phloem cell - specific expression of GUS gene in transgenic tobacco plants. *Proc. Natl. Acad. Sci., USA.* **87**: 4144-4148.

Yarosh, O.K., Charles, T.C.,and Finan. T.M. 1989. Analysis of C_4-dicarboxylate transport genes in *Rhizobium meliloti. Mol. Microbiol.* **3**: 813-823.

Young, J.P.W.,and Johnston, A.W.B. 1989. The evolution of specificity in the legume-*Rhizobium* symbiosis. *Tree.* **4**: 341-349.

Nitrogen Nutrition in Higher Plants, 1995
Editors : H.S. Srivastava & R.P. Singh
Associated Publishing Co., New Delhi, India
pp. 131-143.

Nitrate Versus Ammonium Nutrition and Plant Growth

JOSÉ R. MAGÁLHAES and MANLIO S. FERNANDES

I. Introduction

In most crop systems, the availability of nitrogen is often a more limiting factor influencing plant growth than is any other nutrient. Also in agriculture very often less than 50% of the nitrogen fertilizer applied may ultimately be utilized by crops (Barker and Volk, 1964). Nitrate ions are highly mobile and are not adsorbed by soil colloids. Nitrate leaching, lossess of N_2O, N_2, and other oxides of nitrogen are recognized as major contributors to this ineffective utilization of nitrogen. Application of ammonium fertilizer would seem to offer a potential means of increasing the efficiency of utilization of applied nitrogen fertilizer because the ammonium ion is not as readily subject to losses as is the nitrate ion.

Theoretically, ammonium nitrogen should be the preferred form by plants (Reseinauer, 1978), because it needs not to be reduced before incorporation into organic matter. Consequently, it should be used more efficiently by plants than nitrate. Actually, some plants, such as rice and blueberry, grow better with ammonium than with nitrate as the nitrogen source (Spiers, 1978). However, at certain levels, ammonium ions are toxic to many higher plants. (Magalhaes and Wilcox, 1983a; 1983b; 1983c; 1984b).

In fertile neutral soil, most of the nitrogen is taken up by plants as nitrate into which the ammoniacal form is converted by nitrifying bacteria. These microorganisms are most active at soil pH about 6 to 7. But in most forest soils, in acidic soils, in dry or waterlogged soils, under low temperature, low nitrogen and other limiting factors for nitrification, the nitrifying bacteria are relatively inactive (Middleton and Smith, 1979; Tills and Alloway, 1981), and ammonium is the major nitrogen source (Blevins, 1989). Much of the nitrogen becomes available to plants in the form of ammonium ions, after mineralization of soil organic nitrogen from plant residues by ammonifying bacteria, which operate effectively over a wider range of pH and stressing conditions (Middleton and Smith, 1979). Ammonium is one of the most prevalent forms of nitrogen fertilizers employed in the world. Furthermore, due to the increasing cost of nitrogen fertilizers, nitrification inhibitors have also been used to try to reduce leaching losses and increase the efficiency of nitrogen (Prasad et al., 1971; Tsai et al., 1978; Tells and Alloway, 1981). Actually, all forms of organic N must be elaborated from ammonia, so that NH_3 assimilation must be regarded as an extremely important step in plant nitrogen metabolism (Rhodes, 1987). Substantial capability to assimilate ammonia in leaves is required to prevent the build up of potentially toxic concentration of ammonia resulting from photorespiration and other possible sources, such as:

(1) $NO_3^- \xrightarrow{\text{Nitrate-Reductase}} NO_2^- \xrightarrow{\text{Nitrite-Reductase}} NH_3$

(2) $2(\text{Glycine}) \xrightarrow{\text{Glycine Decarboxylase}} \text{Serine} + CO_2 + NH_3$

(3) $\text{Glutamate} + NAD^+ \xrightarrow{\text{GDH}^-} \text{2-Oxoglutarate} + NH_3 + NADH$

(4) $\text{Threonine} \xrightarrow{\text{Threonine Dehydratase}} NH_3 + \text{2-Oxobutyrate}$

(5) $\text{Asparagine} \xrightarrow{\text{Asparaginase}} \text{Aspartate} + NH_3$

(6) $\text{Urea} \xrightarrow{\text{Urease}} CO_2 + NH_3$

The growth and development of plants under NH_4^+ nutrition may be quite different from plants grown with NO_3^- nutrition. Contrasting NO_3^- nutrition with NH_4^+ nutrition is of special interest in this discussion, relating to different plant species and different environmental conditions.

II.   Contrasting Ammonium and Nitrate Nutrition in Higher Plants

Ammonium ions are adsorbed at the surface of roots in short time but adsorption is not evident in the case of nitrate. After adsorption, ammonium is taken up much more rapidly than is nitrate-nitrogen. (Middleton and Smith, 1979). In addition, in the dark, nitrogen transport to the shoot from nitrate solutions, almost ceases, while that of ammonium-nitrogen does not. When nitrate nitrogen is transported to the shoot, nitrate needs to be reduced to ammonium. Thus, much more energy is required for nitrate than for ammonium assimilation (Mengel and Viro, 1978). Middleton and Smith (1979), studied the energetics of nitrogen assimilation by perennial ryegrass through a thermodynamic model on the basis of the amount of carbohydrate consumed during assimilation of nitrogen (one nitrogen atom assimilated for $1\frac{1}{3}$ moles of glucose consumed), they calculated that for a solar energy input of 911 Kcal, ammonia assimilation returns 427 Kcal of usable free energy, whereas nitrate assimilation returns only 355 Kcal under similar conditions, giving an advantage for the ammonium form.

In contrast to these advantages, even though growing plants need to assimilate nitrogen ultimately in the fully reduced form, ammonium ions are toxic to many plants (Givan, 1979). The toxicity of ammonium to higher plant tissues has been demonstrated repeatedly even though the exact biochemical causes of toxicity, as well as the specific concentrations causing toxicity, are not always clear. In many higher plants, the prolonged application of ammonium as a source of nitrogen may disrupt various aspects of plant metabolism, leading to serious physiological and morphological disorders resulting in chlorosis, restricted growth (Kirkby and Mengel, 1967; Magalhaes and Wilcox, 1984b), uncoupled photophosphorylation, inhibition of ATP formation, reduced CO_2 fixation within the choroplast (Puritch and Barker, 1967; Ikeda and Yamada, 1981), inhibition of $NADP^+$ reduction (Vernon and Zang, 1960), reduced carboxylase enzyme activity, blocked starch synthesis (Matsumoto et al., 1971), low absorption of inorganic cations such as Ca^{++}, Mg^{++}, and K^+ and, in some case resulting in death of the plant (Magalhaes and Wilcox, 1984a; 1984b).

Nitrogen metabolism is complex, and the pathways of N assimilation depend on the form of nitrogen supplied (Raven and Smith, 1976; Magalhaes and Huber, 1989a). There is evidence that nitrate assimilation can occur either in the roots and/or in the shoots of vascular plants (Yoneyama and Kumazawa, 1975); however, most nitrate reduction takes place in shoots and leaf tissues (Raven and Smith, 1976). On the other hand, most of the nitrogen from ammonium sources is assimilated in the roots (Yoneyama and Kumazawa, 1974). Although there are evidences that free ammonium is translocated to the shoot of plants (Wilcox et al., 1977; Magalhaes et al., 1992). The biochemical pH state is a topic that illustrates the response of plants to NO_3^- vs NH_4^+ uptake and assimilation. During NO_3^- reduction, OH^- is produced and the pH of the tissue increases ($NO_3^- + 8H^+ + 8e^- \xrightarrow{\text{nitrate reduction}} NH_3 + 2H_2O + OH^-$) (Raven and Smith, 1976). The increase in pH stimulates phosphoenolpyruvate (PEP) carboxylation by the enzyme PEP carboxylase, producing oxalocetate (OAA), which is rapidly converted to malate by malate dehydrogenase. This sequence of events has been termed "the biochemical pH stat" (Davies, 1973; Raven and Smith, 1976). Labelling leaves with $^{14}CO_2$ resulted in $^{14}CO_2$ (as HCO_3^-) efflux from roots of plants fed with NO_3^- (Ben-Zioni et al., 1970). This indicated that an increase in alkalinity of the rhizosphere with NO_3^- nutrition was due to exchange or efflux of HCO_3^- from the roots. This pH stat eliminates the possibility of alkalinization of the tissues by utilizing OH^- and CO_2 to synthesize malate. Malate is subsequently transported to the root, decarboxylated and HCO_3^- effluxed into the soil. This is an effective means of eliminating into the rhizosphere a potentially dangerous situation for leaf tissues, i.e. the build up of high pH. Under NH_4^+ nutrition the pH stat reacts very differently: the rhizosphere become acid (i.e., pH down to 3.8 in unbuffered solutions (Magalhaes and Wilcox, 1984a, 1984b; Marschner, 1986). Apparently, NH_4^+ uptake involves H^+ pumping as a way of compensating for the decrease in tissue pH (Fernandes and Rossiello, 1986; Raven and Smith 1976). In the roots, keto acids, (i.e. OAA or) α-Ketoglutarate (α-KG) are quickly aminated, depleting the supply of carbon in the root. Ammonium ion, the dominant cation in terms of uptake is thus consumed along with organic anions in the synthesis of amino acids and amides. An understanding of the metabolism associated with NO_3^- vs NH_4^+ nutrition and of the pH stat, is important for the mechanisms involved in root diseases or control of soil born pathogens.

III.   Responses of Ammonium Assimilation Enzymes to Nitrogen Form

Nitrogen assimilation in different species is not expected to be the same since species such as tomato grow very poorly with NH_4^+ as the sole N source (Magalhaes and Wilcox, 1983a; 1983b; 1983c; 1984b), and plants such as blueberry and rice grow better with NH_4^+ than NO_3^- (Dkjkshoorn and Ismunadji, 1972; Spiers, 1978; Tills and Alloway, 1981). Activity of the NH_4^+ assimilating enzymes may regulate N assimilation and growth; and variation in their activity may explain observed differences in the response of plant species to N forms. In general, glutamate dehydrogenase (GDH) activity increases in the presence of NH_4^+ (Barash et al., 1975; Singh and Srivastava, 1982; Handa et al., 1984; Anghinoni et al., 1988; Magalhaes and Huber, 1989a; 1989b; 1991a; 1991b). However, increased activity may not necessarily indicate that GDH plays an active role in assimilation of NH_4^+ (Lewis et al., 1983; Magalhaes et al., 1990). In flowing solution with acidity controlled at pH 5.7 during the uptake of NO_3^- or NH_4^+, ammonia toxicity was reduced although tomato and corn growth were still 30% less with NH_4^+ (Magalhaes and Huber, 1989a). Restricted growth observed for NH_4^+ fed plants in media where rhizosphere pH is not effectively controlled (Kirkby and Mengel, 1967) is an effect of acidification during the uptake of ammonium (Peet et al., 1985). However the mechanism of acidification during uptake of NH_4^+ is not the only cause of NH_4^+ toxicity, since the effect of NH_4^+ in decreasing pH of the solution was similar for rice, tomato, and maize, and rice growth was not affected by NH_4^+ at either pH, 5.7 or 3.5, (Magalhaes and Huber, 1989a).

Glutamine synthetase activity in rice shoots, a plant efficient in NH_4^+ assimilation, was 260% higher than that observed in tomato shoots, a plant sensitive to NH_4^+ toxicity (Magalhaes and Huber, 1989a). This strongly suggests that GS may be a key enzyme to detoxify NH_4^+ in green tissues. It appears from this study that GS activity is important to incorporate NH_4^+ in shoots as well as roots. The higher GDH activity in roots of tomato and maize treated with NH_4^+ compared to NO_3^- was inversely related to plant growth (Magalhaes and Huber, 1991a). The enzyme could function as a deaminating enzyme in shoots (Guadinova, 1983; Srivastava and Singh, 1987; Magalhaes, 1991), to provide carbon skeletons to plants where a low photosynthetic rate results from NH_4^+ toxicity. This may further accelerate the toxicity in green tissues. The reducing power or energy required for NH_4^+ assimilation catalyzed by GDH was not a limiting factor in NH_4^+ treated plants, and the content of NADH was slightly higher in NH_4^+ than in NO_3^- fed plants, (Magalhaes et al., 1992). Thus it is likely that a deficiency of carbon skeletons in tomato is the limiting factor for NH_4^+ assimilation as will be discussed latter. This may be consequence of the accumulation of extremely high concentrations of free NH_4^+ in green tissues, (Table 1), which uncouples photophosphorylation and inhibits photosynthesis to result in NH_4^+ toxicity (Ikeda and Yamada, 1982). In contrast to tomato, green tissue of rice accumulated very little free NH_4^+ (Table 1), possibily as a result of the high GS activity in rice shoots. From these results, it looks reasonable to use the parameters free ammonia and GS activity in leaves for selecting plants for N assimilation efficiency.

Table 1. Free ammonium in tomato and rice shoots as affected by the form of N and pH.

Plant Species	N Form/pH			
	NO_3^-	NH_4^+	NH_4^+	L.S.D.
	5.7	5.7	3.5	(0.05)
	n moles g^{-1}FW			
Tomato	298	9865	19543	386
Rice	246	751	4548	251

*Mean of three determinations.

IV. Light Intensity and Plant Responses to Nitrogen Form

Reduced growth of plants receiving NH_4^+ as compared with those fed NO_3-N under high light intensity, is the general response reported for many plant species (Givan, 1979; Barker and Mills, 1980). However, tomatoes grown with NH_4^+, under shade, by reducing light intensity to 67% of full sunlight, did not show symptoms of NH_4 toxicity (Magalhaes and Wilcox, 1984b). Similar effect of light intensity on plant growth was observed by Ta and Ohira (1981) and Fernandes (1983; 1984). Etiolated radish seedlings also exhibited fewer symptoms of NH_4^+ toxicity than green seedlings (Goyal et al., 1982). In contrast, some authors have suggested that high illumination levels are necessary to avoid NH_4^+ toxicity, based on the obvious influence of light on the rate of photosynthesis and thus the production and availability of carbon skeletons (Givan, 1979; Barker and Mills, 1980). Magalhaes and Wilcox (1984b), have found that ammonium toxicity symptoms were increased with increased light intensity. Differences in levels of free NH_4^+ in shoots, could be a reason for higher NH_4^+ toxicity under high light intensity. Under full sunlight, the uncomplexed NH_4^+ content in tomato shoot tissue was found to be much higher than in shaded plants. Plants under higher light intensity have higher transpiration rate and have been reported to transport complexed NH_4^+ (Barker and Mills, 1980) and free NH_4^+ (Wilcox et al., 1977; Magalhaes et al. data not published) to the shoot in levels sufficient to trigger ammonia toxicity through their uncoupling of photophosphorylation (Ikeda and Yamada, 1981). However, in a study carried out by Magalhaes and Wilcox (1984b), the total amount of absorbed N was similar for NH_4^+-treated plants regardless of light intensity, so reduced absorption of NH_4^+ was not the factor reducing NH_4^+ toxicity under low light intensity. If the plant fails to assimilate the NH_4^+ taken up or generated in leaves, then NH_4^+ accumulates to toxic levels (Miflin and Lea, 1980). In support of this hypothesis, significant levels of free NH_4^+ (4 to 12 mM) were detected in tomato stem exudates by Wilcox et al. (1977). Another mechanism of NH_4^+ accumulation in shoots of plants under high light intensity could be photorespiration and the release of NH_4^+ from amides by deamination. Photorespiration, a physiological defense against light-oxygen toxicity, increases with increasing light intensity and also is greatly affected by water stress (Tolbert, 1980). The conversion of 2 mol glycine to 1 mol serine during photorespiration in the mitochondria, leads to the stoichiometric release of CO_2 and NH_3 (Martin et al., 1983). The NH_3 liberated from photorespiration is assumed to be reassimilated into glutamine via the glutamine synthetase/glutamate synthase pathway; however, the precise fate of NH_3 released during glycine oxidation within the cell remains unclear. It is suggested that this process can contribute to NH_4^+ accumulation found in shoots of plants grown under high light intensity, and our study showed that the glycine content in shoots was significantly lower under full sunlight as compared to low light, (Magalhaes and Wilcox, 1984b). Whatever the mechanism involved, it was clearly evident from the symptoms that NH_4^+ produced more toxicity in tomato plants under the highest light intensity, as was found for rice by Fernandes (1983; 1984). The overall amino acid content in the tomato shoot tissues under high light as reported by Magalhaes and Wilcox (1984b), is also in agreement with several other findings. Characteristically, NH_4^+ fed plants tended to accumulate larger amounts of amino acids, mainly asparagine and glutamine (Beevers, 1976; Givan, 1979; Fernandes, 1983; 1984). It has been reported that soluble amino acids in leaves under low light intensity are increased to the detriment of protein synthesis or due to protein breakdown to produce soluble NH_4^+ (Beevers, 1976). However, in NH_4^+- fed tomato plants both free amino acid and total reduced nitrogen contents increased with shading and indicates that shading did not suppress either continued amino acid incorporation or protein synthesis in NH_4^+-fed tomato plants, (Magalhaes and Wilcox, 1984b).

It is important to recognize the high efficiency of NH_4^+–fed tomato plants in incorporating NH_4^+ into amino acids at low levels of light. Asparagine and glutamine concentration in recently mature and expanding tip portions of NH_4^+-fed tomato plants increased 7- and 3-fold, respectively, under 67% shade; and the percentage of N as uncomplexed NH_4^+ was reduced 3-times as compared to full sunlight (Magalhaes and Wilcox, 1984b).

The NO_3^- accumulation in NO_3^- - fed plants with decreasing light intensity is well documented and could be explained in terms of decreased nitrate reduction in leaves under low light intensity (Canvin and Atkins, 1974). Magalhaes et al. (1974) also reported that the reduction of nitrate to ammonia in chloroplasts is a light-dependent reaction.

Conclusion about the utilization of various nitrogen forms needs careful evaluation because of interactions of environmental factors and plant species. It is suggested that light intensity is an important factor in plant response to NH_4^+-N and expression of toxicity symptoms. Further studies with combinations of the two nitrogen forms and different species under different light levels can provide increased understanding of the observed responses.

V. Simulated Waterlogging and its Effects on Nitrogen Form

The effect of ammonium in depressing plant growth as compared to nitrate has been widely reported (Anghinoni et al., 1988; Fernandes, 1984; Magalhaes and Wilcox, 1984a; Magalhaes and Huber, 1989a; 1989b).

However, no research has been done on the effects of oxygen depletion and ammonium assimilation in plants, although under waterlogging condition with no free oxygen, ammonium is the predominant, if not the sole, form of nitrogen available for plants. In complete anoxia, nitrate is lost by denitrification, and nitrifying bacteria are inactive (Middleton and Smith, 1979; Tills and Alloway, 1981). It is intricate the synergistic effect of O_2 deficiency and NH_4^+ nitrogen form in depressing maize growth, with nitrate enhancing plant growth under low O_2 level and NH_4^+ toxicity being alleviated in presence of high O_2 pressure (Table 2). Actually, in nature ammonium N is associated with anoxia in waterlogging condition. As flooded soil has no free oxygen, plants adapted to waterlogging, like paddy rice, must have an alternative mechanism to overcome ammonium toxicity and O_2 deficiency. It is notable that the maize cv. SARACURA, a genotype selected for waterlogging condition (EMBRAPA-Brazil), seems to be more efficient in ammonium assimilation, compared to the other maize genotypes tested, since SARACURA, showed best growth under NH_4^+ and O_2 stress, showing the least free ammonia accumulation in green tissue (Table 3) (Magalhaes et al., data not published), and less NH_4^+ toxicity (Magalhaes and Huber, 1989a; Magalhaes, 1991).

Table 2. Free ammonia in shoot of maize genotypes as affected by N form and O_2 level in aerated solution culture, with pH adjusted with $CaCO_3$

GENOTYPE	$+O_2$		$-O_2$	
	NO_3^-	NH_4^+	NO_3^-	NH_4^+
		n mol/gfw		
BR 201	730	1784	1098	2240
BR 107	692	1920	1160	2418
SARACURA	618	1410	906	1808
HS 20X22	830	2208	1246	3164
MEAN	717	1830	1102	2407
L.S.D (0.05) 326				

In paddy rice, a plant efficient in assimilating ammonia (Magalhaes and Huber, 1989a), their roots receive oxygen from aerial parts by gaseous diffusion through aerenchyma continuous with roots (Barber et al., 1962), or biochemically by a mechanism of oxygen secretion (the glycolic acid pathway of respiration), which may account for the strong oxidizing power of rice roots (Krizek, 1988). Species adapted to waterlogging, such as rice, have a greater ability to synthesize aminoacids under flooding condition than any other intolerant species (Krizek, 1988). This mechanism is likely to be linked to the higher ammonia assimilation efficiency in rice compared to other plant species. By supplying high level of O_2 to the root growing system, it alleviated ammonia toxicity, resulting in enhanced plant growth and decreased free ammonia content in the green tissues of maize treated with NH_4^+-nitrogen (Table 3). This suggests that oxygen in root systems is involved with ammonium assimilation. The glycolic acid pathway of respiration in roots (Krisek, 1988), not only provides oxygen but also provides carbon skeleton for ammonia incorporation into aminoacids. In this case, it is reasonable to conclude that oxygen is related to ammonia assimilation. However, another hypothesis needs to be involved to explain the effect of O_2 provided exogenously in increasing NH_4^+ assimilation as observed in roots of maize. Regardless of the mechanism involved, the tolerance to waterlogging and NH_4^+ assimilation efficiency appears to be related, although further research is needed for better understanding the physiological mechanisms.

Table 3. **Shoot dry weight of maize genotype as affected by nitrogen form and O_2 level in aerated solution culture with pH adjusted by using $CaCO_3$**

GENOTYPE	High O_2 $^+O_2$			Low O_2 $^-O_2$	
	NO_3^-	NH_4^+		NO_3^-	NH_4^+
		(g/12 plants)			
BR 201	7.13	6,54		6,74	4.72
BR 107	6.78	6,45		5,75	4.90
SARACURA	8.08	7,55		7,18	6.12
HS 20X22	8.05	6,05		6,36	4.12
MEAN	7.51	6,64		6,50	4.96

L.S.D.(0,05)   0.64

VI. Evidence of Exogenous Carbon Skeleton Effects on Increasing $^{15}NH_4^+$ Assimilation.

A high rate of photosynthesis and thus the production and availability of carbon skeletons is necessary to prevent the toxic effect of ammonia (Givan, 1979; Barker and Mills 1980). Reports that plants well supplied with carbohydrate are better able to utilize NH_4^+ (Reisenauer, 1978), are consistent with these ideas. Accordingly, it appears to be logical that supplying carbon skeleton exogenously to the root system would have a great effect in increasing ammonia assimilation by providing substrate for NH_4 incorporation into amino acids. To test this specific point, ammonia assimilation into amino acids and total reduced nitrogen, was monitored in both roots and shoots of two weeks old tomato seedling supplied with 5mM 99% ($^{15}NH_4)_2SO_4$

Table 4. Pool of free amino acids In tomato tissue of $^{15}NH_4^+$ grown plant treated with 1 mM α - keto glutarate (α KG) for 24 hours

Amino acid	ROOT		SHOOT	
	-αKG	+αKG	-αKG	+αKG
		n mols/gfw*		
Ala	1241	2129	280	827
Glu	149	201	175	190
B-Ala	47	58	15	37
Val	183	105	175	227
Thr	166	93	125	264
Ser	347	242	2393	4063
Leu	195	109	57	50
Ile	140	76	60	60
Gaba	1098	2209	206	962
Pro	123	70	48	20
Met	132	125	31	143
Asn	464	940	130	211
Asp	63	166	412	590
Phe	99	32	59	98
Gln	2984	5013	341	1130
Glu	452	1082	1277	2142
Lys	62	38	10	17
Mean	467	747	341	653

*Mean of two determinations

Table 5. TIme courses of changes In ^{15}N abundance of amino acIds, free NH_4^+ and total N In tIssues of $^{15}NH_4^+$ grown tomato treated with 1 mM α-Keto glutarate for 24 h.

Amino Acid	ROOT		SHOOT	
	-αKG	+αKG	-αKG	+αKG
		% labelled ^{15}N		
Ala	79.7*	86.1	37.6	64.5
Glu	35.7	58.6	37.2	59.5
B-Ala	30.5	48.7	27.0	35.5
Val	25.6	57.5	14.1	35.6
Thr	33.5	54.1	17.6	47.5
Ser	31.2	64.4	40.7	67.8
Leu	39.1	66.8	25.4	46.8
Ile	31.6	63.0	28.4	53.2
GABA	82.2	86.0	43.6	69.4
Pro	26.4	50.0	13.7	32.1
Met	33.6	58.9	22.6	42.1
Asn	46.2	75.9	23.1	56.7
Asp	54.2	76.3	35.4	65.6
Gln	71.7	85.5	42.8	69.1
Glu	70.7	83.5	37.1	67.0
Phe	25.3	58.7	27.1	51.3
Lys	21.4	43.4	12.6	30.2
Mean	43.4	65.7	27.0	52.6
NH_4^+	57.0	83.5	39.1	66.5
TN	24.4	40.0	10.5	21.1

TN = Total Nitrogen

*Mean of two determinations

Table 6. **Ammonium content in tomato tissue of NH_4^+ treated plants as affected by α-Keto Glutarate.**

Treatment Period (Days)	SHOOT		ROOT	
	NH_4^+ - αKG	NH_4^+ + αKG	NH_4^+ - αKG	NH_4^+ + αKG
	n mol/gfw*			
1	2847	1988	5232	6539
2	6125	1830	5322	1248
3	13363	1467	5463	1428
Mean	7445	1761	5339	2705

*Mean of four replicatoins.

via the aerated root medium in hydroponic culture, in the presence and absence of 1mM α -keto glutarate (α - KG). The pools of amino acids increased in average 2-fold in α - KG treated plants, compared to control, and the amino acids accumulated in response to α - KG, exhibited an isotope abundance 1.5 and 2-fold higher in roots and shoots respectively (Table 4, 5), (Magalhaes et al., 1992). Free ammonia accumulation decreased sharply in the presence of exogenous α - KG, being about 10-fold lower in the shoots of 13 days treated plants, compared to control (Table 6), (Magalhaes et al., 1992), and the total ^{15}N labelled nitrogen assimilation increased twice, showing a clear increase in $^{15}NH_4$ assimilation due to the effect of exogenous α - ketoglutarate, indicating carbon skeleton as a key limiting factor for ammonia assimilation.

Literature Cited

Anghinoni, I., Magalhaes, J.R., and Barber, S.A. 1988. Enzyme activity, nitrogen uptake and corn growth as affected by ammonium concentration in soil solution. *J. Plant Nutr.*, **11**: 131-144.

Barash, I.; Mor, H., and Sadon, T. 1975. Evidence for ammonium dependent de novo synthesis of glutamate dehydrogenase in detached oat leaves. *Plant Physiol.*, **56**: 856-858.

Barber, D.A.; Ebert, M., and Evans, M.T.S. 1962. The movement of ^{15}O through barley and rice plants. *J. Exp. Bot.* **13**: 397-403.

Barker, A.V., and Mills, M.A. 1980. Ammonium and nitrate nutrition of horticultural crops. In *Horticultural Review (Janick J. Ed.), Vol. 2 AVI, West Port, Conecticut.* pp. 395-423.

Barker, A.V., and Volk, R.J. 1964. Determination of ammonium amide, amino and nitrate nitrogen in extracts by a modified Kjeldahl method. *Anal. Chem.* **36**: 439-444.

Beevers, L. 1976. Nitrogen Metabolism in Plants. Elsevier, New York.

Ben-Zioni, A., Vaadia, Y., and Lips, S.H. 1970 Correlations between nitrate reduction, protein synthesis and malate accumulation. *Physiol. Plant.* **23**: 1039-1047.

Blevins, D.G. 1989. An overview of nitrogen metabolism in higher plants. In *Plant Nitrogen Metabolism.* (Conn. E.E. Ed.), Plenum Press, New York. pp. 1-42.

Canvin, D.T., and Atkins, C.A. 1974. Nitrate, nitrite and ammonia assimilation by leaves: Effect of light, carbon dioxide and oxygen. *Planta* 116: 207-224.

Davies, D.D. 1973. Metabolic control in higher plants. In *Biosynthesis and its Control in Plants*. (Milborrow B.V., ed.), Academic Press, London. pp. 1-20.

Dkjkshoorn, W., and Ismunadji, M. (1972). Nitrogen nutrition of rice plants measured by growth and nutrient content in pot experiments. 3. Changes during growth. *Neth. J. Agric. Sci.* 20: 133-144.

Fernandes, M.S. 1983. N-carriers, light and temperature. Influences on the free-amino acid pool composition of rice plants. *Turrialba.* 33: 297-301.

Fernandes, M.S. 1984. N-carriers, light and temperature. Influences on uptake and assimilation of nitrogen by rice seedlings. *Turrialba*, 34: 9-19.

Fernandes, M.S., and Rossiello, R.O.P. 1986. Aspectos do metablismo e utilizacao do Nitrogenio em gramineas tropicais. In *Calageme Adubacao de Pastagens. Potafos.* Piracicaba. S.P. pp. 93-123.

Gaudinova, A. 1983. The effect of NO_3^- and NH_4^+ ions on enzymes involved in nitrogen assimilation in *Pisum sativum* L. *Biol. Plant.* 25: 440-448.

Givan, C.V. 1979. Metabolic detoxification of ammonia in tissues of higher plants. *Phytochemistry* 18: 375-382.

Goyal, S.S., Huffaker, R.C., and Lorens, O.A. 1982. Inhibitory effects of ammoniacal nitrogen on growth of radish plants. 11. Investigation on the possible causes of ammonium toxicity to radish plants and its reversal by nitrate. *J. Amer. Soc. Hort. Sci.* 107: 130-135.

Handa, S., Warren, H.L., Huber, D.M., and Tasai, C.Y. 1984. Nitrogen nutrition and seedling development of normal and opaque-i genotypes. *Can. J. Plant Sci.* 64: 885-894.

Ikeda, M., and Yamada, Y. 1981. Dark CO_2 fixation in leaves of tomato plants grown with ammonium and nitrate as nitrogen sources. *Plant and Soil* 60: 213-222.

Kirkby, E.A., and Mengel, K. 1967. Ionic balance in different tissue of the tomato plant in relation to nitrate, urea, or ammonium nutrition. *Plant Physiol.* 42: 6-14.

Krizek, D.T. 1988. Plant response to atmospheric stress caused by waterlogging. *In Breeding Plants for Less Favorable Environments* (Christiansen M.N. and Lewis, C.F., Ed.), John Wiley and Sons, New York. pp. 293-325.

Lewis, D.A.M., Chadwick, S., and Withers, J. 1983. The assimilation of ammonium by barley roots. *Planta* 159: 483-486.

Magalhaes, J.R., and Wilcox, G.E. 1983a. Tomato growth and mineral composition as influenced by nitrogen form and light intensity. *J. Plant. Nutr.* 6: 847-862.

Magalhaes, J.R., and Wilcox, G.E. 1983b. Tomato growth, nitrogen fractions and mineral composition in response to nitrate and ammonium foliar sprays. *J. Plant Nutr.* 6: 911-939.

Magalhaes, J.R., and Wilcox, G.E. 1983c. Tomato growth and nutrient uptake patterns as influenced by nitrogen form and light intensity. *J. Plant Nutr.* 6: 941-956.

Magalhaes, J.R., and Wilcox, G.E. 1984a. Growth, free amino acids and mineral composition of tomato plants in relation to nitrogen form and growing media. *J. Amer. Soc. Hort. Sci.* 109: 406-411.

Magalhaes, J.R., and Wilcox, G.E. 1984b. Ammonium toxicity development in tomato plants relative to nitrogen form and light intensity. *J. Plant. Nutr.* 7: 1477-1496.

Magalhaes, J.R., and Huber, D.M. 1989a. Ammonium assimilation in different plant species as affected by nitrogen form and pH control in solution culture. *Fert. Res.* 21: 1-6.

Magalhaes, J.R., and Huber, D.M. 1989b. Maize growth and ammonium assimilation enzyme activity in response to nitrogen forms and pH control. *J. Plant Nutr.* 12: 985-996.

Magalhaes, J.R., Ju, G.C., Rich, P.J., and Rhodes, D. 1990. Kinetics of $^{15}NH_4^+$ assimilation in *Zea mays;* preliminary studies with a glutamate dehydrogenase (GDH1) Nul Mutant. *Plant Physiol.* 94: 647-656.

Magalhaes, J.R., and Huber, D.M. 1991a. Responses of ammonium assimilation enzymes to nitrogen treatment forms in different plant species. J. Plant. Nutr. **14**: 175-185.

Magalhaes, J.R., and Huber, D.M. 1991b. Free ammonia, free amino acids, and enzyme activity
in maize tissue treated with methionine sulfoximine. J. Plant. Nutr. **14**: 883-895.

Magalhaes, J.R. 1991. Kinectics of $^{15}NH_4$ assimilation in tomato plants: Evidence of NH_4^+ assimilation via GDH in tomato roots. J. Plant. Nutr. **14**: 1341-1354.

Magalhaes, J.R., Huber, D.M., and Tsai, C.Y. 1992. Evidence of increased ^{15}N-ammonium assimilation in tomato plants with exogenous α-ketoglutarate. Plant Sci. **85**: 135-141.

Magalhaes, A.R., Neyra, C.A., and Hageman, R.H. 1974. Nitrite assimilation and amino nitrogen
synthesis in isolated spinach chloroplasts. Plant Physiol. **53**: 411-415.

Marchner, H. 1986. Mineral Nutrition in Higher Plants. Academic Press, New York. 674 p.

Martin, F., Winspear, M.J., MacFarlene, J.D., and Oaks, A. 1983. Effect of methionine sulfoximine
on the accumulation of ammonia in C_3 and C_4 leaves. Plant Physiol. **72**: 177-181.

Matsumoto, H., Wakiuchi, N., and Takahashi, E. 1971. Changes of some mitochondrial enzyme
activities of cucumber leaves during ammonium toxicity. Physiol. Plant. **25**: 353-357.

Mengel, K., and Viro, M. 1978. The significance of plant energy status for the uptake and incorporation of NH_4^+-nitrogen by young rice plants. Soil Sci. Plant Nutr. **24**: 407-416.

Middleton, K.R., and Smith, G.S. 1979. A comparison of ammoniacal and nitrate nutrition of perennial ryegrass through a thermodynamic model. Plant Soil **53**: 487-504.

Miflin, B.J., and Lea, P.J. 1980. Ammonia assimilation. In The Biochemstry of Plants, vol. 5. (Stumpf,
P.K. and Conn, E.E. eds.), Academic Press, New York. pp. 169-202.

Miflin, B.J., and Lea, P.J. 1976. The pathway of nitrogen assimilation in plants. Phytochemistry.
15: 873-885.

Peet, N.M., Raper, Jr., C.D., Tolley, L.C., and Robarge, W.P. 1985. Tomato responses to ammonium and nitrate nutrition under controlled root-zone pH. J. Plant Nutr. **8**: 787-798.

Pill, W.G., and Lambeth, V.N. 1977. Effect of NH_4^+ and NO_3^- nutrition with and without pH adjustment on tomato growth, ion composition and water relations. J. Am. Soc. Hort. Sci. **102**: 78-81.

Prasad, R., Rajale, G.B., and Lekhdive, B.A. 1971. Nitrification retarders. Adv. Agron. **23**: 337-383.

Puritch, G.S., and Barker, A.V. 1967. Structure and function of tomato leaf chloroplasts during ammonium toxicity. Plant Physiol. **42**: 1229-1238.

Raven, J.A., and Smith, F.A. 1976. Nitrogen assimilation and transport in vascular land plants in
relation to intracellular pH regulation. New Phytol. **76**: 415-431.

Reisenauer, H.M. 1978. Absorption and utilization of ammonium nitrogen by plants. In Nitrogen in
the environment. Vol 2. (Nielsen D.R. and Macdonald, J.G. eds.), Academic Press, New York.
pp. 157-189.

Rhodes, D. 1987. Nitrogen Metabolism. Hort 6505. Purdue University, West Lafayette, p. 297.

Singh, R.P., and Srivastava, H.S. 1982. Glutamate dehydrogenase activity and assimilation of inorganic nitrogen in maize seedlings. Biochem. Physiol Planzen. **177**: 633-642.

Spiers, J.H. 1978. Effects of pH level and nitrogen source on elemental leaf content of "Tifblue"
rabbit-eye blueberry. J. Amer. Soc. Hort. Sci. **103**: 705-708.

Srivastava, H.S., and Singh, R.P. 1987. Role and regulation of l-glutamate dehydrogenase activity
in higher plants. Phytochemistry **26**: 697-610.

Ta, T.C., Ohira Ta, T.C. and Ohira, K. 1981. Effects of various environmental and medium conditions on the response of Indica and Japonica rice plants to ammonium and nitrate nitrogen.
Soil Sci. Plant Nutr. **27**: 347-355.

Tills, A.R., and Alloway, B.J. 1981. The effect of ammonium and nitrate nitrogen sources on copper
uptake and amino acid status of cereals. Plant and Soil. **62**: 279-290.

Tolbert, N.E. 1980. Photorespiration. In The biochemistry of plants, Vol. 2. (Stumpf, P.K. and Conn,
E.E. eds.), Academic Press, New York. pp. 487-523.

Tsai, C.Y., Huber, D.M., and Warren, H.L. 1978. Relationships of the kernel sink for N to maize productivity. *Crop Science.* **18**: 399-404.

Vernon, L.P., and Zang, W.S. 1960. Photo reduction by fresh and aged choroplasts: Requirements for ascorbate and 2, 6-dichlorophenol endophenol with aged chloroplasts. *J. Biol. Chem.* **235**: 2728-2733.

Wilcox, G.E., Mitchel, C.A., and Hoff, J.E. 1977. Influence of nitrogen form and exudation rate on ammonium amide and cation composition of xylem exudate on tomato. *J. Amer. Soc. Hort. Sci.* **102**: 192-196.

Yoneyama, T., and Kumazawa, K. 1974. A kinetic study of the assimilation of ^{15}N-labelled ammonium in rice seedling. *Plant Cell Physiol.* **15**: 655-661.

Yoneyama, T., and Kumazawa, K. 1975. A kinetic study of the assimilation of ^{15}N-labelled nitrate in rice seedling. *Plant Cell Physiol.* **16**: 21-31.

Nitrogen Nutrition in Higher Plants, 1995
Editors : H.S. Srivastava & R.P. Singh
Associated Publishing Co. New Delhi, India
pp. 145-164.

Nitrate Reductase

H.S. SRIVASTAVA

I. Introduction

Nitrate is the predominant form of soil nitrogen available to most plants. Salt Peter of Chili ($NaNO_3$), is a natural ore of nitrogen, which was mined extensively in early 20th century for extracting nitrogen for various commercial and agricultural uses. Urea and ammonium salts added to the soil as nitrogenous fertilizers increase nitrate content of the soil due to the nitrifying activities of soil bacteria; *Nitrosomonas* and *Nitrobacter* spp. Plants absorb and assimilate soil nitrate to build up their nitrogenous organic molecules. The first step in the assimilation of nitrate is its reduction by the enzyme nitrate reductase (NR, E.C. 1.6.6.1) to nitrite, which is considered to be the rate limiting step in the process. The excess of nitrate absorbed is accumulated inside the vacuoles in the cell. Most plant tissues are able to accumulate upto 1% of nitrate of their dry weight, although canola leaves can accumulate upto 2.2% of nitrate-nitrogen, under certain conditions (Srivastava and Fletcher, 1992). However, the nitrite and the ammonium, the sequential products of nitrate reduction are toxic and therefore, do not accumulate inside the cell under normal conditions. Thus, once the nitrate is reduced, the nitrogen is eventually incorporated into amino acids, proteins and other such nitrogenous molecules. A positive correlation between NR activity and organic nitrogen content in nitrate supplied plants, has been observed in many cases (Srivastava, 1980; Barvo et al., 1991).

The level of NR activity determines seed production also in several crops. It shows a positive correlation with the grain protein and grain nitrogen in wheat (Hernandez et al., 1974; Dalling et al., 1975). The enzyme itself however, represents a very small proportion, 0.02 to 0.05% of the total soluble protein (Calza et al., 1987). Thus, the activity of this enzyme plays a pivotal role in the supply of nitrogen and in the growth and productivity of plants, specially the cereals, that are the world's main food producers. The enzyme also plays an important role in the maintenance of the natural plant communities, as well. It has been estimated that on global basis, nitrate assimilation produces more than 2×10^4 megatons of organic nitrogen per year, as compared to 3×10^2 megatons for bilogical nitrogen fixation (Guerrero et al., 1981). Rightly so, the physiology, biochemistry and molecular biology of NR has been investigated perhaps more extensively than any other plant enzyme in the last 10 years or so. The probability of biotechnological application of this enzyme has also been now expressed. Robert B. Mellor and coworkers (1992) have used the purified immobilized enzyme from maize along with nitrite reductase and nitrous oxide reductase from *Rhodopseudomonas* in a electrobioreactor to reduce nitrate into nitrogen gas. This opens the possibility of application of the enzyme for reducing nitrate contamination of the ground water.

II. Enzyme Extraction and Assay

The presence and the level of enzyme activity in the plant tissues is usually determined by measuring the reduction of nitrate to nitrite, either by intact tissue (*in vivo* method) or by the extract of the tissue (*in vitro* method).

In *in vivo* assay, small segments of plant samples are incubated in a buffer containing appropriate amount of nitrate as substrate of the enzyme. The method was used for the first time by Kumada (1953) and the optimization of various assay conditions was described by Jaworski (1971). Although, various components of the assay may differ in different investigations, the fundamentals remain the same: that is to incubate excised plant organs in nitrate under darkness and to measure the nitrite produced and secreted out in the medium. Propanol or ethanol and Triton x-100 or Neutrony x-600 are added in the incubation medium as wetting and permeation agents. Freezing and thawing are also used sometimes to accelarate the secretion of nitrite from the tissue. The incubation is carried out in darkness and at a desired temperature (usually 25-30°C). Chloramphenicol is added sometimes in the incubation medium, to prevent bacterial contamination. Anaerobicity is usually achieved by one of the following methods: (a) By keeping the plant tissue immersed in the assay medium contained in a tightly stoppered vial. (b) By vacuum infiltration of the tissue or (c) By purging nitrogen gas in the medium. The nitrite secreted out in the medium is measured colorimetrically after diazotization with 1% sulfanilamide solution (in 1.5 N HCl) followed by the addition of 0.2% of aquous solution of naphthalene ethylene diamine dihydrochloride.

Procedure to detect NR activity by *in vitro* method was first described by Evans and Nason (1953). *In vitro* assay of the enzyme is done either with the crude or with purified enzyme preparations. In either case, the freshly harvested samples are usually frozen in liquid nitrogen and then extracted in cold; either with a pestle and mortor or with a tissue homogenizer for 1 to 2 min using a buffered extraction medium of controlled ionic strength and pH. Ethylene diamine tetra-acetic acid (EDTA) is also added in the medium for solubilizing the cell walls and membranes. Some other chemicals are also added sometimes in the extraction medium, to achieve some specific objectives (Table 1). The selection of these chemicals depends primarily on the plant tissue, from which the enzyme is to be extracted. The extract is cleared off by centrifugation in cold (0-4°C). The clear supernatant is used as a crude enzyme preparation for the assay of its activity. The assay mixture consists of a buffer, KNO_3 (substrate), NADH (coenzyme) and the enzyme preparation. The nitrite produced after the desired incubation period at a specified temperature is determined colorimetrically as in *in vivo* method.

Table 1. Some specific additives in the extraction medium for nitrate reductase and their role

Additive	Function
Bovine serum albumin and casein	Stabilize the enzyme by protecting against the action of endogenous proteases
Chymostatin	Inhibits specifically the protease action on NR
Dithiothreitol	Protects sulfhydryl groups of the enzyme
Flavin adenine dinucleotide	Stabilizes the enzyme by protecting against redox changes
Glutathione $+NiCl_2$	Removes inhibitory effects of endogenously present cyanide in *Sorghum*
Polyvinyl pyrrolidine	Adsorbs phenolic compounds in the extract, which otherwise inhibit enzyme activity
Phenyl methyl sulfonyl fluoride	Inhibits serine proteases and stabilizes the enzyme

The major precaution in *in vitro* enzyme assay is of temperature control. The enzyme is very unstable at room temperature and all operations but for the assay, must be performed in cold, preferably below 4°C. Sometimes the centrifugation of the extract is omitted to minimize NR decay between homogenization and assay (Tischler et al., 1978). In some cases, post assay treatment of the assay mixture is also done to remove excess of NADH, which inhibits colour development of nitrite.

Both, *in vivo* and *in vitro* procedures of determining NR activity have their own merits and demerits, but they often yield comparable data. An usual range of total activity is 2 to 5 μmol NO_2^- h^{-1} (g fr. wt)$^{-1}$. But some unusually high values of 44-45 μmol NO_2^- h^{-1} (g fr.wt.)$^{-f}$ have been measured by *in vivo* method in peanut leaves (Silberbush et al., 1988) and by *in vitro* method in maize leaves (Kenis and Campbell, 1989) under optimum conditions of enzyme induction. The selection of the procedure may depend upon the tissues to be analysed and the laboratory facilities. Perhaps it would be better to check the comparativity of the data by the two procedures, in one or two experiments, while performing most of the tests by one particular method.

III. Enzyme Purification

The purification of NR from higher plants had eluded the plant scientists for almost two decades, because of its extremely labile nature. Solomonson (1975) purified NR from the green alga *(Chlorella vulgaris)* by affinity chromatography. He demonstrated that the blue dye ligand Cibacron blue FG 3-A provided a semi specific affinity for NR. This method of using blue sepharose gel as an affinity material was subsequently adopted for enzyme purification from the tissues of higher plants. Zinc-chelate, 5′ AMP affinity chromatography and amphiphilic gel chromatography have also been used to purify the enzyme. In a typical purification procedure (for example, Campbell and Smarelli, 1978; Nakagawa et al., 1984), the enzymic protein from the crude extract is precipitated out with 50% $(NH_4)_2SO_4$. The protein is dissolved in a buffer, dialysed and then subjected to column chromatography, generally in a series of two to three columns. The first chromatography is done on hydroxypatite column and the second one on blue sepharose column (affinity column). Elution is performed with NADH or KNO_3 for which the enzyme has stronger affinity than for the column material. The selection of the column material, and of the equilibration and elution buffer is a meticulous exercise, aimed to achieve the maximum purification and recovery of the enzyme. Nakagawa et al. (1985) have used another column containing blue Toyopearl-M, along with hydroxy-patite brushite and blue sepharose columns to achieve a satisfactory purification of the spinach enzyme. Blue toyopearl is an amphiphilic gel, which was manufactured initially as an adsorbent for membrane proteins.

Electrophoretic purification of the enzyme, with an aim to separate its isozymes has also been attempted in a few cases (Heath- Pagliuso et al., 1984; Nakagawa et al., 1985). The electrophoresis is carried out in cold (0-4°C) using either acrylamide or hydrolysed potato starch as a gel. The location of the enzymic protein (s) in the gel is spotted by assaying the enzyme activity. The gel is incubated in a solution containing phosphate buffer, KNO_3 and reduced methyl viologen or NADH. The nitrate is reduced to nitrite by the enzyme and then the nitrite is determined colorimetrically. However, if only the protein bands are to be detected, and no enzyme activity, Coomasie brilliant blue is used as a staining agent.

Immunochemical purification and quantification of NR protein has also been achieved in many cases (Hyde et al. 1991). Columns loaded with monoclonal NR antibody, can be used for the purification of NR to apparent homogenity in a single step. The antibody is immobilized in the column by using blue sepharose. In the crude enzyme preparations, the enzyme can be quantified also by using antibodies. Schuster and Mohr (1990) have observed that NR activity in mustard cotyledons was strictly proportional to the amount of immunorespective material.

IV. Physico-chemical Nature

The enzyme from higher plants was first characterized by H.J. Evans and A. Nason (1953), which was shown to require reduced NAD (P) for a two electron reduction of nitrate to nitrite. The reaction as known today is as follows:

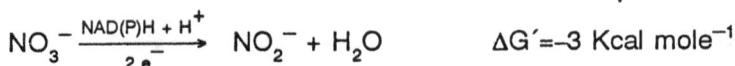

$$NO_3^- \xrightarrow[2e^-]{NAD(P)H + H^+} NO_2^- + H_2O \qquad \Delta G' = -3 \text{ Kcal mole}^{-1}$$

However, biochemical characterization of the enzyme could be achieved only in the 1980's when the enzyme could be purified and quantified using immunochemical methods. Three isoforms of NR have been described from soybean tissues (Streit et al., 1985; 1987). These are:

(i) A nitrate inducible NADH:NR (E.C. 1.6.6.1) with a pH optimum of 7.5.
(ii) A constitutive bispecific NADH/NADPH:NR (E.C. 1.6.6.2) with a pH optimum of 6.5 and
(iii) A constitutive NADH:NR (E.C. number not yet assigned) with a pH optimum of 6.5

The first two isoforms are usually found in close association and perhaps both contribute to the reduction and assimilation of nitrate. The constitutive NADH:NR consitutes about 12 to 20% of the total NR activity, but its absence does not impede nitrate utilization (Ryan et al., 1983)

There are evidences for NR polymorphism in other species as well. The monocots maize, barley and rice contain both, a NADH specific and a NAD(P)H bispecific NR (Kleinhofs et al., 1989). In barley and maize, the NAD(P)H:NR is present in roots only. In tree species *Erythrina senegalensis* and *Betula pendula*, there is no NADH:NR only NAD(P)H:NR, is present (Stewart and Orebamjo, 1979; Friemann et al., 1991). Different isoforms of NR are apparently controlled by different genes. In barley, while *nar*1 gene codes for NADH:NR, *nar*7 codes for NAD(P)H:NR (Warner et al., 1987).

The NADH:NR is the most ubiquitous and extensively studied isoform of the enzyme. Unless stated otherwise, the physicochemical characteristics described in the following paragraphs relate to this isoform.

The enzyme has been purified from several species of higher plants including barley, spinach, tobacco, squash and maize. It is present in almost all kinds of tissues, such as in aleurone layers, endosperm, scutellum, cotyledons, stems, petioles, ovules, fruits, developing pods of some legumes and so on (Srivastava, 1992). But the most active tissues are of roots and leaves. The relative contribution of the root or leaf in nitrate reduction depends upon the species (Andrews, 1986). Age of the plant, development of competing sinks and the light intensity can also modify the relative contributions of root or shoot (Oaks, 1992). There seems to be a great deal of similarity between the root and the leaf enzyme. In spinach, the comparison of root to leaf enzyme, using the Ouchterlony double diffusion technique reveals a high degree of similarity between the two (Nakagawa et al., 1986). However, Ferrario et al. (1983) have concluded that root and leaf enzymes were not the identical proteins.

Nitrate reductase from higher plants is a homodimer, with monomer size of 100-120 kD (Caboche and Rouze, 1990; Campbell, 1988). However, the *Chlorella* NR is a homotetramer with a dihedral symmetry (Howard and Solomonson, 1982). The amino acid sequence of monomeric unit is now available from several species. There are about 881 (bean) to 926 (spinach) amino acid residues in each unit (Hoffe et al., 1991). Acidic amino acids are more abundant than basic ones and the isoelectric point of the squash NR has been determined to be 5.7 (Redinbaugh and Campbell, 1981). The C-terminal sequence is more conserved among different speices than the N-terminal, indicating that the N-terminal part was not involved in the catalytic action of the enzyme. The enzyme from spinach has an interchain disulfide bond, which stabilizes the dimeric structure of the enzyme (Hyde et al., 1989). There are 9 to 19 cysteine residues in each sub unit. In bean, three cysteine residues, cys-167, cys-221 and cys-406 are probably involved in forming disulfide bridge (Hoff et al., 1992). The holoenzyme contains three prosthetic groups FAD, heme (cytochrome b_{557}) and Mo-pterin in a 1:1:1 stoichiometry per sub unit (Solomonson and Barber, 1990). The Mo atom is linked to pterin (a heterocyclic compound) via thiol linkage (Rajagopalan, 1989). It has been possible to assign three different regions in the protein, with respect to the binding with different prosthetic groups (Calza et al., 1987). The Moco factor domain of the protein is assigned to N-terminal half and the heme binding domain to the C-terminal part of the protein (Fig.1).

Tungsten, a metal classified with Mo in the periodic table can compete with Mo for incorporation into the enzyme protein, and thus the enzyme activity is suppressed in its presence (Heimer et al., 1969), although the synthesis of enzymic protein and the corresponding m-RNA is rather activated (Deng et al., 1989).

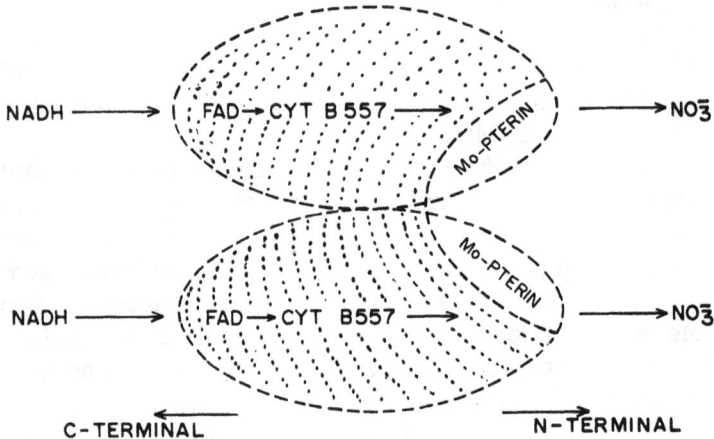

Fig. 1. Model of NADH nitrate reductase and electron transport.

A. Electron Transfer and Partial Activities

Nitrate reductase acts as an electron transport system, transferring electrons from NAD(P)H to nitrate through its various components, in the following sequence:

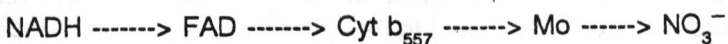

$$NADH \text{ -------> } FAD \text{ -------> } Cyt\ b_{557} \text{ -------> } Mo \text{ ------> } NO_3^-$$

Reduced NAD(P) donates two electrons to the initial acceptor FAD. Then the electrons flow through cytochrome b_{557} site to MoCo site and finally to nitrate, which is reduced to nitrite. Each unit seems to be acting independently as a catalyst for reducing nitrate to nitrite.

Evidences for the oxidation-reduction of the enzymic components involved in electron transfer are available. For example, the absorption spectrum of purified enzyme preparations from barley (Somers et al., 1982) and squash (Smarelli and Campbell, 1983) changes during NADH oxidation and nitrate reduction, indicating the involvement of cytochrome b_{557} in the process. Barber and Solomonson (1986) by measuring EPR spectra of Mo in NR from *Chlorella vulgaris* demonstrated that the metal was undergoing reduction-oxidation during the enzymic process. They observed that the reduction of Mo(VI) to Mo (V) and then to Mo (IV) involved addition of one proton in each step.

The source of reducing power (NADH) for *in situ* reduction of nitrate may be (a) the oxidation of glyceraldehyde 3-phosphate by the cytosolic glyceraldehyde 3-phosphate dehydrogenase (Klepper et al., 1971) (b) the oxidation of malate by the cytosolic malate dedydrogenase (Mann et al., 1978) or (c) the reduction of cytosolic NAD^+ via a malate-oxalacetate shuttle between the mitochondrial matrix and cytosol (Woo et al., 1980).

The enzyme also shows partial activities involving one or two of its prosthetic groups. One of the partial activities is dehydrogenase (diaphorase) action, which is perhaps catalysed by the proximal part (C-terminal) of the polypeptide, encompassing the flavin domain of the electron transport chain (Brown et al., 1981). The electron from NAD(P)H can be passed on to some acceptors other than the redox system of the enzyme and thus the enzyme can act as a NADH: ferricyanide reductase, a NADH: dichloroindophenol reductase and a NADH: cytochrome c reductase (Guerrero et al., 1981). The enzyme can reduce nitrate to nitrite using distal Mo-pterin part (N-terminal of the polypeptide) of the enzyme, employing reduced flavin or reduced methyl viologen as a reductant. Proteolytic cleavage of the holoenzyme by proteinases such as trypsin or V8 protease produces at least two enzyme fragments of 59 and 49 kD, which have partial enzyme activities (Kuo et al., 1980). In spinach, the target amino acid cleavage action of trypsin is arginine while that for V8 protease is the glutamate (Shiraishi et al., 1991). These two amino acids are apparently involved in linking the three functional domains of NR; FAD binding site, cytochrome b_{557} and Mo-pterin sites. The smaller segment (diaphorase segment) contains FAD and may use NADH as electron donor for the reduction of ferricyanide and other electron acceptors. The larger segment (terminal segment) contains Mo-pterin and may reduce nitrate to nitrite using artificial electron donors. The partial activities of the holoenzyme may or may not assume any physiological significance in higher plants (Srivastava, 1992). In spinach however, all partial activities are faster than the full NADH:NR activity (Barber and Notton, 1990). Cytochrome c reductase activity has been shown to be functional in many crude and purified enzyme preparations (Wray and Filner, 1970; Kuo et al., 1980). Daniel-Vedele et al. (1989) have suggested that NR was evolved from the fusion of three gene sequences coding each for one redox centre protein. Thus, there might be some wild ancestors of modern day plants, with functional components of NR.

B. Some Kinetic Properties

A two site ping pong steady state kinetic mechanism has been proposed for NADH: NR (Campbell and Smarelli, 1978). The transfer of electrons from heme to Mo is not the rate limiting step. As mentioned earlier, all partial activities of the enzyme are faster than the full NADH:NR activity. Histidine residue is believed to be involved in the binding of NADH; modification of histidines using diethyl pyrocarbonate inhibits the dehydrogenase activity, but not the terminal NR activity in the *Amaranthus* NR (Baijal and Sane, 1987). Apparently arginine residues are involved in the catalytic activity of FAD and MoCo domains of the enzyme (Baizal and Sane, 1988). Arginine and lysine residues are also involved in NADH binding with the enzyme (Sato et al., 1992).

In almost all the plants examined, NADH:NR has an optimum pH of 7.5. The K_m value for nitrate varies according to the species, but the normal range with purified enzyme preparations is 50 to 100 μM (Table 2). Higher values have been reported for other isoforms of the enzyme. The NAD(P)H:NR of *Erythrina senegalensis* has a K_m value of 10,000 μM (Stewart and Orebamjo, 1979), while the constitutive form of NADH:NR in soybean has a K_m NO_3^- of 15,000 μM (Li and Gresshoff, 1990). In soybean the K_m value for constitutive NADH:NR increased to 60,000 μM in the presence of 200 μM allantoic acid (a competitive inhibitor of NR) (Li and Gresshoff, 1990).

Table 2. Some kinetic properties of higher plant nitrate reductases

Plant material	Enzyme purification	K_m, μM NO_3^-	NADH	Vmax, μmol NO_2^- min^{-1} mg^{-1} protein	Reference
Cucurbita maxima	Affinity chromatography	40	5	-	Redinbaugh and Campbell, 1985.
Dunaliella parva	Crude	10	-	-	Heimer, 1976.
Glycine max	Crude	9500	-	-	Beevers et al., 1964.
Hordeum vulgare					
-roots	Affinity chromatography	130	2.6	6.2	Oji et al., 1988.
-leaves	Affinity chromatography	2.4-270	3.8	8.0	Kuo et al., 1980; Oji et al., 1988.
Nicotiana tabaccum	Affinity chromatography	0.2	7.5	-	Mendel and Muller, 1980.
Spinacea oleracea	Affinity chromatography	-	1.4	2.98	Sanchez and Heldt, 1989.
	Immunoaffinity chromatography	13	7.0	-	Barber and Naffon, 1990.
Suaeda maritima	Crude	90-110	50	-	Billard and Boucaud, 1982.
Zea mays	Affinity chromatography	70	-	-	Redinbough and Campbell, 1981.

Several natural and synthetic molecules, unrelated to the nitrate assimilation pathway metabolites affect kinetic properties of the enzyme. Synthetic compound phenylglyoxal, an antagonist of proton gradient, causes non-competitive inhibition of the spinach enzyme (Hucklesby and Blanke, 1991). The *in vitro* enzyme preparations from spinach is inhibited by ATP; but the activity is restored back after the removal of ATP by gel filtration and subsequent incubation of the enzyme with AMP (Kaiser and Spill, 1991). The enzyme preparation from barley leaves is inhibited by Ca^{2+} and to some extent by Mg^{2+} and is protected by EDTA and inorganic phosphate (De Cires et al., 1993). Thus, there is a strong possibility of the enzyme being regulated by phosphorylation-dephosphorylation mechanism, in *in vivo* conditions.

V. Sub-Cellular Location

The intra-cellular location of NR in higher plants is uncertain, although evidences favouring its cytosolic location seem to be rather overwhelming. But, there are reports of its location in plastids or in plasma membrane as well. Most early studies involving cell fractionation methods and differential centrifugation, indicated that the enzyme was located in the cytosol (Dalling et al., 1972; Harel et al., 1977). Immunogold localization tests of Vaughn and Campbell (1988) also confirmed the cytosolic location of maize NR. Suzuki et al. (1981) by using differential centrifugation and density gradient separation techniques demonstrated that the NR in the roots of five species namely barley, bean, maize, pea and rice was cytosolic. Vaughn et al. (1984) performed immunofluorescence localization tests in soybean cotyledons using anti-NR from *Chlorella fusca*. The enzyme appeared to be distributed in clusters throughout the cytoplasm. The cytosolic location of the enzyme may be circumstantially advantageous for the cell, as the physiological electron donor, the NADH, is also produced in the cytosol.

However, the non-cytosolic localization of the enzyme has also been demonstrated in a few cases. Kamachi et al. (1987) by employing immunogold localization technique demonstrated that the NR of spinach leaves was chloroplastic. Still in a few investigations, the membranal location of the enzyme has been indicated. Ward et al. (1989) for example, found about 19% of the total NR activity to be associated with the microsomal fraction of the maize roots.

VI. Gene Cloning

Genes coding for NR have been cloned and sequenced from fungi, algae and higher plants. In *Arabidopsis thaliana*, two genes for NR, nia 1 and nia 2 have been identified (Cheng et al., 1988; Wilkinson and Crawford, 1991). In bean, a root specific NR gene, PV*nr1* has been identified and cloned; and the possibility of another gene present in the leaf has been suggested (Hoffe et al., 1991). The complete sequence of gene is known from *Arabidopsis*, birch [NAD(P)H:NR], barley, bean, rice, tobacco, tomato, spinach and squash (Table 3). The gene contains about 3 to 5 kbp and has two to three introns. As in most other eukaryotic genes, all three introns in tomato NR gene begin with the dinucleotide GT and end with the dinuc-leotide AG (Daniel-Vedele et al., 1989). The average A+T content of these regions is also high, about 73%. The A+T content of the coding sequence is about 47.3%, as has been reported in bean (Hoffe et al., 1991).

Table 3. Some characteristics of nitrate reductase gene cloned from higher plants

Plant species and organ	Clone	Some characteristics	Reference
A. *Complete NADH : NR gene*			
Arabidopsis thaliana leaves	Genomic (*nia* 2)	3829 bp, two introns coding for 917 amino acids	Wilkinson and Crawford (1991).
Cucurbita maxima (squash)	cDNA	2754 bp plus a untranslatable region of 135 nucleotides at 3' end and a poly A tail of 16 nucle-otides, coding for 918 amino acids	Hyde et al., (1989).
Hordeum vulgare (barley) leaves	Genomic/ cDNA	7300 bp; one large intron of 2700 bp, coding for 915 amino acids	Schnorr et al., (1991).
Lycopersicum esculentum (tomato) leaves	Genomic	5300 bp, three introns, coding for 911 amino acids	Deniel-Vedele et al., (1989).
Nicotiana tabaccum (tobacco)	Genomic	6000 bp, three introns coding for 904 amino acids	Vaucheret et al., (1989).
Oryza sativa (rice) leaves	Genomic	5400 bp, three introns, coding for 916 amino acids	Choi et al., (1989).
Phaseolus vulgaris (bean) roots	Genomic	4600 bp, three introns coding for 881 amino acids	Hoffe et al., (1991).
Spinacea oleracea (spinach) leaves	cDNA	3284 bp, two introns, coding for 926 amino acids	Prosser and Lázarus, (1990).
B. *Partial NADH : NR gene*			
Cucurbita maxima (squash)	cDNA	1200 bp	Crawford et al., (1986).
Nicotiana tabaccum (tobacco)	cDNA	coding for 554 amino acids	Calza et al., (1987).
Spinacea oleracea (spinach)	cDNA	2324 bp, three introns, coding for 640 amino acids	Shiraishi et al., (1991).
C. *Complete NAD(P)H : NR gene*			
Betula pendula (birch) leaves	cDNA	3031 bp, coding for 898 amino acids	Friemann et al., (1991).
Hordeum vulgare (barley) leaves	Genomic	3538 bp, two introns, coding for 891 amino acids	Miya Zaki et al., (1991).

There is a high degree of similarity (63 to 91%) between nucleotide sequence from different species of higher plants. The plant NRs show about 40% homology to NR from fungus *Aspergillus nidulans* (Kinghorn and Campbell, 1989). There is also a great deal of similarity in nucleotide sequence among various related enzymes. Hyde et al. (1991) have suggested that enzyme such as NADH:NR, Cytochrome b_5 reductase, NADH: cytochrome b_5 reductase and sulfite oxidase, which have a great deal of homology, should be grouped as "flavoprotein pyridine nucleotide cytochrome reductase".

As is apparent from table 3, NAD(P)H:NR gene isolated from barley (Miyazaki et al., 1991) and birch (Friemann et al., 1991) is slightly smaller than the NADH: NR. In barley, which contains both NADH as well as NAD(P)H:NR genes, the NAD(P)H:NR gene is 3538 bp long as compared to 7300 bp in NADH:NR gene. Further, it encodes for 891 amino acids as compared to 915 amino acids in NADH: NR. The gene is however, larger than the *Aspergillus nidulans* NR gene, which encodes for 873 amino acids. In barley, the NAD(P)H:NR gene has 70% homology with the NADH:NR. It would be pertinent to mention that changes in only a few amino acids of another enzyme lipoamide dehydrogenase converts it from NADH specific enzyme to NADPH specific isoform (Scrutton et al., 1990). Perhaps, we should visualize a similar scenario for NR as well.

With successful gene cloning, genetic transformation with respect to NR expression has also been achieved. A *nia* gene consisting of a full length NR c-DNA from tobacco could be introduced into NR deficient mutants and the mutants made to grow on nitrate as sole source of nitrogen (Vincentz and Caboche, 1991; Gabard et al., 1987). Dorbe et al. (1992) have successfully complemented a tomato gene with tobacco. A cloned tomato NR gene was introduced into a plasmid vector, and then it was used for transforming a mutant of *Nicotiana plumbaginifolia* lacking NR, via *Agrobacterium* mediated gene transfer. The transgenic *N. plumbaginifolia* plants expressed from undetectable to 17% of the control NR activity in their leaves. Further, the introduced NR gene was regulated by light, and nitrate and showed circadian rythm as in tomato plants. However, a cloned *nia* 2 gene from *Nicotiana tabacum* when integrated in yeast, retained only cytochrome c reductase activity, but not NR activity (Truong et al., 1991). Elisa and Western blot analysis and other biochemical data show that the inserted gene in yeast has functional heme and FAD domains, but no Mo cofactor.

VII. Substrate Inducibility of the Enzyme

NADH:NR is a substrate inducible enzyme. Its activity is very low in the tissues from plants grown without nitrate; although in some cases other forms of inorganic nitrogen such as ammonium, support an appreciable enzyme activity. In intact seedlings, as well as in excised plant organs nitrate supply readily increases the enzyme activity, and as nitrate is depleted or removed the activity diminishes (Srivastava, 1980). This is one of the very few examples of substrate induction in higher plants, although this phenomenon is quite common in heterotrophs.

Several lines of evidences indicate that nitrate induces *de novo* synthesis of the enzyme. The inducer apparently acts at the transcription level. Increase in NR specific m-RNA following nitrate treatment has been recorded in barley and other species (Cheng et al., 1986; Meltzer et al., 1989). This appearance of m-RNA in the presence of nitrate appears to be independent of protein synthesis. In roots, scutella and leaves of maize plants, pretreatment of organs with cycloheximide prior to nitrate exposure does not prevent the appearance of m-RNA (Gowri et al., 1992). The appearance of m-RNA is followed by the appearance of NR-apoprotein. The decline in NR m-RNA occurs while the NR activity is increased. Further, the translation inhibitors, such as cycloheximide and puromycin prevent nitrate induced increase in NR activity. In *in vitro* experiments, the m-RNA from maize and tobacco could be translated into NR apoenzyme in cell free systems (Commemere et al., 1986; Calza et al., 1987). The availability of NR specific antibodies has lead to the immunological demonstration of NR protein formation during substrate induction of the enzyme. Western blotting and immunoelectrophoresis have shown that polypeptides which cross react with NR antibodies appear after a few hours of exposure of plants to nitrate (Somers et al., 1983; Remmler and Campbell, 1986). There is a good correlation between the amount of NR protein and the enzyme activity (Maki et al., 1986). The conversion of apoprotein to active enzyme molecule requires insertion of Mo atoms. Tungsten competes with the Mo for incorporation into enzyme complex and results in enzyme inactivation. But it does not prevent the expression of NR gene to form NR m-RNA and NR protein. In fact, in tobacco leaves, the supply of tungstate increases NR protein several folds and it has been suggested that tungsten causes over-expression of NR gene (Deng et al., 1989)

Nitrate appears to be involved at post-transcriptional stages of enzyme expression as well. As nitrate is depleted or removed from the medium, NR protein and NR activity decline (Galangau et al., 1988). In a few instances, the decline in activity has been reported to be faster than the decline in NR protein; which may indicate that the nitrate was involved also in the post translational modification of the enzyme. But the mechanism of the modification is unknown.

Chlorate, bromate and iodate, the structural analogues of nitrate, also induce the expression of NR gene. Chlorate supply to ammonium grown *Arabidopsis thaliana* plants increases NR m-RNA in leaves, but has no effect on NR protein and inhibits NR activity (LaBrie et al., 1991). Perhaps. It induces the NR gene by binding with the same receptor to which nitrate binds. However, chlorate acts as a competitive inhibitor of NR activity (Nakagawa and Yamashita, 1986). All the three halogenates chlorate, bromate and iodate are reduced by NR, although the reduction of bromate or iodate is less efficient than that of chlorate. Reduction of chlorate to chlorite is the chemical basis of its toxic action on plants. Chlorate resistant mutants of plants have either absent or impaired nitrate reduction.

VIII. Regulation by Ammonium and Amino Acids

Ammonium as an end product of nitrate reduction is expected to inhibit NR activity. This expectation is realized in micro-organisms (Marzluff, 1981), but not in higher plants. Moreover, in many higher plants, addition of ammonium in the nutrient solution causes a noticeable increase in enzyme activity (Srivastava, 1992). In maize roots and shoots, there is only a mild enhancement of NADH:NR, but the bispecific NAD(P)

H:NR in the roots in substantially enhanced by the addition of ammonium (Oaks and Long, 1991). The exact mechanism and the physiological significance of such an increase in NR activity is not understood. In some cases, specially where unsterilized culture and treatment conditions are employed, the increase may be linked to the oxidation of ammonium to nitrate by the contaminating nitrifying bacteria. In rice seedlings cultured in non sterile conditions, there is a substantial increase in leaf NR activity and also some in nitrate content during ammonium supply; and when nitrapyrin, an inhibitor of nitrification, is added in the culture medium the increase in nitrate content and in NR activity is reduced (Uhel et al., 1989). Further, ammonium causes very little increase in NR activity in sterile culture conditions. Poulle and Rambour (1985) by employing immunochemical technique have demonstrated that NR protein was present in ammonium cultured *Silene alba* cells also, although it was inactive. Perhaps, a small amount of nitrate was required to trigger the activation of this protein.

It is not known whether the NR protein formed in the presence of ammonium is NADH specific or NAD(P)H bispecific. Schuster et al. (1989) have separated as many as four isozymes of NR from *Sinapis alba* cotyledons, using anion exchange chromatography; namely NR_1, NR_2, NR_3, and NR_4. The ammonium supply induces NR_3 and NR_4 forms, which are not represented in the presence of nitrate.

Amino acids are other potential regulators of NR activity. In bean seedlings, when cotyledons, the source of amino acids are removed, the substrate induction of NR activity in the leaves is higher as compared to when cotyledons are left intact (Srivastava, 1975). The repression of NR activity by exogenously supplied amino acids has been demonstrated in quite a few cases. But the effect of individual amino acid on enzyme activity seems to be varying with the species and the tissues. Further, the effects have been more clear cut in cultured tissues and cells than in intact plant organs. Glutamine seems to be the critical regulator of the enzyme as it inhibits enzymes induction in cultured protoplasts of tobacco (Marion-Poll et al., 1984), cells of soybean (Curtis and Smarelli Jr., 1987) and so on. The other amide asparagine also inhibits the induction of NR in a few cases, although the inhibition is not as pronounced as with glutamine (Langendorfer et al., 1988). Diurnal rythm in glutamine and ammonium contents have also been reported, besides the one reported for NR activity and NR m-RNA in tobacco (Deng et al., 1991). There is an inverse correlation between the glutamine content and the NR m-RNA level.

The inhibition of NR activity by glutamine may be manifested at one or more of the following steps: (a) nitrate uptake (b) translocation of nitrate from storage to the metabolic pool (c) NR gene expression and (d) post translational modification of the NR protein. Heimer and Filner (1970) demonstrated that most amino acids which inhibited NR activity in cultured tobacco cells, inhibited nitrate uptake also. Inhibition of nitrate uptake by glutamine and many other amino acids has been demonstrated in squash seedlings (Langendorfer et al., 1988). The decrease in NR transcript in the presence of glutamine has also been reported (Langendorfer et al., 1900; Callaci and Smarelli Jr., 1991). In soybean, NR m-RNA starts increasing after 2 h of nitrate addition to the medium and reaches to a maximum level in 48 h. The glutamine supply suppresses this increase in m-RNA (Callaci and Smarelli Jr., 1991).

IX. The Possibilities of Alternate Functions of the Enzyme

Functions of NR other than the reduction of nitrate also appear to be important under certain conditions (Srivastava, 1992). Reasonably purified NR preparations from squash cotyledons and maize leaves are able to catalyse the reduction of ferric citrate (Campbell and Smarelli Jr., 1978; Redinbaugh, 1984). That NR and ferric citrate reductase were the functions of the same protein, was demonstrated by immunochemical tests and inhibitor studies. However, the optimum pH for Fe(III) reductase was 6.2 (Campbell and Smarelli Jr., 1978; Redinbaugh, 1984) which may suggest that perhaps the constitutive isoform of NR,which has a pH optimum of 6.5, was involved in Fe (III) reduction.

The dissimilatory function of NR has been indicated in several micro-organisms including green algae. These organisms can use nitrate as electron sink to unload their excess of reductant, and thus they can grow anaerobically. Garcia-Novo and Crawford (1973) and Reggiani et al. (1985) have suggested a dissimilatory role of NR in flood tolerant rice plants. During flooding, annoxia is created, but the oxygen generated from the reduction of nitrate to nitrite is able to oxidise NADH to NAD^+ for the continued operation of glycolysis and tricarboxyllic acid cylcle.

Reduction of chlorate to chlorite is another function of the enzyme, which has been exploited for selecting NR deficient mutants in bacteria, algae and higher plants. Apparently, the structural similarities between the chlorate and nitrate allow the former to compete for the active site on the enzyme. Although this function of NR may not have any physiological significance under normal environmental conditions, it may be an important factor in determining the form of inorganic nitrogen fertilizer for growing crops in soils contaminated with chlorate.

The possibility of NR being involved in the acquisition of nitrate from the medium was raised by Butz and Jackson (1977). This suggestion was based primarily on the kinetic similarities between the induction of NR and of nitrate uptake; although the nitrate transporter was never isolated. But for the involvement of NR in ion acquisition, the enzyme must be located on the membranes. As discussed earlier, although the sub-cellular location of NR is not fully resolved, the bulk of the data suggest its cytosolic location.

X. Conclusions and Future Prospects

Much advancement has been made in the understanding of the biochemistry and molecular biology of NR, during last 10 years or so. Characterization of NR deficient mutants, specially in *Arabidopsis thaliana*, barley and tobacco has contributed a great deal to the understanding. Recently developed techniques of immuno-characterization, the use of molecular probes and of gene cloning have helped a lot in unravelling the molecular biology of the enzyme. But there are still a few unanswered questions about the enzyme. One such aspect is the physical configuration of the enzyme. Lu et al. (1992) have been able to crystallize FAD domain of maize NADH:NR expressed in the bacterium *Escherichia coli* and have made a few crystallographic observations. Such observations with other domains of the enzyme are also required. The sub-cellular location of the enzyme is another aspect, which has to be investigated and the enigma resolved. The distribution, characterization and physiological role of the constitutive isoform (s) of the enzyme are also to be investigated. Certainly these and many other facets of the enzyme investigation may keep NR experts involved for many more years to come.

Literature Cited

Andrews, M. 1986. The partitioning of nitrate assimilation between roots and shoots of higher plants. *Plant Cell Environ.* **9**: 511-519.

Baijal, M., and Sane, P.V. 1987. Chemical modification of nitrate reductase from *Amaranthus. Indian J. Biochem. Biophys.* **24**: 75-78.

Baijal, M., and Sane, P.V. 1988. Arginine residue (s) at the active site (s) of the nitrate reductase complex from *Amaranthus. Phytochemistry* **27**: 1969-1972.

Barber, M.J., and Notton, B.A. 1990. Spinach nitrate reductase. Effects of ionic strength and pH on the full and partial enzymic activities. *Plant Physiol.* **93**: 537-540.

Barber, M.J., and Solomonson, L.P. 1986. Properties of the molydbenum domain of nitrate reductase. *Polyhedron* **5**: 577-580.

Barvo, F., Fontes, A.G., and Maldonado, J.M. 1991. Organic nitrogen content and nitrate and nitrite reductase activities in tritordeum and wheat grown under nitrate or ammonium. *Plant Soil* **135**: 251-256.

Beevers, L., Flesher, D., and Hageman, R.H. 1964. Studies on the pyridine nucleotide specificity of nitrate reductase in higher plants and its relationship to sulfhydryl levels. *Biochim. Biophys. Acta* **89**: 453-464.

Billard, J.P., and Boucaud, J. 1982. Effect of sodium chloride on the nitrate reductase from *Saueda maritima* var. marcocarpa. *Phytochemistry* **21**: 1225-1228.

Brown, J., Small, I.S., and Wray, J.L. 1981. Age dependent conversion of nitrate reductase to cytochrome c reductase species in barley. *Phytochemistry* **20**: 389-398.

Butz, R.G., and Jackson, W.A. 1977. A mechanism for nitrate transport and reduction, *Phytochemistry* **13**: 409-417.

Caboche, M., and Rouze, P. 1990. Nitrate reductase: a target for molecular and cellular studies in higher plants. *Trends Genet.* **6**: 187-192.

Callaci, J.J., and Smarelli, J. Jr. 1991. Regulation of inducible nitrate reductase isoform from soybeans. *Biochim, Biophys. Acta* **1088**: 127-130.

Calza, R., Huttner, E., Vincentz, M., Rouze, P., Gangau, F., Vaucheret, H., Cherel, I., Meyer, C., and Kronenberger, 1987. Complimentary to tobacco nitrate reductase m-RNA and encoding epitopes common to the nitrate reductase from higher plants. *Mol. Gen. Genet.* **209**: 552-562.

Campbell, W.H. 1988. Nitrate reductase and its role in nitrate assimilation in higher plants. *Physiol. Plant.* **74**: 214-219.

Campbell, W.H., and Redinbaugh, M.G. 1984. Ferric citrate reductase activity of nitrate reductase and its role in iron assimilation by plants. *J. Plant Nutr.* **7**: 799-806.

Campbell, W.H., and Smarelli, J. Jr. 1978. Purification and kinetics of higher plant nitrate reductase. *Plant Physiol.* **61**: 611-616.

Cheng, C., Dewdeny, J., Kleinhofs, A., and Goodman, H.M. 1986. Cloning and nitrate induction of nitrate reductase m-RNA. *Proc. Natl. Acad. Sci. U.S.A.* **83**: 6825-6828.

Cheng, C., Dewdeny, J., Nam, H., Denboer, B.G.W., and Goodman, H.M. 1988. A new locus (Nia 1) in *Arabidopsis thaliana* encoding nitrate reductase. *EMBO J.* **7**: 3309-3314.

Choi, H.K., Kleinhors, A., and An, G. 1989. Nucleotide sequence of the rice nitrate reductase gene. *Plant Mol. Biol.* **13**: 731-733.

Commemere, B., Cherel, I., Kronenberger, J., Galangau, F., and Caboche, M. 1986. *In vitro* translation of nirate reductase messenger RNA from maize and tobacco and detection with an antibody directed against the enzyme of maize. *Plant Sci.* **44**: 191-203.

Crawford, N.M., Campbell, W.H., and Davis, R.W. 1986. Nitrate reductase from squash: C-DNA cloning and nitrate regulation. *Proc. Natl. Acad. Sci. U.S.A.* **93**: 8073-8076.

Curtis, L.T., and Smarelli, J. Jr. 1987. Metabolic control of nitrate reductase activity in cultured soybean cells. *J. Plant Physiol.* **127**: 31-39.

Dalling, M.J., Halloran, G.M., and Wilson, J.H. 1975. The relation between nitrate reductase activity and grain nitrogen productivity in wheat. *Aust. J. Agric. Res.* **26**: 1-10.

Dalling, M.J., Tolbert, N.E., and Hageman, R.H. 1972. Intracellular location of nitrate reductase and nitrite reductase I. Spinach and tobacco leaves. *Biochim. Biophys. Acta* **283**: 505-512.

Daniel-Vedele, F., Dorbe, H.F., Caboche, M., and Rouze, P. 1989. Cloning and analysis of tomato nitrate reductase encoding gene: Protein domain structure and amino acid homologues in higher plants. *Gene* **85**: 371-380.

De Cires, A., De la Torre, A., Delgado, B., and Lara, C. 1993. Role of light and CO_2 fixation in the control of nitrate-reductase activity in barley leaves. *Planta* **190**: 277-283

Deng, M.D., Moureaux, T., and Caboche, M. 1989. Tungstate, a molybdate analog inactivating nitrate reductase, deregulates the expression of the nitrate reductase structural gene. *Plant Physiol.* **91**: 304-309.

Deng, M.D., Moueaux, T., Cherel, I., Boutin, J.P., and Caboche, M. 1991. Effect of nitrogen metabolites on the regulation and circadian expression of tobacco nitrate reducaste. *Plant Physiol. Biochem.* **29**: 239-247.

Dorbe, M-F., Caboche, M., and Daniel-Vedele, F. 1992. The tomato nia gene complements a *Nicotiana plumbaginifolia* nitrate reductase deficient mutant and is properly regulated. *Plant Mol. Biol.* **18**: 363-375.

Evans, H.J., and Nason, A. 1953. Pyridine nucleotide nitrate reductase from extracts of higher plants. *Plant Physiol.* **28**: 233-234.

Ferrario, S., Hirel, B., and Gadal, P. 1983. Immunochemical characterization of nitrate reductase from spinach leaves and roots. *Biochem. Biophys. Res. Commn.* **113**: 733-737.

Friemann, A., Brinkmann, K., and Hachtel, W. 1991. Sequence of a C-DNA encoding the bispecific NAD(P)H-nitrate reductase from the tree *Betula pendula* and identification of conserved protein regions. *Mol. Gen. Genet.* **227**: 97-105.

Gabard, J., Marion-Poll, A., Cherel, I., Meyer, C., Muller, A.J., and Caboche, M. 1987. Isolation and characterization of *Nicotiana plumabiginifolia* nitrate reductase deficient mutants. Genetics and biochemical analysis of the NIA complementation group. *Mol. Gen. Genet.* **209**: 596-606.

Galangau, F., Daniel-Vede, F., Moureaux, T., Dorbe, M-F., Leydecker, M.T., and Caboche, M. 1988. Expression of leaf nitrate reductase genes from tomato and tobacco in relation to light dark regimes and nitrate supply. *Plant Physiol.* **88**: 383-388.

Garcia-Novo, F., and Crawford, R.M.M. 1973. Soil aeration, nitrate reduction and flooding tolerance in higher plants. *New Phytol.* **72**: 1031-1039.

Gowri, G., Kenis, J.D., Ingemarsson, B., Redinbaugh, M.G., and Campbell, W.H. 1992. Nitrate reductase transcript is expressed in the primary response of maize to environmental nitrate. *Plant Mol. Biol.* **18**: 55-64.

Guerrero, M.G., Vega, J.M., and Losada, M. 1981. The assimilatory nitrate reducing system and its regulation. *Annu. Rev. Plant Physiol.* **32**: 169-204.

Harel, E., Lea, P.J., and Miflin, B.J. 1977. The localization of enzymes of nitrogen assimilation in maize leaves and their activities during greening. *Planta* **134**: 195-200.

Heath-Pagluiso, S., Huffaker, R.C., and Allard, R.W. 1984. Inheritance of nitrite reductase and regulation of nitrate reductase, nitrite reductase and glutamine synthetase isozymes. *Plant Physiol.* **76**: 353-358.

Heimer, Y.M. 1976. Specificity for NAD and NADP of nitrate reductase from the salt tolerant alga *Dunaliella parva*. *Plant Physiol.* **57**: 58-59.

Heimer, Y.M., and Filner, P. 1970. Regulation of nitrate assimilation pathway of cultured tobacco cells II. Properties of a variant cell line. *Biochim. Biophys. Acta* **215**: 152-165.

Heimer, Y.M., Wray, J.L., and Filner, P. 1969. The effect of tungstate on nitrate assimilation in higher plant tissues, *Plant Physiol.* **44**: 1197-1199.

Hernandez, H.H., Walsh, D.E., and Bauer, A. 1974. Nitrate reductase of wheat, its relation to nitrate fertilization. *Cereal Chem.* **51**: 330-336.

Hoffe, T., Stumman, B.M., and Henningsen, K.W. 1991. Cloning and expression of a gene encoding a root specific nitrate reductase in bean (*Phaseolus vulgaris*). *Physiol. Plant.* **82**: 197-204.

Hoffe, T., Stumman, B.M., and Henningsen, K.W. 1992. Structure, function and regulation of nitrate reductase in higher plants. *Physiol. Plant.* **84**: 616-624.

Howard, W.D., and Solomonson, L.P. 1982. Quarternary structure of assimilatory NADH: nitrate reductase from *Chlorella*. *J. Biol. Chem.* **257**: 10243-10250.

Hucklesby, D.P., and Blanke, M.M. 1991. Limitation of nitrogen assimilation in plants III. Nitrate assimilation in spinach as affected by a nitrate transport inhibitor. *Gartenbauwissenchaft* **56**: 264-266.

Hyde, G.E., Crawford, N.M., and Campbell, W.H. 1991. The sequence of squash NADH: nitrate reductase and its relationship to other flavoprotein oxidoreductases. A family of flavoprotein pyridine nucleotide cytochrome reductases. *J. Biol. Chem.* **266**: 23542-23547.

Hyde, G.E., Wilberding, J.A., Meyer, A.L., Campbell, E.R., and Campbell, W.H. 1989. Monoclonal antibody based immunoaffinity chromatography for purifying corn and squash NADH: nitrate reductases. Evidence for an interchain disulfide bond in nitrate reductase. *Plant Mol. Biol.* **13**: 233-246.

Jaworski, E.G. 1971. Nitrate reductase assay in intact plant tissues. *Biochem. Biophys. Res. Commun.* **43**: 1274-1279.

Kaiser, W.M., and Spill, D. 1991. Rapid modulation of spinach leaf nitrate reductase activity by photosynthesis II. *In vitro* modulation by ATP and AMP. *Plant Physiol.* **96**: 368-375.

Kamachi. K., Amemiya, Y., Ogura, N., and Nakagawa, H. 1987. Immunogold localization of nitrate reductase in spinach (*Spinacea oleracea*) leaves. *Plant Cell Physiol.* **28**: 333-338.

Kenis, J.D., and Campbell, W.H. 1989. Oxygen inhibition of nitrate reductase biosynthesis in detached corn leaves via inhibition of total soluble protein synthesis. *Plant Physiol.* **91**: 883-888.

Kinghorn, R., and Campbell, E.I. 1989. Amino acid sequence relationship between bacterial, fungal and plant nitrate reductase and nitrite reductase proteins. In *Molecular and Genetic Aspects of Nitrate Assimilation*, (J.L. Wray and J.R. Kinghorn, eds.) Oxford University Press, New York. pp. 285-403.

Kleinhofs, A, Warner, R.L., and Melzer, J.M. 1989. Genetics and molecular biology of higher plant nitrate reductases. In *Recent Advances in Phytochemistry* Vol 23 *Plant Nitrogen Metabolism*, J.E. Poulton, J.T. Romoeo and E.C. Conn, eds. Plenum Press, New York. pp. 117-155.

Klepper, L., Flesher, D., and Hageman, R.H. 1971. Generation of nicotinamide adenine dinucleotide for nitrate reduction in green leaves. *Plant Physiol.* **48**: 580-590.

Kumada, H. 1953. The nitrate utilization in seed embryos of *Vigna sesquipedalis*. *J. Biochem.* **440**: 439-450.

Kuo, T., Kleinhofs, A., and Warner, R.L. 1980. Purification and partial characterization of nitrate reductase from barley leaves. *Plant Sci. Lett.* **17**: 371-381.

LaBrie, S.T., Wilkinson, J.Q., and Crawford, N.M. 1991. Effect of chlorate treatment on nitrate reductase and nitrite reductase gene expression in *Arabidopsis thaliana*. *Plant Physiol.* **97**: 873-879.

Langendorfer, R.L. Walters, M.T., and Smarelli, J. Jr. 1988. Metabolite control of squash nitrate reductase. *Plant Sci.* **57**: 119-125.

Li, Z.Z., and Gresshoff, P.M. 1990. Developmental and biochemical regulation of constitutive nitrate reductase activity in leaves of nodulating soybean. *J. Expt. Bot.* **41**: 1231-1236.

Lu, G., Campbell, W.H., Lindqvist, Y., and Schneider, G. 1992. Crystallization and preliminary crystallographic studies of the FAD domain of corn NADH: nitrate reductase. *J. Mol. Biol.* **224**: 277-279.

Maki, H., Yamagishi, K., Sato, T., Ogara, N., and Nakagawa, H. 1986. Regulation of nitrate reductase activity in cultured spinach cells as studied by an enzyme linked immuno-absorbent assay. *Plant Physiol.* **82**: 739-741.

Mann, A.F., Hucklesby, D.P., and Hewitt, E.J. 1978. Sources of reducing power for nitrate reduction in spinach leaves. *Planta* **140**: 261-263.

Marion-Poll, A., Huet, J.C., and Caboche, M. 1984. Regulation of nitrate reductase in protoplast derived cells. Influence of exogenously supplied nitrate, ammonium and amino acids. *Plant Sci. Lett.* **34**: 61-72.

Marzluff, G.A. 1981. Regulation of nitrogen metabolism and gene expression in fungi. *Microbiological Reviews* **45**: 437-461.

Mellor, R.B., Ronnenberg, J., Campbell, W.H., and Diekmann, S. 1992. Reduction of nitrate and nitrite in water by immobilized enzymes. *Nature* **355**: 717-819.

Meltzer, J.M., Kleinhofs, A., and Warner, R.L. 1989. Nitrate reductase regulation: Effects of nitrate and light on nitrate reductase m-RNA accumulation. *Mol. Gen. Genet.* **217**: 341-346.

Mendel, R.R., and Muller, A.J. 1980. Comparative characterization of nitrate reductase from wild type and molybdenum cofactor defective cell cultures of *Nicotiana tabacum. Plant Sci. Letters* **18**: 277-288.

Miyazaki, J., Juricek, M., Angelis, K., Schnorr, K.M., Kleinhofs, A., and Warner, R.L. 1991. Characterization and sequence of a novel nitrate reductase from barley. *Mol. Gen. Genet.* **228**: 329-334.

Nakagawa, H., Poulle, M., and Oaks, A. 1984. Characterization of nitrate reductase from corn leaves (*Zea mays* L. cv. 64A x W182E). Two molecular forms of the enzyme. *Plant Physiol.* **75**: 285-289.

Nakagawa, H., Yamagishi, K., Yamashita, N., Sato, T., Ogura, N., and Oaks, A. 1986. Immunological chatacterization of nitrate reductase in different tissues of spinach seedlings. *Plant Cell Physiol.* **27**: 627-633.

Nakagawa, H., and Yamashita, N. 1986. Chlorate reducing activity of spinach nitrate reductase. *Agric. Biol. Chem.* **50**: 1893-1894.

Nakagawa, H., Yonemura, Y., Yamamoto, H., Sato, T., Ogura, N., and Sato, R. 1985. Spinach nitrate reductase. Purification, molecular weight and sub-unit composition. *Plant Physiol.* **77**: 124-128.

Oaks, A. 1992. A re-evaluation of nitrogen assimilation in roots. *Bioscience* **42**: 103-111.

Oaks, A., and Long, D.M. 1991. NO_3^- assimilation in root systems with special reference to *Zea mays* L. (cv. W64A x W 182E). In *Nitrogen Metabolism in Plants*, (K. Mengel and D.J. Pilbeam, eds.) Oxford Univ. Press, pp. 91-102.

Oji, Y., Takahashi, M., Nagai, Y., and Wakiuchi, N. 1988. NADH dependent nitrate reductase from two row barley roots. Purification, characteristics and comparison with leaf enzyme. *Physiol. Plant.* **72**: 311-315.

Poulle, M., and Rambour, S. 1985. Regulation of nitrate reductase in suspension cultures of *Silene alba*. Immunochemical appraoch. *Plant Sci.* **40**: 111-115.

Prosser, I.M., and Lazarus, C.M. 1990. Nucleotide sequence of a spinach nitrate reductase cDNA. *Plant Mol Biol.* **15**: 187-190.

Rajagopalan, K.W. 1989. Chemistry and biology of the molybdenum cofactor. In *Molecular and Genetic Aspects of Nitrate Assimilation*, J. Wray and J. Kinghorn, Eds. Oxford Univ. Press, New York. pp. 212-226.

Redinbaugh, M.G., and Campbell, W.H. 1981. Purification and characterization of NAD(P)H: nitrate reductase and NADH: nitrate reductase from corn roots. *Plant Physiol.* **88**: 115-20.

Redinbaugh, M.G., and Campbell, W.H. 1985. Quarternary structure and composition of squash nitrate reductase. *J. Biol. Chem.* **260**: 338-385.

Reggiani, R., Brambilla, I., and Bertani, A. 1985. Effect of exogenous nitrate on anaerobic metabolism in excised rice roots I. Nitrate reduction. *J. Expt. Bot.* 36: 1193-1198.

Remmler, J.L., and Campbell, W.H. 1986. Regulation of corn leaf nitrate reductase II. Synthesis and turnover of the enzyme activity and protein. *Plant Physiol.* 80: 440-442.

Ryan, S.A., Nelson, R.S., and Harper, J.E. 1983. Soybean mutants lacking constitutive nitrate reductase activity II. Nitrogen assimilation, chlorate resistance and inheritance. *Plant Physiol.* 72: 510-514.

Sanchez, J., and Heldt, H.W. 1989. On the regulation of spinach nitrate reductase. *Plant Physiol.* 92: 684-689.

Sato, Y., Shiraishi, N., Sato, T., Ogura, N., and Nakagawa, H. 1992. Arginine and lysine residues as NADH binding sites in NADH nitrate reductase from spinach. *Phytochemistry* 31: 2259-2262.

Schnorr, K.M., Juricek, M., Huag, C., Culley, D., and Kleinhofs, A. 1991. Analysis of barley nitrate reductase cDNA and genomic clones. *Mol. Gen Genet.* 227: 411-416.

Schuster, C., and Mohr, H. 1990. Photooxidative damage to plastids affects the abundance of nitrate reductase m-RNA in mustard cotyledons. *Planta* 181: 125-128.

Schuster, C., Schmidt, S., and Mohr, H. 1989. Effects of nitrate, ammonium, light and plastidic factor on the appearance of multiple forms of nitrate reductase in mustard (*Sinapis alba* L.) cotyledons. *Planta* 177: 74-83.

Scrutton, N.S., Berry, A., and Perham, R.N. 1990. Redesign of the coenzyme specificity of a dehydrogenase by protein engineering. *Nature* 343: 38-43.

Shiraishi, N., Kubo, Y., Takebo, G., Kiyoto, S., Sakano, K., and Nakagawa, H. 1991. Sequence analysis of cloned cDNA and proteolytic fragments for nitrate reductase from *Spinacea oleracea* L. *Plant Cell Physiol.* 32: 1031-1038.

Silberbush, M., Golau-Goldhirsh, A., Heimer, Y., Ben-Asher, J., and Lips, S.H. 1988. Response of peanuts (*Arachis hypogaea* L.) grown in saline nutrient solution to potassium nitrate. *J. Plant Physiol.* 132: 229-233.

Smarelli, J. Jr., and Campbell, W.H. 1983. Heavy metal inactivation and chelator stimulation of higher plant nitrate reductase. *Biochim. Biophys. Acta* 742: 435-445.

Solomonson, L.P. 1975. Purification of NADH nitrate reductase by affinity chromatography. *Plant Physiol.* 56: 853-855.

Solomonson, L.P., and Barber, M.J. 1990. Assimilatory nitrate reductase, functional properties and regulation. *Annu. Rev. Plant Physiol. Plant Mol. Biol.* 41: 225-253.

Somers, D.A., Kuo, T.M., Kleinhofs, A., and Warner, R.L. 1982. Barley nitrate reductase contains a functional cytochrome b $_{557}$. *Plant Sci. Lett.* 24: 261-265.

Somers, D.A., Kuo, T.M., Kleinhofs, A., Warner, R.L., and Oaks, A. 1983. Synthesis and degradation of barley nitrate reductase. *Plant Physiol..* 72: 949-952.

Srivastava, H.S. 1975. Distribution of nitrate reductase in ageing bean seedlings. *Plant Cell Physiol.* 19: 995-999.

Srivastava, H.S. 1980. Regulation of nitrate reductase activity in higher plants. *Phytochemistry* 19: 725-733.

Srivastava, H.S., 1992. Multiple functions and forms of nitrate reductase in higher plants. *Phytochemistry* 31: 2941-2947.

Srivastava, H.S., and Fletcher, R.A. 1992. Triademenol treatment of canola seeds increases nitrate reductase activity and nitrate content of leaves. *J. Expt. Bot.* 43: 1267-1271.

Stewart, G.R., and Orebamjo, J.O. 1979. Some unusual characteristics of nitrate reduction in *Erythrina senegalensis* D.C. *New Phytol.* 83: 311-319.

Streit, L., Martin, B.A., and Harper, J.E. 1987. A method for the separation and partial purification of the three forms of nitrate reductase present in wild type soybean leaves. *Plant Physiol.* 84: 654-657.

Streit, L., Nelson, R.S., and Harper, J.E. 1985. Nitrate reductase from wild type and nr1 mutant soybean (*Glycine max* (1.) Merr) leaves I. Purification, kinetics and physical properties. *Plant Physiol.* **78**: 80-84.

Suzuki, A., Gadal, P., and Oaks, A. 1981. Intracellular distribution of enzymes associated with nitrogen assimilation in roots. *Planta* **151**: 457-461.

Tischler, C.R., Purvis, A.C., and Jordon, W.R. 1978. Factors involved in *in vitro* stabilization of nitrate reductse from cotton (*Gossypium hirsutum* L.) cotyledons, *Plant Physiol.* **61**: 714-717.

Truong, H.N., Meyer, C., and Daniel-Vedele, P. 1991. Characteristics of *Nicotiana tabaccum* nitrate reductase protein produced in *Sachromyces cerevisie*. *Biochem. J.* **278**: 383-392.

Uhel, C., Roument, C., and Salsac, D. 1989. Inducible nitrate reductase of rice plants as a possible indicator for nitrification in water logged paddy soils. *Plant and Soil* **116**: 197-206.

Vaucheret, H., Vincentz, M., Kronenberger, J., Caboche, M., and Rouze, P. 1989. Molecular cloning and characterization of the two homologous genes coding for nitrate reductase in tobacco. *Mol. Gen. Genet.* **216**: 10-15.

Vaughn, K.C., and Campbell, W.H. 1988. Immunogold localization of nitrate reductase in maize leaves. *Plant Physiol.* **88**: 1354-1357.

Vaughn, K.C., Duke, S.O., and Funkhouser, E.A. 1984. Immunochemical characterization and localization of nitrate reductase in nor-flurazon treated cotyledons. *Physiol. Plant.* **62**: 481-484.

Vincentz, M., and Caboche, M. 1991. Constitutive expression of nitrate reductase allows normal growth and development of *Nicotiana plumbaginifolia* plants. *EMBO J.* **10**: 1027-1035.

Ward, M.R., Grimes, G.D., and Huffaker, R.G. 1989. Latent nitrate reductase activity is associated with the plasma membrane of corn roots. *Planta* **177**: 470-475.

Warner, R.L., Narayanan, K.R., and Kleinhofs, A. 1987. Inheritance and expression of NAD(P)H nitrate reductase in barley. *Theor. Appl. Genet.* **74**: 714-717.

Wilkinson, J.Q., and Crawford, N.M. 1991. Identification of the *Arabidopsis* CHL$_3$ gene as the nitrate reductase structural gene NIA2. *The Plant Cell* **3**: 461-471

Woo, K.C., Joniken, M., and Canvin, D.T. 1980. Reduction of nitrate via a dicarboxylate shuttle in a reconstituted system of supernatant and mitochondria from spinach leaves. *Plant Physiol.* **65**: 433-436.

Wray, J.L., and Filner, P. 1970. Structural and functional relationships of enzyme activities induced by nitrate in barley. *Biochem. J.* **119**: 715-725.

Nitrogen Nutrition in Higher Plants, 1995
Editors : H.S. Srivastava & R.P. Singh
Associated Publishing Co., New Delhi, India
pp. 165-188.

Nitrite Reduction

S.K. SAWHNEY

Abbreviations:—ENDOR, Electron nuclear double resonance spectroscopy; EPR, Electron paramagnetic resonance; NiR, Nitrite reductase, NR. Nitrate reductase.

I. Introduction

Nitrite reduction is the second step in the pathway of nitrate assimilation. Nitrite is, infact, the only known free intermediate formed during this eight electron reductive process of conversion of nitrate to ammonia. Being highly toxic, nitrite is rapidly reduced further to ammonia and, consequently, it rarely accumulates in any appreciable amounts in plant tissues (Hewitt et al., 1976; Beevers and Hageman, 1980). More than a century ago, Meyer and Schulze (1884) proposed that the overall process of nitrate reduction to ammonia entails four distinct reactions, each involving transfer of 2 electrons with nitrite, hyponitrite and hydroxylamine as the intermediates. Subsequent demonstration of activities of NR (Evans and Nason, 1953), nitrite reductase (Hageman et al., 1962) and hydroxylamine reductase (Nason et al., 1954) in plant extracts was, infact, consistent with such a multi-step mechanism. However, in early 1960's serious doubts began to arise about hydroxylamine as an intermediate in this pathway. Some of the observations in this regard were: (i) Cell free extracts were found to produce ammonia from nitrite at much faster rates than compared with hydroxylamine (Hageman et al., 1962; Hewitt et al., 1968; Hucklesby and Hewitt, 1970); (ii) The rate and stoichiometry of conversion of nitrite to ammonia was not affected at all by exogenously added hydroxylamine (Cresswell et al., 1962; Hewitt et al., 1968) and (iii) Reduction of hydroxylamine was susceptible to severe inhibition by nitrite (Cresswell et al., 1962; Hewitt et al., 1968). All these strongly argued against involvement of hydroxylamine as an intermediate during reduction of nitrite to ammonia. Finally, highly purified preparations of NiR, which were almost completely devoid of hydroxylamine reductase activity, were shown to rapidly and stoichiometrically convert nitrite to ammonia (Hewitt et al., 1968; Hucklesby and Hewitt, 1970). This provided a convincing evidence that a hydroxylamine reductase activity was not involved in reduction of nitrite to ammonia.

Though occurrence of hydroxylamine reductase has been reported by several workers (Nason et al., 1954; Hageman et al., 1962; Cresswell et al., 1965; Hucklesby and Hewitt, 1970; Sawhney and Nicholas, 1975) its role in plant metabolism remains obscure. It is also significant that till date neither hyponitrite, another intermediate proposed in the scheme by Meyer and Schulze (1884), nor its reductase has been detected in plant tissues. It is now well established that complete reduction of nitrite to ammonia is accomplished without formation of any free intermediates and that this remarkable transformation is mediated by a single enzyme viz. nitrite reductase (Hewitt et al., 1976; Beevers and Hageman, 1980). The present paper aims at presenting an overview of our current understanding of this facet of inorganic nitrogen metabolism in higher plants.

II. Uptake and Enzymic Reduction of Nitrite

A. Uptake

Nitrate and ammonium are the two principal forms of inorganic nitrogen available to plants from soil. Under natural field conditions, situations where nitrite is likely to be the predominant source of nitrogen, are virtually non-prevalent. Traces of nitrite formed in the soil during mineralization of organic nitrogen may, however, be absorbed by roots. Like nitrate, nitrite is also taken up by plants via a carrier-mediated, energy dependent and inducible system (Jackson et al., 1974; Aguera et al., 1990). Nitrogen starved seedlings on being transferred to nitrite-containing media initially absorb nitrite at very slow rates and after sometime this is followed by a phase of accelerated uptake of nitrite (Jackson et al., 1974; Aguera et al., 1990). Such a lag period is characteristic of an induced system and implies that development of the uptake system is initiated only in response to nitrite ions. This is supported by the observation that previous exposure of seedlings to either nitrate or nitrite ions eliminates the lag phase and such plants exhibit high initial rate of nitrite uptake. Furthermore, this stimulatory effect of pre-treatment with nitrate or nitrite depends on *de novo* synthesis as it is suppressed by cycloheximide (Aguera et al., 1990; Maldonado et al., 1990). Involvement of a specific carrier has been deduced from saturation kinetics and inhibition of nitrite uptake by a sulphydryl group reagent, N-ethylmaleimide (Aguera et al., 1990; Maldonado et al., 1990). Since nitrite absorption is significantly impeded under anaerobic conditions (Aguera et al., 1990), this process is apparently energy dependent. Available evidence also indicates that nitrite is absorbed through a system distinct from the one for uptake of nitrate. For example, in sunflower, ammonium interferes with absorption of nitrite without influencing that of nitrate (Maldonado et al., 1990). On the other hand, in these seedlings uptake of nitrate, but not of nitrite, is impaired by tungstate.

B. Enzymic Reduction

In plant tissues nitrite is rapidly reduced to ammonia by action of nitrite reductase. In most of the earlier studies on purification, the enzyme from diverse plant sources was recovered as a single symmetrical peak during ion exchange chromatography. However, two forms of the enzyme were isolated from scutella (Hucklesby et al., 1972) and from roots and etiolated leaves (Hucklesby et al., 1972; Hucklesby and Hageman, 1976) of maize. Green maize leaves, however, contained only one species of the enzyme. The two forms from scutella of maize seeds were identical with respect to molecular weight, electron donors, inducibility, pH optimum K_m for nitrite, products of reaction and sensitivity to chelators but differed in heat stability, isoelectric focussing and chromatographic behaviour (Hucklesby et al., 1972). More recent studies suggest that the phenomenon of occurrence of multiple forms of the enzyme in plant tissues might be more widespread than believed hither-to. Thus, presence of two distinct forms of NiR has been demonstrated in maize, *Triticum aestivum, Gallium aparine* (Kutscherra et al., 1987), wild oats (Heath-Pugliuso et al., 1984) and mustard cotyledons (Schuster and Mohr, 1990). DNA analysis of maize genome also indicates existence of at least two genes for NiR (Lahners et al., 1988). Interestingly, in *T. aestivum* and *G. aparine* (Kutscherra et al., 1987) and in mustard cotyledons (Schuster and Mohr, 1990) activity of one of the forms (NiR_1) was significant only during early stages of growth and its level was not influenced by either nitrate supply or light. On the other hand, NiR_2 was the predominant form during later stages of development and its activity was markedly enhanced by nitrate and light treatments.

Another interesting development in recent years is the discovery that in young soybean leaves nitrite can also be reduced to nitrogen oxides by the constitutive bispecific NR. The mutants which lack constitutive NR do not produce nitrogen oxides

(Nelson et al., 1983). Unlike normal inducible NR in plants, the constitutive enzyme from soybean leaves has a pH optima of 6.5 and reduces both nitrate as well as nitrite with either NADH or NADPH as an electron donor. It has a low K_m of 0.1 mM for nitrite and the rate of production of nitrogen oxides from nitrite far exceeds than that from nitrate. Reduction of nitrite appears to occur at the molybdenum site since nitrite-dependent evolution of nitrogen oxides is severely inhibited by azide (Dean and Harper, 1988; Campbell, 1990).

III. Properties of Nitrite Reductase

A. *Comparison of Enzyme from Different Sources*

Nitrite reductase from leaves and roots exhibit very similar properties and there is no evidence for the enzyme from these tissues being different proteins (Hirasawa et al., 1987a). Thus, NiR from barley roots and spinach leaves are almost similar with respect of their molecular weight, isoelectric point, pH stability and affinity for substrates and inhibitors (Ida et al., 1974). Comparison of purified enzyme from roots and leaves of pea also did not reveal any significant differences in their sensitivity to inhibitors, in light absorption spectra and electroparamagnetic resonance (Bowsher et al., 1988) and these were also immunologically indistinguishable in their cross reactivity with antibodies prepared against purified nitrite reductase from leaves of *Cucurbita pepo* (Bowsher et al., 1988; Wray et al., 1990). Similarly, the enzyme from roots of barley (Wray et al., 1990) and spinach (Hirasawa et al., 1987a) gave positive cross reaction with antibodies raised against the one from leaves of these plant species thereby implying that NiR from these tissues have common antigenic determinants.

Though the enzyme from different plant species exhibit almost identical physico-chemical and kinetic properties, there seem to be some structural variations between them as revealed by their immunological reactivities. During rocket immunoelectrophoresis, using monospecific antiserum against spinach ferredoxin NiR, the enzyme from all the eight examined dicotyledonous species behaved identically whereas that from monocotyledonous plants responded differently (Ida, 1987). Moreover, much larger amount of antiserum was needed to effect 50% inhibition of the enzyme from monocots than from the dicots. In immunoprecipitation tests, Ishiyama et al. (1985b) also found that antiserum against nitrite reductase from *Phaseolus angularis* inhibited the enzyme from leaves of spinach, comfrey, corn, barley and leek to varying extents. In immunodiffusion tests, the enzyme from spinach and comfrey leaves gave discrete precipitin lines but no such reaction was discernible with NiR from corn, barley and leek (Ishiyama et al., 1985b). The enzyme from *Cucurbita pepo, Zea mays, Hordeum vulgare* and *Brassica* rape was reported to be immunochemically non-reactive with antibodies for spinach NiR (Hirasawa et al., 1987a). These observations evidently suggest prevalence of significant differences in antigenic determinant regions of the enzyme from diverse plant species.

B. Physio-chemical Properties

According to most of the reports NiR from photosynthetic as well as non-photosynthetic tissues is composed of a single polypeptide chain of a molecular weight ranging from 60-64 kD (Hewitt et al., 1976; Losada and Gurrero, 1979; Lahners et al., 1988; Bowsher et al., 1988; Wray et al., 1990), with a sedimentation coefficient value around 4.65 (Losada and Guerreo, 1979; Ida and Mikami, 1986; Wada et al., 1986). The native enzyme from etiolated leaves of Phaseolus angularis, however, has been reported to comprise of two dissimilar sub-units of 35 and 64 kD (Ishiyama and Tamura, 1985). According to Hirasawa and Tamura (1980) the enzyme from spinach leaves has a molecular weight 85 kD and during purification it gets dissociated into two components each of 24 and 61 kD. It was proposed that the extensively documented nitrite reductase in the earlier literature, with molecular weight of around 61 kD, might infact, be a sub-unit of a larger native enzyme molecule (Hirasawa and Tamura, 1980; Hirasawa et al., 1982, 1987b). It was observed that the putative 85 kD enzyme has 3-4 times higher activity with ferredoxin than with methyl viologen as the reductant. On dissociation into 61 and 24 kD components during purification, the larger sub-unit continues to function efficiently with methyl viologen but exhibits low activity with ferredoxin. However, in careful studies on the enzyme from leaves of peas (Bowsher et al., 1988) and barley (Wray et al.,1990), no evidence could be obtained for 61 kD nitrite reductase being part of a larger complex. Obviously, critical and detailed investigations are needed to resolve this discrepancy.

Nitrite reductase from leaves of spinach (Ida and Mikami, 1986), Cucurbita pepo (Hucklesby et al., 1976) and barley (Wray et al., 1990) have quite similar amino acid composition and all of them show preponderance of acidic amino acids. The isoelectric point of the enzyme from maize scutellum (Hucklesby et al., 1972) and leaves (Wray et al., 1990) and roots of barley (Ida et al., 1974) is consistent with its acidic nature. Investigations on NiR cDNA clones isolated from spinach (Back et al., 1988) and maize (Lahners et al., 1988) indicate that the enzyme in spinach leaves is synthesized as a larger precursor having 594 amino acids out of which 32 amino acids, which constitute the leader sequence, get cleaved during post-translational processing. The enzyme contains about 10 half cysteine residues per mole and four of these are highly conserved and are believed to be involved in the formation of iron-sulphur clusters and one in binding of the prosthetic group siroheme (Mc Ree et al., 1986; Ostrowski et al., 1987; Wray et al., 1990).

Early studies on inhibitory action of certain metal chelators (Hucklesby et al., 1970) and on the influence of deficiency of micronutrients on level of nitrite reductase (Cresswell et al., 1965) implicated iron in the process of nitrite reduction. Subsequently, chemical analysis of purified NiR confirmed it to be an iron-containing enzyme. The reported values for iron content vary from 2-4.2 atoms of Fe per mole (Cardenas et al., 1972; Vega and Kamin 1977; Hirasawa and Tamura, 1980). These results have been interpreted as indicating 3 Fe atoms per mole with one atom associated with siroheme and the remaining two with labile sulphide to form a binuclear iron-sulphur centre (Hucklesby et al., 1976; Vega and Kamin, 1977). However, according to some of the reports the enzyme might, infact, contain a tetranuclear iron-sulphur centre (Lancaster et al., 1979; Ida and Mikami, 1986). The enzyme thus has two prosthetic groups, namely, iron-sulphur centre and siroheme (Hucklesby et al., 1976; Krueger and Siegel, 1982; Mc Ree et al., 1986) both of which contain iron. The only other enzyme known to contain siroheme is ferredoxin-sulphite reductase (Murphy et al., 1974) which shares a common property with NiR cat-alyisng a six electron reductive reaction. This unusual heme has been character-ized as iron tetrahydroporphyrin of isobacteriochlorin type with eight carboxylic acid containing side chains (Murphy et al., 1974). The exact number of siroheme in the enzyme remains doubtful with the earlier estimates of one siroheme per mole of enzyme (Ida and Mikami, 1986). However, according to Hirasawa et al. (1987b) the native ferredoxin NiR with molecular weight of 85 kD contains two moles of sir-oheme. As mentioned earlier, this enzyme comprises of two polypeptides of 61 and 24 kD and it is quite likely that the second siroheme molecule might be associated with the smaller sub-unit.

Purified NiR is reddish brown in colour (Hucklesby et al., 1976; Losada and Guer-rero, 1979). Purified enzyme from barley leaves exhibits absorption peaks at 280, 392, 574 and 693 nm with shoulders at 291, 530 and 638 nm (Wray et al., 1990). Almost similar spectral features have been reported for the enzyme from roots (Ishiyama et al., 1985b; Bowsher et al., 1988, Hucklesby et al., 1990) and leaves (Murphy et al., 1974; Hucklesby et al., 1976) of other plants.

Electron paramagnetic resonance spectra of the enzyme from leaves of pea (Bow-sher et al., 1988) and *Cucurbita pepo* (Cammack et al., 1978) and pea roots (Bow-sher et al., 1988) are almost identical with prominent signals at around g = 6.9, 5.0 and 1.95 which correspond to gz, gy and gx of a rhombically distorted high spin ferric heme. On reaction of dithionite-reduced enzyme with cyanide, the signals due to ferric heme are eliminated and another appears at g = 1.94, which is typical of a iron-sulphur protein (Bowsher et al., 1988; Hucklesby et al., 1990). During turn-over of the enzyme in presence of dithionite-reduced methyl viologen and nitrite, a third type axial spectra emerged with signals at g = 2.00 and 2.6 which corres-pond to nitrosyl-siroheme signals (Cammack et al., 1978; Bowsher et al., 1988; Huck-lesby et al., 1990; Wray et al., 1990). These EPR spectral studies provide a definitive proof of the siroheme and iron-sulphur nature of NiR from leaves as well as roots. The cysteine residues at positions 473, 479, 514 and 518 (spinach numbering) are highly conserved in maize and spinach enzyme and are believed to be involved in the formation of iron-sulphur clusters. Another cysteine residue appears to par-ticipate in binding of siroheme to the enzyme (Mc Ree et al., 1986; Ostrowski et al., 1987; Wray et al., 1990).

C. Catalytic Properties

Progress on characterization of nitrite reductase from higher plants was hamp-ered initially due to lack of proper understanding about the nature of electron donor as both pyridine nucleotides and flavin nucleotides were found to be ineffective (Hewitt et al., 1976; Beevers and Hageman, 1980). A major break through was achieved in 1962, when Hageman and his associates discovered that leaf extracts could reduce nitrite to ammonia in the presence of benzyl viologen, reduced either chemically or enzymatically with NADPH and a diaphorase (Cresswell et al., 1962; Hageman et al., 1962). The enzyme could also function with reduced methyl viologen (Cresswell et al., 1965; Hucklesby and Hewitt, 1970; Cardenas et al., 1972; Sawhney and Naik, 1973b). Subsequently, several groups (Hewitt and Betts, 1963; Huzisige and Satoh 1961; Losada, et al., 1963) almost simultaneously demonstrated that reduced ferredoxin was an effective electron donor and established that in photosynthetic tissues it functioned as a physiological reductant for the reduction of nitrite to ammonia. The enzyme from non-photosynthetic tissues can also utilize reduced viologen dyes or ferredoxin (Hucklesby et al., 1972; Beevers and Hageman, 1980). It is believed that the immediate reductant is probably a ferredoxin-like non-heme iron protein (Beevers and Hageman, 1980). A component, which is reduced by NADPH plus ferredoxin-NADP reductase system and is capable of replacing methyl viologen as an electron donor for nitrite reduction, has been isolated from extracts of etiolated bean stems (Ishiyama et al., 1985a). This electron carrier had a molecular weight of 34 kD and absorption maxima at 280 and 415 nm with shoulders at 340, 375, 525 and 565 nm. On reduction, the band at 415 nm shifted to 423 nm and a small peak appeared at 555 nm. Apparently properties of this protein differ from that of spinach ferredoxin. Cultured tobacco cells have also been shown to contain a carrier which upon reduction with NADPH plus ferredoxin-NADP reductase can reduce nitrite (Ninamiya and Sato, 1984). Its absorbance spectra, with peaks at 419, 459, 469 and shoulders at 552, 553 and 587 nm, is very similar to that of ferredoxin. On reduction, the peak at 419 nm gets diminished. EPR studies on reduced form of the carrier gave signal at g = 1.93 which conforms to that of iron-sulphur group. However, its molecular weight of 19,500 is distinctly higher than that of ferredoxin. Occurrence of a ferredoxin-like natural electron carrier which undergoes enzymic reduction with either NADH or NADPH has been demonstrated in maize roots (Suzuki et al., 1985). This carrier, on reduction with dithionite, was effective in nitrite reduction as well as in reaction catalysed by ferredoxin dependent glutamate synthase. More recently, Hucklesby et al., (1990) have reported presence of two electron carriers in pea root plastids which cross-react with antibodies against cucurbita leaf ferredoxin and have molecular weights of 13.9 and 14.9 kD. Interestingly, these carriers were detected only in the roots of pea plants fed with nitrate. Their possible role in nitrite metabolism is yet to be ascertained.

The probable sequence of events during ferredoxin-dependent reduction of nitrite has been described (Beevers and Hageman, 1980; Singh and Sawhney, 1989; Hageman, 1990). According to the available evidence from Mossbauer, ENDOR and EPR investigations, discussed in detail by Siegel and Wilkerson (1989), there is an electronic lap between iron-sulphur cluster and siroheme and that these two prosthetic groups are linked together via S atom of cysteine residue. Both of them participate in transfer of electrons from the reductant to nitrite. According to the postulated mechanism, first siroheme undergoes reduction by accepting an electron from ferredoxin and this enables binding of nitrite to form a NO_2^- siroheme complex. Formation of this complex then facilitates reduction of tetranuclear iron-sulphur centre and these electrons are then passed on to NO_2^- siroheme complex resulting in the formation of NO-siroheme. EPR analysis and spectral data are in conformity with formation of this NO-siroheme complex (Vega and Kamin, 1977; Bowsher et al., 1988; Hucklesby et al., 1990; Wray et al., 1990). The further sequence of addition of five electrons and hydrogen required to reduce nitrite to ammonia and also the nature of enzyme bound intermediates is not known (Hageman, 1990). However, the observation that complete reduction of iron-sulphur centre occurs only in the presence of inhibitors such as carbon monoxide or cyanide indicate that this prosthetic group functions as an intermediate carrier of reducing equivalents from ferredoxin to siroheme (Losada and Guerrero, 1979). Its higher negative mid-point redox potential than siroheme (Cammack et al., 1978) is in accord with the above postulated path of electron flow during nitrite reduction.

Cyanide (Cresswell et al., 1965, Hucklesby et al., 1970, Bowsher et al., 1988) and carbon monoxide (Murphy et al., 1974; Hucklesby et al., 1976; Hewitt et al., 1976; Vega and Kamin, 1977) are two very potent inhibitors of NiR activity. Activity of purified pea root enzyme is, infact, completely abolished by as low as 12.5 µM cyanide (Bowsher et al., 1988). In absence of nitrite, the reduced enzyme is extremely sensitive to carbon monoxide and its inhibitory effect is overcome by light (Ida et al., 1974; Murphy et al., 1974; Vega and Kamin, 1977). On reaction with carbon monoxide, the dithionite-reduced enzyme gives a derivative with absorbance bands 398, 589, 543 nm and on oxidation the peak at 589 shifts to 584 nm whereas the other two peaks change slightly in altitude but not in wavelength (Hucklesby et al., 1976). The loss of CO-spectra after oxidation, is typical of hemoproteins. It is believed that carbon monoxide forms an adduct with siroheme of the reduced enzyme (Murphy et al., 1974; Vega and Kamin 1977). That cyanide and carbon monoxide interfere with the enzyme activity by interacting with siroheme component is consistent with the observations that neither of them inhibits the activity in presence of nitrite and that cyanide acts as a competitive inhibitor with respect to nitrite (Ida et al., 1974).

Other inhibitors of NiR include reagents such as p-chloromercuric benzoate (Cresswell et al., 1965; Hucklesby et al., 1972; Ida et al., 1974b), phenylmercuric acetate (Ida et al., 1974) and mersalyl (Cresswell et al., 1965; Hucklesby et al., 1976) which block the sulphydryl groups. The adverse effect of phenylmercuric acetate and mersalyl can readily be overcome by glutathione provided it is added within 3 min but the enzyme gets irreversibly inactivated after 30-60 min treatment with these inhibitors (Hucklesby et al., 1976). An extended treatment with these inhibitors probably causes degradation of siroheme as the loss of enzyme activity is associated with irreversible alteration in absorption bands attributable to this prosthetic group (Vega and Kamin, 1977; Hucklesby et al., 1976). The activity of NiR from higher plants in general is not sensitive to metal chelators. The enzyme from roots of peas (Bowsher et al. 1988) and barley (Ida et al., 1974) and from maize scutellum (Hucklesby et al., 1972) has been reported to be inhibited by 8-hydroxyquinoline and diethyldithiocarbamate.

IV. Synthesis of Nitrite Reductase

A. Site of Synthesis

Nitrite reductase is a nuclear encoded chloroplastic enzyme (Heath-Pugliuso et al., 1984; Rajasekhar and Oelmuller, 1987; Redinbaugh and Campbell, 1991). In earlier studies with maize (Schrader et al., 1967) and rice leaves (Sawhney and Naik, 1973a) chloramphenicol, which interferes with protein synthesis on 70 S chloroplastic ribosomes but not on 80 S cytoplasmic ribosomes, was reported to preferentially suppress the formation of NiR without significantly affecting that of NR. Consequently, it was deduced that NiR in leaves was synthesized in chloroplasts. In rice leaves, in addition to chloramphenicol, the production of NiR as well as NR was impeded to the same extent by cycloheximide (Sawhney and Naik, 1972; 1973a). The results were interpreted to indicate that the formation of NiR required protein synthesis both on cytoplasmic and chloroplastic ribosomes. However, in subsequent investigations the appearance of NiR has been shown to be markedly suppressed by cycloheximide (Sluister- Scholten, 1973; Aguera et al., 1990; Maldonado et al., 1990; Sopory and Sharma, 1990), but not by chloramphenicol (Sluister-Scholten, 1973; Heath-Pugliuso et al., 1984; La Haba et al., 1988; Maldonado et al., 1990). The cause for this disparity in response of induced synthesis of NiR is not known. It is quite likely that in the earlier studies on maize (Schrader et al., 1967) and rice (Sawhney and Naik, 1973a) relatively high concentrations of the inhibitor used might have indirectly affected synthesis of the enzyme by interfering with other metabolic processes. The observation that plants with impaired chloroplastic protein synthesis, such as heat-treated 70 S ribosome deficient rye seedlings (Feierabend, 1986) or barley and maize mutants (Borner et al., 1986) which are deficient in chloroplastic ribosomes, retain capacity to form NiR lends further support to the view that this enzyme is synthesized in cytoplasm. In this context it is also relevant that analysis of cDNA clones has revealed that in spinach leaves a precursor of NiR, which comprises of 594 amino acids and has a leader sequence of 32 amino acids at the N-terminal end, is believed to be synthesized in cytoplasm (Back et al., 1988). This leader sequence of about 4 kD (Small and Grey, 1984; Rajasekhar and Oelmuller, 1987) most probably functions as a transit peptide and facilitates the transport of cytoplasmically synthesized NiR into chloroplasts. In pea leaves also two polypeptides which differ in their molecular mass but give positive cross reaction with antibodies for purified NiR have been detected and the one with higher molecular weight is probably a precursor of the enzyme (Gupta and Beevers, 1985).

B. Regulation of Synthesis

Nitrite reductase activity in plant tissues varies quite considerably in response to a number of factors. Effect of two of these namely, nitrogen source and light, which have a profound influence on level of the enzyme is discussed below.

Nitrogen Source:

Like NR, NiR is also an adaptive enzyme. Tissues of plants grown on nitrogen-free media or those supplied with reduced forms of nitrogen such as ammonium or urea exhibit very low enzyme activity (Aguera et al., 1987; La Haba et al., 1988; Maldonado et al. 1990). Nitrite reductase activity begins to increase quite markedly on supplying nitrate (Ingle et al., 1966; Hucklesby et al., 1972; Sawhney and Nicholas, 1975) and recedes again on its withdrawal (Gupta and Beevers, 1983; Hucklesby et al., 1990). Even in soybean leaves, which in addition to nitrate-inducible NR is exceptional in having a constitutive NR, very little NiR activity is detectable in seedlings grown on nitrate-free media (Nelson et al., 1984).

Whether nitrate is the true inducer of NiR or its effect is indirect due to generation of nitrite upon its reduction by NR, has been examined by several workers. In initial studies the observed lag phase in appearance of NiR with nitrate, in contrast to a linear increase in response to nitrite (Ingle et al., 1966), was considered to imply that nitrite rather than nitrate is the actual inducer. Other investigations, however, indicate that the nitrate-mediated increase in NiR activity can occur in the absence of functional NR. Thus, NO_3^- induced synthesis of NiR is not affected in tungstate-treated plants (Kelker and Filner, 1971; Maldonado et al., 1990). Secondly, NR deficient mutants have been found to possess comparable or even higher NiR activity than the wild type (Feenstra and Jacobsen, 1980; Warner et al., 1982). Thirdly, nitrate ions are not effective in inducing synthesis of NR in embryos of wheat seeds during first 24 h of germination due to prevalence of non-inducible phase, yet it causes considerable enhancement in activity of NiR (Gupta et al., 1983). All these observations lend support to the view that nitrate ions *per se* function as inducers of this enzyme. Infact, in the leaves of nitrite fed-barley seedlings nitrite ions have been proposed to promote synthesis of NiR indirectly after being oxidized to nitrate (Aslam and Huffaker, 1989). The results obtained on induced synthesis of mRNA in *Arabidiopsis thaliana*, however, seem to be at variance with direct action of nitrate in promoting synthesis of NiR. Exposure of ammonium-grown seedlings to chlorate significantly elevated m-RNA for NR in leaves whereas there was no such increase in mRNA for NiR (La Brie et al., 1992).

As mentioned earlier, occurrence of multiple forms of NiR in tissues of higher plants has been reported. In a few cases effect of externally supplied nitrate on levels of different forms of the enzyme has been investigated. In maize scutella, synthesis of both the forms was induced by nitrate (Hucklesby et al., 1972). In *Triticum aestivum* and *Gallium aparine* (Kutscherra et al., 1987) and in cotyledons of mustard (Schuster and Mohr, 1990) NiR_1, which showed significant activity during the initial stages of seedling development, was found to be synthesized constitutively, whereas the level of NiR_2, which was the predominant isoform during later stages was profoundly influenced by nitrate supply. Using clones of maize and spinach NiR, to monitor mRNA for NiR, it has been confirmed that uninduced leaves of both the species contain minute quantities of NiR mRNA and its amount increases significantly on induction with nitrate (Kramer et al., 1990). From these investigations, it was inferred that two distinct genes code for NiR and one of these is expressed constitutively.

Only a limited information is available about effect of other sources of nitrogen on nitrate or nitrite induced synthesis of NiR. In cotyledons of mustard (Rajasekhar and Mohr, 1986) nitrate-induced formation of the enzyme was suppressed by ammonium ions but no discernible change was recorded in the leaves of barley (Wray et al., 1990), maize (Rothstein et al., 1989) and cotyledons of sunflower (La Haba et al., 1988). However, in sunflower cotyledons, glutamine depressed development of the enzyme (La Haba et al., 1988).

Several workers have demonstrated that nitrate-dependent increase in activity of NiR is prevented by the inhibitors of transcription and translation (Ingle et al., 1966; Schrader et al., 1967; Sawhney and Naik, 1972; Hucklesby and Hageman, 1976; Gupta et al., 1983, Small and Gray, 1984) thereby implying that enhancement in enzyme activity involves its *de novo* synthesis. This has been confirmed by immunological studies. Cotyledons of mustard (Weber et al., 1990) and leaves of maize (Kramer et al., 1989), barley (Wray et al., 1990) and pea (Gupta and Beevers, 1985) from the seedlings raised in nitrate-free media contain very minute quantities of material capable of cross-reacting with antibodies against purified NiR and their amount as well as the enzyme activity increases steadily on transferring the seedlings to a nitrate or nitrite (Lahners et al., 1988) containing media, thus indicating that the increase in activity in response to either nitrate or nitrite is associated with elevated level of the enzyme protein. By using *in vitro* translational technique or cDNA clones coding for NiR, it has also been demonstrated that nitrate causes enhancement in level of mRNA for NiR (Small and Gray, 1984; Gupta and Beevers, 1987; Lahners et al., 1988; La Brie et al., 1992). Thus, the stimulatory effect of nitrate or nitrite, in general appears to be manifested at the level of gene expression. However, the observation that unlike in spinach leaves (Seith et al., 1991), nitrate had no effect on level of transcripts in mustard cotyledons (Schuster and Mohr, 1990), has led to the suggestion that the control of expression of the gene for NiR may vary in different plant species (Neininger et al., 1992). In wheat seeds also, during the initial stages of germination nitrate appears to act at translational level since the nitrate-induced formation of the enzyme is impaired by cycloheximide but not by 6-methyl purine (Gupta et al., 1983).

Light

In photosynthetic tissues light has a profound stimulatory influence on induced synthesis of nitrite reductase. Leaves of nitrate-supplied etiolated seedlings (Sawhney and Naik, 1972; Sluister-Scholten, 1975; Sawhney and Nicholas, 1975; Ogawa and Ida, 1987) or greened seedlings kept in darkness (Sawhney and Naik, 1972;

Gupta and Beevers, 1983; Ogawa and Ida, 1987) have very low activity of nitrite reductase. The enzyme activity gets markedly elevated on exposure of these seedlings to light and begins to decline when these are transferred back to darkness (Gupta and Beevers, 1983). This light mediated increase in activity appears to involve *de novo* synthesis of the enzyme and is abolished by inhibitors of protein synthesis (Sawhney and Naik, 1972; Sluister-Scholten, 1973; Small and Gray, 1984). This has been confirmed by immunological techniques as enhancement in enzyme activity is accompanied by appearance in higher quantities of material which cross-reacts with antibodies against purified nitrite reductase (Gupta and Beevers, 1984; Small and Gray, 1984; Schuster and Mohr, 1990; Weber et al., 1990). The precise role of light in induction of nitrite reductase has not been fully resolved. Three possible mechanisms have been postulated by various workers and these include: involvement of phytochrome, dependence on photosynthetic reactions and finally association with development of chloroplasts. According to several reports a brief exposure of either etiolated or green seedlings to red light of low fluence density causes substantial enhancement of nitrite reductase activity and, as expected of phytochrome-mediated response, this effect of red light is readily reversed by a brief irradiation with far-red light (Schuster et al., 1987; Sopory and Sharma, 1990; Neininger et al., 1992). In etiolated maize seedlings, nitrite reductase fully escapes photoreversibility by far red light 2 h after treatment with red light (Sopory and Sharma, 1984). Both LIR and HIR mechanisms have been implicated in this response (Sopory and Sharma, 1984; Rajasekhar and Oelmuller, 1987; Sharma and Sopory, 1990; Neininger et al., 1992). The latter effect is elicited upon continuous exposure of plants to far red light and is manifested through maintenance of low level of phytochrome in its Pfr form for longer periods. Activity of nitrate reductase is also elevated by continuous, but not short pulses of blue light (Rajasekhar and Sopory, 1985; Sopory and Sharma, 1990). The response to blue light is not mediated via phytochrome because brief irradiation with red light causes further enhancement in the activity (Rajasekhar and Sopory, 1985).

Available evidence indicates that nitrate and phytochrome act independently of each other in inducing the synthesis of nitrite reductase. For example, after attainment of maximum activity in response to nitrate in dark, an irradiation with red light causes a still further enhancement in the activity (Sharma and Sopory, 1984). Conversely, red light treatment given prior to nitrate supply, promotes nitrate-induced formation of the enzyme in the dark. These observations evidently suggest that phytochrome initiates a biochemical signal distinct from that by nitrate ions. The molecular basis for effects of light and nitrate seemingly vary with plant species. In cotyledons of mustard (Schuster and Mohr, 1990) and pea leaves (Gupta and Beevers, 1987) the level of *in vitro* translatable mRNA for nitrite reductase increased in response to irradiation but not by nitrate. The latter was nonetheless essential for the appearance of the enzyme protein. Thus, in these tissues transcription of the gene was regulated by phytochrome, and nitrate ions somehow or the other promoted the translational process. In contrast, in spinach the amount of mRNA for nitrite reductase was influenced by nitrate and not by light (Seith et al., 1991). Yet in another study on tobacco cotyledons (Neininger et al., 1992), for a substantial increase in transcripts for nitrite reductase both light and nitrate were required and neither of them alone was effective.

According to another view, action of light on nitrite reductase is linked with capacity of leaves to carry out photosynthesis. Thus, the effect of light on the enzyme is hardly noticeable in plants lacking functional chloroplasts such as etiolated seedlings greened in presence of chloramphenicol (Sawhney and Naik, 1972; Sluisters-Scholten 1975) or albino mutants (Sawhney et al., 1972; Warner and Kleinhofs, 1974) and is also significantly suppressed when photosynthetic activity is impaired by treatments such as DCMU or simazine application (Sawhney and Naik, 1972; Sluister-Scholten, 1975) or on placing the plants under CO_2-free atmosphere (Sawhney and Naik, 1972; Gandhi et al., 1974). Since in rice seedlings exogenously supplied sugars were ineffective in substituting the requirement of light, it was deduced that the formation of nitrate reductase as well as nitrite reductase was perhaps influenced by Hill activity-mediated changes in redox state of photosynthetic tissue (Sawhney and Naik, 1972). Contradictory results were obtained in *Phaseolus vulgaris* where exogenous supply of sugars led to restoration of the enzyme activity in DCMU-treated seedlings (Sluisters-Scholten, 1975). Schuster et al. (1987) have, however, concluded that in mustard cotyledons the action of light does not require active photosynthesis since light treatment given prior to nitrate application strongly promotes NO_3^- induced synthesis of the enzyme during subsequent dark period and this response to light remains cryptic upto 12 h in darkness. As emphasized elsewhere (Sawhney and Naik, 1990), the nature of tissue, its developmental stage and physiological state may be important factors and may account for the differences in the obtained results. This is illustrated by the observation that even though the competence to respond to light is retained in aged mustard cotyledons, the inducibility of nitrate reductase as well as nitrite reductase in dark by stimulus from prior light treatment diminishes with age (Schuster et al., 1987).

Even though nitrite reductase is a nuclear encoded enzyme and is synthesized in cytoplasm, its response to light requires presence of undamaged chloroplasts. Thus, leaves of nitrate-supplied albino mutants grown in light, which obviously lack normal chloroplasts, possess very low activity of nitrite reductase (Sawhney et al. 1972; Warner and Kleinhofs, 1974). Retardation of chloroplast development by pretreating etiolated seedlings with chloramphenicol, an inhibitor of protein synthesis on 70 S ribosomes, prior to their being transferred to light also suppresses synthesis of nitrite reductase to a far greater extent than the effect of the antibiotic on light-grown seedlings (Sawhney and Naik, 1972). Similarly, chloramphenicol prevented formation of the enzyme when provided to etiolated *Phaseolus vulgaris* seedlings at the beginning of light treatment but it was not effective if given to the plants after 24 h of exposure to light (Sluisters-Scholten, 1975). A more convincing evidence is provided by the failure of light to evoke stimulation of enzyme in cotyledons of tobacco (Neininger et al., 1992), mustard (Rajasekhar and Mohr, 1986; Schuster and Mohr, 1990) and spinach (Seith et al., 1991) in which chlo-ʾ roplasts were photo-xidatively damaged by treatment with non-fluorozan. The need for undamaged chloroplasts has been ascribed to generation of a plastidic signal-factor (Schuster and Mohr, 1990; Seith et al. 1991) which is postulated to exert a positive control over transcription of nuclear encoded genes of the proteins des-tined for chloroplasts (Oelmuller, 1989). This plastidic signal/factor is presumed to emanate concomitantly with onset of chloroplast development. However, the exact nature of this postulated factor and its molecular mechanism of action remains to be established.

V. Nitrite Reduction in Photosynthetic and Non-Photosynthetic Tissues

Both chlorophyllous and non-chlorophyllous tissues are capable of reducing nitrite (Beevers and Hageman, 1980; Singh and Sawhney, 1989). In leaves of C_4 plants, bulk of nitrite reductase is contained in mesophyll cells, though a small activity is detected in bundle sheath cells as well (Mellor and Tregunna, 1971; Neyra and Hageman, 1974; Hirel et al., 1977; Schnutz and Brunold, 1985). However, in *Panicum maximum* most of the enzyme activity has been reported to be associated with bundle sheath cells (Rathnam and Edwards, 1976). Cell fractionation studies on leaves of C_3 (Sawhney and Nicholas, 1975; Wallsgrove et al., 1979; Wada et al., 1986) and C_4 (Neyra and Hageman, 1974; Rathnam and Edwards, 1976; Hirel et al., 1977) reveal that NiR is located almost exclusively in chloroplasts. The pos-sibility that a small proportion of the enzyme might be in cytoplasm, however, cannot be ruled out. In this context it is pertinent to note that out of the two forms of the enzyme viz. NiR_1 and NiR_2 in mustard cotyledons, formation of NiR_2 but not NiR_1 gets abolished on photoxidative damage to chloroplasts by nor-fluorozan treat-ment. It has been suggested that NiR_2 is chloroplastic whereas NiR_1, which is syn-thesized constitutively during early seedling development, is an extrachloroplastic enzyme (Schuster and Mohr, 1990). In roots (Dalling et al., 1972; Washitani and Sato, 1977; Emes and Fowler, 1979) and cultured cells of tobacco (Washitani and Sato, 1977) the enzyme activity is recovered almost exclusively in the proplastidic fraction.

Though NiR in both the types of plant organs is located in plastids, the metabolic systems for supplying physiological reductant for nitrite reduction in non-photo-synthetic tissues and in photosynthetic tissues in darkness, is distinctly different from that in actively photosynthesizing tissues.

A. Photosynthetic Tissues

The profound stimulatory effect of ight on the rate of nitrite reduction in leaves is well documented (Losada and Guerrero, 1979; Beevers and Hageman, 1980;

Sawhney and Naik, 1990). This is evident from the observation that nitrite accumulated during *in vivo* assay of NR in dark, disappears rapidly on transferring the leaves to light (Jones and Sheard, 1978; Reed and Canvin, 1982; Prakash et al., 1984; Watt et al., 1987). The direct role of light in nitrite reduction in the photosynthetic tissues became apparent with the establishment of chloroplastic location of the enzyme and more importantly with the discovery that it requires reduced ferredoxin for its activity (see Section III C). Reduction of nitrite in illuminated leaves is thus directly coupled to photochemical reactions of photosynthesis. This is well supported by the observation that photoreduction of nitrite is markedly retarded by inhibitors of photosynthetic electron transport chain such as DCMU and simazine (Atkins and Canvin, 1975; Neyra and Hageman, 1974; Klepper, 1978) but is not affected by uncouplers of oxidative phosphorylation (Reed and Canvin, 1982). Nitrite, infact, acts as a typical Hill reagent in photochemical reactions of photosynthesis as denoted by stoichiometric relationship of 1: 1.5: 3:1 between nitrite reduced, oxygen evolved, ATP produced and ammonia formed (Losada and Guerrero, 1979). It has also been established that reduction of nitrite is functionally associated with photosystem I since this process is impeded by the presence of electron carriers such as phenazine methonosulphonate and methyl viologen (Neyra and Hageman, 1974) and disalicylidene propanediamine (Trebst and Burba, 1967). Furthermore, chloroplast grana with heat inactivated photosystem II readily photoreduce nitrite in the presence of dichlorophenol indophenol and ascorbate (Hewitt et al., 1968; Hucklesby and Hewitt, 1970). Nitrite also competes for reducing equivalents from ferredoxin as photoreduction of NADP and CO_2 fixation got depressed in its presence (Hewitt et al., 1968; Magalhaes et al., 1974; Baysdorfer and Robinson, 1985). However, in studies (Robinson, 1986; 1988) with isolated mesophyll cells or chloroplasts from spinach it has been shown that even at limiting light intensity, reduction of nitrite is not repressed by CO_2 and *vice versa*. It hence appears that even under low irradiation, reduced ferredoxin is generated *in situ* in amounts sufficient to meet the demands of both the processes of CO_2 assimilation and nitrite reduction.

Contrary to the earlier belief that nitrite reduction in photosynthetic tissues proceeds only in light (Canvin and Atkins, 1974), it is now accepted that these tissues do reduce nitrite even in complete darkness, though, at much slower rates than in light (Jones and Sheard, 1978; Yoneyama, 1981; Reed et al., 1983; Prakash, et al., 1984). The requirements for reduction of nitrite in dark are distinctly different from the light dependent reactions. Thus, whereas, nitrite is reduced equally well both under aerobic and anaerobic conditions in light, its reduction in dark proceeds only under aerobic conditions (Jones and Sheard, 1978; Mann et al., 1979; Cresswell et al., 1990). Secondly, inhibitors of photosynthetic electron transfer chain impede its reduction in light but these show no discernible effect in dark (Ben-Shalom et al., 1983). Finally, light dependent reduction of nitrite is not impaired by uncouplers but its reduction in dark is severely retarded in their presence (Mann et al., 1979; Ben-Shalom et al., 1983). The requirement for oxygen and ATP imply that nitrite reduction in dark is coupled to respiratory metabolism and, more particularly, to an ATP-dependent reaction. This provides a probable explanation for the observation that the rate of nitrite reduction in leaves in dark is greatly influenced by carbohydrate status of the tissue (Aslam and Huffaker, 1984; Watt et al., 1987; Cresswell et al., 1990). It has been proposed that in dark, reduced ferredoxin is generated from NADPH by reversal of the reaction catalysed by ferredoxin: NADP reductase and NADPH for this step is supplied via oxidative pentose phosphate pathway. Glucose-6-phosphate for oxidative pentose phosphate pathway is produced by the action of hexokinase and this accounts for dependence of the process of nitrite reduction in dark on ATP supply and consequently on respiratory metabolism. In conformity with this postulated mechanism, coupling of leaf NiR with glucose-6-phosphate dehydrogenase and ferredoxin: NADP reductase in presence of glucose-6-phosphate and NADP in a reconstituted chloroplastic system has been demonstrated (Hucklesby et al., 1990). It was also shown that this system is much slower than nitrite reduction *in vivo* in light or the system based on illuminated chloroplasts. It is conceivable that this mechanism of NADPH-linked dark reduction of nitrite ceases to operate in light because on illumination glucose-6-phosphate dehydrogenase, the key enzyme of oxidative pentose phosphate pathway, is rendered inactive via light mediated reactions (Sawhney and Naik, 1990).

B. Non-photosynthetic Tissues

As indicated earlier, the exact identity of the physiological electron donor for the activity of NiR in non-chlorophyllous tissues has not been established unequivocally but it is perhaps a ferredoxin-like protein. Since nitrite reduction in these tissues is also inhibited by uncouplers of oxidative phosphorylation and anaerobic conditions (Dry et al., 1981; Watt et al., 1987), the system for reduction of this electron carrier appears to be similar to the one operating during dark reduction of nitrite in photosynthetic tissues and hence is dependent on oxidative pentose phosphate pathway for supply of NADPH (Emes and Fowler, 1979). NADPH-dependent reduction of this carrier is catalysed by a diaphorase. Proplastids of pea roots have, infact, been shown to contain NAD(P)H linked ferredoxin reductase which is 7 times more active with NADPH than with NADH and its activity is 4-5 times higher in plants supplied with nitrate (Bowsher et al., 1989). Several workers have noted enhanced metabolism of glucose via oxidative pentose phosphate pathway in roots of nitrate-fed plants (Lee et al., 1979; Hucklesby et al., 1990) and increase in activity of glucose-6-phosphate dehydrogenase in roots but in the leaves (Weismann, 1972). It has also been shown that nitrite reduction in roots or root proplastids is accelerated by glucose-6-phosphate or fructose-6-phosphate and the K_m for glucose-6-phosphate and fructose-6-phosphate for nitrite reduction are similar for glucose-6-phosphate dehydrogenase and hexose phosphate isomerase (Dry et al., 1981; Hucklesby et al., 1990). Retardation of nitrite reduction in plastids of pea roots on treatment with uncouplers of oxidative phosphorylation is accompanied by depleted levels of glucose-6-phosphate. It is also pertinent that nitrite reduction in these tissues is restored by exogenous supply of glucose-6-phosphate (Dry et al., 1981). In conformity with this view is the observation of Stulen and Lanting (1976) that *in vivo* reduction of nitrite in radish cotyledons declines rapidly after 19 h of germination in dark in spite of high activity of nitrite reductase. Significantly, around the same period there is a marked decrease in activities of glucose-6-phosphate dehydrogenase and 6-phosphagluconate dehydrogenase. All these observations implicate oxidative pentose phosphate pathway in supply of NADPH which is then utilized for reduction of the immediate electron donor for NiR in the non-chlorophyllous tissues.

VI. Conclusions and Future Prospects

A considerable progress has undoubtedly been made over the years in understanding of the process of nitrite reduction in higher plants. However, there is a need to obtain further detailed information about this process and also to resolve some of the uncertainities. For example, it needs to be resovled unambiguously whether the much studied 61 kD NiR exists as such *in situ* or is it a component of a larger enzyme. It also needs to be established whether multiple forms of the enzyme in plant tissues is a universal phenomenon or is it restricted to a few plant species and organs. Evidently, in such studies it will be essential to ensure that techniques employed are free of artefacts and are adequate enough to resolve multiple forms. It is also necessary to determine whether all these forms are plastidic enzymes and, if not, what is source of reductant for the extraplastidic form(s) of the enzyme. Though a number of workers have isolated and characterized electron carriers in non-photosynthetic tissues which are capable of mediating nitrite reduction with NADPH as the reductant, the true electron carrier in these tissues still awaits to be unquivocally established. It would also be desirable to obtain information about constitution and topography of the active site of the enzyme. Light has been known to promote synthesis of NiR in photosynthetic tissues since a long time. However, the precise mechanism for its action has remained enigmatic. One of the fruitful areas for future work would be to understand the molecular biology of induced synthesis of the enzyme both with regard to induction by nitrate and nitrite ions and also the stimulatory effect of light on its formation.

Literature Cited

Aguera, E., de La Haba, P., and Maldonado, J.M. 1987. Induction of nitrate reductase, nitrite reductase and glutamine synthetase by nitrate and/or nitrite in germinating sunflower cotyledons. Effect of plant hormones. *Plant Physiol.* (Life Sci. Adv.). **6**: 255-258.

Ague, E., de La Haba, P., Fontes, A.G., and Maldonado, J.M. 1990. Nitrate and nitrite uptake and reduction by intact sunflower plants. *Planta* **182**: 149-154.

Aslam, M., and Huffaker, R.C. 1984. Dependency of nitrate reduction on soluble carbohydrates in primary leaves of barley under aerobic conditions. *Plant Physiol.* **75**: 623-628.

Aslam. M., and Huffaker, R.C. 1989. Role of nitrate and nitrite in induction of nitrite reductase in leaves of barley seedlings. *Plant Physiol.* **91**: 1152-1156.

Atkins, C.A., and Canvin, D.T. 1975. Nitrate, nitrite and ammonia assimilation by leaves. Effect of inhibitors. *Planta.* **123**: 41-51.

Back, E., Burkhart, W., Moyer, M., Privalle, L., and Rothstein, S. 1988. Isolation of cDNA coding for spinach nitrite reductase: Complete sequence and nitrate induction. Mol. *Gen. Genet.* **212**: 20-26.

Baysdorfer, C., and Robinson, J.M. 1985. Metabolic interactions between spinach leaf nitrite reductase and ferredoxin: NADP reductase. *Plant Physiol.* **77**: 318-320.

Beevers, L.,and Hageman, R.H. 1980. Nitrate and nitrite reduction. *In 'The Biochemistry of Plants',* (Miflin, B.J. ed), Vol. 5, Academic Press, New York. pp. 116-168.

Ben Shalom, N., Huffaker, R.C., and Rappaport, L. 1983. Effect of photosynthetic inhibitors and uncouplers of oxidative phosphorylation on nitrate and nitrite reduction in barley leaves. *Plant Physiol.* **71**: 63-66.

Borner, T., Mendel, R.R., and Schiemann, J.J. 1986. Nitrite reductase is not accumulated in chloroplast deficient mutants of higher plants. *Planta* **169**: 202-207.

Bowsher, C.G., Emes, M.J., Cammack, R., and Hucklesby, D.P. 1988. Purification and properties of nitrite reductase from roots of pea (*Pisum sativum* cv. Meteor). *Planta* **175**: 334-340.

Bowsher, C.G., Hucklesby, D.P., and Emes, M.J. 1989. Nitrite reduction and carbohydrate metabolism in plastids purified from roots of *Pisum sativum* L. *Planta* **177**: 359-366.

Canvin, D.T., and Atkins, C.A. 1974. Nitrate, nitrite and ammonia assimilation by leaves. Effect of light, carbon dioxide and oxygen. *Planta* **116**: 207-224.

Cammack, R., Hucklesby, D.P., and Hewitt, E.J. 1978. Electron paramagnetic resonance studies of the mechanism of leaf nitrite reductase. *Biochem. Jour.* **171**: 519-526.

Campbell, W.H. 1990. Purification, characterization and immunochemistry of higher plant nitrate reductase. *In Nitrogen in Higher Plants,* (Abrol, Y.P., ed.), John Wiley and Sons, New York. pp. 65-88.

Cardenas, J., Barea, J.L., Rivas, J., and Moreno, C.G. 1972. Purification and properties of nitrite reductase from spinach leaves. *FEBS Letts.* **23**: 113-135.

Cresswell, C.F., Hageman, R.H., and Hewitt, E.J. 1962. The reduction of nitrite and hydroxylamine by enzymes of vegetable marrow (*Cucurbita pepo*). *Biochem. J.* **83**: 38p.

Cresswell, C.F., Hageman, R.H., Hewitt, E.J., and Hucklesby, D.P. 1965. The reduction of nitrate, nitrite and hydroxylamine to ammonia by enzymes from *Cucurbita pepo* in presence of benzyl viologen as electron donor, *Biochem. J.* **94**: 40-53.

Cresswell, C.F., Watt, M.P., Amory, A.M., and Whittaker, A. 1990. The regulation of reduction of nitrogen in chlorophyllous tissues: Uptake and reduction of nitrite by intact chloroplasts. *In Inorganic Nitrogen in Plants and Micro Organisms,* (Ullrich, W.R., Rigano, C., Fuggi, A. and Aparicio, P.J., eds.), Springer-Verlag, Berlin. pp. 149-156.

Dalling, M.J., Tolbert, N.E., and Hageman, R.H. 1972. Intracellular location of nitrate reductase and nitrite reductase. II. Wheat roots. *Biochim. Biophys. Acta.* **283**: 513-519.

Dean, J.V., and Harper, J.E. 1988. The conversion of nitrite to nitrogen oxide (s) by constitutive NAD(P)H -nitrate reductase from soybean. *Plant Physiol.* **88**: 389-395.

Dry, I.B., Wallace, W.,and Nicholas, D.J.D. 1981. Role of ATP in nitrite reduction in roots of wheat and peas. *Planta* **152**: 234-238.

Emes, M.J., and Fowler, M.W. 1979. Intracellular interactions between pathways of carbohydrate oxidation and nitrate assimilation in plant roots. *Planta* **145**: 287-292.

Evans, H.J.,and Nason,A. 1953. Pyridine nucleotide nitrate reductase from extracts of higher plants. *Plant Physiol.* **28**: 233-254.

Feenstra, W.J.,and Jacobsen, E. 1980. Isolation of nitrate reductase deficient mutants of *Pisum sativum* by means of selection for chlorate resistance. *Theor. Appl. Genet.* **58**: 39-42.

Feierabend, J. 1986. Investigation of site of synthesis of chloroplast enzymes of nitrogen metabolism by use of heat-treated 70 S ribosome deficient rye leaves. *Physiol. Plant.* **67**: 145-150.

Gandhi, A.P., Sawhney, S.K., Mehta, S.L.,and Naik, M.S. 1974. Effect of inhibitors of Hill reaction on polyribosome level and on synthesis of nitrate and nitrite reductases in rice seedlings. *Indian J. Biochem. Biophys.* **11**: 40-42.

Gupta, A., Disa, S., Saxena, I.M., Sarin, N.B., Guha-Mukherjee, S.,and Sopory, S.K. 1983. Role of nitrate in induction of nitrite reductase activity during wheat seed germination. *J. Expt. Bot.* **34**: 396-404.

Gupta, S.C., and Beevers, L. 1983. Environmental influences on nitrite reductase activity in *Pisum sativum* L. seedlings. *J. Expt. Bot.* **34**: 1455-1462.

Gupta, S.C., and Beevers, L. 1984. Synthesis and degradation of nitrite reductase in pea leaves. *Plant Physiol.* **75**: 251-252.

Gupta, S.C., and Beevers, L. 1985. Regulation of synthesis of nitrite reductase in pea leaves. *In vivo* and *in vitro* studies. *Planta* **166**: 89-95.

Gupta, S.C., and Beevers, L. 1987. Regulation of nitrite reductase. Cell free translation and processing. *Plant Physiol.* **83**: 750-754.

Hageman, R.H. 1990. Historical perspectives of the enzymes of nitrate assimilation by crop plants and potential for biotechnological applications. *In Inorganic Nitrogen in Plants and Microorganisms* (Ullrich, W.R., Rigano, C., Fuggi, A. Aparicio, P.J., eds.), Springer-Verlag, Berlin, pp. 3-11.

Hageman, R.H., Cresswell, C.F., and Hewitt, E.J. 1962. Reduction of nitrate, nitrite and hydroxylamine to ammonia by enzymes extracted from higher plants. *Nature* **193**: 247-250.

Harel, F., Lea, P.J., and Miflin, B.J. 1977. Localization of enzymes of nitrate assimilation in maize leaves and their activities during greening. *Planta* **134**: 195-200.

Heath-Pugliuso, S., Huffaker, R.C.,and Allard, R.W. 1984. Inheritance of nitrate reductase and regulation of nitrate reductase, nitrite reductase and glutamine synthetase isozymes. *Plant Physiol.* **76**: 353-358.

Hewitt, E.J., and Betts, C.F. 1963. The reduction of nitrite and hydroxylamine by ferredoxin and chloroplast grana from *Cucurbita pepo*. *Biochem. J.* **89**: 20 p.

Hewitt, E.J., Hucklesby, D.P., and Betts, C.F. 1968. Nitrogen assimilation in higher plants. *In Recent Aspects of Nitrogen Metabolism in Plants*, (Hewitt, E.J. and Cutting, C.V., eds.), Academic Press, New York. pp. 47-81.

Hewitt, E.J., Hucklesby, D.P., and Notton, B.A. 1976. Nitrate Metabolism. *In Plant Biochemistry*, (Bonner, J. and Varner, J.E, eds.), Academic Press, New York. pp. 663-682.

Hirasawa, K., and Tamura, G. 1980. Ferredoxin-dependent nitrite reductase from spinach leaves. *Agric. Biol. Chem.* **44**: 749-758,

Hirasawa, M., Horie, S.,and Tamura, G. 1982. Further characterization of ferredoxin-nitrite reductase and the relationship between the enzyme and methyl viologen nitrite reductase. *Agric. Biol. Chem.* **46**: 1319-1328.

Hirasawa, M., Fukushima, M., Tamura, G., and Knaff, D.B. 1987a. Immunochemical characterization of nitrite reductase from spinach leaves, spinach roots and other higher plants. *Biochim. Biophys. Acta* **791**: 145-154.

Hirasawa, M., Shaw, R.W., Palmer, G., and Knaff, D.B. 1987b. Prosthetic group content and ligand binding properties of spinach nitrite reductase. *J. Biol. Chem.* **262**: 12428-12433.

Hucklesby, D.P., and Hageman, R.H. 1976. Hydroxylamine reductase enzymes from scutellum and their relationship to nitrite reductase. *Plant Physiol.* **57**: 693-698.

Hucklesby, D.P., and Hewitt, E.J. 1970. Nitrite and hydroxylamine reduction in higher plants. *Biochem. J.* **119**: 615-627.

Hucklesby, D.P., Hewitt, E.J., and James, D.M. 1970. Possible active sites in nitrite reductase and hydroxylamine reductase from vegetable marrow. (*Cucurbita pepo*). *Biochem. J.* **117**: 30 p.

Hucklesby, D.P., Dalling, M.J., and Hageman, R.H. 1972. Some properties of two forms of nitrite reductase from corn (*Zea mays*) scutellum. *Planta* **104**: 220-223.

Hucklesby, D.P., Emes, M.J., Bowsher, C.G., and Cammack, R. 1990. Nitrite reduction in the roots and leaves of *Pisum sativum. In Inorganic Nitrogen in Plants and Micro-organisms*. (Ullrich, W.R., Rigano, C., Fuggi, A., Aparicio, P.J., eds), Springer-Verlag, Berlin. pp. 210-215.

Hucklesby, D.P., James, D.M., Banwell, M.J., and Hewitt, E.J. 1967. Properties of nitrite reductase from *Cucurbita pepo. Phytochemistry.* **15**: 599-603.

Huzisige, H., and Satoh, K. 1961. Photosynthetic nitrite reductase. I. Partial purification and properties of the enzyme from spinach leaves. *Bot. Mag.* (Tokyo). **74**: 175-185.

Ida, S. 1987. Immunological comparisons of ferredoxin-nitrite reductases from higher plants. *Plant Sci. Letts.* **49**: 111-116.

Ida, S., and Mikami, B. 1986. Spinach ferredoxin-nitrite reductase: A purification procedure and characterization of chemical properties. *Biochim. Biophys. Acta.* **871**: 167-176.

Ida, S., and Morita, Y. 1973. Purification and properties of spinach leaf nitrite reductase. *Plant Cell Physiol.* **14**: 661-671.

Ida, S., Mori, E.,and Morita, Y. 1974. Purification, stabilization and characterization of nitrite reductase from barley roots. *Planta* **121**: 213-224.

Ingle, J., Joy, K.W., and Hageman, R.H. 1966. Regulation of activity of enzymes involved in assimilation of nitrate in higher plants. *Biochem. J.* **100**: 577-588.

Ishiyama, Y., and Tamura, G. 1985. Isolation and partial characterization of homogenous nitrite reductase isolated form etiolated bean leaves and *Phaseolus angularis. Plant Sci. Letts.* **37**: 251-256.

Ishiyama, M., Shinoda, I., and Tamura, G. 1985a. Electron donor for nitrite reductase from etiolated bean leaves. *Agric. Biol. Chem.* **49**: 2223-2224.

Ishiyama, M., Shinoda, I, Fukushima, K., and Tamura, G. 1985b. Some properties of ferredoxin-nitrite reductase from green shoots of *Phaseolus angularis* and an immunological comparison with nitrite reductase from roots and etiolated shoots. *Plant Sci. Letts.* **39**: 89-95.

Jackson, W.A., Johnson, R.E., and Volk, R.J. 1974. Nitrite uptake patterns in wheat seedlings as influenced by nitrate and ammonium. *Physiol. Plant.* **32**: 108-114.

Jones, R.W., and Sheard, R.W. 1978. Accumulation and stability of nitrite in intact aerial leaves. *Plant Sci. Letts.* **11**: 285-291.

Kelker, H.C.,and Filner, P. 1971. Regulation of nitrite reductase and its relationship to the regulation of nitrate reductase in cultured tobacco cells. *Biochim. Biophys. Acta.* **252**: 69-82.

Klepper, L.A. 1979. Effect of certain herbicides and their combinations on nitrate and nitrite reduction. *Plant Physiol.* **64**: 273-275.

Kramer, V., Lahners, K., Back, E., Privalle, L.S., and Rothstein, S. 1989. Transient accumulation of m-RNA for nitrite reductase in maize following the addition of nitrate. *Plant Physiol.* **90**: 1214-1220.

Krueger, R.J.,and Siegel, L.M. 1982. Spinach siroheme enzymes: Isolation and characterization of ferredoxin-sulphite reductase and comparison with ferredoxin-nitrite reductase. *Biochemistry* **21**: 2892-2909.

Kutscherra, M., Jost, W.,and Schlee, D. 1987. Isoenzymes of nitrite reductase in higher plants: Occurrence, purification, properties and alterations during ontogenesis. *J. Plant Physiol.* **129**: 383-393.

La Brie, S.T., Wilkinson, J.Q., and Crawford, N.M. 1992. Effect of chlorate treatment on nitrate reductase and nitrite reductase gene expression in *Arabidiopsis thaliana. Plant Physiol.* **97**: 873-879.

La Haba, P., Aguera, E.,and Maldonado, J.M. (1988). Development of nitrogen assimilating enzymes in sunflower cotyledons during germination as affected by nitrogen source. *Planta* **173**: 52-57.

Lahners, K., Kramer, V., Back, E., Privalle, L., and Rothstein, S. 1988. Molecular cloning of complementary DNA encoding maize nitrite reductase: Molecular mechanisms and nitrate induction. *Plant Physiol.* **88**: 7141-7146.

Lancaster, J.R., Vega, J.M., Kamin, H. Orme-Johnson, N.R., Orme-Johnson, W.H., Krueger, R.J., and Siegel, L.M. 1979. Identification of iron-sulphur centre of spinach ferredoxin-nitrite reductase as a tetranuclear centre and preliminary EPR studies of mechanism. *J. Biol. Chem.* **254**: 1268-1272.

Lee, R.B. 1979. The release of nitrite from barley roots in response to metabolic inhibitors, uncoupling agents and anoxia. *J. Expt. Bot.* **30**: 119-133.

Losada, M., and Guerrero, G.M. 1979. Photosynthetic reduction of nitrate and its regulation. In *Photosynthesis in Relation to Model Systems* (Barber, J., ed.), Elsevier-North Holland Biomedical Press, Amsterdam. pp. 366-407.

Losada, M., Paneque, A., Ramirez, J.M., and Del Campo, F.F. 1963. Mechanism of nitrite reduction in chloroplasts. *Biochem. Biophys. Res. Commun.* **10**: 298-303.

Magalheas, A.C., Neyra, C.A., and Hageman, R.H. 1974. Nitrite assimilation and amino acid synthesis in isolated spinach chloroplasts. *Plant Physiol.* **53**: 411-415.

Mann, A.F., Hucklesby, D.P., and Hewitt, E.J. 1979. Effect of aerobic and anaerobic conditions on in vivo nitrate reductase assay in spinach leaves. *Planta* **146**: 83-89.

Maldonado, J.M., Aguera, E., and La Haba, P. 1990. Nitrate and nitrite utilization by sunflower. In *Inorganic Nitrogen in Plants and Micro-organisms* (Ullrich, W.R., Rigano, C., Fuggi, A. and Aparicio, P.J., eds.), Springer-Verlag, Berlin. pp. 159-164.

Mc Ree, D.E., Richardson, D.C., Richardson, J.S., and Siegel, L.M. 1986. The heme and $Fe_4 S_4$ cluster in crystallographic structure of *Escherichia coli* sulphite reductase. *J. Biol. Chem.* **261**: 10277-10281.

Mellor, G.E., and Tregunna, E.B. 1971. The localization of nitrate assimilating enzymes in leaves of plants with C_4 pathway of photosynthesis. *Can. J. Bot.* **49**: 137-142.

Meyer, V., and Schulze, E. 1894. Uebeur die einwirkung von hydroxylamisalzen auf pflansen. *Ber Dtsch. Chem. Ges.* **17**: 1554-1558.

Murphy, J.M., Siegel, L.M., Tove, S.R., and Kamin, H. 1974. Siroheme: A new prosthetic group participating in six electron reduction reactions catalysed by both sulphite and nitrite reductase. *Proc. Nat. Acad. Sci.* **71**: 612-616.

Nason, A., Abraham, R.G., and Averbach, B.C. 1954. The enzymic reduction of nitrite to ammonia by reduced pyridine nucleotides. Biochim. *Biophys. Acta.* **15**: 159-161.

Neininger, A., Kronenberger, J., and Mohr, H. 1992. Coaction of light, nitrate and plastidic factor in controlling nitrite reductase gene expression in tobacco. *Planta* **187**: 381-387.

Nelson, R.S., Ryan, S.A., and Harper, J.E. 1983. Soybean mutants lacking constitutive nitrate reductase activity. II. Nitrogen assimilation, chlorate resistance and inheritance. *Plant Physiol.* **72**: 503-509.

Nelson, R.S., Streit, L., and Harper, J.E. 1984. Biochemical characterization of nitrate reductase and nitrite reductase in wild type and a nitrate reductase mutant of soybean. *Physiol. Plant.* **61**: 384-390.

Neyra, C.A., and Hageman, R.H. 1974. Dependence of nitrate reduction on electron transport in chloroplasts. *Plant Physiol.* **54**: 480-483.

Ninamiya Y., and Sato, S. 1984. A ferredoxin - like electron carrier from non-green cultured tobacco cells. *Plant Cell Physiol.* **25**: 453-458.

Oelmuller, R. 1989. Photoxidative destruction of chloroplasts and its effect on nuclear gene expression and extra-plastidic enzymes. *Photochem. Photobiol.* **49**: 229-239.

Ogawa, M., and Ida, S. 1987. Biosynthesis of ferredoxin nitrite reductase in rice seedlings. *Plant Cell Physiol.* **28**: 501-508.

Ostrowski, J., Back, E.W., Madden, J., Siegel, L.M.,and Kredich, N.M. 1987. Comparison of deduced amino acid sequences of and α-sub units of *S. typhimurium* sulfite reductase (SR) with those . of spinach nitrite reductase (NR) and cytochrome P-450 oxidoreductase (P 450 R). *Fed. Proc. Abst. No. 1780.* p. 2231.

Prakash, S., Singh, P., Sawhney, S.K.,and Naik, M.S. 1984. Regulation of nitrate assimilation in plants in light and dark. *Plant Sci. Letts.* 34: 25-34.

Rajasekhar, V.K., and Mohr. M. 1986. Appearance of nitrite reductase in cotyledons of mustard (*Sinapis alba*) seedling as affected by nitrate, phytochrome and photoxidative damage of chloro-plasts. *Planta* 168: 369-376.

Rajasekhar, V.K., and Oelmuller, R. 1987. Regulation of induction of nitrate reductase and nitrite reductase in higher plants. *Physiol. Plant.* 71: 517-521.

Rajasekhar, V.K., and Sopory, S.K. 1985. The blue light effect and its interaction with phytochrome in control of nitrite reductase in *Sorghum bicolor* wild. *New Phytol.* 101: 251-258.

Rathnam, C.K.M., and Edwards, G.E. 1976. Distribution of nitrate assimilating enzymes between mesophyll protoplasts and bundle sheath cells in leaves of three groups of C_4 plants. *Plant Physiol.* 57: 881-885.

Redinbaugh, M.G., and Campbell, W.H. 1991. Higher plant responses to environmental nitrate. *Physiol. Plant.* 82: 640-650.

Reed, A.J., and Canvin, D.T. 1982. Light and dark controls of nitrate reduction in wheat (*Triticum aestivum*) protoplasts. *Plant Physiol.* 69: 508-513.

Reed, A.J., Canvin, D.T., Sherrard, J.H., and Hageman, R.H. 1983. Assimilation of (^{15}N)-nitrate and (^{15}N) - nitrite in leaves of five plant species under light and dark conditions. *Plant Physiol.* 71: 29-1-294.

Robinson, J.M. 1986. Carbon dioxide and nitrite photo assimilatory processes do not inter-compete for reducing equivalents in spinach and soybean leaf chloroplasts. *Plant Physiol.* 80: 674-684.

Robinson, J.M. 1988. Spinach leaf chloroplasts: Carbon dioxide and nitrite photoassimilation do not complete for reductant. *Plant Physiol.* 88: 1373-1380.

Rothstein, S., Lahners, K., Kramer, V., Privalle, L., and Back, E. 1989. Expression of corn nitrite reductase in hydroponically grown seedlings and cell suspensions. In *Abstracts Advance Course in Inorganic Nitrogen Metabolism.* Seiano. p. 103.

Sawhney, S.K., and Naik, M.S. 1972. Role of light in synthesis of nitrate reductase and nitrite reductase in rice seedlings. *Biochem. J.* 130: 475-485.

Sawhney, S.K., and Naik, M.S. 1973a. Effect of chloramphenicol and cycloheximide on synthesis of nitrate reductase and nitrite reductase in rice leaves. *Biochem. Biophys. Res. Commun.* 51: 61-73.

Sawhney, S.K., and Naik, M.S. 1973b. Some properties of nitrate reductase and nitrite reductase from rice seedlings. *Indian. J. Biochem. Biophys.* 11: 40-42.

Sawhney, S.K., and Naik, M.S. 1990. Role of light in nitrate assimilation in higher plants. In *Nitrogen in Higher Plants,* (Abrol Y.P., ed.), John Wiley and Sons, New York. pp. 93-128.

Sawhney, S.K., and Nicholas, D.I.D. 1975. Nitrite, hydroxylamine and sulphite reductases in wheat leaves. *Phytochemistry.* 14: 1499-1503.

Sawhney, S.K., Prakash, V., and Naik, M.S. 1972. Nitrate and nitrite reductase activities in induced chlorophyl mutants of barley. *FEBS Letts.* 22: 200-202.

Schnutz, D., and Brunold, C. 1985. Localization of nitrite and sulphite reductase in bundle sheath and mesophyll cells of maize leaves. *Physiol. Planta.* 64: 423-428.

Schrader, L.E., Beevers, L., and Hageman, R.H. 1967. Differential effects of chloramphenicol on induction of nitrate and nitrite reductase in green leaf tissue. *Biochem. Biophys. Res. Comm.* 26: 14-17.

Schuster, C., and Mohr, H. 1990. Appearance of nitrite reductase m-RNA in mustard seedling cotyledons is regulated by phytochrome. *Planta.* **181**: 324-327.

Schuster, C., Oelmuller, R., and Mohr, H. 1987. Signal storage in phytochrome action on nitrate mediated induction of nitrite reductase in mustard seedling cotyledons. *Planta* 171: 136-143.

Seith, B., Schuster, C., and Mohr, H. 1991. Coaction of light, nitrate and plastidic factor in controlling nitrite reductase gene expression in spinach. *Planta* **184**: 78-80.

Sharma, A.K., and Sopory, S.K. 1984. Independent effects of phytochrome and nitrate on nitrate reductase and nitrite reductase activities in maize. *Photochem. Photobiol.* **39**: 491-493.

Siegel, L.M., and Wilkerson, 1989. In *Molecular and Genetic Aspects of Nitrate Assimilation,* (Wray, J.L. and Kinghorn, J.R., eds), Oxford University Press, Oxford. pp. 262-283.

Singh P., and Sawhney, S.K. 1989. Nitrate reduction in plants. In *Recent Advances in Plant Biochemistry,* (Mehta, S.L., Lodha, M.L. and Sane, P.V., eds.), ICAR Publication, New Delhi, pp. 143-172.

Sluister-Scholten, C.M. Th. 1973. The effect of chloramphenicol and cycloheximide on induction of nitrate reductase and nitrite reductase in bean leaves. *Planta* **113**: 229-240.

Sluister-Scholten, C.M. Th. 1975. Photosynthesis and induction of nitrate reductase and nitrite reductase in bean leaves. *Planta* **123**: 175-184.

Small, I.S., and Gray, J.C. 1984. Synthesis of wheat leaf nitrite reductase *de novo* following induction with nitrate and light. *Eur. J. Biochem.* **145**: 291-297.

Sopory, S.K., and Sharma, A.K. 1990. Special quality of light, hormones and nitrate assimilation. In *Nitrogen in Higher Plants,* (Abrol, Y.P., ed.), John Wiley and Sons. New York. pp. 129-157.

Stulen I., and Lanting, L. 1976. Nitrate reductase and nitrite reductase in radish seedlings: Relation to supply of NADPH. *Physiol. Plant.* **37**: 139-142.

Suzuki, A., Oaks, A., Jaquet, J.P., Vidal, P., and Gadal, P. 1985 An electron transport system in maize roots for reactions of glutamate synthase and nitrite reductase. *Plant Physiol.* **78**: 374-378.

Trebst, A., and Burba, M. 1967. Inhibition of photosynthetic reductions in isolated chloroplasts and in Chlorella by disalicylidene propanediamine. *Z. Pflanzenphysiol.* **57**: 419-423.

Vega, J.M., and Kamin, M. 1977. Spinach nitrite reductase: Purification and properties of siroheme containing iron-sulphur enzyme. *J. Biol. Chem.* **252**: 896-909.

Wada, K., Onda, M., and Matsubara, H. 1986. Ferredoxin isolated from plant non-photosynthetic tissues. Purification and properties. *Plant Cell Physiol.* **27**: 407-415.

Wallsgrove, R.M., Key, A.J., Lea P.J., and Miflin, B.J. 1983. Photosynthesis, photorespiration and nitrogen metabolism. *Plant Cell Environ.* **6**: 301-309.

Wallsgove, R.M., Lea, P.J., and Mikflin, B.J. 1979. Distribution of enzymes of nitrate assimilation within pea leaf cells. *Plant Physiol.* **63**: 232-236.

Warner, R.L., and Kleinhofs, A. 1974. Relationship between nitrate reductase, nitrite reductase and ribulose diphosphate carboxylase activities in chlorophyll deficient mutants of barley. *Crop Sci.* **14**: 654-658.

Warner, R.L., Kleinhofs, A., and Muehibauer, F.J. 1982. Characterization of nitrate reductase deficient mutants in peas. *Crop Sci.* **22**: 389-393.

Washitani, I., and Sato, S. 1977. Studies on function of proplastids in metabolism of *in vitro* cultures tobacco cells. II. Source of reducing power for amino acid synthesis from nitrite. *Plant Cell Physiol.* **18**: 1235-1241.

Watt, M.P., Gray, V.M., and Cresswell, C.F. 1987. Control of nitrate and nitrite assimilation by carbohydrate reserves, adenosine nucleotides and pyridine nucleotides in leaves of *Zea mays* under dark aerobic conditions. *Planta* 172: 548-554.

Weber, M., Schmidt, S., Schuster, C., and Mohr, H. 1990. Factors involved in coordinate appearance of nitrite reductase and glutamine synthetase in mustard (*Sinapis alba* L.) seedlings. *Planta* **180**: 429-434.

Weismann, G.S. 1972. Influence of ammonium and nitrate nutrition on enzymatic activity in soy-
      bean and sunflower. *Plant Physiol.* **49**: 138-141.
Wray, J.L., Ip. S.M., Duncanson, E., Gilkes, A.F., and Kirk, D.W. 1990. Biochemistry, regulation
      and genetics of nitrite reductase. *In Inorganic Nitrogen in Plants and Micro-organisms*,
      (Ullrich, W.R., Rigano, C., Fuggi, A. and Aparicio, P.J., eds.), Springer-Verlag, Berlin.,
      pp. 201-209.
Yoneyama, T.E. 1981. [15]N Studies in *in vivo* assay of nitrate reductase in leaves. Occurrence of
      under estimation of activity due to dark assimilation of nitrate and nitrite. *Plant Cell Physiol.*
      22: 1507-1520.

Nitrogen Nutrition in Higher Plants, 1995
Editors : H.S. Srivastava & R.P. Singh
Associated Publishing Co., New Delhi, India
pp. 189-203.

Ammonia Assimilation

RANA P. SINGH

Abbreviations: NH$_3$, Ammonia; Fd red, Ferrodoxin reduced; GDH, L-Glutamate dehydrogenase; GS, Glutamine synthetase; GOGAT, Glutamate synthase; MSO, Methionine sulfoximine

I. Introduction

In plants only ammonium and no other form of inorganic nitrogen gets incorporated into the organic nitrogen cycle. Irrespective of various primary and secondary origin of the molecule (Fig. 1) it usually does not accumulate in the plant tissues possibly because of its toxic nature and it gets assimilated rapidly into glutamine, glutamate and some times in asparagine or even in other amino acids (Miflin and Lea, 1980).

Assimilation of ammonium, in addition to its consideration for one of the most important processes of nitrogen metabolism in plants, has attracted attention of the plant scientists during last two decades because of its changing concepts too. Prior to 1970 major route of entry of ammonium into organic form had been considered via glutamate dehydrogenase (GDH), an ubiquitously available reversible enzyme catalyzing reductive ammination of 2-oxoglutarate and *vice versa* (eg. 1)

$$\text{2-oxoglutarate} + NH_3 + NAD(P)H \xrightarrow{\text{GDH}} \uparrow \text{ L-glutamate} + H_2O + NAD(P)^+ ...(eq. 1)$$

Glutamine synthetase catalyzed entry of ammonia into l-glutamate to form l-glutamine, an amide of glutamate (eq 2) was another known route for ammonia assimilation in plants. But, despite of rapid formation of glutamine in the ammonia assimilating tissues there was no known route showing transfer of amide nitrogen to the α-amino position to give amino acids till 1970.

PRIMARY SOURCES   **SECONDARY SOURCES**

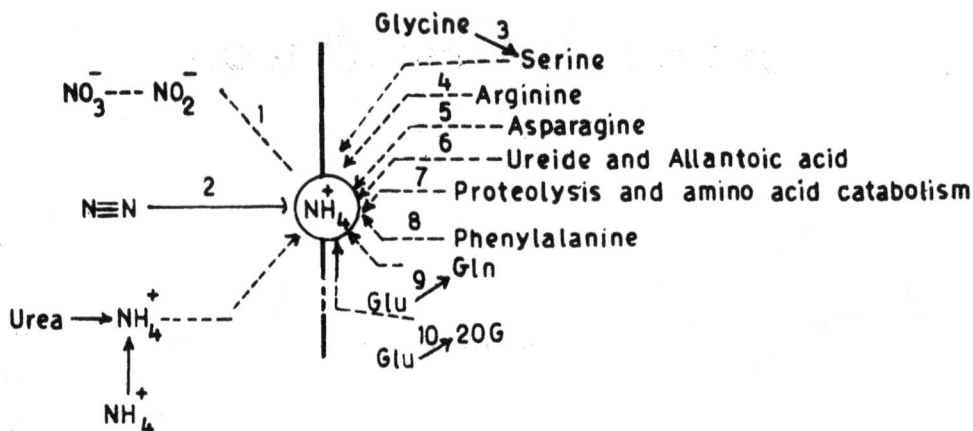

Fig. 1. Origin and source of ammonia in plant cell. 1. Nitrite reduction
2. Dinitrogen fixation. 3. Photorespiration 4. Arginase 5. Asparaginase
6. Catabolism of ureides in legumes 7. Catabolism primarily during senescence and stresses 8. Phenylalanine ammonia lyase
9. Glutaminase (?) 10. Catabolic GDH.

    Break-through was, however, achieved by D.W. Tempest, J.L. Meers and C.M. Brown (1970) who discovered a new enzyme, glutamate synthase in bacterium *Klebsiella aerogenes* grown on limited amount of ammonium which was subsequently reported in higher plants by Dougall (1974) and Lea and Miflin (1974). This could facilitate the transfer of amide nitrogen to α-position of 2-oxoglutarate to produce l-glutamate. It is interesting to note that glutamate acts as both the acceptor and product of ammonia assimilation by this newly described route catalyzed by the joint action of glutamine synthetase and glutamate synthase enzymes (eq. 2-4; Fig. 2) which is often designated as GS-GOGAT cycle (Miflin and Lea, 1980; Rhodes et al., 1980) or the GOGAT cycle (Rhodes et al., 1980; Lea et al., 1990; Robinson et al., 1991; 1992).

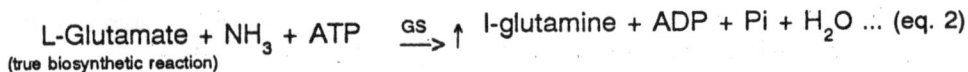

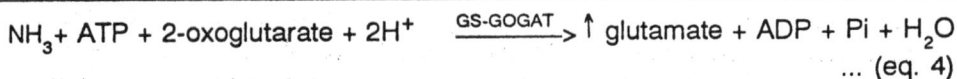

L-Glutamate + NH$_3$ + ATP $\xrightarrow{\text{GS}}$ ↑ l-glutamine + ADP + Pi + H$_2$O ... (eq. 2)
(true biosynthetic reaction)

L-Glutamine + 2-oxoglutarate + 2H$^+$ [NAD (P)H or red Fd] $\xrightarrow{\text{GOGAT}}$ ↑ 2 (l-glutamate) + NAD(P)$^+$ or Fd (oxidized)   ... (eq. 3)

NH$_3$+ ATP + 2-oxoglutarate + 2H$^+$ $\xrightarrow{\text{GS-GOGAT}}$ ↑ glutamate + ADP + Pi + H$_2$O
                                                       ... (eq. 4)

Several evidences from recent works suggest that GS-GOGAT pathway is the major route for ammonia assimilation in the plants under normal growth conditions (Kumar and Abrol, 1990; Lea et al., 1990). The actual role of l-glutamate dehydrogenase in plants, on the other hand, is not fully resolved as yet and some recent reports, focus the attention towards a reassessment of the role of this unique enzyme in the process (Srivastava and Singh, 1987; Yamaya and Oaks, 1987; Lea et al., 1990; Bhadula and Shargool, 1991; Magalhaes and Huber, 1991; Robinson et al., 1991, 1992; Frick and Pahlich, 1992). Whichever route is availed for ammonia assimilation, a minimum of two reducing equivalents are required for the net formation of one glutamate molecule and in addition an extra ATP molecule is needed in GS catalyzed entry if GS-GOGAT route is under operation. A carbon skeleton is also required to accept the ammonia assimilated which is usually a five carbon TCA-cycle product of 2-oxoglutarate if nitrogen is maintained as glutamate. However, if ammonia is incorporated to produce a storage amide asparagine then only two carbon atoms are required for each molecule of ammonia incorporated (Lea and Fowden, 1975).

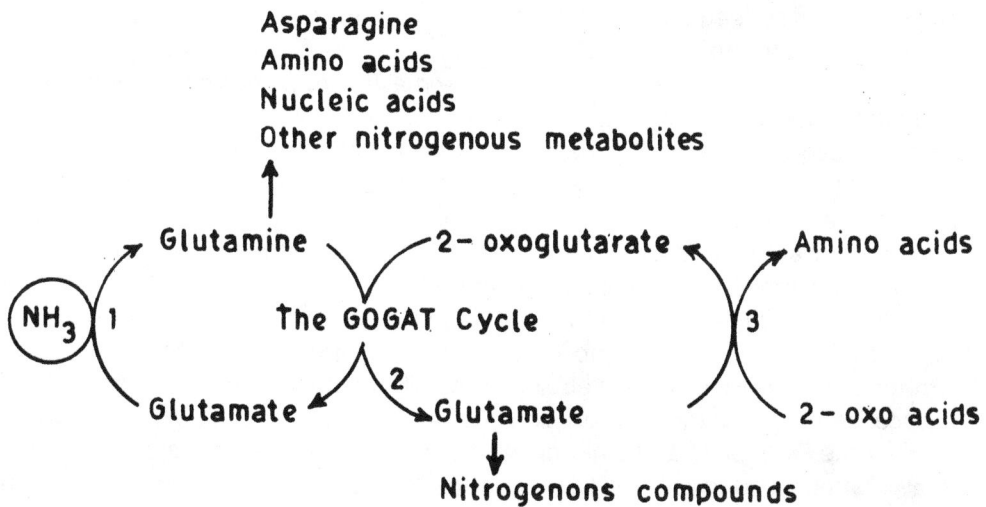

Fig. 2. Ammonia assimilation by the GOGAT cycle

1. GS 2. GOGAT 3 Transaminases

The enzymes catalyzing assimilation of ammonia i.e. GS-GOGAT and GDH have been extensively discussed elsewhere in this volume (chapters 8-10). This chapter is designed for a comparative analysis of the two routes of ammonium assimilation in relation to their physiological significance in the various plant parts.

II. Origin and Source of Ammonia

Ammonia available in the surrounding of a plant is easily taken up to the plants by roots and to some extent by leaves also (Singh et al., 1984). It can readily be available as NH_3 to the plants if urea or some other NH_3 containing fertilizers are supplied and even it occurs in the natural habitat under certain conditions as major source of inorganic nitrogen (see chapters 1 and 4 for details). There is a concensus, however, that nitrate is the predominant form of nitrogen in most of the natural habitats and it can be available to the plants as NH_3 after its reduction by the sequential action of nitrate reductase and nitrite reductase enzymes. In the nitrogen fixing organisms, fixation of dinitrogen (N_2) is another major source of inorganic nitrogen (Fig. 1).

In addition to these primary sources of NH_3 there are various secondary sources too (Fig. 1) which are very significant and contribute a major share in the internal pool of NH_3 in the plants. Photorespiration, for example, releases ammonia during the conversion of glycine to serine in mitochondria upto 80 μ moles g^{-1} fresh weight h^{-1} in plant leaves; several fold higher than the calculated rate of net ammonia assimilation (Keys et al., 1978; Miflin and Lea, 1980; Joy, 1988). The origin of this photorespiratory ammonia is phosphoglycolate in the chloroplast which is metabolized to glycolate and then enters into the peroxisomes. There it is metabolised into glycine and subsequently after its entry in mitochondria it is converted to serine accompanied with the release of NH_3.

Amides, glutamine and asparagine are other major nitrogenous compounds transported in the temperate legumes and cereals (Lea and Miflin, 1980). Asparagine is metabolized either via deamination or transamination to yield NH_3 in the cytosol or in peroxisomes (Sieciechawicz et al., 1988; Lea et al., 1990). In the tropical legumes like soybean, on the other hand, the main nitrogen transport forms are ureides like allantoin and allantoic acid. Allantoin is converted to allantoic acid via allantoinase in the leaves, pods and maturing seeds (Tazima et al., 1977; Rao and Singh, 1988) which is in turn assimilated in to glyoxylate and urea. Urea releases NH_3 and CO_2 by the catalytic action of urease (Shelp et al., 1985)

Proteolysis and amino acid catabolism specially during seed germination and leaf senescence are other measurable sources of NH_3 in the cells. A large amount of enzymes like RUBISCO and other proteins in the leaves and other storage organs are also catabolized to release NH_3 under certain conditions (Thomas, 1978; Thimann, 1980). Certain amino acids like arginine, cysteine, serine, tryptophan and phenylalanine, mono, di and polyamines and several other nitrogenous molecules are also degraded and ammonia released may be recycled through its assimilation into amino acids. The list of ammonia liberating reactions is very extensive and describing each is beyond the scope of this chapter. Nevertheless a multifacial origin and source of ammonia in the various plant tissues, as indicated, may be realized to understand its significance in the plants.

III. Sites of Ammonia Assimilation

The assimilation of NH_3 in a particular tissue or the cell organelle depends upon the activity of the relevant enzymes. Root GS is primarily active in cytosol although plastidial activity of GS has also been reported by several workers (see Lea et al., 1990). Leaves contain two GS forms GS_1 and GS_2 distributed in cytosol and chloroplasts respectively. In root nodules GS accounts for about 2 % of total soluble proteins (Mcfarland et al., 1976), where it is located within the inner cortical cells of alder (*Alnus glutinosa*) root nodules (Hirel et al., 1982) and a part of enzyme is present in the bacteroid (Planke et al., 1977). Two forms of nodule GS, GS_{n1} and GS_{n2} have been reported in which GS_{n1} has been shown to be nodule specific and GS_{n2} is similar to root GS (Lara et al., 1983). The enzyme is also reported in the storage organs like seeds, awns, glumes, seed coats etc. (Kumar and Abrol, 1990). Two isoforms of glutamate synthase (NADH and Fd-dependent) have been reported from most of the tissues and now it has been shown that two activities of the enzyme are carried out by separate proteins (see Lea et al., 1990; and Chapter 10). It has been shown that NADH-dependent GOGAT is predominant in roots and non-green tissues and is more active during early growth phase where as mature green leaves have a more active Fd-dependent GOGAT activity (Matoh and Takahashi, 1982). The localization of the enzyme has been described in detail elsewhere in this volume (Chapter 10).

L-Glutamate dehydrogenase; the enzyme of another pathway is also distributed in various plant parts and upto 17 isozymes of GDH have been reported in young *Pisum* shoots preincubated with tap water (Nauen and Hartmann, 1980). Isozymic number of the enzyme varies with plant species as well as other nutritional and environmental conditions and usual number reported are 7 and 14 although the number may increase by the addition of ammonium and amino acids and decrease by the addition of sucrose (Srivastava and Singh, 1987 and references therein; Chapter 8). This enzyme is predominantly localized in mitochondria and it has been shown to be loosely bound to the mitochondrial membrane using modern techniques (Yamaya et al., 1984). However, a separate enzyme differing from the mitochondrial enzyme has been characterized from the chloroplasts of many plants, where it is shown to be tightly bound to the chloroplast lamellae (Srivastava and Singh, 1987 and references there in). A plastidial localisation of GDH has been shown in soybean suspension culture (Bhadula and Shargool, 1991) and in roots (Srivastava and Singh, 1987). In root nodule, GDH is present both in cytosol and bacterial fractions and it has also been reported in the storage plant organs (Srivastava and Singh, 1987).

IV. Physiological Significance of the Pathways in Different-Plant Organs

Recent studies on the primary NH_3 assimilation, reassimilation of enormous amount of NH_3 released during photorespiration of C_3 leaves, and NH_3 derived from the secondary sources such as the catabolism of nitrogenous transport and storage metabolites (Fig. 1) indicate that the GOGAT cycle is the major route for

NH_3 assimilation in higher plants (Fig. 2). This new concept on the subject has been developed during last two decades on the basis of pulse labelling studies, use of specific inhibitors and genetic studies, for example identification and location of genes for the enzymes and the regulation of the gene expression and studies with mutants lacking specific enzyme system (Stewart et al., 1980; Miflin and Lea 1980; 1982; Srivastava and Singh, 1987; Oaks, 1986; Yamaya and Oaks, 1987; Rhodes et al., 1989; Lea et al., 1990; Kumar and Abrol, 1990).

Regarding the status of the ubiquitously occurring plant GDH in NH_3 assimilation, the opinion is divided. Lea et al. (1990) have ruled out the possibility of any aminating role of the enzyme *in situ* and they have proposed that the enzyme is only involved in the catabolism of l-glutamate to provide carbon skeleton for the intermediary metabolism and for the NH_3 assimilation by the GOGAT cycle under carbon-limiting conditions. This hypothesis has recently been supported experimentally by Robinson et al. (1991; 1992) in carrot cell suspension cultures and it has been demonstrated that NH_3 is not incorporated into glutamate in the presence of methionine sulfoximine even when the activity of NADH-GDH is increased. The enzyme is suggested to apparently play a somewhat similar role identified for GDH in animal cells (Frieden, 1963).

Another opinion emerging from the pertinent literature emphasizes for the need of a reassessment of aminating role of GDH, in spite of the consensus for the GOGAT cycle as major pathway for NH_3 assimilation in plants (Oaks, 1986; Srivastava and Singh, 1987; Yamaya and Oaks, 1987; Rhodes et al., 1989). The facts favouring this opinion are :-

(1)   A high K_m value for NH_3 and other substrates for GDH may be affected by various external' and internal factors, and the values estimated may not be a true reflection of what might be happening *in situ* (Srivastava and Singh, 1987 and references therein). Pahlich and Gerlitz (1980) have demonstrated a biphasic K_m for NH_3 to plant GDH which is dependent upon the concentration of NH_3. Further, the catalytic properties of GDH may be modified by the availability of effectors such as Ca^{+2} (Nagel and Hartmman, 1980; Pahlich and Joy, 1981; Yamaya et al., 1984; Yamaya and Oaks, 1987; Das et al., 1989; Loulakakis and Roubelakis-Angelakis, 1990 a) and may change its affinity for binding with the substrate (s).

(2)   The aminating and deaminating activities of the enzyme respond in a different manner to the nitrogen sources in maize (Singh and Srivastava 1982; 1983; Loyola-Vargas and de Jimenez, 1984, Table 1); to the developmental stages and dark stress in *Arabidopsis* (Cammaerts and Jacobs, 1985), to Ca^{+2} in grapevine (Loulakakis and Roubelakis-Angelakis, 1990) and it would be rather unrealistic to assume that only the deaminating role of enzyme was physiologically significant (Table 1). Loulakakis and Roubelakis-Angelakis (1991) have shown that anabolic reaction of GDH is correlated with α -polypeptide and catabolic reaction of the enzyme with β -polypeptide and both functions are apparently related with isoenzyme composition in the tissue. Further isoforms related with different functions of GDH have been demonstrated by differential electrophoretic mobility towards the opposite poles (Loulakakis and Roubelakis-Angelakis, 1991; Watanabe et al., 1992).

(3)   During pulse labelling studies, glutamate is also labelled alongwith glutamine. Thus it is difficult to exclude a role of GDH in glutamate formation.

(4)   Magalhaes et al. (1990) have demonstrated a 40-50 % lower rate of root NH_3 assimilation in GDH-1 null mutant of maize. Magalhaes (1991) has also shown the incorporation of ^{15}N into glutamate in tomato roots in spite of pretreatment with MSO, an inhibitor of GS.

(5)   Supply of MSO alongwith NO_3^-/NH_4^+ in some cases show very low accumulation of NH_3 in roots of several C_3 and C_4 plants and it also increases the activities of GS and GDH if supplied with inorganic N (Martin et al., 1983; Cammaerts and Jacobs, 1985).

Perhaps it would be worthwhile to compare the aminating and deaminating efficiency of GDH under various nutritional and environmental conditions which are known to influence ammonia assimilation, to gain some insight into the possible role of GDH in plants.

Table 1. Differential response of aminating and deaminating activities of GDH to some exogenous factors.

Species	Organs	Factor/ Cause	Activity	Response	Reference
Arabidopsis thaliana	Cultured plantlets	Darkness	Aminating	Increase	Cammaerts and Jacobs (1985).
			Deaminating	Decrease	-do-
		Darkness + light	Aminating	No effect	-do-
			Deaminating	Decrease	-do-
	Leaves	$NH_4^+/NH_4^+ +$ Sucrose	Aminating/	Increase	-do-
			Deaminating	Decrease	-do-
	Roots	$NH+_4/NH+_4^+$ Sucrose	Aminating/ Deaminating	Increase	-do-
Brasica napus	Leaves	Electrophoretic separation	Aminating	Anodal isozyme	Watanabe et al. (1992).
			Deaminating	Cathodal isozyme	-do-
Vitis vinifera	Leaves	Ca^{+2}	Aminating	Increase	Loulakakis and Roubelakis-Angelakis (1990).
			Deaminating	No effect	-do-
Zea mays	Root, Shoot, Leaves	NH_4^+/NO_3^-	Aminating	Increase	Singh and Srivastava (1982). -do-
			Deaminating	Decrease	
	Root	Amino-acids	Aminating	Decrease	Singh and Srivastava (1983).
			Deaminating	Increase	-do-
	Leaves		Aminating	No effect	-do-
			Deaminating	Increase	-do-
	Callus	Glutamate/ urea	Aminating	No effect	Loyola-Vargas and de Jimenej (1984).
			Deaminating	Decreasee	-do-
	Root	Glutamate	Aminating	Increase	-do-
			Deaminating	Decrease	-do-

A. Roots and Cell Cultures

Plant roots and cell cultures are not exactly similar in nature but both of them are usually non-chlorophyllous and directly exposed to the nutrients and the environment. The response of enzymes related to NH_3 assimilation have been found largely similar in two systems (Miflin and Lea, 1982). Roots and cell cultures contain a large amount of GS, GOGAT and GDH activities and it is a general understanding that the major pool of NH_3 irrespective of its origin and source is assimilated/reassimilated via the GOGAT cycle as in other plant organs. Though the role of plant GDH in other tissues is doubtful (Lea et al., 1990), roots are still considered as the organ where GDH might be involved in the synthesis of glutamate in addition to its role in glutamate catabolism (Oaks, 1986; Yamaya and Oaks, 1987; Srivastava and Singh, 1987; Loyola-Vargas et al., (1987); Rhodes et al., 1989; Sengar and Srivastava, 1990; Sechley et al., 1991; Magalhaes, 1991; Magalhaes and Huber, 1992).

The evidences demonstrated in favour of the aminating role of root GDH include those by Loyola-Vargas et al. (1987) in *Canavalia ensiformis*, Cammaerts and Jacobs (1983; 1985) in *Arabidopsis*, Sechley et al., (1991) in *Cichorium intybus*, Magalhaes et al. (1990) in GDH-1 null mutant of maize and Magalhaes (1991) and Magalhaes and Huber (1992) in tomato. Magalhaes and coworkers have shown incorporation of NH_3 into glutamate in tomato roots even in the presence of MSO. A general increase in NADH-GDH by exogenously supplied NH_3, which has been demonstrated to be at the level of enzyme synthesis (see Srivastava and Singh, 1987) and more recently at mRNA level (see Bhadula and Shargool, chapter 8 this volume) is believed to be an adaptive feature of the tissue so that it can detoxify higher level of NH_3. But Robinson et al. (1991, 1992) have shown in carrot cell suspension cultures, using NMR and other modern techniques that even the elevated level of GDH under inducible conditions were actually not playing any role in the detoxification of tissue NH_3 and the enzyme was rather involved in deaminating l-glutamate to produce 2-oxoglutarate.

Another belief that GDH may play a key aminating role in senescing tissues and during stresses when the GOGAT cycle is inefficient, has also not found much support recently. It should be, however, noted that response of the enzymes of NH_3 assimilation including NADH-GDH is not uniform to each stress. For example, Miranda-Ham and Loyola-Vargas (1988) have shown a switch on for NH_3 assimilation in *Canavalia* root from GDH pathway to the GOGAT cycle under water sress but not under the salt stress. An increase in GDH and GS in horsegram roots (Sudhakar and Veeranjaneyulu, 1988) and GDH in rice callus cultures (Subhashini and Reddy, 1990) have been observed whereas Misra and Dwivedi (1990) have shown a negative response of GDH in *Phaseolus aureus* roots under salt stress. Studying a seasonal fluctuation in the roots of the persistent weedy perennial *Cichorium intybus*, Sechley et al. (1991) have shown that during winter NR, GS and Fd-GOGAT activities increase where as GDH shows an elevated activity during summer months.

A hypothesis presented by Oaks (1986) that aminating role of GDH in plant roots is necessary to maintain the level of glutamate, which is required if the glutamine is exported from the organ, appears to be valid.

B. Leaves and Shoots

Reduction of nitrate is the major source of NH_3 in the shoots and leaves of C_4 plants. On the other hand, the rate of NH_3 production in the glycine-serine conversion reaction during photorespiration in a C_3 plant leaf is approximately 10 times more than the normal rate of nitrate assimilation (Keys et al., 1978; Lea et al., 1990). Deamination and reassimilation of the secondary sources such as arginine, asparagine and other amino acids and storage nitrogenous forms are also important NH_3 generating reactions in the shoots and leaves of various plants especially during maturation and senescence of the organ (see Srivastava and Singh, 1987 and references there in). Several recent studies have demonstrated that the GOGAT cycle (catalysed by the cytosolic and chloroplastic isoforms of GS and NADH and Fd-dependent GOGAT enzymes) is the major route for NH_3 assimilation in leaves (see Stewart et al., 1980; Miflin and Lea, 1980; 1982; Oaks, 1986; Srivastava and Singh, 1987; Yamaya and Oaks, 1987; Rhodes et al., 1989; Lea et al., 1990; Kumar and Abrol, 1990 for reviews). It has also been suggested that GDH may play a significant role in the net synthesis of glutamate in the leaf mitochondria especially to supply glutamate required for the normal protein synthesis (Oaks, 1986; Yamaya and Oaks, 1987). On the other hand, Day et al. (1988) have shown glutamate oxidation by the mitochondrial preparations from soybean cotyledon and leaf and have suggested a role of mitochondria in protein reserve mobilization and glutamate metabolism during seedling growth.

Lea et al. (1990), however, have stated that most of the NH_3 liberated during photorespiratory nitrogen-cycle in the mitochondria passes directly into the chloroplasts where it gets converted to glutamine by the chloroplastic isoforms of GS; as in most C_3 plants, the cytosolic GS is not efficient enough to carry out the high rate of NH_3 recycling required. A major role of the GOGAT cycle in the reassimilation of photorespiratory NH_3 in many plants (Lea et al., 1990; Yu and Woo, 1991) and in the active assimilation and reassimilation of remobilized NH_3 in pre and post senescing phases of wheat leaves during grain development (Garg et al., 1991) has been suggested. A major change in the understanding of NH_3 assimilation during senescence and environmental stresses can be visualized on the basis of recent studies. The studies now indicate that the GOGAT cycle may be important pathway for assimilation of NH_3 under such conditions as well (Berger et al., 1985; Miranda-Ham and Loyola-Vargas, 1988; Misra and Dwivedi, 1990; Kamachi et al., 1991; Peeters and Van Laere, 1992; Yamaya et al., 1992), although the operation of alternate pathways may also depend upon the nature of stress and tissue being studied (Miranda-Ham and Loyola-Vargas, 1988; Sudhakar and Veeranjaneyulu, 1988). The physiological significance of increased activity of the plant GDH during various stresses such as darkness (Cammaerts and Jacobs, 1985; Peeters and Van Laere, 1992), and heavy metal pollution (Chugh et al., 1992; S. Dabas and R.P. Singh, unpublished data) are now being re-examined following the study of Robinson et al., (1991, 1992). Peeters and Van Laere (1992) have shown a gradual decrease in the levels of free amino acids and NH_3 at the later stages of senescence which could be in part due to volatilization of NH_3 from the shoot parts. They have suggested, however, that the induced level of GDH might as well cause NH_3 assimilation to reduce its accumulation (Peeters and Van Laere 1992).

Yamaya and his associates have shown a role for GS in the remobilization of leaf nitrogen during the natural senescence in rice leaves and suggested that GS-1 in senescing leaf blades is responsible for the synthesis of glutamine which is then transferred to the growing tissues in the rice plants (Kamachi et al., 1991, 1992 a, b). They have further demonstrated that NADH-GOGAT is important for the synthesis of glutamate from the glutamine which is transported from senescing source tissues through the phloem to the non-green sink tissues in rice leaves (Yamaya et al., 1992). It is shown that GS-1 mRNA increased gradually during senescence when GS, Fd-GOGAT and RUBISCO proteins decreased (Kamachi et al., 1991).

Observations of Kumar et al. (1984) for a higher level of the endogenous free NH_3 in the C_4 than in the C_3 plants in the absence of MSO and a reserve trend in the presence of the inhibitor indicate some differences in the mechanism for NH_3 assimilation in two groups of plants which needs a thorough investigation to evaluate the actual path of *in situ* NH_3 assimilation.

C. Root Nodules

When legume roots are infected with *Rhizobium* there is a dramatic increase in the activities of the enzymes of NH_3 assimilation i.e. GS, GOGAT and GDH and also in phosphoenolpyruvate carboxylase (Duke and Ham, 1976; Gadal, 1983; Vance et al., 1983). The faster rate of import and utilization of photosynthate permits a faster export of reduced nitrogen. Miflin and Lea (1980, 1982) have presented evidences for the GOGAT cycle operation into the assimilation of symbiotically fixed NH_3 in the root nodules. Ortega et al., (1992) have pointed out that the C/N balance within the nodule and not NH_3 derived from the bacteroid *per se* is the a primary modulating factor of nodule GS-γ isoform, in bean.

Srivastava and Singh (1987) have suggested for a possible aminating role of GDH under certain nutritional and environmental conditions which depends upon plant species and type of symbiotic association.

D. Developing Seeds and Storage Tissues

The major sources of NH_3 in the developing seeds and storage tissues are glutamine, arginine, asparagine, ureides and other secondary nitrogenous compounds. Though the assimilation of NH_3 in such tissues has not been studied extensively, evidences are available in the favour of the operation of the GOGAT cycle as major route for the glutamate synthesis (Miflin and Lea, 1980, 1982; Lea et al., 1990). Garg et al. (1991) have shown relatively higher activities of GS and GOGAT in Shera wheat leaves compared with C-306 wheat leaves and suggested that it might lead to more active NH_3 assimilation. They have also shown an increased GDH activity during grain filling period which was substantially higher in the second leaf of Shera. Ta (1991) has demonstrated that stalk of maize has an important function as a temporary reservoir for nitrogen in maize during the ear development.

V. Conclusions and Future Prospects

Studies of Robinson et al. (1991, 1992), using sophisticated techniques provide evidences for the operation of the GOGAT cycle in the assimilation of NH_3 in higher plants, and for a catabolic role of the GDH, involving deamination of glutamate to 2-oxoglutarate. Although, glutamate catabolism by GDH is of common occurrence in animals, the physiological significance of such a process in plants, specially in the green photosynthesizing leaves is not clear. The observations questioning the role of GDH in the amination process include; high K_m for NH_3, labelling of glutamine first during pulse feeding of $^{15}NH_4^+$, and the cessation of NH_3 assimilation in the presence of the inhibitors of GS and GOGAT enzymes. But there have been exceptions also to these observations; and the presence of high GDH levels in most plant tissues indicates that more rigorous experimentation, specially at the molecular level are required to assess the role of GDH in amination process. Perhaps the development of mutants of various enzymes of NH_3 assimilation pathway, and methodologies to estimate *in vivo* concentration of NH_3, glutamate and glutamine and of enzyme proteins, may help in this assessment.

Acknowledgements

The assistance provided by my research students Ms. Nisha Bharti and Mr. Gulshan Kumar is acknowledged.

Literature Cited

Berger, M.G., Woo, K.C., Wong, S.C., and Fock, H.P. 1985. Nitrogen metabolism in senescent flag leaves of wheat (*Triticum aestivum* L.) in the light. *Plant Physiol.* **78**: 779-783.

Bhadula, S.K., and Shargool, P.D. 1991. A plastidial localization and origin of l-glutamate dehydrogenase in a soybean cell culture. *Plant Physiol.* **95**: 258-263.

Cammaerts, D., and Jocobs, M. 1985. A study of the role of glutamate dehydrogenase in the nitrogen metabolism of *Arabidopsis thaliana*. *Planta* **163**: 517-526.

Chugh, L.K., Gupta, V.K., and Sawhney, S.K. 1992. Effect of cadmium on enzymes of nitrogen metabolism in pea seedlings. *Phytochemistry*. **31**: 395-400.

Das, R., Sharma, A.K., and Sopory, S.K. 1989. Regulation of NADH-glutamate dehydrogenase activity by phytochrome, calcium and calmodulin in *Zea mays*. *Plant Cell Physiol.* **30**: 317-323.

Day, D.A., Caroline, L.S., Azcon-Bieto, J., Dry, I.B., and Wiskich, J.T. 1988. Glutamate dehydro-
     genase by soybean cotyledon and leaf mitochondria. *Plant Cell Physiol.* **29**: 1193-1202.
Dougall, D.K. 1974. Glutamate synthase in extracts of carrot cell cultures. *Biochem. Biophys. Res.
     Comm.* **58**: 639-646.
Duke, S.H., and Ham, G.E. 1976. The effect of nitrogen additions on N_2-fixation and on glutamate
     dehydrogenase and glutamate synthase activities in nodules and roots of soybeans inoc-
     ulated with various strains of *Rhizobium japonica. Plant Cell Physiol.* **17**: 1037-1044.
Frick, W., and Pahlich, E. 1992. Malate : A possible source of error in the NAD-glutamate dehydro-
     genase assay. *J. Exp. Bot.* **43**: 1515-1518.
Frieden, C. 1963. Glutamate dehydrogenase. In *The Enzymes* (Boyers, P.D., Lardy H., and Myr-
     back K., eds). Vol 7. Academic Press, New York. pp. 3-24.
Gadal. P. 1983. Phosphoenolpyruvate carboxylase and nitrogen fixation. *Physiol. Veg.* **21**: 1069-1074.
Garg, N., Singh, R., and Batra, V.I.P. 1991. Enzymes of ammonia assimilation in leaves of high
     and low protein wheat during grain development. *Proc. Indian Nat. Sci. Acad.* **B57**: 63-68.
Hirel, B., Perrot-Richenmann, C., Mandinas. B., and Gadal, P. 1982. Glutamine synthetase in alder
     (*Alnus glutinosa*) root nodules. *Physiol. Plant.* **35**: 197-203.
Joy, K.W. 1988. Ammonia, glutamine and asparagine : a carbon nitrogen interface. *Can. J. Bot.*
     66: 2103-2109.
Kamachi, K., Yamaya, T., Mae, T., and Ojima, K. 1991. A role for glutamine synthetase in the remo-
     bilization of leaf nitrogen during natural senescence in rice leaves. *Plant Physiol.* **96**: 411-417.
Kamachi, K., Yamaya, T., Hayakawa, T., Mae, T., and Ojima, K. 1992 a. Changes in cytosolic gluta-
     mine synthetase polypeptide and its mRNA in a leaf blade of rice plants during natural sene-
     scence. *Plant Physiol.* **98**: 1323-1329.
Kamachi, K., Yamaya, T., Hayakawa, T., Mae, T., and Ojima, K. 1992 b. Vascular bundle specific
     localization of cytosolic glutamine synthetase in rice leaves. *Plant Physiol.* **99**: 1481-1486.
Keys, A.J., Bird, I.F., Cornelius, M.J., Lea, P.J., Wallsgrove, R.M., and Miflin, B.J. 1978. Photo-
     respiratory nitrogen cycle. *Nature* **275**: 741-643.
Kumar, P.A., Nair, T.V.R., and Abrol. Y.P. 1984. Effect of photorespiratory metabolites, inhibitors
     and methionine sulfoximine on the accumulation of ammonia in the leaves of mungbean and
     Amaranthus. Plant Sci. Lett. **33**: 303-307.
Kumar. P.A., and Abrol. Y.P. 1990. Ammonia assimilation in higher plants. In *Nitrogen in Higher
     Plants* (Abrol, Y.P. ed.). John Wiley and Sons. New York. pp. 159-180.
Lara, M., Cullimore, J.V., Lea, P.J., Miflin, B.J., Johnson, A.W.B., and Lamb, J.W. 1983. Appear-
     ance of a novel form of plant glutamine synthetase during nodule development in *Phaseolus
     vulgaris* L. *Planta* **157**: 254-258.
Lea, P.J., and Miflin, B.J. 1974. An alternative route for nitrogen assimilation in higher plants. *Nature*
     251: 614-616.
Lea, P.J., and Fowden, L. 1975. The purification and properties of glutamine-dependent aspara-
     gine synthetase isolated from *Lupinus albus. Proc. Royal. Soc. (London) B* **192**: 13-26.
Lea, P.J., and Miflin, B.J. 1980. Transport and metabolism of asparagine and other nitrogen com-
     pounds within the plant. In *The Biochemistry of Plant. A Comprehensive Treatise: Amino
     Acids and Their Derivatives* (Miflin, B.J. ed.), Vol 5. Academic Press, New York. pp. 569-607.
Lea, P.J., Robinson, S.A., and Stewart, G.R. 1990. The enzymology and metabolism of glutamine,
     glutamate and asparagine. In *The Biochemistry of Plants. A Comprehensive Treatise. Inter-
     mediary Nitrogen Metabolism,* (Miflin, B.J. and Lea, P.J. ed.), Vol **16**: Academic Press, New
     York. pp. 121-159.
Loyola-Vargas, V.M., and de Jimenez, E.S. 1984. Differential role of glutamate dehydrogenase in
     nitrogen metabolism of maize tissues. *Plant Physiol.* **76**: 536-540.
Loyola-Vargas, V.M., Yanez, A., Caldera, J., Oropeza, C., Robert, M.L., Quriroz, J., and Scorer,
     K.N. 1988. Nitrogen metabolism in *Canavalia ensiformis* L. Dc II. Changing activities of nitro-
     gen assimilating enzymes during growth. *J. Plant Physiol.* **132**: 289-293.

Loulakakis, C.A., and Roubelakis-Angelakis, K.A. 1990. Intercellular localization and properties of NADH-glutamate dehydrogenase from *Vitis vinifera* L. Purification and characterization of the major leaf isoenzyme. *J. Exp. Bot.* **41**: 1223-1230.

Loulakakis. C.A., and Roubelakis-Angelakis, K.A. 1991. Plant NAD(H)-glutamate dehydrogenase consists of two subunit polypeptides and their purification in the seven isoenzymes occurs in an ordered ratio. *Plant Physiol.* **97**: 104-111.

Magalhaes, J.R., Patrick, G.C.J., and Rhodes, D. 1990. Kinetics of $^{15}NH_4^+$ assimilation in *Zea mays*. Preliminary studies with a glutamate dehydrogenase (GDH 1) null mutant. *Plant Physiol.* **94** : 646-656.

Magalhaes, J.R. 1991. Kinetics of $^{15}NH_4^+$-assimilation in tomato plants. Evidence of NH_4^+-assimilation via GDH in tomato roots. *J. Plant Nutr.* **14**: 1341-1354.

Magalhaes, J.R., and Huber, D.M. 1991. Response of ammonium assimilation enzymes to nitrogen treatments in different plant species. *J. Plant Nutr.* **14**: 175-185.

Martin, F., Winspear, M.J., McFarlane, J.D., and Oaks, A. 1983. Effect of methionine sulfoximine on the accumulation of ammonia in C_3 and C_4 leaves. The relationship between NH_4^+ accumulation and photorespiratory activity. *Plant Physiol.* **71**: 177-181.

Matoh, T., and Takahashi, E. 1982. Changes in the activities of ferredoxin and NADH-glutamate synthase during seedling development of peas. *Planta* **54**: 289-294.

McFarland, R.H., Guevarra, J.G., Becker, R.R., and Evans, H.J. 1976. The purification of glutamine synthetase from the cytosol of soybean root nodules. *Biochem. J.* **53**: 411-415.

Miflin, B.J., and Lea, P.J. 1980. Ammonia assimilation. *In The Biochemistry of Plants : A Comprehensive Treatise; Amino Acids and Their Derivatives* (Miflin, B.J. ed.), Vol 5: Academic Press, New York. pp. 169-202.

Miflin, B.J., and Lea, P.J. 1982. Ammonia Assimilation and amino acid metabolism. *In 'Encyclopedia of Plant Physiol". New Series* (Pirson, A. and Zimmerman, M.H. eds.), Vol. 14A. Springer Verlag Berlin, Heidelberg. pp. 5-64.

Miranda-Ham, M.L., and Loyola Vargas, V.M. 1988. Ammonia assimilation in *Canavalia ensiformis* plants under water and salt stress. *Plant Cell Physiol.* **29**: 747-753.

Misra, N., and Dwivedi, U.N. 1990. Nitrogen assimilation in germinating *Phaseolus aureus* seeds under saline stress. *J. Plant Physiol.* **135**: 719-724.

Nagel, M., and Hartmann. T. 1980. Glutamate dehydrogenase from *Medicago sativa* L. Purification and comparative kinetic studies of organ specific multiple forms. *Z. Naturforsch.* **35** *c*: 406-415.

Nauen, W., and Hartmann, T. 1980. Glutamate dehydrogenase from *Pisum sativum* L. Localization of the multiple forms and glutamate formation in isolated mitochondria. *Planta* **148**: 7-16.

Oaks, A. 1986. Biochemical aspects of nitrogen metabolism in a whole plant context. *In Fundamental, Ecological and Agricultural Aspects of Nitrogen Fixation in Higher Plants* (Lambers, H., Neeteson, J.J. and Stulen, I. eds.), Martinus Nijhoff Publishers, Dordrecht. pp. 133-151.

Ortega, J.L., Sanchez, F., Soberon, M., and Flores, M.L. 1992. Regulation of nodule glutamine synthetase by CO_2 levels in bean (*Phaseolus vulgaris* L.). *Plant Physiol.* **98** : 584-587.

Pahlich, E., and Joy, K.W. 1981. Glutamate dehydrogenase from pea roots : Purification and properties of the enzymes. *Can. J. Biochem.* **49**: 127-138.

Pahlich, E., and Gerlitz, C. 1980. Deviations from Michaelis-Menten behaviour of plant glutamate dehydrogenase with ammonium as the variable substrate. *Phytochemistry.* **19** : 11-13.

Planke, K., Kennedy, I.R., Quispel, A., and Van Brussel, A.A.N. 1977. Location of nitrogenase and ammonium assimilatory enzymes in bacteriods of *Rhizobium leguminosarum* and *Rhizobium lupini*. *J. Gen. Microbiol.* **102** : 95-104.

Peeters, K.M.U., and Van Laere, A.J. 1992. Ammonium and amino acid metabolism in excised leaves of wheat (*Triticum aestivum*) senescing in dark. *Physiol. Plant.* **84**: 243-249.

Rao, A.S., and Singh, R. 1988. Ureides metabolism in nodulated legumes. *In Advances in Frontier Areas of Plant Biochemistry*. (Singh, R. and Sawhney, S.K. eds.), Prentice Hall (India), New Delhi. pp. 281-316.

Rhodes, D., Sims, A.P., and Fokes, B.F. 1980. Pathway of ammonia assimilation in illuminated *Lemna minor. Phytochemistry* 19: 357-365.

Rhodes, D., Rich, P.J., and Brunk, D.G. 1989. Amino acid metabolism of *Lemna minor* L. IV. ^{15}N labelling kinetics of the amide and amino groups of glutamine and asparagine. *Plant Physiol.* 89: 1161-1171.

Robinson, S.A., Slade, A.P., Fox, G.G., Phillips, R., Ratcliffe, R.G., and Stewart, G.R., 1991. The role of glutamate dehydrogenase in plant nitrogen metabolism. *Plant Physiol.* 95: 509-516.

Robinson, S.A., Stewart, G.R., and Phillips, R. 1992. Regulation of glutamate dehydrogenase activity in relation to carbon limitation and protein catabolism in carrot cell suspension cultures. *Plant Physiol.* 98: 1190-1195.

Sechley, K.A., Oaks, A., and Bewley, J.D. 1991. Enzymes of nitrogen assimilation undergo seasonal fluctuations in roots of the persistent weedy perennial *Cichorium intybus. Plant Physiol.* 97: 322-329.

Sengar, R.S., and Srivastava, H.S. 1990. Effect of age on NADH-glutamate dehydrogenase activity and ammonia assimilation in maize seedlings. *Indian J. Plant Physiol.* 33: 40-45.

Sieciechawicz, K., Joy, K.W., and Ireland, R.J. 1988. The metabolism of asparagine in plants. *Phytochemistry* 27: 663-671.

Singh. R.P., and Srivastava, H.S. 1982. Glutamate dehydrogenase activity and assimilation of inorganic nitrogen in maize seedlings. *Biochem. Physiol. Pflanzen.* 177 : 633-642.

Singh, R.P., and Srivastava, H.S. 1983. Regulation of glutamate dehydrogenase activity by amino acids in maize seedlings. *Physiol. Plant* 57: 549-554.

Singh, R.P., Mehta, P., and Srivastava, H.S. 1984. Characteristics of ammonium absorption by excised root and leaf tissues of maize. *Physiol. Plant.* 60: 119-129.

Shelp, B.J., Sieciechowicz, K., Ireland, R., and Joy, K.W. 1985. Determination of urea and ammonia in leaf extracts: Application to ureide metabolism. *Can J. Bot.* 63: 1135-1140.

Srivastava, H.S., and Ormrod, D.P. 1984. Effects of nitrogen dioxide and nitrate nutrition on growth and nitrate assimilation in bean leaves. *Plant Physiol.* 76: 418-423.

Srivastava, H.S., and Singh, R.P. 1987. Role and regulation of l-glutamate dehydrogenase activity in higher plants. *Phytochemistry.* 26: 597-610.

Stewart, G.R., Mann. A., and Fentem, P.A. 1980. Enzymes of glutamate formation: glutamate dehydrogenase, glutamine synthetase and glutamate synthase. *In The Biochemistry of Plants. A comprehensive Treatise; Amino Acids and Their Derivatives* (Miflin, B.J. ed.), Vol. 5 Academic Press, New York. pp. 271-327.

Subhashini, K., and Reddy, G.M. 1990. Effect of salt stress on enzyme activities in callus cultures of tolerant and susceptible rice cultivars. *Indian J. Exp. Biol.* 28 : 277-279.

Sudhakar, C.,and Veeranjaneyulu, K. 1988. Effect of salt stress on enzymes of nitrogen metabolism in Horsegram *Dolichos biflorus* L. *Indian J. Exp. Biol.* 26: 618-620.

Ta, C.T. 1991. Nitrogen metabolism in the stalk tissue of maize. *Plant Physiol.* 97: 1375-1380.

Tazima, S., Yatazawa, M., and Yamamoto, Y. 1977. Allantoin production and its utilization in relation to nodule formation in soybean. *Soil Sci. Plant Nutr.* 23: 225-235.

Tempest, D.W., Meers, J.L., and Brown, C.M. 1970. Synthesis of glutamate in *Aerobactor aerogens* by a hitherto unknown route. *Biochem. J.* 117: 405-407.

Thimann, K.V. 1980. The senescence of leaves. *In Senescence in Plants* (Thimann, K.V. ed.), C.R.C. Press, Boca Raton, pp. 85-115.

Thomas, H. 1978. Enzymes of nitrogen mobilization in detached leaves of *Lolium temulentum* during senescence. *Planta ;* 142: 161-169.

Vance, C.R., Slade, S., and Maxwell, C.A. 1983. Alfalfa root nodule carbon dioxide fixation. I. Association with nitrogen fixation and incorporation into amino acids. *Plant Physiol.* 72: 469-473.

Yamaya, T., Oaks, A., and Matsumoto, H. 1984. Characteristics of glutamate dehydrogenase in mitochondria prepared from corn shoots. *Plant Physiol.* 76: 1009-1013.

Yamaya, T., and Oaks, A. 1987. Synthesis of glutamate by mitochondria—An anaplerotic function for glutamate dehydrogenase. *Physiol. Plant.* 79: 749-756.

Yamaya, T., Hayakawa, T., Tanasawa, K., Kamachi, K., Mae, T., and Ojima, K. 1992. Tissue distribution of glutamate synthase and glutamine synthetase in rice leaves. *Plant Physiol.* 100: 1427-1432.

Yu, J., and Woo, K.C. 1991. Correlation between the development of photorespiration and the change in activities of NH_3 assimilation enzymes in greening oat leaves. *Aust. J. Plant Physiol.* 18: 583-588.

Watanabe, M., Nakayama, H., Watanabe, Y., and Shimada, N. 1992. Induction of a specific isoenzyme of glutamate dehydrogenase during isolation and the first 48 h of culture of *Brassica napus* leaf protoplast. *Physiol. Plant.* 86: 231-235.

Nitrogen Nutrition in Higher Plants, 1995
Editors : H.S. Srivastava & R.P. Singh
Associated Publishing Co., New Delhi, India
pp. 205-217.

Glutamate Dehydrogenase: Purification, Properties and Regulation

SHAILENDRA K. BHADULA and PETER D. SHARGOOL

I. Introduction

The enzyme l-glutamate dehydrogenase (GDH, EC 1.4.1.2-4) catalyzes the reductive amination of 2-oxoglutarate or the oxidative deamination of glutamate in the reverse direction and is found in almost all living organisms. The GDH aminating reaction was long considered to be the primary route of ammonia assimilation in higher plants. However, since the discovery of glutamate synthase (GOGAT), GDH has been a subject of controversy and its role in ammonia assimilation has been questioned (Miflin and Lea, 1976; Stewart et al., 1980; Srivastava and Singh, 1987). It is now well accepted that under normal conditions, the major pathway for ammonia assimilation in plants involves the combined action of the enzymes glutamine synthetase (GS) and glutamate synthase, commonly known as GS/GOGAT cycle. This conclusion is based primarily on the high K_m values of GDH for ammonia and the inhibition of ammonium assimilation in the presence of inhibitors of GS/GOGAT pathway (Jain and Shargool, 1987; Srivastava and Singh, 1987). L-Glutamate dehydrogenase is thought not to function in the normal metabolic assimilation of ammonia, but rather in glutamate catabolism, providing carbon skeletons for the tricarboxylic acid cycle and thus, acting as a link between carbon and nitrogen metabolism (Miflin and Lea, 1976; Goodwin and Mercer, 1983; Robinson et al., 1991). A little evidence, however, still exists supporting the idea that GDH may have an important role to play in the assimilation/detoxification of the ammonia produced as a result of the tissue damage and degradation and under certain environmental and nutritional stress conditions (Srivastava and Singh, 1987). GDH is widely distributed throughout the plant kingdom (Stewart et al., 1980; Goodwin and

Mercer, 1983; Srivastava and Singh, 1987) except for some species of *Sphagnum* (Meade, 1984). Although in most cases, GDH in higher plants is shown to be a mitochondrial enzyme, it has also been reported to exist in the chloroplasts, non-green plastids and supernatant fractions (Stewart et al., 1980; Srivastava and Singh, 1987; Bhadula and Shargool, 1991). The enzyme has been purified and characterized from a few species of higher plants (Pahlich and Joy, 1971; King and Wu 1971; Mazurova et al., 1980; Nagel and Hartman, 1980; Scheild et al., 1980; Yamaya et al., 1984; Itagaki et al., 1986; Shargool and Jain, 1989; Loulakakis and Roubelakis-Angelakis, 1990). The aim of the present article is to discuss the general purification procedure for GDH from higher plants and cell cultures, and to briefly review its properties, possible functions and regulation.

II. Extraction of the Enzyme

GDH can be extracted from plant tissues (leaves, shoots, roots etc.) or plant cell cultures using a low molarity buffer (tricine, phosphate buffer, tris-HCl buffer). Addition of a proteolytic inhibitor to the extraction buffer is also helpful when dealing with tissues containing high proteolytic activity to minimize the loss of GDH activity. The enzyme can also be purified form cell organelles which involves (a) isolation and purification of the desired cell organelles e.g. mitochondria, chloroplasts or non-green plastids from the plant tissues and (b) extraction and purification of the enzyme from the organelles. In our laboratory, we used 0.2 M tris-HCl buffer (pH 7.5) containing 1mM EDTA and 1mM-β mercaptoethanol for GDH extraction from SB1 and SB3 cell suspension cultures of soybean (*Glycine max* L., cv. Mandarin).

The cells were grown as suspension cultures in 1000 ml B5 medium and were subcultured after 96 h in the growth medium. The enzyme extracts were prepared according to the methods described previously by Chiu and Shargool (1979). The cells were harvested and homogenized for 3 min using a bead beater and the extraction buffer described above. The ratio of cells to beads to buffer was 1:3:2. The homogenate was filtered through two layers of miracloth and centrifuged at 12,000 x g for 20 min using a Sorvall SS-34 rotor. The supernatant was termed as "crude extract" and was used for enzyme activity and purification of GDH. Protein content in the samples was determined using the method of Bradford (1976).

III. Assay of the Enzyme

The reversible reaction catalyzed by GDH can be measured in either direction by monitoring the oxidation or reduction of the coenzyme at 340 nm. In our studies, GDH activity in the aminating direction was measured by monitoring NAD(P)H oxidation at 340 nm according to the method modified from Dougall (1974). The assay mixture contained 100 mM tris-HCl buffer, pH 7.5, 10 mM 2-oxoglutarate, 100 mM ammonium chloride, 1.0 mM calcium chloride, 0.2 mM NAD(P)H and enzyme in a final volume of 0.5 ml. The activity in the deaminating direction was measured using glutamate (30mM) as substrate and 1.0 mM NAD(P) as coenzyme. One unit of GDH activity was defined as the amount of enzyme catalyzing the oxidation or reduction of 1.0 μmol of coenzyme min^{-1} at 30°C.

IV. Purification

Depending on the type of plant tissues/cell cultures/organelles used for GDH extraction, various methods have been used to purify the enzyme. In general, the purification of GDH can be achieved by using the following steps: (a) acid precipitation and/or ammonium sulfate fractionation of the crude supernatant, (b) desalting by dialysis or sephadex column chromatography, (c) DEAE-cellulose chromatography, (d) affinity chromatography, (e) gel filtration and (f) concentration of the purified GDH protein. Acid precipitation of the crude extracts has been shown to be an important step in the purification of GDH from roots (Pahlich and Joy, 1971) and seeds (Scheild et al., 1980).

In our studies, GDH was purified to homogeneity from the crude extracts of soybean SB1 and SB3 cells using the method modified from Yamaya et al (1984). This procedure for GDH purification from soybean cell cultures represents a typical method for plant enzyme, involving enzyme extraction from the cultures, ammonium sulfate fractionation, dialysis, DEAE-cellulose, sephadex, and affinity chromatography. With some modifications, this method for GDH purification should also be applicable to other plant tissues.

The crude enzyme extracts prepared as described above were subjected to ammonium sulfate precipitation and the 30-60% fraction suspended in a small volume of 50% extraction buffer was dialyzed against the same buffer overnight at 4°C. The dialysed preparation was centrifuged at 12,000 x g for 10 min. The pellet was discarded and the supernatant was loaded onto a DEAE cellulose column (1.6 x 35 cm) previously equilibrated with 50 mM tricine buffer (pH 8.5) containing 5 mM 2-oxoglutarate, 1 mM β-mercaptoethanol and 0.1 mM EDTA. The column was washed with 10 ml of the same buffer. GDH bound to the column was then eluted using a linear gradient derived from 250 ml of tricine buffer and 250 ml of 0.5 M KCl dissolved in tricine buffer. Serveral 4.0 ml fractions were collected and the fractions containing GDH activity were pooled. Solid ammonium sulfate (70% saturation) was then added to precipitate the protein. After centrifugation at 12,000 x g for 15 min, the pellet was suspended in 2 ml 50 mM tris-HCl buffer (pH 7.0) containing 1 mM $CaCl_2$, 0.5 mM EDTA, and 1 mM β-mercaptoethanol. The samples were passed through a sephadex G-200 column (3.2 x 45 cm) and the GDH eluted with the same buffer. Fractions containing maximal GDH activity were pooled and concentrated using an Amicon ultrafiltration cell (W.R. Grace and Co., Danvers, MA.).

Further purification of GDH was achieved by passing the samples through an affinity column (1 x 8 cm) of NAD-agarose (Sigma Chemical Co., St. Louis, MO.) previously equilibrated with tris-HCl buffer (pH 7.0) containing 1 mM $CaCl_2$, 0.5 mM

EDTA, and 1 mM β-mercaptoethanol. The column was washed with 50 ml of the same buffer and GDH activity was then eluted in a single peak with a linear gradient derived from 25 ml of tris buffer (pH 7.0) and 25 ml of 0.5 mM NADH dissolved in tris buffer.

To check the purity of the GDH protein, the fraction containing GDH peak eluted from the NAD-agarose column was loaded onto a 7% native polyacrylamide gel slab following the method of Davis (1964). Fig. 1B shows the coomassie blue R-250 stained gel of GDH obtained from soybean SB1 (lane 1) and SB3 (lane 2) cells. At this stage, a single band was observed indicating that the enzyme preparation was homogeneous.

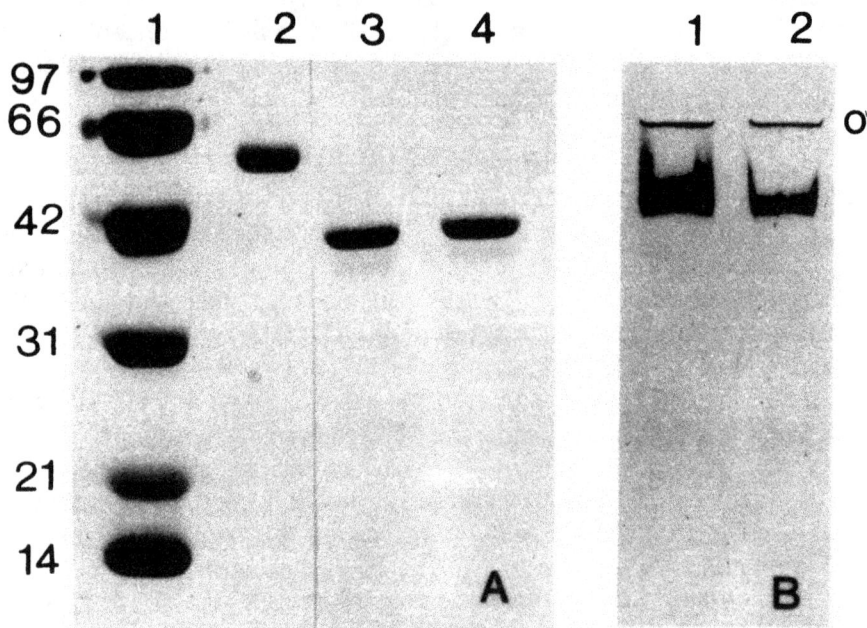

Fig. 1. Polyacrylamide gel electrophoresis (PAGE) of purified glutamate dehydrogenase. A. SDS-PAGE. Lane 1: Standard molecular weight markers. Numbers on the left indicate molecular weight. Lane 2: Purified GDH from rat liver (sigma). Lane 3: Purified GDH from SBI cells. Lane 4: Purified GDH from SB3 cells. B. Native PAGE of GDH. Lane 1: GDH from SB1 cells. Lane 2: GDH from SB3 cells. O=Origin.

V. Molecular Weight Determination

The molecular weight of the native enzyme was determined by gel filtration using a column of sephadex G-200 as described by Andrews (1964). Using standard molecular weight markers, the molecular weight of GDH protein was found to be 263,000 ± 12,000. The molecular weight of the GDH subunit was determined by sodium dodecyl sulfate-polyacrylamide gel electrophoresis (SDS-PAGE) following the method of Laemmli (1970). The purified enzyme was mixed with SDS-sample buffer and loaded onto a 10% SDS containing slab gel. Low molecular weight markers were used to determine the molecular weight of GDH-polypeptide. Fig. 1A shows the coomassie blue stained SDS-PAGE gel of purified GDH from soybean SB1 (lane 3) and SB3 (land 4) cells. A single band of approximately 41 ± 2 kD appeared in the SDS-gels suggesting a homohexameric configuration of soybean GDH.

VI. Properties

As described above, the purified enzyme from soybean cell cultures appeared to be a hexamer of six identical subunits with a molecular weight of 263,000 and the subunit molecular weight of approximately 41 kD. Structural studies on GDH from algae and other plant sources have shown that the enzyme is a hexamer with mw in the range of 200, 300, kD (Stewart et al., 1980; Srivastava and Singh, 1987). The GDH from bovine liver was also shown to be a hexamer (for reference, see Stewart et al., 1980). In yellow lupine root nodules, Ratajczak et al.(1988) also reported a hexameric configuration of GDH. Whereas the root enzyme was shown to be a hexamer of six identical subunits, the nodular enzyme was shown to be a hexamer with seven isoenzymes, five of which are suggested to be made up of two types of subunits. A similar pattern of the isoenzymes of GDH has been reported in *Arabidopsis thaliana* (Cammaerts and Jacob, 1983) Thus, in higher plants, both homohexameric and heterohexameric forms of GDH have been reported. In addition, tetrameric forms of the enzyme have also been reported in some cases. (Stewart et al., 1980; Scheild et al., 1980; Srivastava and Singh, 1987).

GDH from a number of higher plants has been shown to exist in multiple isoenzymic forms. In our studies, GDH purified from soybean cells showed a single band in native and SDS-polyacrylamide gels. A major band of GDH activity was also observed after isoelectric focussing of the crude soybean cell extracts, with an isoelectric point of 5.9. However, cells grown in high ammonium medium as sole source of nitrogen showed the induction of GDH activity associated with the appearance of a new isoenzyme (S.K. Bhadula and P.D. Shargool, unpublished results). It is well known that high ammonium concentration generally stimulates the activity/synthesis of the enzyme and also increases the number of isoenzymes in many species of higher plants (Kanamori et al., 1972; Barash et al., 1975; Lauriere and Daussant, 1983; Jain and Shargool, 1987; Bhadula and Shargool, 1991). In addition to ammonium, some amino acids also stimulate GDH isoenzymes whereas carbohydrates are known to inhibit GDH activity and isoenzymes (Rataczak et al., 1977, Maurova et al., 1980; Srivastava and Singh, 1987).

In many species, seven isoenzymes arising from the random association of two subunits have been described (Cammaerts and Jacob, 1983; Ratajczak et al., 1988). However, the total number of isoenzymes and the relative abundance of individual isoenzyme(s) varies with plant species and within the different plant parts, subcellular compartments, developmental stages, and environmental and nutritional conditions (Nauen and Hartmann, 1980; Srivastava and Singh, 1987). As many as seventeen isoenzymes were reported in detached pea shoots floated on tap water

(Nauen and Hartmann, 1980). In *Arabidopsis thaliana,* Cammaerts and Jacobs (1983) have reported seven anodal migrating isozymes. Two electrophoretic variants of GDH, namely slow (S) and normal (N) were discovered by these workers in 85 geographical races of *Arabidopsis* and 22 isoenzymes of GDH were identified in the heterozygotes between the two variants. Isoenzymes of GDH were also shown to be affected by light-dark period (Lauriere et al., 1981; Lauriere and Daussant, 1983) and during nodule development (Mazurova et al., 1980; Srivastava and Singh, 1987; Ratajczak et al., 1988).

GDH isoenzymes with different relative coenzyme specificities have also been reported (Stewart et al., 1980; Rataczak et al., 1988). A detailed discussion of the coenzyme specificity of different glutamate dehydrogenases from various plant species is given in previous reviews on GDH (Stewart et al., 1980; Srivastava and Singh, 1987). GDH from both the cell cultures used in this study showed high activity with NADH and NADPH in the aminating direction (Jain and Shargool, 1987; Bhadula and Shargool, 1991).The rate of the deaminating reaction was found to be very low with NAD^+ (about 10% of the aminating reaction) and was hardly detectable when $NADP^+$ was used as a coenzyme (Jain and Shargool, 1987). However, using mixed substrate methods, it has also been shown that NAD(H) and NADP(H) bind to the same active site in the chloroplastic and mitochondrial GDH enzymes, thus suggesting a true dual cofactor specificity (Stewart et al., 1980).

The pH optima for the aminating and deaminating GDH catalyzed reactions in some higher plants have been reported (Stewart et al., 1980). In general, the pH optimum for the aminating reaction varies from 7.5 to 8.2 whereas for the deaminating reaction, the values are generally higher (8.9-10.0). The enzyme from pea roots, however, was shown to have an optimal reaction at pH 8.0 in aminating as well as deaminating direction (Pahlich and Joy, 1971). The possible implications of the differences in the optimum pH for aminating and deaminating reactions of GDH on the reaction mechaninsm have been discussed by Stewart et al (1980).

The K_m values for the various substrates of GDH enzymes isolated from different species of higher plants are shown in Table 1. It is clear that these values differ considerably for GDH enzymes isolated from different sources. The most important point of interest is that in most species studied, the K_m values for ammonia are several times higher (4.76 - 70.0 mM, Table 1) than the normal tissue concentrations of ammonia. This is, in fact the major objection to the role of GDH in ammonia assimilation. Some exceptions to this may occur and the tissue concentrations of ammonia may be higher, especially for plants grown under high ammonium conditions or when ammonium is the sole source of nitrogen as in tissue/cell culture experiments (Stewart et al., 1980; Srivastava and Singh, 1987; Jain and Shargool, 1987). Even under the high ammonium conditions, inhibition of glutamine synthetase by methionine sulfoximine (MSX) results in accumulation of ammonia in the culture medium suggesting that GDH is not involved in ammonia assimilation (Jain and Shargool, 1987). It should also be noted that K_m values are highly dependent on the concentrations of other substrates of the GDH reaction. Also, calcium has a very significant effect on the apparent K_m values for (Hadzi-Taskovic, 1990). One particular point of interest, however, is that in almost all the studies on plant GDH, the reaction equilibrium favours the synthesis of glutamate in the aminating direction. In our own laboratory, the deaminating activity of soybean GDH was only 10% of that in the aminating direction (Jain and Shargool 1987). However, in the light of recent advances on GDH reaction mechanism, it is hard to assume that the favoured aminating reaction *in vitro* reflects the actual *in vivo* picture, especially when the glutamine synthetase/glutamate synthase route of ammonia assimilation is widely accepted.

Table 1. K_m **Values (mM) for substrates of l-glutamate dehydrogenase isolated from some species of higher plants**

Plant species	Glutamate	2-Oxoglutarate	Ammonium	NAD$^+$	NADH	Reference
Carthamus tinctorius (cotyledon)	8.0	4.0	35.4	0.26	0.065	Errel et al., 1993.
Glycine max (cotyledon)	8.0-25.0	1.20	9.4	0.21-1.2	0.015-0.039	King and Yu, 1971.
Lectuca sativa (chloroplast)			5.2		0.12 *(0.118)	Lea and Thurman, 1972.
Lemna minor	2.5	1.40	33.0	0.2	0.017	Stewarts and Rhodes, 1977.
Pisum sativa (epicotyl)	3.5	0.62	70.0	0.59	0.026 *(0.66)	Davies and Texiera, 1975.
Pisum sativa (root)	7.3	3.30	38.0	0.65	0.86	Pahlich and Joy, 1971.
Turnip (mitochondria)	28.6	2.0	22.2	0.25	0.09	Itagaki et al., 1988.
Vitis vinifera (leaf)	18.0	2.1	13.0	0.069	0.195	Loulakis and Roubelakis-Angelakis, 1990.
Zea mays (shoot)		2.22	4.76-12.5		0.063	Yamaya et al., 1984.

*Values in parentheses indicate K_m values for NADPH

Higher plant GDH enzymes have been shown to be metalloproteins. Calcium is a well known activator of GDH. The inhibitory effect of EDTA on higher plant GDH is also shown to be reversed by Ca^{2+} and Zn^{2+} Other divalent cations e.g. Mn^{2+}, Co^{2+}, Cu^{2+} and Fe^{2+} are also known to activate GDH from various sources. However, inhibitory effects of some of these ions on aminating or deaminating reactions of GDH isolated from various species have also been reported (Stewart et al., 1980). A more detailed discussion on the effects of metal ions on GDH from various tissues has been discussed by other workers (Stewart et al., 1980; Srivastava and Singh, 1987). The aminating reaction for soybean GDH was also calcium dependent, the activity being much higher when 1 mM calcium was present in the assay mixture. Thus, GDH from both of these cell cultures appeared to have properties analogous to the GDH enzymes from other plant sources.

Preliminary experiments in our laboratory to crystallize the enzyme from soybean cells for X-ray crystallographic studies were partially successful (Shargool and Delbaere, 1991). The method described by Ollis and White (1990) was used with pure GDH protein dissolved in polyethylene glycol-4000 containing phosphate, tris and azide ions. Small crystals of GDH were obtained. Further work is now in progress to produce GDH crystals for X-ray diffraction studies. Once the crystal structure of this soybean GDH is obtained, future structural studies on GDH will be done on the three dimensional structure to develop a better understanding of its active site and the role of calcium in allosteric activation of the enzyme.

VII. Regulation and Function

GDH activity levels in various plant tissues have been shown to be affected by age, nutritional and environmental conditions. Several workers have shown an induction of GDH activity by high levels of ammonia (Kanamori et al., 1972; Shepard and Thurman, 1973; Singh and Srivastava, 1982; Lauriere and Daussant, 1983; Prunkard et al., 1986; Srivastava and Singh, 1987; Jain and Shargool, 1987; Shargool and Jain, 1989; Bahadula and Shargool, 1991) . In soybean cell cultures, it was shown that the increase in GDH activity during the growth of cells in high ammonium medium was not due to the activation of pre-existing inactive enzyme, but due to the synthesis of the enzyme protein (Shargool and Jain, 1989) as evidenced by rocket immunoelectrophoresis. Using inhibitors of protein synthesis, Bhadula and Shargool (1991) showed that the ammonia-induced synthesis of GDH was inhibited by chloramphenicol but not by cycloheximide suggesting that the enzyme was not synthesized in the cytosol. Using purified plastids for protein synthesis, GDH was shown to be synthesized in the plastids (Bhadula and Shargool, 1991). However, the synthesis of ammonium-induced GDH on 80S ribosomes has been reported for *Lemna gibba* (Shepard and Thurman, 1973).

The induction of GDH isoenzyme (s) in response to high ammonium is well documented (Srivastava and Singh, 1987 and references therein). The induction of a specific GDH isoenzyme during isolation and culture of *Brassica napus* leaf protoplasts was reported recently by Watanabe et al. (1992). These workers also found that the high specific activity of NAD-GDH (deaminating) in leaves correlated with cathodal enzymes whereas the aminating NADH-GDH in the protoplasts was correlated with ancdal isoenzymes. A similar pattern was also reported in *Arabidopsis thaliana* (Cammaerts and Jacob, 1985). It has also been suggested that GDH isoenzymes have a varying anabolic and catabolic function (Loulakakis and Roubelakis-Angelakis, 1990)

An increase in GDH activity has also been reported during senescence of plant tissues. A.A.K. Hussain et al. (unpublished data) have recently shown that during ethylene induced senescence in *Brassica*, the levels of ammonia are increased accompanied by an increase in GDH mRNA levels and the *de-novo* synthesis of GDH. These results support the concept that the increase in GDH activity seen on exposure to high levels of ammonia is initiated at the gene level. It should, however, be noted that different species may respond differently to high concentrations of ammonium and in some cases, either no effect and in a few species an inhibitory effect of high ammonium on GDH has also been reported (Srivastava and Singh, 1987).

GDH also seems to play an important role during plant development when nitrogen compounds are being mobilized (Srivastava and Singh, 1987). An increase in growth temperature, salinity and water stress also lead to an increased level of GDH in the tissues stressed. It has been suggested that under stress conditions, GDH is a more stable enzyme than glutamine synthetase and may play a very important role in determining the plant response to potentially toxic levels of ammonia (Srivastava and Singh, 1987).

Since GDH is a ubiquitous enzyme and is often found in high levels in certain plant tissues, and is induced during senescence and by high levels of ammonia, it seems to have an important and specific function. As discussed earlier, the GS/GOGAT cycle is now believed to be the primary route of ammonia assimilation in plants under normal conditions. Although the thermodynamically favoured direction of the GDH reaction is the production of glutamate, GDH from various plant species has been shown to have a high K_m for ammonia and this is one of the major objections to its role in ammonia assimilation. However, some controversy still exists and it has been suggested by some workers that since GDH is found in high levels, it can account for assimilation of ammonia even if the affinity for ammonia is low (Murray and Kennedy, 1980; Loyola-Vargas and Sanchez de Jimenez, 1984). A role for mitochondrial GDH in ammonia assimilation has also been suggested Davies and Teixeira (1975) and by Yamaya and co-workers (Yamaya et al., 1984; Yamaya and Oaks, 1987). They have shown that the affinity of GDH for ammonia is increased by the addition of physiologically occurring levels of calcium ion. These workers have proposed that the mitochondrial GDH functions in glutamate synthesis utilizing some of the ammonia released during photorespiration (Yamaya and Oaks, 1987). It has also been proposed that GDH aminating activity is adaptive, and is involved in maintaining intracellular level of glutamate when it cannot be maintained by the glutamate synthase cycle (Munoz-Blanco and Cardenas, 1989).

Many studies on plant tissues and cell cultures have suggested a role for GDH in the oxidative deamination of glutamate. In soybean cell cultures, Jain and Shargool (1987) reported an induction of GDH activity when cells were grown in high ammonium medium. However, addition of compounds which inhibit the glutamine synthetase/glutamate synthase pathway such as MSX and azaserine to the culture medium inhibited ammonia utilization by the cells. Also, the addition of 2-oxoglutarate or glutamate to the ammonia rich medium completely inhibited the induction of GDH. Based on these observations, the authors suggested that GDH may in fact play a role in the oxidative deamination of glutamate. Using automated $^{15}N/^{13}C$ mass spectrometry, Robinson et al. (1991) have also shown that in carrot cell cultures, GDH functions in the oxidation of glutamate and not in the reductive amination of 2-oxoglutarate. They found that GDH was derepressed in carbon limited cells, where its role appeared to be the oxidation of glutamate to provide carbon skeletons for the TCA cycle. One thing that still remains unclear is that if GDH is not involved in ammonia detoxification during growth in high ammonia medium (Jain and Shargool, 1987), then why is the enzyme induced and synthesized in very high amounts under such conditions?

Although in most cases, the inhibition of glutamine synthetase by MSX leads to an accumulation of ammonia and inhibition of ^{15}N-labelling of glutamate (Jain and Shargool, 1987; Magalhaes, 1991), suggesting a catabolic function for GDH, tomato seems to be an exception (Magalhaes, 1991; Magalhaes et al., 1992). Magalhaes (1991) has recently shown that in tomato, some assimilation of ammonia continues even when the GS/GOGAT pathway is inhibited. It was also reported that the labelling of glutamate with ^{15}N continued whereas the labelling of the amino-N moiety of glutamine was completely inhibited in presence of MSX (Magalhaes, 1991). Megalhaes et al. (1992) also observed an increase in ^{15}N-ammonium assimilation in tomato with exogenous 2-oxoglutarate. Based on their results, these workers have concluded that in tomato roots, GDH catalyzes the synthesis of glutamate in presence of MSX and that this glutamate synthesis would also account for the low percentages of ammonia assimilation when GS/GOGAT pathway is active.

It should be noted that GDH exists as a number of isoenzymes which vary in number according to environmental conditions, and developmental stages. GDH isoenzymes have been found localized in the mitochondrial, chloroplastic, plastidial and cytoplasmic fractions (Stewart et al., 1980; Srivastava and Singh, 1987; S.K. Bhadula and P.D. Shargool, unpublished data). Since GDH in different subcellular compartments is subjected to a particular microenvironment, it is difficult to explain how the activity of different isoenzymes is coordinated within the plant cell.

The mitochondrial enzyme is thought to be the product of a nuclear gene. The majority of studies carried out with protein synthesis inhibitors have suggested that the *de novo* synthesis of GDH obtained upon induction with ammonia, is dependent upon 80S ribosomes (Kanamori et al., 1972; Shepard and Thurman, 1973). This fact suggests a cytosolic origin for GDH, and a nuclear site for the GDH gene (or genes). The chloroplastic GDH of *Chlorella sorokiniana* has also been shown to be synthesized by a nuclear gene (Prunkard et al., 1986). However, Bhadula and Shargool (1991) have recently shown that GDH is not only localized in the plastids of soybean cells, but is also synthesized there. The changes in GDH isoenzyme pattern under environmental and developmental conditions and the compartmentalisation of GDH suggest that there are distinct GDH genes present in plant tissues. Changes in GDH isoenzyme pattern under different growth conditions have also been reported for other isoenzymic systems associated with nitrogen metabolism such as glutamine synthetase (Gebharlt et al., 1986).

VIII. Conclusion and Future Prospects

GDH purified from soybean cell cultures appeared to be a hexamer of six identical subunits and showed properties similar to GDH enzymes isolated from other plant tissues. Although evidence from our studies and from many other laboratories indicates that GDH mainly functions in glutamate catabolism, the possibility still exists that it may have a role in ammonia assimilation under certain environmental and nutritional conditions. Based on their studies on *Arabidopsis thaliana*, Cammaerts and Jacobs (1983) have concluded that NADH-GDH was involved in the detoxification of high nitrogen levels supplied exogenously or endogenously produced, whereas NAD-GDH was responsible for the supply of energy to the cell during active assimilation. These workers have also suggested that two genes (GDH1 and GDH2) involved in GDH synthesis control the expression of GDH enzymes with different metabolic functions (Cammaerts and Jacobs, 1983; 1985).

The studies on the subcellular localization and biosynthetic origin of GDH suggest that GDH enzymes have a plastidial as well as a cytosolic biosynthetic origin within plant tissues. Future studies dealing with the molecular mechanisms involved in the expression of GDH genes under different environmental, nutritional or stress conditions should be helpful to understand the role that GDH plays in plant nitrogen metabolism. Studies on the crystal structure of the plant enzyme should be helpful to understand the mechanism of the GDH reaction and to see how the plant enzyme compares with the *Clostridium* enzyme (Rice et al., 1987). Initial attempts to crystallize soybean GDH for X-ray diffraction studies have been successful (P.D. Shargool and L.T.J. Delbaere, unpublished data) and further work in this direction is needed.

Literature Cited

Andrews, P. 1964. Estimation of the molecular weights of proteins by sephadex gel filtration. *Biochem. J.* **91**: 222-233.

Barash, I., Mor, H., and Sadon, T. 1975. Evidence for ammonium-dependent *de novo* synthesis of glutamate dehydrogenase in detached oat leaves. *Plant Physiol.* **56**: 856-858.

Bhadula, S.K., and Shargool, P.D. 1991. A plastidial localization and origin of l-glutamate dehydrogenase in a soybean cell culture. *Plant Physiol.* **95**: 258-263.

Bradford, M.M. 1976. A rapid and sensitive method for quantitation of microgram quantities of protein utilizing the principle of protein-dye binding. *Anal. Biochem.* **72**: 248-254.

Cammaerts, D., and Jacobs, M. 1983. A study of the polymorphism and the genetic control of the glutamate dehydrogenase isozymes in *Arabidopsis thaliana*. *Plant Sci. Lett.* **31**: 65-73.

Cammaerts, D., and Jacobs, M. 1985. A study of the role of glutamate dehydrogenase in the nitrogen metabolism of *Arabidopsis thaliana*. *Planta* **163**: 517-526.

Chiu, J.Y., and Shargool, P.D. 1979. Importance of glutamate synthase in glutamate synthesis by soybean cell suspension cultures. *Plant Physiol.* **63**: 409-415.

Davies, D.D., and Teixeira, A.N. 1975. The synthesis of glutamate and the control of glutamate dehydrogenase in pea mitochondria. *Phytochemistry* **14**: 647-656.

Davis, B.J. 1964. Disc electrophoresis. II. Method and application to human serum proteins. *Annals NY Acad. Sci.* **121**: 404-427.

Dougall, D.K. 1974. Evidence for the presence of glutamate synthase in extracts of carrot cell cultures. *Biochem. Biophys. Res. Comm.* **58**: 639-646.

Errel, A., Mor, H., and Barash, I. 1973. The isozymic nature and kinetic properties of glutamate dehydrogenase from safflower seedlings. *Plant Cell Physiol.* **14**: 39-50.

Gebhardt, C., Oliver, J.E., Forde, B.G., Saarelainen, R., Miflin, B.J., Idler, K.B., and Baker, R.F. 1986. Primary structure and differential expression of glutamine synthetase gene in nodules, roots and leaves of *Phaseolus vulgaris. EMBO J.* **5**: 1429-1435.

Goodwin, T.W., and Mercer, E.I. 1983. Introduction to Plant Biochemistry. 2nd Edn., Pergamon Press, pp. 333-334

Hadzi-Taskovic Sukalovic V. 1990. Properties of glutamate dehydrogenase from developing maize endosperm. *Physiol. Plant.* **80**: 238-242.

Itagaki, T., Dry, I.B., and Wiskich, J.T. 1988. Purification and properties of NAD-glutamate dehydrogenase from turnip mitochondria. *Phytochemistry* **27**: 3373-3378.

Jain, J.C., and Shargool, P.D. 1987. Use of an aneuploid soybean cell culture to examine the relative importance of the GS:GOGAT system and GDH in ammonia assimilation. *J. Plant Physiol.* **130**: 137-146.

Kanamori, T., Konishi, S., and Takahashi, E. 1972. Inducible formation of glutamate dehydrogenase in rice roots by the addition of ammonia to the media. *Physiol. Plant.* **26**: 1-6.

King, J., and, Wu, W. Y-F. 1971. Partial purification and kinetic properties of glutamic dehydrogenase from soybean cotyledons. *Phytochemistry.* **10**: 915-928.

Laemmli, U.K. 1970. Cleavage of structural proteins during the assembly of the head of bacteriophage T4. *Nature* **227**: 680-685.

Lauriere, C., and Daussant, J. 1983. Identification of the ammonium-dependent isoenzyme of glutamate dehydrogenase as the form induced by senescence or darkness stress in the first leaf of wheat. *Physiol. Plant.* **58**: 89-92.

Lauriere, C., Weisman, N., and Daussant, J. 1981. Glutamate dehydrogenase in the first leaf of wheat. II. De novo synthesis upon darkness stress and senescence. *Physiol. Plant.* **52**: 151-155.

Lea, P.J., Thurman, D.A. 1972. Intracellular location and properties of plant L-glutamate dehydrogenase. *J. Expt. Bot.* **23**: 440-449.

Loulakakis, C.A., and Roubelakis-Angelakis, K.A. 1990. Intracellular localization and properties of NADH-glutamate dehydrogenase from *Vitis vinifera* L.: Purification and characterization of the major leaf isoenzyme. *J. Expt. Bot.* **41**: 1223-1230.

Loyola-Vargas, V.M., and Sanchez de Jimenez, E. 1984. Differential role of glutamate dehydrogenase in nitrogen metabolism of maize tissues. *Plant Physiol.* **76**: 536-540 .

Magalhaes, J.R. 1991. Kinetics of $^{15}NH_4$ assimilation in tomato plants: Evidence for $^{15}NH_4$ assimilation via GDH in tomato roots. *J. Plant Nutr.* **14**: 1341-1353.

Magalhaes, J.R., Huber, D.M., and Tsai, C.Y. 1992. Evidence of increased ^{15}N-ammonium assimilation in tomato plants with exogenous α-ketoglutarate. *Plant Sci.* **85**: 135-141.

Mazurova, H., Ratajczak, W., and Ratajczak, L. 1980. Glutamate dehydrogenase characteristics in the organ and root nodules of *Lupinus luteus* L. *Acta Physiol. Plant.* **2**: 167-177.

Meade, R. 1984. Ammonia assimilating enzymes in bryophytes. *Physiol. Plant.* **60**: 305-308.

Miflin, B.J., and Lea, P.J. 1976. The pathway of nitrogen assimilation in plants. *Phytochemistry* **15**: 873-885.

Munoz-Blanco, J., and Cardenas. J. 1989. Changes in glutamate dehydrogenase activity of *Chlamydomonas reinhardtii* under different trophic and stress conditions. *Plant Cell Environ.* **12**: 173-182.

Murray, D.R., and Kennedy, I.R. 1980. Changes in activities of enzymes of nitrogen metabolism in seedcoats and cotyledons during embryo development in pea seeds. *Plant Physiol.* **66**: 782-785.

Nagel, M., and Hartmann, T. 1980. Glutamate dehydrogenase from *Medicago sativa* L.: Purification and comparative kinetic studies of the organ-specific multiple forms. *Z. Naturforsch* **35c**: 406-415.

Nauen, W., and Hartmann, T. 1980. Glutamate dehydrogenase from *Pisum sativum* L. Localization of the multiple forms and of glutamate formation in isolated mitochondria. *Planta* **148**: 7-16.

Ollis, D., and White, S.. 1990. Protein crystallization. *In Methods in Enzymology.*, (Deutscher, M.P. ed), Academic Press Inc. New York. **182**: 646-659.

Pahlich, E., and Joy, K. 1971. Glutamate dehydrogenase form pea roots: Purification and pro-perties of the enzyme. *Can. J. Biochem.* **49**: 127-138.

Prunkard D.E., Bascomb. N.F., Robinson, R. W., and Schmidt, R.R. 1986. Evidence for chloro-plastic localization of an ammonium-inducible glutamate dehydrogenase and synthesis of its subunit from a cytosolic precursor-protein in *Chlorella sorokiniana. Plant Physiol.* **81**: 349-355.

Ratajczak, L., Pecikiewics, M., Ratajczak, W., Mazurowa, H., and Koroniak, D. 1988. Glutamate dehydrogenase isoenzymes in yellow lupine root nodules. I. Developmental pattern and coen-zyme requirements. *Acta Phyiol. Plant* **10**: 41-48.

Ratajczak, L., Ratajczak, W., Koroniak, D, Przybylska, M., and Mazurowa, H. 1988. Glutamate dehydrogenase isoenzymes in yellow lupine root nodules. II. subunits composition. *Acta Phys-iol. Plant* **10**: 49-55.

Ratajczak, L., Ratajczak, W., and Mazurowa, H. 1977. Isoenzyme pattern of glutamate dehydro-genase as a reflection of nitrogen metabolism in *Lupinus albus. Acta Soc. Bot. Pol.* **46**: 347-356.

Rice, D.W., Baker, P.J., Farrants, G.W., and Hornby, D.P. 1987. The crystal structure of glutamate dehydrogenase from *Clostridium symbiosum* at 0.6 nm resolution. *Biochem. J.* **242**: 789-795.

Robinson, S.A., Slade, A.P., Fox, G.G., Phillips, R., Ratcliffe, R.G., and Stewart, G.R. 1991. The role of glutamate dehydrogenase in plant nitrogen metabolism. *Plant Physiol.* **95**: 509-516.

Scheild, H-W., Ehmke, A., and Hartmann, T. 1980. Plant NAD-dependent glutamate dehydroge-nase. Purification, molecular properties and metal ion activation of the enzymes from *Lemna minor* and *Pisum sativum. Z. Naturforsch.* **35c**: 213-221.

Shargool, P.D., and Jain, J.C. 1989. Purification and immunological properties of an NAD(H) depen-dent glutamate dehydrogenase from soybean cells *(Glycine max* L.), *Plant Sci.* **61**: 173-179.

Shepard, D.V., and Thurman, D.A. 1973. Effect of nitrogen sources upon the activity of L-glutamate dehydrogenase of *Lemna gibba. Phytochemistry* **12**: 1937-1946.

Srivastava, H.S., and Singh, R.P. 1987. Role and regulation of L-glutamate dehydrogenase activity in higher plants. *Phytochemistry,* **26**: 597-610.

Singh, R.P., and Srivastava, H.S. 1982. Glutamate dehydrogenase activity and assimilation of inor-ganic nitrogen in maize seedlings. *Biochem. Physiol. Pflanzen.* **177**: 633-642.

Stewart, G.R., and Rhodes, D. 1977. A comparison of the characteristics of glutamine synthetase and glutamate dehydrogenase from *Lemna minor* L. *New Phytol.* **79**: 257-268.

Stewart, G.R., Mann, A.F., and Fentem, P.A., 1980. Enzymes of glutamate formation: Glutamate dehydrogenase, glutamine synthetase and glutamate synthase. *In Biochemistry of Plants,* Vol. 5 (Miflin, B.J. ed), Academic Press, New York. pp. 271-327.

Watanabe, M., Nakayama, H., Watanabe, Y., and Shimada, N. 1992. Induction of a specific iso-enzyme of glutamate dehydrogenase during isolation and the first 48 h of culture of *Brassica napus* leaf protoplasts. *Physiol. Plant.* **86**: 231-235.

Yamaya, T., and Oaks, A. 1987. Synthesis of glutamate by mitochondria. An anaplerotic function for glutamate dehydrogenase. *Physiol. Plant.* **70**: 749-756.

Yamaya, T., Oaks, A. and Matsumoto, H. (1984) Characteristics of glutamate dehydrogenase in mitochondria prepared from corn shoots. *Plant Physiol.* **76**: 1009-1013.

Nitrogen Nutrition in Higher Plants, 1995
Editors : H.S. Srivastava & R.P. Singh
Associated Publishing Co., New Delhi, India
pp. 219-228.

Glutamine Synthetase

TOMOYUKI YAMAYA

I. Introduction

Prior to 1974, glutamate dehydrogenase (GDH; EC 1.4.1.2) and glutamine synthetase (GS; EC 6.3.1.2) were considered to be the major ports of entry of NH_4^+ in higher plants. However, in 1970, Tempest et al. discovered NADPH- dependent glutamate synthase (NADPH-GOGAT: EC 2.6.1.53) in NH_4^+-limited chemostat cultures of *Aerobacter aerogenes* (Tempest et al., 1970). In higher plants, Dougall discovered NADH-dependent GOGAT (EC 1.4.1.14) in carrot cell cultures in 1974 (Dougall, 1974), while Lea and Miflin found a ferredoxin (Fd) -dependent GOGAT (EC 1.4.7.1) in pea leaves (Lea and Miflin, 1974). The significance of the discovery of GOGAT is that, in conjunction with GS, it provides an alternative route for the net synthesis of glutamate from NH_4^+ and 2-oxoglutarate. The conjugated action of GS and GOGAT was termed as the GOGAT cycle by Rhodes et al. (1980). With this cycle, two glutamate molecules are produced, one of which could be used to regenerate glutamine and the other could be used in various ways, such as incorporation into protein, an amino donor in transamination reactions, and so on (Rathnam et al., 1976; Sechley et al., 1992). Evidence in favour of GS activity as the major reaction in the assimilation of NH_4^+ are: (1) GS has much lower K_m for NH_4^+ than GDH (Rathnam et al., 1976) (2) glutamine is the primary product after feeding $^{15}NH_4^+$ or $^{15}NO_3^-$ (Arima, 1979; Yoneyama and Kumazawa, 1974) (3) $^{15}NH_4^+$ is incorporated into amide group of glutamine (Rhodes et al., 1989), (4) methionine sulfoximine, an inhibitor of GS, blocks the NH_4^+ assimilation (Rhodes et al., 1980), and (5) mutants lacking either GS (Wallsgrove et al., 1987) or Fd-GOGAT (Somerville and Ogren, 1980) in chloroplasts cannot survive under conditions which permit photorespiration. Because of this evidence it is assumed that the NADH-GDH in the mitochondria functions primarily in the oxidation of glutamate (Robinson et al., 1991).

Because of the central importance of GS in the assimilation of NH_4^+ derived from such reactions as nitrate reduction, photorespiration, protein degradation, and nitrogen fixation, GS has been extensively studied in higher plants. There have been many studies with various higher plants on occurrence of GS isoforms and their localization (Hirel et al., 1982; Botella et al., 1988) as well as their biochemical characteristics (Hirel and Gadal, 1980), gene structures and expressions (Lightfoot et al., 1988; Sakamoto et al., 1989; Mc Grath and Coruzzi, 1991; Becker et al., 1992) and physiological functions (Wallsgrove et al., 1987; Edwards et al., 1990; Kamachi et al., 1992). In this chapter, GS in plant leaves and roots will be described. A recent review is available for a more complete bibliography on GS in nodules of legumes (Vance, 1990).

II. Enzyme Extraction

All procedures are carried out at 0 to 4°C. Either the fresh plant tissues or tissues frozen with liquid nitrogen can be used as a source of GS. Various types of electric blenders and mixers can be used for homogenization of relatively soft plant tissues, such as spinach leaves, while a chilled mortar and pestle in the presence of washed sand is recommended for extraction of GS from hard tissues, such as rice leaves and roots. For protection of the enzyme from proteolytic breakdown during extraction, protease inhibitors, i.e. 10-50 μM leupeptin, 1-5 mM phenylmethylsulfonyl fluoride (PMSF), and 1-10 mM EDTA, are usually included in the extraction buffer. When plant tissues contain polyphenolic substances, addition of insoluble polyvinylpyrrolidone (PVPP) to the extraction buffer is recommended for removal of phenolics. The preparation of GS from rice leaves is described below (Kamachi et al., 1991b).

Fully expanded rice leaves are harvested, weighed, and frozen in liquid nitrogen. Leaves frozen with liquid nitrogen (about 4.0 g fresh weight) are ground to a fine powder in a mortar and pestle in the presence of washed quartz sand. The powder is then homogenized in an extraction buffer containing 25 mM tris-HCl (pH 7.6), 10 mM $MgCl_2$, 10 mM 2-mercaptoethanol, and 5% (w/v) PVPP. Five ml of the buffer per g fresh weight of plant materials were used. The homogenate is filtered through four layers of gauze and the filtrate is centrifuged at 27,000 x g for 20 min. The supernatant fraction (crude extract) is used for assay of GS activity.

To reduce the background, the crude extract should be filtered through a sephadex G-25 column or be dialyzed for the removal of amino acids and other low molecular substances.

III. Enzyme Assay

GS catalyzes the following reaction in the presence of Mg^{2+} (a true biosynthetic reaction).

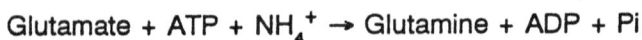

$$\text{Glutamate} + \text{ATP} + NH_4^+ \rightarrow \text{Glutamine} + \text{ADP} + \text{Pi}$$

The isolated enzyme also enhances the formation of γ-glutamylhydroxamate through either a semibiosynthetic reaction in the presence of Mg^{2+} or a tansferase reaction in the presence of Mn^{2+}, ADP, and arsenate.

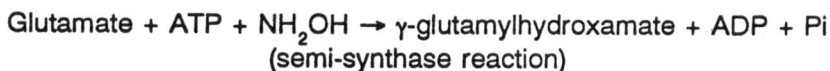

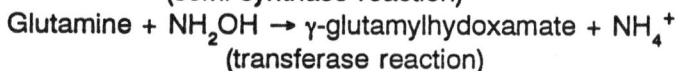

Glutamate + ATP + NH_2OH → γ-glutamylhydroxamate + ADP + Pi
(semi-synthase reaction)
Glutamine + NH_2OH → γ-glutamylhydoxamate + NH_4^+
(transferase reaction)

The latter two reactions, in which the activity can be measured spectrophotometrically, are frequently used in the analysis of GS activity, since the reaction rate is both fast and sensitive. The transferase assay generally produces higher activity than the semi-synthase assay, so that the transferase assay is most commonly used. However, it should be noted that the tansferase: semi-synthase ratio is not a fixed parameter, the best fitting relationship is quadratic rather than linear (Rhodes et al., 1975). The following are examples for measuring the GS activity.

A.   True Biosynthetic Assay (Vezina et al., 1987)

The reaction mixture contained: 50 μmol imidazole-HCl, 50 μmol $MgCl_2$, 12 μmol ATP, 14 μmol NH_4Cl, 80 μmol glutamate, and 250 μL of enzyme solution partially purified on sephadex G-25. The final volume of the reaction mixture was 500 μl and had a pH of 8.0. The reaction was initiated by the addition of enzyme solution and was incubated at 30°C for 30 min. The reaction was stopped by addition of 100 μl sulfosalicylic acid (30 mg) and centrifuged at 14, 000xg. Samples were neutralized and diluted in 10% (v/v) methanol. The synthesized glutamine was determined by HPLC. Reactions with boiled enzyme are used for the blank test.

B.   Semi-synthetase Assay (Rhodes et al., 1975)

The reaction mixture contained: 36 μmol ATP, 90 μmol $MgSO_4$, 12 μmol NH_2OH, 184 μmol glutamate, and 100 μmol imidazole-HCl. The final volume of the reaction mixture was 2 ml and had a pH of 7.2. The reaction was initiated by the addition of enzyme solution and was incubated at 30°C for 10 to 30 min. The reaction was terminated by addition of 1 ml ferric chloride regent (0.37 M $FeCl_3$, 0.67 N HCl, 0.2 M trichloroacetic acid) and the denatured proteins were removed by centrifugation at 10,000xg for 15 min. The synthesized γ-glutamylhydroxamate was determined by measurement of absorbency at 540 nm. Commercially available γ-glutamylhydroxamate was used to develop the standard curve. Either 0 time reaction or minus ATP is used for the blank test.

C.   Transferase Assay (Rhodes et al., 1975)

The reaction mixture contained: 0.34 μmol ADP, 5 μmol $MnCl_2$, 34 μmol hydroxylamine, 130 μmol glutamine, 66 μmol Na-arsenate, and 200 μmol tris-acetate, pH 6.4. Total volume was 2 ml. Other methods were the same as described in "semi-synthetase assay".

IV.   Localization

In 1979, occurrence of two distinct isoforms of GS, designated as GS1 and GS2, was shown by ion-exchange chromatography in extracts obtained from plant leaves and shoots (Guiz et al., 1979; Mann et al., 1979). Extracts from spinach leaves contained only GS2 and its localization in chloroplasts was established by immuno-cytochemical studies (Hirel et al., 1982). Observations localizing GS2 in the stroma of tomato chloroplasts have been confirmed by immunogold techniques (Botella et al., 1988). Although cytosolic localization of GS1 was only indicated by cell fractionation studies (Hirel and Gadal, 1981), recent studies using molecular technics support to the biochemical observation. The native GS isoproteins comprise eight subunits and are coded by nuclear genes (Sechley et al., 1992). The primary translation products of GS2 mRNA contain an N-terminal transit peptide that is removed after transport into the chloroplasts (Lightfoot et al., 1988; Tingey et al., 1988; Sakamoto et al., 1989), whereas those of GS1 mRNA contain no transit sequences (Tingey et al., 1988; Sakamoto et al., 1989; Miao et al., 1991). The proportion of GS1 and GS2 isoproteins varies with plant species (Mc Nally et al., 1983). Recent studies with monospecific antibody for GS1 show that the GS1 protein is localized specifically in large and small vascular bundles of rice leaf blades, whereas GS2 protein is in mesophyll cells (Kamachi et al., 1992). In maize leaves, where the assimilation of photorespiratory NH_4^+ is in bundle sheath cells and that of NH_4^+ from NO_3^- is in the mesophyll cells (Rathnam and Edwards, 1976), similar proportions of GS1 and GS2 are present in bundle sheath strand and mesophyll cells (Yamaya and Oaks, 1988).

In roots of many plant species, GS occurs predominantly in the cytosol (Hirel and Gadal, 1980; Kamachi et al., 1991a). On the other hand, in root of *Pisum sativum* cv little Marvel, about 50% of GS is in the plastids, when the plants are grown in the presence of nitrate (Vézina et al., 1987). There are also cytosolic and plastidic GS isoforms in chicory roots (Sechley et al., 1991). Cytosolic GS in rice roots (Sakamoto et al., 1989) and pea roots (Tingey et al., 1988) is encoded by a nuclear gene homologous but distinct from the genes for the leaf GS1 and GS2.

V.   Properties of the Enzyme

The native GS protein has a molecular weight of about 350 kD and it is comprised of eight identical subunits with a molecular weight of 38 to 43 kD for cytosolic GS and that of 44 to 45 kD for chloroplastic isoform (Sechley et al., 1992). Summary for the molecular mass of GS subunits in various plant leaves and roots is shown in Table 1. These two native isoforms in leaves can be separated by

Table 1. Molecular mass (in kD) of GS subunits in various plants

Plant species	Leaf		Root	
	Chloroplast (GS2)	Cytosol (GS1)	Plastid	Cytosol
Non legumes				
Spinacea oleracea	44	-	ND	ND
Nicotiana plumbagnifolia	44(49)	-	-	38(38)
Oryza sativa "Sasanishiki"	44(49)	41	-	41(41)
Oryza sativa "Delta"	45	38	-	45
Legumes				
Pisum sativum "Sparkle"	44(49)	38	-	38
P. sativum "Little Marvel"	-	-	44	38
Phaseolus vulgaris	45	43	-	43
Cichorium intybus	ND	ND	42	39

This table was adapted from Sechley et al. (1992).
ND, Not determined.
Value in parentheses is relative molecular mass of polypeptides translated *in vitro*.

DEAE-Sephacel column chromatography (Hirel and Gadal, 1981). Although GS in rice roots was eluted from the column at the position same to the leaf GS1, they can be resolved by hydroxyapatite column chromatography (Hirel and Gadal, 1980). Generally, GS2 is a heat labile protein and is sensitive to inactivation by sulfhydryl reagents (Hirel and Gadal, 1980). The pH optimum of GS2 is about 8.0, while that of GS1 and root GS is around 7.5 in the synthetic reactions (Hirel and Gadal, 1980). Cytosolic GS isoforms, i.e. GS1 and GSr, show normal Michaelis-Menten kinetics with respect to l-glutamate, while GS2 shows negative cooperativity (Guiz et al., 1979). Because of the highly homologous amino acid sequences among GS isoforms, polyclonal antibodies raised against either a purified GS1 (Kawakani and Watanabe, 1988) or a purified GS2 (Tingey et al., 1988; Kamachi et al., 1991) recognize all GS isoforms. GS2 is coded for by the single nuclear gene (Lightfoot et al., 1988; Tingey et al., 1988; Sakamoto et al., 1989; Becker et al., 1992). For cytosolic GS, molecular studies have revealed the presence of a number of distinct isoforms in several plant species (Tingey et al., 1988; Kamachi et al., 1991). In bean plants, GS is encoded by four differentially expressed nuclear genes (Swarup et al., 1990). One gene encodes a precursor to the GS2 (Lightfoot et al., 1988) and the other three encode cytosolic GS polypeptides which appear able to assemble into a variety of different isoforms (Bennett and Cullimore, 1989). In pea, it has been shown that two classes of genes encode homologous but distinct cytosolic GS genes (Tingey et al., 1988).

VI. Function and Regulation

Barley mutants lacking GS2 in chloroplasts clearly show that a major role of GS2 is the reassimilation of NH_4^+ released from photorespiration (Blackwell et al., 1987; Wallsgrove et al., 1987). The Fd-GOGAT in chloroplasts is also essential for the assimilation of photorespiratory NH_4^+ (Somerville and Ogren, 1980). Using a different approach, both GS2 activity and protein disappear in parallel with the loss of RUBISCO and Fd-GOGAT proteins during natural senescence of rice leaves (Kamachi et al., 1991b). GS1 protein, on the other hand, remains at a high level during the senescence period. Since the mutation lacking GS2 is lethal, GS1 and GSr apparently do not replace completely the function of GS2. As with the Fd-GOGAT mutants, the GS2 mutants are able to grow normally when photorespiration is repressed. Therefore, GS1 and GSr could be important in the synthesis of glutamine for normal growth and development. Promoters isolated from the nuclear genes for GS2 and cytosolic GS in root nodules of pea are able to direct cell-specific expression of the β-glucuronidase reporter gene in transgenic plants: GS2 is expressed in cells with chloroplasts whereas cytosolic GS is found in phloem elements (Edwards et al., 1990). These observations support the results with barley mutants which suggest that GS2 plays a major role in the reassimilation of photorespiratory NH_4^+. Using monospecific antibody raised against a synthetic 17-residue peptide following the deduced amino acid sequence for rice GS1 and tissue print immunoblot experiments, GS1 protein is indeed localized in the large and small vascular bundles of rice leaves (Kamachi et al., 1992b). These studies strongly suggest that GS1 is involved in the synthesis of glutamine that is destined for transport. Thus, GS1 and GS2 in leaves probably do not have overlapping functions. There is an example to suggest some overlap in their functions. When GS2 and Fd-GOGAT mutants are transferred from high CO_2 to normal air, deficiency symptoms appear earlier in the Fd-GOGAT mutants (Wallsgrove et al., 1987). One interpretation of this observation is that the GS1 in vascular tissues can be active to a limited extent in the reassimilation of photorespiratory NH_4^+.

In pea leaves, light affected the expression of GS2, and light-induced accumulation of GS2 mRNA was caused in part by the action of phytochrome as well as light-induced changes in chloroplast metabolism (Edwards et al., 1990). A phytochrome-mediated light stimulation of GS2 gene expression was also indicated with tomato cotyledons during greening (Becker et al., 1992). In bean plants, the GS2 gene was highly expressed in tissues containing chloroplasts (Lightfoot et al., 1988) and the GS2 mRNA was detected only in light-grown cotyledons (Swarup et al., 1990). The promoter for pea GS2 was expressed specifically within pho-tosynthetic cells of transgenic tobacco plants and retained the ability to confer a light-regulated gene expression (Edwards et al., 1990). The light-dependent expression of GS2 was consistent with its function in reassimilating photorespiratory NH_4^+. Another example with transgenic tobacco showed that the promoter for rice GS2 was also regulated by light (Kozaki et al., 1992). This promoter, however, was active in both leaf and root tissues. The expression of GS1 was unaffected by light (Tingey et al., 1988).

Leaf age also influences differently on the abundance of GS1 and GS2 proteins. For example, the amounts of GS2 polypeptide on a unit of fresh weight of attached rice leaves declined continuously during natural senescence (Kamachi et al., 1991b). Other stroma enzymes in chloroplasts, such as RUBISCO and Fd-GOGAT, declined in parallel with GS2. In contrast, the GS1 polypeptide remained constant throughout the senescence period (Kamachi et al., 1991b). However, changes in the contents of GS1 and GS2 polypeptides was not parallel with the change in the steady state levels of corresponding mRNAs (Kamachi et al., 1992a). Thus, the accumulation of these GS mRNAs did not determine the abundance of GS1 and GS2 polypeptides in senescing rice leaves.

It has been suggested that genes encoding cytosolic GS in soybean roots and nodules are directly induced by NH_4^+ either supplied externally or made available as a result of nitrogen fixation (Hirel et al., 1987). This observation has been confirmed in transgenic *Lotus corniculatus* and in transgenic tobacco plants (Miao et al., 1991). In these experiments, chimeric genes, consisting of the cytosolic GS promoter fused with a reporter gene (GUS) were expressed in a root-specific fashion in both species of plants. Treatment with NH_4^+ increased the expression of this chimeric gene in the legume background, but it caused no induction in tobacco roots (Miao et al., 1991). In bean plants, on the other hand, neither NH_4^+ (Cock et al., 1990) nor NO_3^- (Swarup et al., 1990) has an effect on the expression of genes for the plastidic and cytosolic GS isoforms. Promoters of the genes for both GS1 and GS2 (Kozaki et al., 1992) from rice responded to externally added NH_4^+ in transgenic tobacco plants. It appears, therefore, that differences in the promoter regions of various genes encoding GS polypeptides lead to important differences in their expression regulated by nitrogen.

VII. Conclusion and Future Prospects

From those studies with GS2 minus mutants, transgenic plants, and tissue localization cited above, it is clear that GS2 in chloroplasts of C_3 green leaves plays a major role in the assimilation of photorespiratory NH_4^+. However, the function of GS2 in non-photosynthetic organs and that in mesophyll chloroplasts in C_4 leaves are not clearly understood. The transgenic studies as well as the studies with tissue print immunoblot strongly suggest that GS1 in the vascular tissues is involved in the synthesis of glutamine that is destined for transport. However, there is no information on the availability and generation of substrates for the GS1 reaction, i.e. NH_4^+, glutamate, and ATP, in such tissues. This problem together with the precise localization of the GS1 protein among the different cell types of the vascular tissue, should be resolved to fully understand the function of GS1. Thus, recent studies indicate that GS1 and GS2 in leaves have nonoverlapping functions.

Analyses of gene expression for GS isoforms and the regulation by environmental factors have been advanced with several different plants. To understand the regulation of gene expression more precisely, further studies on the finding of *cis*-acting regulatory elements and *trans*-acting nuclear factors are required. For example, a nuclear protein factor, which bound to a 17 bp DNA element within the promoter for cytosolic GS gene of pea, has recently been identified and it is required for expression (Brears et al., 1991).

Literature Cited

Arima, Y. 1979. ^{15}N-nitrate assimilation in association with glutamine synthesis in rice seedling roots. *Soil Sci. Plant Nutr.* **25**: 311-322.

Becker, T.W., Caboche, M., Carrayol, E., and Hirel, B. 1992. Nucleotide sequence of a tobacco cDNA encoding plastidic glutamine synthetase and light inducibility, organ specificity and diurnal rhythmicity in the expression of the corresponding genes of tobacco and tomato. *Plant Mol. Biol.* **19**: 367-379.

Bennett, M.J., and Cullimore, J.V. 1989. Glutamine synthetase isoenzymes of *Phaseolus vulgaris* L.: Subunit composition in developing root nodules and plumules. *Planta* **179**: 433-440.

Blackwell, R.D., Murray, A.J.S., and Lea, P.J. 1987. Inhibition of photosynthesis in barley with decreased levels of chloroplastic glutamine synthetase activity. *J. Exp. Bot.* **38**: 1799-1807.

Botella, J.R., Verbelen, J.P., and Valpuesta, V. 1988. Immunocytolocalization of glutamine synthetase in green leaves and cotyledons of *Lycopersicon esculentum*. *Plant Physiol.* **88**: 943-946.

Brears, T., Walker, E.L., and Coruzzi, G.M. 1991. A promoter sequence involved in cell-specific expression of the pea glutamine synthetase GS3A gene in organs of transgenic tobacco and alfalfa. *Plant. J.* **1**: 235-244.

Cock, J.M., Mould, R.M., Bennett, M.J., and Cullimore, J.V. 1990. Expression of glutamine synthetase genes in roots and nodules of *Phaseolus vulgaris* following changes in the ammonium supply and infection with various *Rhizobium* mutants. *Plant Mol. Biol.* **14**: 549-560.

Dougall, D.K. 1974. Evidence for the presence of glutamate synthase in extracts of carrot cell cultures. *Biochem. Biophys. Res. Commun.* **58**: 639-646.

Edwards, J.W., Walker, E.L., and Coruzzi, G.M., 1990. Cell-specific expression in transgenic plants reveals nonoverlapping roles for chloroplast and cytosolic glutamine synthetase. *Proc. Natl. Acad. Sci. USA* **87**: 3457-3463.

Guiz, C., Hirel, B., Shedlofsky, G., and Gadal, P. 1979. Occurrence and influence of light on the relative proportions of two glutamine synthetases in rice leaves. *Plant Sci. Lett.* **15**: 271-277.

Hirel, B., and Gadal, P. 1980. Glutamine synthetase in rice. A comparative study of the enzymes from roots and leaves. *Plant Physiol.* **66**: 619-623.

Hirel, B., and Gadal, P. 1981. Glutamine synthetase isoforms in pea leaves. Intracellular localization. *Z. Pflanzenphysiol.* **102**: 315-319.

Hirel, B., Perrot-Rechenmann, C., Suzuki, A., Vidal, J., and Gadal, P. 1982. Glutamine synthetase in spinach leaves. Immunological studies and immunocytochemical localization. *Plant Physiol.* **69**: 983-987.

Hirel, B., Bouet, C., King, B., Layzell, D., Jacobs, F., and Verma, D.P.S. 1987. Glutamine synthetase genes are regulated by ammonia provided externally or by symbiotic nitrogen fixation. *EMBO J.* **6**: 1167-1171.

Kamachi, K., Yamaya, T., Mae, T., and Ojima, K. 1991a. Multiple polypeptides of glutamine synthetase subunit in rice roots *in vivo* and *in vitro*. *Agric. Biol. Chem.* **55**: 887-888.

Kamachi, K., Yamaya, T., Mae, T., and Ojima, K 1991b. A role for glutamine synthetase in the remobilization of leaf nitrogen during natural senescence in rice leaves. *Plant Physiol.* **96**: 411-417.

Kamachi, K., Yamaya, T., Hayakawa, T., Mae, T., and Ojima, K. 1992a. Changes in cytosolic glutamine synthetase polypeptide and its mRNA in a leaf blade of rice plants during natural senescence. *Plant Physiol.* **98**: 1323-1329.

Kamachi, K., Yamaya, T., Hayakawa, T., Mae, T., and Ojima, K. 1992b. Vascular bundle-specific localization of cytosolic glutamine synthetase in rice leaves. *Plant Physiol.* **99**: 1481-1486.

Kawakami, N., and Watanabe, A. 1988. Senescence-specific increase in cytosolic glutamine synthetase and its mRNA in radish cotyledons. *Plant Physiol.* **88** : 1430-1434.

Kozaki, A., Sakamoto, A., and Takeba, G. 1992. The promoter of the gene for plastidic glutamine synthetase (GS2) from rice is developmentally regulated and exhibits substrate-induced expression in transgenic tobacco plants. *Plant Cell Physiol.* **33**: 233-238.

Lea, P.J., and Miflin, B.J. 1974. Alternative route for nitrogen assimilation in higher plants. *Nature* **251**: 614-616.

Lightfoot, D.A. Green, N.K., and Cullimore, J.V. 1988. The chloroplast located glutamine synthetase of *Phaseolus vulgaris* L.: Nucleotide sequence, expression in different organs and uptake into isolated chloroplasts. *Plant Mol. Biol.* **11**: 191-202.

Mann, A.F., Fentem, P.A., and Stewart, G.R. 1979. Identification of two forms of glutamine synthetase in barley (*Hordeum vulgare*). *Biochem. Biophys. Res. Commun.* **88**: 515-521.

McGrath R.B., and Coruzzi G.M. 1991. A gene network controlling glutamine and asparagine biosynthesis in plants. *Plant J.* **1**: 275-280.

McNally, S.F., Hirel, B., Gadal, P., Mann, A.F., and Stewart, G.R. 1983. Glutamine synthetases of higher plants. Evidence for a specific isoform content related to their possible physiological role and their compartmentation within the leaf. *Plant Physiol.* **72**: 22-25.

Miao, G.-H., Hirel, B., Marsolier, M.C., Ridge, R.W., and Verma, D.P.S. 1991. Ammonia-regulated expression of a soybean gene encoding cytosolic glutamine synthetase in transgenic *Lotus corniculatus*. *Plant Cell* **3**: 11-22.

Miflin, B.J., and Lea, P.J. 1977. Amino acid metabolism. *Annu. Rev. Plant Physiol.* **28**: 299-329.

Rathnam, C.K.M., and Edwards, G.E. 1976. Distribution of nitrate-assimilating enzymes between mesophyll protoplasts and bundle sheath cells in leaves of three groups of C_4 plants. *Plant Physiol.* **57**: 881-885.

Rhodes, D., Sims, A.P., and Folkes, B.F. 1980. Pathway of ammonia assimilation in illuminated *Lemna minor*. *Phytochemistry* **19**: 357-365.

Rhodes, D., Rendon, G.A., and Stewart, G.R. 1975. The control of glutamine synthetase level in *Lemna minor* L. *Planta* **125**: 201-211.

Rhodes, D., Rich, P.J., and Brunk, D.G. 1989. Amino acid metabolism of *Lemna minor* L. IV. ^{15}N-labelling kinetics of the amide and amino groups of glutamine and asparagine. *Plant Physiol.* **89**: 1161-1171.

Robinson, S.A., Slade, A.P., Fox, G.G., Phillips, R., Ratcliffe, R.G., and G. Stewart, G.R. 1991. The role of glutamate dehydrogenase in plant nitrogen metabolism. *Plant Physiol.* **95**: 509-516.

Sakamoto, A., Ogawa, M., Masumura, T., Shibata, D., Takeba, G., Tanaka, K., and Fujii, S. 1989. Three cDNA sequences coding for glutamine synthetase polypeptides in *Oryza sativa* L. *Plant Mol. Biol.* **13**: 611-614.

Sechley, K.A., Oaks, A., and Bewley, J.D. 1991. Enzymes of nitrogen assimilation undergo seasonal fluctuations in the roots of the persistent weedy perennial *Cichorium intybus*. *Plant Physiol.* **97**: 322-329.

Sechley, K.A., Yamaya, T., and Oaks, A. 1992. Compartmentation of nitrogen assimilation in higher • plants. *Inter. Rev. Cytol.* **134**: 85-163.

Somerville, C.R., and Ogren, W.L. 1980. Inhibition of photosynthesis in mutants of *Arabidopsis* lacking glutamate synthase activity. *Nature* **286** : 257-259.

Swarup, R., Bennett, M.J., and Cullimore, J.V. 1990. Expression of glutamine-synthetase genes in cotyledons of germinating *Phaseolus vulgaris* L. *Planta* **183**: 51-56.

Tempest, D.W., Meers, J.L., and Brown, C.M. 1970. Synthesis of glutamate in *Aerobacter aerogenes* by a hitherto unknown route. *Biochem. J.* **117**: 405-407.

Tingey, S.V., Tsai, F.-Y., Edwards, J.W., Walker, E.L., and Coruzzi, G.M. 1988. Chloroplast and cytosolic glutamine synthetase are encoded by homologous nuclear genes which are differentially expressed *in vivo*. *J. Biol. Chem.* **263**: 9651-9657.

Vance, C.P. 1990. Symbiotic nitrogen fixation: recent genetic advances. In *The Biochemistry of Plants: A Comprehensive Treatise. Intermediary Nitrogen Metabolism*. Vol. 16 (Miflin, B.J.,and Lea, P.J., eds.). Academic Press, New York. pp. 43-88.

Vezina, L.-P., Hope, H.J., and Joy, K.W. 1987. Isoenzymes of glutamine synthetase in roots of pea *Pisum sativum* L. cv Little Marvel) and alfalfa (*Medicago media* Pers. cv Saranac). *Plant Physiol.* **83**: 58-62.

Vezina, L.-P., and Langlois, J.R. 1989. Tissue and cellular distribution of glutamine synthetase in roots of pea *(Pisum sativum)* seedlings. *Plant Physiol.* **90**: 1129-1133.

Wallsgrove, R.M., Lea, P.J., and Miflin, B.J. 1979. Distribution of the enzymes of nitrogen assimilation within the pea leaf cell. *Plant Physiol.* **63**: 232-236.

Wallsgrove, R.M., Turner, J.C., Hall, N.P., Kendall, A.C., and Bright, S.W.J. 1987. Barley mutants lacking chloroplast glutamine synthetase; biochemical and genetic analysis. *Plant Physiol.* **83**: 155-158.

Yamaya, T., and Oaks, A. 1988. Distribution of two isoforms of glutamine synthetase in bandle sheath and mesophyll cells of corn leaves. *Physiol. Plant.* **72**: 23-28.

Yoneyama, T., and Kumazaya, K. 1974. A kinetic study of the assimilation of ^{15}N-labelled ammonium in rice seedling roots. *Plant Cell Physiol.* **15**: 655-561.

Nitrogen Nutrition in Higher Plants, 1995
Editors : H.S. Srivastava & R.P. Singh
Associated Publishing Co., New Delhi, India
pp. 229-242.

Glutamate Synthase

REKHA M. PURANIK and H.S. SRIVASTAVA

I. Introduction

Ammonium in higher plants is believed to be assimilated primarily by the glutamine synthetase-glutamate synthase pathway, although the glutamate dehydrogenase pathway may also contribute significantly to the process, under certain environmental conditions. Ammonium, either absorbed from the soil or produced endogenously from the reduction of nitrate or from photorespiratory deamination process, is rapidly fixed up into glutamine by an ATP dependent GS enzyme, which has a high affinity for the ammonium. The enzyme GOGAT transfers the amide nitrogen of the glutamine to 2-oxoglutarate to yield glutamate. Thus, a two step process, called glutamate synthase cycle by Rhodes et al. (1980), carries out the principal process of ammonium assimilation.

Glutamate synthase, also known as glutamate oxoglutarate amino transferase (GOGAT), was first discovered in the bacterium *Aerobacter aerogens* by Tempest et al. (1970), and was later found to be ubiquitous in higher plants. The electron donor for the enzyme is NADPH (E.C. 1.4.1.13), NADH (E.C. 1.4.1.14) or reduced ferredoxin (E.C. 1.4.7.1), depending upon the plant species, and organs of the plant, although in some cases, more than one isoforms are active in the same organ. NADH and ferredoxin dependent GOGAT are of common occurrence in the tissues of higher plants, and are involved in the assimilation of primary ammonia, as well as of photorespiratory ammonia. Thus, this enzyme seems to be playing a pivotal role in linking the inorganic nitrogen to the organic compounds.

II. Organ Specific Spectra and Cellular Distribution

The presence of GOGAT activity has been demonstrated in most vegetative and a few reproductive organs of higher plants. Some organs have both NAD(P)H- as well as Fd-specific enzymes, while others have only one of these isoforms. The higher plant GOGAT, first described from cultured cells, (Dougall, 1974) and roots (Fowler et al., 1974), was found to be specific for NAD(P)H. Subsequently, Miflin and Lea (1975) demonstrated that both NAD(P)H- as well as Fd-specific isoforms were present in pea roots. However, the enzyme present in rice roots was active with ferredoxin only; and no activity was detected with NAD(P)H by Arima and Kumazawa (1977). The enzyme has been detected in shoot tissues also; the enzyme from the shoots of etiolated pea seedlings is active with both types of reductants (Matoh et al., 1980). The young and developing leaves have generally a higher level of NADH-GOGAT (Yamaya et al., 1992), while mature green leaves have higher Fd-GOGAT activity, as compared to that of NADH-GOGAT (Wallsgrove et al., 1982; Avilla et al., 1987; Hecht et al., 1988). One or both the isoforms of the enzyme have been detected in soybean cotyledons also (Storey and Reporter, 1978), developing maize endosperm (Misra and Oaks, 1985) and in endosperm, embryo, pedicel and pericarp of maize kernels (Muhitch, 1991). The enzyme, specially the NADH-GOGAT, is also present in root nodules, where it plays a major role in ammonia assimilation (Boland and Benny, 1977; Awonaike et al., 1981; Suzuki et al., 1988; Anderson et al., 1989; Chen et al., 1990). Immunochemical characterization of the enzyme from alfalfa root nodules shows that the same protein is not present in the roots or shoots (Anderson et al., 1989).

Cellular localization studies of Rathnam and Edwards (1976) with some C_4 species revealed that the enzyme was exclusively localized in the mesophyll cells. However, Hirel et al., (1977) reported the presence of enzyme both in the mesophyll as well as in bundle sheath cells. In leaves and cotyledons of tomato seedlings, localization tests for Fd-GOGAT by immunochemical methods established that the enzyme was a constituent of chloroplast stroma of mesophyll cells and also of xylem parenchyma and epidermis (Botella et al., 1988). In root nodules, it is distributed in both cortex as well as medulla region (Chen and Cullimore, 1989).

The sub-cellular location of the enzyme is determined by its isoformic nature, in some cases. In most studies, Fd-GOGAT has been found to be located in the. chloroplasts of the green tissues (Botella et al., 1988; Matoh and Takahashi, 1981; Murrilo and Jimenez, 1985; Wallsgrove et al., 1982). In non-chlorophyllous organs such as roots also this isoform is located in plastids (Kubik-Dobos, 1989). The NADH-GOGAT may be located in both cytosol and plastids (Awonoike et al., 1981; Matoh and Takahashi, 1981; Suzuki et al., 1981; Boland et al., 1982; Hecht et al., 1988).

III. Enzyme Purification and Assay

The activity of GOGAT is determined by the measurement of glutamate formed, or in the case of NAD(P)H:GOGAT, also by determining the amount of NAD(P)H oxidized. In a typical protocol (for example, Puranik and Srivastava, 1990), the enzyme is extracted from the freshly harvested plant material in cold (1-4°C) in a phosphate buffer containing EDTA, KCl, mercaptoethanol and Triton x-100. The concentration of these additives in the extraction medium is usually, although not always, similar for various kinds of tissues. The extract is cleared off by centrifugation in cold, and the enzymic protein from the clear extract is precipitated out by adding 55% $(NH_4)_2SO_4$. The protein is dissolved in small quantity of buffer solution (pH 7.5) and filtered through sephadex G-75 column. Various filtrate fractions are assayed for enzyme activity, and the fraction showing maximum enzyme activity is used for further purification and/or enzyme characterization.

The NAD(P)H:GOGAT is usually assayed spectrophotometrically by observing the decrease in absorbance at 340 nm (Matoh et al., 1980). The assay mixture consists of phosphate buffer (pH 7.5), KCl, NAD(P)H, 2-oxoglutarate, glutamine, NAD(P)H and enzyme preparation. The reaction is usually started by the addition of glutamine and the decrease in absorbance is recorded at 340 nm, after every 2 min, and as long as the decrease in absorbance is linear with respect to time. In control sets, usually both the substrates, 2-oxoglutarate and glutamine are excluded, so that the error arising due to non specific oxidation of NAD(P)H, if any, is eliminated.

Ferredoxin dependent GOGAT can be assayed by colorimetric measurement of the amount of glutamate formed. In a typical assay (see Lea, 1987),the assay mixture consists of phosphate buffer (pH 7.5), 2-oxoglutarate, glutamine, enzyme prepartion and reduced ferredoxin. Often, the reduced methyl viologen is used as a substitute for Fd, as it is cheaper and stable on storage. When crude enzyme preparation is used, 10 mM amino-oxyacetate is added in the assay mixture, so that the contaminating transaminase activity is inhibited. After incubation of the assay mixture at about 25-30°C, for 60-90 min, the glutamate formed is separated out, usually by paper chromatography and its amount evaluated after eluting the spot and measuring its absorbance. However, a more reliable method of glutamate separation is ion exchange chromatography, using Dowex-1 acetate columns, or high performance liquid chromatography.

Purification of glutamate synthase beyond $(NH_4)_2SO_4$ precipitation and gel filtration has been achieved by using ion exchange chromatography in a few cases. Anderson et al. (1989) purified NADH-GOGAT from alfalfa root nodules up to 208 fold by ion exchange chromatography. Hirasawa and Tamura (1984) achieved 100 fold purification of the spinach leaf enzyme, by using ferredoxin sepharose 4B as an affinity material. In this case, the enzyme preparation was filtered through a DEAE-cellulose or DEAE-sephacel and a sephacryl S-300 or ultrogel ACA 34 gel filtration column before loading on an affinity column. The elution from the affinity column is usually done with a gradient of NaCl/KCl solution in a concentration range 0 to 600 mM. Using blue sepharose as affinity column, a 430 fold purification of NADH(P)H:GOGAT has been achieved for soybean cell culture enzyme (Chiu and Shargool, 1979). However, the tobacco leaf enzyme could be purified upto 160 fold only, but the enzyme was electrophoretically homogenous (Zehnacker et al., 1992).

Immuno detection of GOGAT protein has also been achieved in quite a few cases (For example, Commere et al., 1986; Hayakawa et al., 1992; Suzuki et al., 1982). Western blotting and immunoprecipitation tests have been performed for tobacco GOGAT and Fd-specific polyclonal antibody has also been obtained (Zehnacker et al., 1992).

IV. The Isoforms

Two isozymic forms of GOGAT have been identified from different organs of higher plants. One isoform requires reduced ferredoxin (Fd-GOGAT) as the reductant, while the other works with reduced pyridine nucleotides (NAD(P)H:GOGAT).

Ferredoxin: GOGAT is the major isoform in green tissues and is localized in chloroplasts. NAD(P)H:GOGAT is present in roots, shoots and nodules. The co-existence of the two isoforms is a general phenomenon in higher plants, and is probably related to different physiological functions of the enzyme. However, mutants lacking Fd-GOGAT have been identified in *Arabidopsis thaliana* (Somerville and Ogren, 1980) and barley (Kendall et al., 1986). In these mutants, NADH:GOGAT and other enzymes of nitrogen assimilation are as active, as in wild types, in both roots and leaves. This indicates that the Fd-dependent and NADH-dependent isoforms are under separate gene control. Analysis of genomic DNA indicates the presence of a single copy gene for Fd-GOGAT in maize leaves (Sakakibara et al., 1991), and in diploid tobacco leaves (Zehnacker et al., 1992), although there is a strong possibility of two genes in amphidiploid tobacco (Zehnacker et al., 1992). The two isoforms of GOGAT have also been separated by using gel filtration and ion exchange chromatography techniques, indicating differences in their physicochemical characteristics. Immunochemical identification of the isoforms have also been achieved in some cases. Antibodies to both isoforms from several plants, have been raised. Antibodies of one isoform do not recognize the antibodies of the other, indicating that the two proteins were quite distinct (Suzuki et al., 1982). Further, Fd-GOGAT from maize roots is inactivated by the antibody of the same isoform from leaves (Suzuki et al., 1982), indicating the presence of organ specific GOGAT in roots or leaves. The antibody of the alfalfa root nodule NADH:GOGAT, cross reacts with the nodule NADH:GOGAT from a number of legume species, but lacks reactivity with the NADH:GOGAT of alfalfa roots and leaves (Anderson et al., 1989). Thus, distinct isoforms appear to be induced in roots in response to the infection with the *Rhizobium*.

The two isoforms of GOGAT show different developmental patterns in early stages of seedling growth (Matoh and Takahashi, 1982), and their expressions are influenced differently by light and exogenous nitrogen sources (Watanabe et al., 1985; Hecht et al., 1988). In wheat plants, NADH:GOGAT is the predominant isoform in young expanding leaves, while Fd:GOGAT is present mostly in fully mature leaves (Cincerova et al., 1991). Yamaya et al., (1992) have also demonstrated that NADH:GOGAT is the principal isoform in non-green and growing leaf blades, while

Fd:GOGAT is abundant in green leaves of rice plants. From their data, they have concluded that NADH:GOGAT was important for the synthesis of glutamate from the glutamine that is transported from senescing source tissues through the phloem in the non-green sink tissues.

Sakakibara et al., (1991) have cloned the c-DNA genome of the Fd:GOGAT from maize leaves. The genome is 5617 bp long and codes for 1616 amino acids; and 42% of the amino acid sequence is identical to the sequence in NADH: GOGAT from *E. coli*. A 450 bp Fd:GOGAT DNA has been cloned from tobacco also (Zehnacker et al., 1992). These authors have identified two isozymic forms of Fd:GOGAT in tobacco leaves. Similarly two isozymic forms of NADH: GOGAT, I and II also exist in the root nodules of *Phaseolus vulgaris* (Chen and Cullimore, 1989; Chen et al., 1990), which have been separated by ion exchange chromatography. The two forms show different kinetic characteristics and developmental patterns during nodulation. Although both isoforms increase in activity during nodulation, the major increase is due to NADH:GOGAT II, which increases over time course similar to nitrogenase activity. This enzyme is twice as active as NADH:GOGAT I, and thus it is the most likely isozyme involved in the assimilation of ammonia derived from fixed atmospheric nitrogen (Chen and Cullimore, 1988).

V. Physico-chemical and Catalytic Characteristics

Glutamate synthases from different sources exhibit differences in molecular size, subunit composition and absence or presence of prosthetic groups. Ferredoxin dependent GOGAT, which may represent upto 1% of the total leaf protein is an iron sulfur flavoprotein (Lea et al., 1992). Evidences obtained from EPR, magnetic circular dichroism and Raman resonance spectroscopic investigations have established the presence of a single 3Fe-4S cluster in the spinach enzyme (see Hirasawa et al., 1992). Also, there is one FMN and one FAD per enzyme molecule, although there is no evidence of flavin in partially purified *Vicia faba* enzyme (Wallsgrove et al., 1977).

The enzyme from green leaves is a monomeric protein with a molar mass ranging from 141 to 270 kD (Table 1). However, Fd:GOGAT from rice leaves has been reported to be a homodimer having molar masses of the native enzyme and subunits 224 and 115 kD respectively (Suzuki and Gadal, 1982). But the sub unit molar mass of Fd:GOGAT from plant cell fractions of alfalfa nodules is 68.2 kD (Suzuki et al., 1988).

The absorption spectrum of the purified NADH:GOGAT from some plants shows the presence of flavin moieties (Boland and Benny, 1977). This isoform has a single polypeptide of molar mass in the range of 200 to 270 kD (Chen and Cullimore, 1988; Anderson et al., 1989).

As mentioned earlier, GOGAT from higher plants require either reduced pyridine nucleotides or reduced ferredoxin as electron donors. Pyridine nucleotide specific enzyme appears to exhibit activity with both, NADH and NADPH, the ratio of the activities wiith two coenzymes varying in different species (Miflin and Lea, 1975; Dougall and Bloch, 1976; Lee and Stewart, 1978). The two isoforms of NADH: GOGAT (I and II) reported in the root nodules of bean (Chen and Cullimore, 1988) exhibit slight difference in their affinity for NADH, as well. In a few cases, reduced methyl viologen has also been found to act as electron donor (Suzuki and Gadal, 1982). In each case, there is an absolute specificity for l-glutamine as amino donor and 2-oxoglutarate as acceptor. No activity is reported when glutamine is replaced by either l-asparagine or ammonia and 2-oxoglutarate is replaced by oxaloacetate, pyruvate or oxobutyrate.

Table 1. **Occurrence and molar masses of glutamate synthase isoforms from higher plants.**

Plant material	Izozymic form	Mol. Wt. kD	Reference
Alfalfa (*Medicago sativa*)			
Nodules	Fd:GOGAT	68.2 per sub unit	Suzuki et al., (1988).
	NADH:GOGAT	235	Anderson et al., (1989).
Beans (*Phaseolus vulgaris*)			
Nodules	NADH:GOGAT I	200	Chen and Cullimore, (1988).
	NADH:GOGAT II	200	- do -
Lupine (*Lupinus augustifolius*)			
Nodules	NADH:GOGAT	235	Boland and Benny, (1977).
Maize (*Zea mays*)			
Endosperm	Fd:GOGAT	171	Misra and Oaks, (1985).
	NADH:GOGAT	270	
Roots	NADH:GOGAT	255	Ferreira and Teixeira, (1985).
Leaves	Fd:GOGAT	145	Commere et al., (1986).
		160	Matoh et al., (1979a).
Pea (*Pisum sativum*)			
Cotyledons	Fd.GOGAT	155	Matoh et al., (1979b).
	NADH:GOGAT	220	- do -
Etiolated shoots	NADH:GOGAT	220	Matoh et al., (1980).
Leaves	Fd:GOGAT	165	Wallsgrove et al., (1977).
Rice (*Oryza sativa*)			
Leaves	Fd:GOGAT	224	Suzuki and Gadal, (1982).
Spinach (*Spinacea oleracea*)			
Leaves	Fd:GOGAT	115	Tamura et al., (1980).
		160	Hirasawa et al.,(1992)
Tobacco (*Nicotiana tabaccum*)			
Leaves	Fd:GOGAT	164	Zehnacker et al., (1992).
Tomato (*Lycopersicon esculentum*)			
Leaves	Fd:GOGAT	141	Avilla et al., (1987).
	NAD (P)H:GOGAT	158	- do -

There is a considerable variation with respect to the apparent pH optimum of gluta-mate synthase which ranges from 6.9 to 8.5. However, the majority of the higher plant NADH:GOGAT have optimum pH around 7.5, while Fd:GOGAT have around 7.0 (Table 2). The dependence of the kinetic constants of the enzyme on pH was demonstrated by Boland and Benny (1977). They found that V_{max} was essentially pH dependent between 6.5 and 9.5 but the K_m values for both substrates were lowest at the optimum pH of 8.5 and increased at lower pH values. The K_m for glut-amine increased markedly above pH 9.0 suggesting that the protonated α-amino group of glutamine (pKa 9.13) could be involved in binding.

Table 2. Kinetic characteristics of glutamate synthases

Plant material	Electron donor	Optimum pH	Apparent K_m, μM					References
			NADH	NADPH	Fd$_{red}$	Gln	20G	
Alfalfa (*Medicago sativa*) Nodules	NADH	7.8	4.2	-	-	466	33	Anderson et al., (1989).
Bean (*Phaseolus vulgaris*) Nodules	NADH I	8.0	14	-	-	770	22	Chen and Cullimore, (1988).
	NADH II	8.5	5.2	-	-	240	87	- do -
Brooad bean (*Vicia faba*) Chloroplasts	Fd	7.4	-	-	2.0	300	150	Wallsgrove et al., (1977).
Maize (*Zea mays*) Endosperm	NADH	7.5	1.0	-	-	850	19	Muhitch, (1991).
	NADH	7.5	7.0	-	-	1350	570	Sodek and de Silva, (1977).
Roots	NADH	7.4	4.7	-	-	620	130	Ferreira and Teixeira, (1985).
	NADH	7.5	2.6	-	-	2570	625	Singh and Srivastava, (1992).
Leaves	NADH	7.5	13	-	-	2850	200	- do -
	Fd	6.9	-	-	1.7	1100	240	Matoh et al., (1979a).
Lupine (*Lupinus augustifolius*) Nodules	NADH	8.5	1.3	-	-	400	39	Boland and Benny, (1977).
Pea (*Pisum sativum*) Cotyledons	NAD(P)H	-	13.3	27.7	-	1430	960	Beevers and Storey, (1976).
Roots	NADH	-	7.0	-	-	1000	500	Matoh and Takahashi, (1982).
Shoots	NADH	7.6	4.0	-	-	400	37	Matoh et al., (1980).
Rice (*Oryza sativa*) Leaves	Fd	7.3	-	-	5.5	270	330	Suzuki and Gadal, (1982).
						570		
Soybean (*Glycine max*) Cultured cells	NADH	8.0	9.0	-	-	630	64	Chiu and Shargool, (1979).
	NAD(P)H	7.5	-	57	-	410	88	- do -
Tomato (*Lycopersicon esculentum*) Leaves	NAD(P)H	7.3	1.7	6.0	-	500	200	Avila et al., (1987).
	Fd	7.8	-	-	0.2	500	300	Avila et al., (1984).

The catalytic activity of GOGAT from various sources has been found to be unstable (Miller and Stadtman, 1972; Wallsgrove et al., 1977). Mercaptoethanol (Boland and Benny, 1977) or dithiothreitol (Suzuki and Gadal, 1982) are often used during enzyme preparation to stabilize it. The enzyme is more stable in phosphate as compared to tris buffer. The general instability, particularly of the NAD(P)H specific isoform may in part explain the difficulty in detecting this enzyme in some plant tissues (Duffus and Rozie, 1978).

A number of mono - and divalent cations appear to inhibit GOGAT activity (Wallsgrove et al., 1977; Suzuki and Gadal, 1982). Calcium a potent activator of GDH has no effect at 1.0 mM level, on Fd: GOGAT from rice leaves (Suzuki and Gadal, 1982). However, Mg^{2+} and Mn^{2+} activate enzyme activity in maize tissues (Oaks et al., 1979). The root NADH:GOGAT requires additional KCl for maximum activity but not the leaf one (Singh and Srivastava, 1992). Ammonium and phosphate ions activate NADH:GOGAT activity in pea shoots (Matoh et al., 1980) but Fd:GOGAT from tomato leaves is inhibited by 3-phosphoserine and 3-phosphoglyceric acid (Avila et al., 1987). The enzyme is inhibited competitively with respect to 2-oxoglutarate by glutamate, oxoglutarate, aspartate and asparagine (Boland and Benny, 1977). However, NAD(P)H:GOGAT from tomato leaves is activated by aspartate (Avila et al., 1987). The dicarboxylic acid glutarate and the trimeric dihydrate glyoxal, whose binding sites are similar to 2-oxoglutarate inhibit NADH:GOGAT (Singh and Srivastava, 1987b). Glutamate analogs, methionine sulfoximine and methionine sulfone also inhibit GOGAT reaction (Buchanan, 1973; Singh and Srivastava, 1987b), but the reaction is insensitive to α-amino oxyacetate (Buchanan, 1973). The glutamine analogs azaserine and 6-diazo 5-oxo L-norvaline have been shown to inhibit NAD(P)H: and Fd:GOGAT activities (Buchanan, 1973: Suzuki and Gadal, 1982). Azaserine is known to alkylate cysteine residues in some enzymes, and it is likely that the inhibition of higher plant GOGAT by azaserine is due to the presence of a catalytically active cysteine residue in the enzyme molecule. The presence of a cysteine residue and its involvement in catalytic activity of the enzyme is demonstrated by the sensitivity of the enzyme towards sulfhydryl reagents also (Matoh et al., 1979a, 1980; Hirasawa and Tamura, 1984). However, alfalfa nodule NADH: GOGAT was not inhibited by 2-2 'dipyridyl and N-ethylmaleimide, suggesting that neither Fe sulfhydryl nor lysine residues were involved in catalysis (Anderson et al., 1989). Metal chelating agents also inhibit enzyme activity in some cases (Matoh et al., 1979a, 1980; Hirasawa and Tamura, 1984), although the metallic ion required for enzyme activity has not been identified.

The specific inhibitors of ammonia assimilating enzymes, such as methionine sulfoximine and azaserine, have been used extensively by the researchers to ascertain the role of GS-GOGAT pathway in ammonium assimilation. In the presence of GS specific inhibitor methionine sulfoximine there is accumulation of ammonium; and in the presence of GOGAT specific inhibitors such as azaserine, albizzine and 6-diazo 5-oxonorleucine, there is an accumulation of glutamine inside the plant tissues (see Lea et al., 1992).

Glutamate synthases from various sources exhibit classical Michaelis-Menten type kinetics with respect to substrate (s) and coenzymes. The K_m value for glutamine varies from 240 to 2850 µM and that for 2-oxoglutarate from 19 to 960 µM (Table 2). From detailed kinetic analysis of lupine nodule NADH:GOGAT, it has been shown that the steady state kinetic behaviour of the enzyme is consistent with an ordered A, random BC type of mechanism with NADH as the leading substrate and glutamine and 2-oxoglutarate as the following substrates (Boland, 1979). A model for the active site of the enzyme from *E. coli* has been proposed by Rendina and Orme-Johnson (1978), showing active site comprising of two distinct catalytic sites.

VI. Some Regulatory Aspects

The supply of inorganic nitrogen in the nutrient solution seems to be a critical factor in the expression of various enzymes involved in nitrogen assimilation in higher plants. Its effects on GOGAT activity depends upon the isoform of the enzyme, the form of the inorganic nitrogen and the plant species. Nitrate supply increases NADH:GOGAT activity in mustard cotyledons (Hecht et al., 1988) and maize seedlings (Singh and Srivastava, 1986), and both NADH: as well as Fd:GOGAT activities in cultured cells of rice (Hayakawa et al., 1990) and *Ipomaea* (Zink, 1989). In mustard cotyledons, the stimulation by KNO_3 and $NaNO_3$ are similar and hence it has been concluded that the nitrate and not the accompanying cation is responsible for enzyme increase (Hecht et al., 1988). Further, in the same system, the increase in enzyme activity by red light is strongly stimulated by NH_4NO_3 and not by ammonium alone. Attempts have been made to understand the possible mechanism of enzyme stimulation by nitrate. In maize root and leaf segments, the increase in NADH:GOGAT activity is apparently through induced synthesis of the enzyme as well as through the protection against degradation/inactivation, or by some physico-chemical modulation of the enzyme molecule (Singh and Srivastava, 1986). Ammonium nitrogen has variable effects on GOGAT activity. The Fd-GOGAT activity increases with NH_4NO_3 supply in radish leaves (Ota and Yamamoto, 1990). Ammonium nitrogen supply increases NADH:GOGAT activity also in maize seedlings (Singh and Srivastava, 1986) and cultured *Ipomaea* cells (Zink, 1989), although the Fd:GOGAT isoform is inhibited by either ammonium or nitrate supply in maize (Loyola-Vergas and Jimenez, 1986). Similarly, in cultured rice cells also, the ammonium supply increases NADH:GOGAT activity but not the Fd:GOGAT one (Hayakawa et al., 1990). On the other hand, in mustard cotyledons, while NADH:GOGAT is unaffected, Fd:GOGAT is inhibited by ammonium supply (Hecht et al., 1988). In bean roots, ammonium supply has no effect on NADH:GOGAT II activity, but it increases NADH:GOGAT I slightly, although in the nodules none of the NADH:GOGAT isoforms are induced (Chen et al., 1990). In the leaves of crassulacean acid metabolism plant, *Kalanchoe lateritta*, when the NH_4NO_3 level in the nutrient solution is reduced from full strength Hoagland's solution to 1/5 or 1/10, the Fd:GOGAT activity decreases considerably, while NADH:GOGAT activity is increased (Santos and Salema, 1992).

Amino acids and amides are also important regulators of GOGAT activities. Glutamine is the substrate and glutamate is the product of GOGAT activity; and the biosynthesis of many other nitrogen metabolites including amino acids are closely linked with glutamine/glutamate content of the plant tissues. The supply of glutamine and glutamate and also of aspartate and asparagine to excised root and leaf segments of maize seedlings, is known to increase NADH:GOGAT activity (Singh and Srivastava, 1987a). Further, in the roots while cysteine has no effect, arginine, lysine, proline, leucine, phenylalanine and tyrosine inhibit enzyme activity; but in the leaves significant inhibition is observed with lysine and phenylalanine only (Singh and Srivastava, 1987a). The inhibition of Fd:GOGAT activity by glutamate has been observed in *Lemna minor* (Rhodes et al., 1976), although these observations do not rule out the possible role of NADH:GOGAT in ammonium assimilation during the abundance of glutamate; as even during the inhibition of Fd:GOGAT by exogenous glutamate, there is an increase in internal glutamate content of the *Lemna minor* tissues (Rhodes et al., 1976). Muhitch (1991) has reported the inhibition of both Fd: and NADH:GOGAT by aspartate and phosphoserine in maize kernels. In the same study, serine and 3-phosphoglycerate inhibited Fd:GOGAT, but had no effect on NADH:GOGAT activities.

Among sugars, 1% glucose has been shown to increase GOGAT activity in excised maize roots during 5 h incubation of the root tips in nutrient solution (Oaks et al., 1980).

The effect of certain environmental factors on GOGAT activity has also been examined. High summer temperature has been demonstrated to be inhibitory for enzyme activity in the needles of pine plants (Prietila et al., 1989). However, an increase in enzyme activity has been reported during salinity (Mishra and Dwivedi, 1990) and desiccation (Gamboa et al., 1991) induced stresses in germinating green gram seeds and *Amaranthus* plants, respectively.

VII. Conclusions and Future Prospects

Although the role of GOGAT in ammonium assimilation is significant, and there is an adequate amount of literature on some aspects of its physio-chemical and catalytic nature and reliable techniques to measure its amount and activity are available, its physiological and molecular biological aspects are relatively unexplored. The role of Fd:GOGAT and the origin of reduced ferredoxin in the roots, are also not well understood, although there is some evidence of the presence of a ferredoxin: NADP reductase in the roots, which could transfer electrons from NAD(P)H to Fd (Suzuki et al., 1985). A full length gene for NADH(P)H:GOGAT is yet to be cloned and characterized. It is so necessary for understanding the molecular basis for coenzyme specificity of the two isoforms of the enzyme. The active site chemistry, and the involvement of metallic cofactor in catalysis, if any, is also to be investigated.

Literature Cited

Anderson, M.P., Vance, C.P., Heichel, G.H., and Miller, S.S. 1989. Purification and charcterization of NADH - glutamate synthase from alfalfa root noduels. *Plant Physiol.* **90**: 351-358.

Arima, Y., and Kumazawa, K. 1977. Evidence of ammonium assimilation via glutamine synthetase-glutamate synthase system in rice seedling roots. *Plant Cell Physiol.* **18**: 1121-1129.

Avilla, C., Botella, J.R., Canovas, F.M., Nunez de Castrol, I.N., and Valpuesta, V. 1987. Different characteristics of the two glutamate synthases in the green leaves of *Lycopersicon esculentum. Plant Physiol.* **85**: 1036-1039.

Avilla, C., Canovas, F., Nunez de Castro, I., and Valpuesta, V. 1984. Separation of two forms of glutamate synthases in leaves of tomato (*Lycopersicon esculentum*) *Biochem. Biophys. Res. Commn.* **122**: 1125-1130.

Awonoike, K.O., Lea, P.J., and Miflin, B.J. 1981. The localization of the enzymes of ammonia assimilation in root nodules of *Phaseolus vulgaris* L. *Plant Sci. Lett.* **23**: 189-195.

Beevers, L., and Storey, R. 1976. Glutamate synthase in developing cotyledons of *Pisum sativum. Plant Physiol.* **57**: 862-866.

Boland, M.J. 1979. Kinetic mechanism of NADH dependent glutamate synthase from lupine nodules. *Eur. J. Biochem.* **99**: 531-539.

Boland, M.J., and Benny, A.G. 1977. Enzymes of nitrogen metabolism in legume nodules. Purification and properties of NADH dependent glutamate synthase from lupin nodules. *Eur. J. Biochem.* **79**: 355-362.

Boland, M.J., Hanks, J.F., Reynolds, P.H.S., Blevins, D.G., Tolbert, N.E., and Staubert, K.R. 1982. Sub-cellular organization of ureide biogenesis from glycolate intermediate and ammonium in nitrogen fixing soybean nodules. *Planta* **155**: 45-51.

Botella, J.R., Verbelen, J.P., and Valpuesta, V. 1988. Immunocytolocalization of ferredoxin - GOGAT in the cells of green leaves and cotyledons of *Lycopersicon esculentum. Plant Physiol.* **87**: 255-257.

Buchanan, J.M. 1973. The amidotransferases. *Advances in Enzymol.* **39**: 91-183.

Chen, F-L., Bennett, M.J., and Cullimore, J.V. 1990. Effect of nitrogen supply on the activities of isoenzymes of NADH dependent glutamate synthase and glutamine synthetase in root nodules of *Phaseolus vulgaris* L. *J. Expt. Bot.* **41**: 1215-1221.

Chen, F-L., and Cullimore, J.V. 1988. Two isoenzymes of NADH dependent glutamate synthase in root nodules of *Phaseolus vulgaris* L. Purification, properties and activity changes during nodule development. *Plant Physiol.* **88**: 1411-1417.

Chen, F-L., and Cullimore, J.V. 1989. Location of two isoenzymes of NADH dependent glutamate synthase in root nodules of *Phaseolus vulgaris* L. *Planta.* **179**: 441-447.

Chiu, J.Y., and Shargool, P.D. 1979. Importance of glutamate synthase in soybean cell suspension cultures. *Plant Physiol.* **63**: 409-415.

Cincerova, A., Novotna, D., and Dvorak, M. 1991. NADH- and ferredoxin dependent glutamate synthase in the life span of the second leaf of wheat plant under conditions of senescene induced by nitrogen deficiency and natural senescence. *Biol. Plant.* **33**: 317-324.

Commere, B., Vidal, J., Suzuki, A., Gadal, P., and Caboche, M. 1986. Detection of messenger RNA encoding for the ferredoxin dependent glutamate synthase in maize leaf. *Plant Physiol.* **80**: 859-862.

Dougall, D.K. 1974. Evidence for the presence of glutamate synthase in extracts of carrot cell cultures. *Biochem. Biophys. Res. Commun.* **58**: 639-646.

Dougall, D.K., and Bloch, J. 1976. A survey of presence of glutamate synthase in plant cell suspension culture. *Can. J. Bot.* **54**: 2924-2927.

Duffus, C.M., and Rosie, R. 1978. Metabolism of ammonium ion and glutamate in relation to nitrogen supply and utilization during grain development in barley. *Plant Physiol.* **62**: 570-574.

Ferreira, R.M., and Teixeira, A.R. 1985. Some physical and catalytic properties of glutamate synthase from maize roots. *Cienc. Biol.* **10**: 15-30.

Fowler, M.W., Jessup, W., and Sarkissian, G.S. 1974. Glutamate synthetase type activity in higher plants. *FEBS Lett.* **46**: 340-342.

Gamboa, A., Valenzuela, E.M., and Murilla, E. 1991. Biochemical changes due to water loss in leaves of *Amaranthus hypochondriacus*. *J. Plant Physiol.* **137**: 586-590.

Hayakawa, T., Kamachi, K., Oikawa, M., Ojima, K., and Yamaya, T. 1990. Responses of glutamine synthetase and glutamate synthase isoforms to nitrogen sources in rice cell cultures. *Plant Cell Physiol.* **31**: 1070-1077.

Hayakawa, T., Yamaya, T., Kamachi, K., and Ojima, K. 1992. Purification, characterization and immunological properties of NADH -dependent glutamate synthase from rice cell cultures. *Plant Physiol.* **98**: 1317-1322.

Hecht, U., Oelmuller, R., Schmidt, S., and Mohr, H. 1988. Action of light, nitrate and ammonium on the levels of NADH- and ferredoxin dependent glutamate synthase in the cotyledons of mustard seedlings. *Planta* **175**: 130-138.

Hirasawa, M., Robertson, D.E., Ameyibor, E., Johnson, M.K., and Knaff, D.B. 1992. Oxidation reduction properties of the ferredoxin linked glutamate synthase from spinach leaf. *Biochim. Biophys. Acta* **1100**: 105-108.

Hirasawa, M., and Tamura, G. 1984. Flavin and iron sulfur containing ferredoxin linked glutamate synthase from spinach leaves. *J. Biochem.* **95**: 983-994.

Hirel, E., Lea, P.J., and Miflin, B.J. 1977. The location of enzymes of nitrogen assimilation in maize leaves and their activities during greening. *Planta* **134**: 195-200.

Kendall, A.C., Wallsgrove, R.M., Hall, N.P., Turner, J.C., and Lea, P.J. 1986. Carbon and nitrogen metabolism in barley (*Hordeum vulgare* L.) mutants lacking ferredoxin dependent glutamate synthase. *Planta* **168**: 316-323.

Kubik-Dobosz, G. 1989. The activity of NADH-, NADPH- and Fd-dependent glutamate synthase in the plastids and cytosol of *Pisum arvense* L. root cells. *Acta Soci. Botan. Poloniae* **58**: 253-261.

Lea, P.J. 1987. Ammonia assimilation and amino acid biosynthesis. In *Techniques in Bioproductivity and Photosynthesis* (Coombs. J., Hall D.C., Long S.P.,and Scurlock J.M.O. eds.),Pergamon Press, England.

Lea, P.J., Blackwell, R.D., and Joy, K.W. 1992. Ammonia assimilation in higher plants. *In Nitrogen Metabolism of Plants* (Mengel K.,and Pilbeam D.J. eds.), Clarendon Press, England. pp. 153-186.

Lee, J.A., and Stewart, G.R. 1978. Ecological aspects of nitrogen metabolism. *Adv. Bot. Res.* **6**: 1-43.

Loyola-Vergas, V.M., and Jimenez, E.S. 1986. Regulation of glutamine synthetase/glutamate synthase cycle in maize tissues. Effects of the nitrogen source. *J. Plant Physiol.* **124**: 147-154.

Matoh, T., Ida, S., and Takahashi, E. 1980. Isolation and characterization of NADH glutamate synthase from pea (*Pisum sativum* L.). *Plant Cell Physiol.* **21**: 1461-1474.

Matoh, T., Suzuki, F., and Ida, S. 1979a. Corn leaf glutamate synthase. Purification and properties of the enzyme. *Plant Cell Physiol.* **20**: 1329-1340.

Matoh, T., and Takahashi, E. 1981. Glutamate synthase in greening pea shoots. *Plant Cell Physiol.* **22**: 727-731.

Matoh, T., and Takahashi, E. 1982. Changes in the activities of ferredoxin and NADH-glutamate synthase during seedling development of peas. *Planta* **154**: 289-294.

Matoh, T., Takahashi, E., and Ida, S.1979b. Glutamate synthase in developing pea cotyledons. Occurrence of NADH dependent enzyme. *Plant Cell Physiol.* **20**: 1455-1459.

Miflin, B.J., and Lea, P.J. 1975. Glutamine and asparagine as nitrogen donors for reductant dependent glutamate synthesis in pea roots. *Biochem. J.* **149**: 403-409.

Miller, R.E., and Stadtman, E.R. 1972. Glutamate synthase from *E. coli*: an iron sulfide flavoprotein. *J. Biochem.* **247**: 7407-7419.

Mishra, N., and Dwivedi, U.N. 1990. Nitrogen assimilation in germinating *Phaseolus aureus* seeds under saline stress. *J. Plant Physiol.* **135**: 719-724.

Misra, S., and Oaks, A. 1985. Ferredoxin and pyridine nucleotide dependent glutamate synthase activities in maize endosperm tissue. *Plant Sci.* **39**: 1-5.

Muhitch, M.J. 1991. Tissue distribution and developmental patterns of NADH-dependent and ferredoxin dependent glutamate synthase activities in maize (*Zea mays*) kernels. *Physiol. Plant.* **81**: 481-488.

Murrilo, E., and Jimenez, E.S. 1985. Glutamate synthase in greening callus of *Bouvardia ternifolia*. *Planta* **163**: 448-452.

Oaks, A., Jones, K., and Misra, S. 1979. A comparison of glutamate synthase obtained from maize endosperms and roots. *Plant Physiol.* **63**: 793-795.

Oaks, A., Stulen, I., Jones K., Winsper, M.J., Misra, S., and Boesel, I.L. 1980. Enzymes of nitrogen assimilation in maize roots. *Planta* **148**: 477-484.

Ota, K., and Yamamoto, Y. 1990. Effects of different nitrogen sources on glutamine synthetase and ferredoxin dependent glutamate synthase activities and on free amino acid composition in radish plants. *Soil Sci. Plant Nutr.* **36**: 645-652.

Pietilla, M., Kusipuro, P., Pietilainen, P., and Lahdesmaki, P. 1989. Specificity and seasonal variation of arginase, glutamate synthase and nitrate reductase activities in Scots pine needles. *Plant Sci.* **64**: 153-160.

Puranik, R.M., and Srivastava, H.S. 1990. Increase in NADH-glutamate synthase activity in bean leaf segments by light. *Curr. Sci.* **59**: 1001-1003.

Rathnam, C.K.M., and Edwards, G.E. 1976. Distribution of nitrate assimilating enzymes between mesophyll protoplasts and bundle sheath cells in leaves of three groups of C_4 plants. *Plant Physiol.* **57**: 881-885.

Rendina, A.R., and Orme-Johnson, W.H. 1978. Glutamate synthase: On the kinetic mechanism of the enzyme from *E. coli. Biochem. J.* **17**: 5388-5399.

Rhodes, D., Rendon, G.A., and Stewart, G.R. 1976. The regluation of ammonia assimilating enzymes in *Lemna minor. Planta* **129**: 203-210.

Rhodes, D., Sims, A.P., and Folkes, B.F. 1980. Pathway of ammonia assimilation in illuminated *Lemna minor. Phytochemistry* **19**: 357-365.

Sakakibara, H., Watanabe, M., Hase, T., and Sugiyama, T. 1991. Molecular cloning and characterization of complimentary DNA encoding for ferredoxin dependent glutamate synthase in maize Leaf. *J. Biol. Chem.* **266**: 2028-2035.

Santos, I., and Salema, R. 1992. Effect of nitrogen nutrition on nitrate and nitrite reductase, glutamine synthase, glutamate synthase and glutamate dehydrogenase in CAM plant *Kalanchoe lateritia. Plant Sci.* **84**: 145-152.

Singh, R.P., and Srivastava, H.S. 1986. Increase in glutamate synthase (NADH) activity in maize seedlings in response to nitrate and ammonium nitrogen *Physiol. Plant.***66**: 413-416.

Singh, R.P., and Srivastava, H.S. 1987a. Increase in glutamate synthase activity (NADH-dependent) in excised root and leaf of maize seedlings in response to acidic amino acids and amides. *Biochem. Physiol. Pflanzen.* **182**: 497-500.

Singh, R.P., and Srivastava, H.S. 1987b. *In vivo* effects of some metabolic inhibitors on glutamate dehydrogenase and glutamate synthase activities in excised maize tissues. *Curr. Sci.* **56**: 93-94.

Singh, R.P., and Srivastava, H.S. 1992. Comparative characteristics of NADH:glutamate synthase from root and leaf tissues of maize seedlings. *Proc. Nat. Acad. Sci.* (India) B. **62**: 109-113.

Sodek, L., and da Silva, W.J. 1977. Glutamate synthase: a possible role in nitrogen metabolism of the developing maize endosperm. *Plant Physiol.* **60**: 602-605.

Somerville, C.R., and Orgen, W.L. 1980. Inhibition of photosynthesis in *Arabidopsis* mutants lacking leaf glutamate synthase activity. *Nature* **286**: 257-259.

Storey, R., and Reporter, M. 1978. Amino acid metabolism in developing soybean (*Glycine max*). Glutamate synthase in the cotyledons. *Can. J. Bot.* **56**: 1349-1356.

Suzuki, A., Carrayol, E., Zehnacker, C., and Deroche, M.E. 1988. Glutamate synthase in *Medicago sativa* L. Occurrence and properties of the Fd-dependent enzyme in plant cell fraction during nodule development. *Biochem. Biophys. Res. Commn.* **156**: 1130-1138.

Suzuki, A., and Gadal, P. 1982. Glutamate synthase from rice leaves. *Plant Physiol.* **69**: 848-852.

Suzuki, A., Gadal, P., and Oaks, A. 1981. Intracellular distribution of enzymes associated with nitrogen assimilation in roots. *Planta* **151**: 457-461.

Suzuki, A., Oaks, A., Jacquot, J-P, Vidal, J., and Gadal, P. 1985. An electron transport system in maize (*Zea mays* L.) roots for reaction of glutamate synthase and nitrite reductase. Physiological and Immunological properties of electron carrier and pyridine nucleotide reductase. *Plant Physiol.* **78**: 374-378.

Suzuki, A., Vidal, J., and Gadal, P. 1982. Glutamate synthase isoforms in rice. Immunological studies of enzymes in green leaf, etiolated leaf and root tissues. *Plant Physiol.* **70**: 827-832.

Tamura, G., Oto, M., Hirasawa, M., and Aketagawa, J. 1980. Isolation and partial characterization of homogenous glutamate synthase from *Spinacea oleracea*. *Plant Sci. Lett.* **19**: 209-215.

Tempest, D.W., Meers, J.L., and Brown, C.M. 1970. Synthesis of glutamate in *Aerobacter aerogens* by hitherto an unknown route. *Biochem. J.* **117**: 405-407.

Wallsgroove, R.M., Hirel, E., Pea, P.J., and Miflin, B.J. 1977. Studies on glutamate synthase from leaves of higher plants. *J. Expt. Bot.* **28**: 588-596.

Wallsgrove, R.M., Lea, P.J., and Miflin, B.J. 1982. The development of NAD(P)H dependent and ferredoxin dependent glutamate synthase in greening barley and pea leaves. *Planta* **154**: 473-476.

Watanabe, M., Hayashi, M., and Sugiyama, T. 1985. Effects of supplemental nitrate activation on the activity of some nitrogen assimilation enzymes and leaf productivity in maize seedlings. *Soil Sci. Plant Nutr.* **31**: 573-580.

Yamaya, T., Hayakawa, T., Tanasawa, K., Kamchi, K., Mae, T., and Ojima, K. 1992. Tissue distribution of glutamate synthase and glutamine synthetase in rice leaves. Occurrence of NADH-dependent glutamate synthase protein and activity in the unexpanded non green leaf blades. *Plant Physiol.* **100**: 1427-1432.

Zehnacker, C., Becker, T.W., Suzuki, A., Carroyl, E., Caboche, M., and Hirel, B. 1992. Purification and properties of tobacco ferredoxin-dependent glutamate synthase, and isolation of corresponding cDNA clones. Light inducibility and organ specificity of gene transcription and protein expression. *Planta* **187**: 266-274.

Zink, M.W. 1989. Regulation of ammonia assimilating enzymes by various nitrogen sources in cultured *Ipomaea* spp. *Can. J. Bot.* **67**: 3127-3133.

Nitrogen Nutrition in Higher Plants, 1995
Editors : H.S. Srivastava & R.P. Singh
Associated Publishing Co., New Delhi, India
pp. 243-272.

Amino Acid Biosynthesis

BIJAY K. SINGH

Abbreviations: AHAS, acetohydroxyacid synthase; AK, aspartate kinase; AS, anthranilate synthase; CM, chorismate mutase; DAHP, 3-deoxyarabinoheputlosonate-7-phosphate; DHAD, dihydroxyacid dehydratase; DHQ, dehydroquinate; DHPS, dihydrodipicolinate synthase; EPSP, 5-enolpyruvylshikimate-3-phoshate; GSA, glutamate semialdehyde; HK, homoserine kinase; HSDH, homoserine dehydrogenase; IAA, indole acetic acid; IGP, indole glycerol phosphate; KARI, ketoacid reductoisomerase; 5-MA, 5-methyl anthranilate; PAT, anthranilate phosphoribosyltransferase; P-5-C, pyrroline-5-carboxylate; PEP, phosphoenolpyruvate; SAM, S-adenosylmethionine; SDS, sodium dodecyl sulfate; S-3-P, shikimate-3-phoshate; TD, threonine dehydratase; TS, tryptophan synthase

I. Introduction

Nitrogen, derived mainly from nitrate and nitrogen gas, is reduced to ammonia before its incorporation into organic matter. The predominant form of organic nitrogen is amino acids. Due to this reason, amino acid biosynthesis is an important component of nitrogen nutrition in plants. Amino acid biosynthetic pathways provide not only the building blocks for protein biosynthesis but the intermediates and end products of these pathways are utilized for a large number of compounds with myriad of functions, such as, folic acid, auxin and ethylene for plant growth; lignin for structural support; phytoalexins and cyanogenic compounds for plant defense; flavonoids for protection from UV light, etc.

In 1806, the first crystals of amino acids were obtained from concentrated and evaporated asparagus juice which were later identified as asparagine and aspartate (Meister, 1992). Although this discovery was made 188 years ago, most of the progress in understanding amino acid biosynthesis in plants has been made in the last two to three decades. Initial interest in this area focussed on identification of the intermediates formed in metabolism of amino acids. Subsequently, enzymes and the pathways involved became the center of attention. With the discovery of intermediates and enzymes in different sub-cellular fractions, the significance of compartmentalization of the pathways became apparent. Regulation of carbon flux through the pathways was realized. In the more recent years, genetic studies involving gene isolation and use of auxotrophic mutants have greatly aided in our understanding of amino acid biosynthesis in plants. Despite a great deal of progress made in this area, pathways leading to several amino acids (e.g. histidine, lysine, leucine etc.) have not yet been elucidated in plants.

Reviews covering the progress made in different areas of amino acid biosynthesis have appeared from time to time (Jensen, 1986a, b; Kishore and Shah, 1988; Bryan, 1990; Singh et al., 1990). The purpose of this review is to provide the current understanding of pathways of amino acid biosynthesis and their regulation in plants. Rather than giving a detailed description of all the knowledge about different pathways, I have attempted to focus on the recent literature.

II. Aromatic Amino Acids

Condensation of erythrose-4-phosphate and phosphoenolpyruvate is the starting point of the biosynthesis of aromatic amino acids phenylalanine, tyrosine and tryptophan. There are seventeen enzymatic reactions leading to the biosynthesis of these amino acids (Fig. 1-3). All of the enzymes catalyzing these reactions have been demonstrated in plants. These amino acids and a number of intermediates of this pathway are precursors for the biosynthesis of numerous secondary metabolites. It has been estimated that as much as 25% of the carbon fixed by green plants flows through this route.

A. Biosynthesis of Chorismate

The first enzyme of the shikimate pathway, 3-deoxyarabinoheptulosonate-7-phosphate synthase, catalyzes the condensation reaction that produces 3-deoxyarabinoheptulosonate-7-phosphate (Fig. 1). Two isoforms of DAHP synthase have been demonstrated in a number of species (Ganson et al., 1986; Jensen, 1986a, b; Bryan, 1990; Singh et al., 1991). A Mn^{2+} requiring form (designated DS-Mn) is present in the chloroplast while a Co^{2+}/Mn^{2+} requiring form (designated DS-Co) is present in the cytosol. A great deal of interest has been generated in this enzyme with the discovery that expression of this enzyme is highly regulated in response to environmental stimuli (discussed later).

Phosphoenolpyruvate + Erythrose-4-phosphate

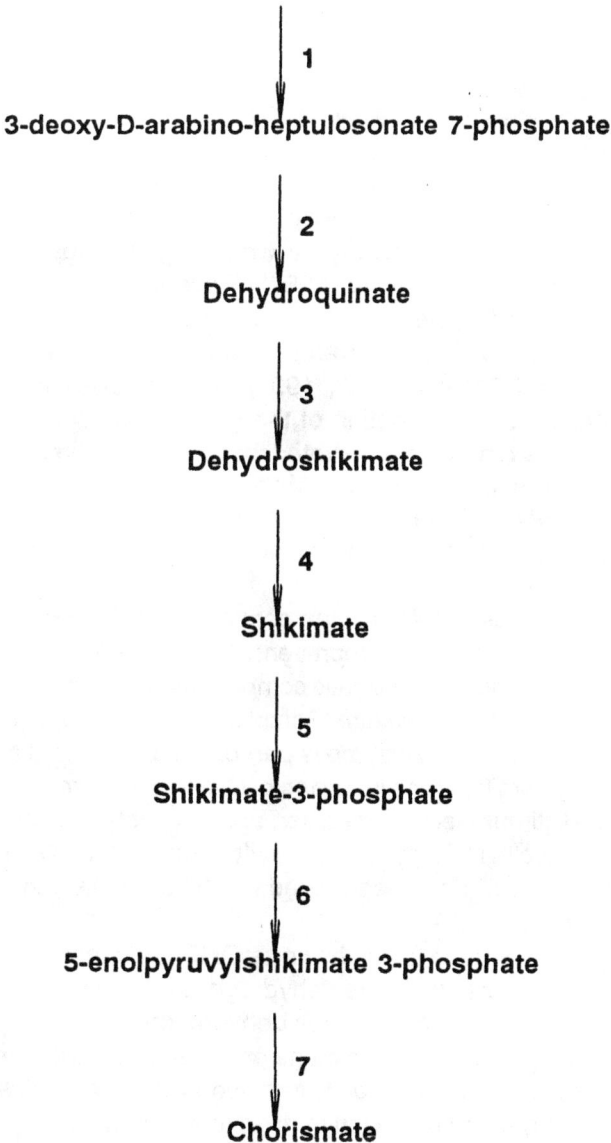

1

3-deoxy-D-arabino-heptulosonate 7-phosphate

2

Dehydroquinate

3

Dehydroshikimate

4

Shikimate

5

Shikimate-3-phosphate

6

5-enolpyruvylshikimate 3-phosphate

7

Chorismate

Fig. 1. Biosynthesis of chorismate. Enzymes: 1, DAHP synthase; 2, dehydroquinate synthase; 3, dehydroquinate dehydratase; 4, shikimate oxidoreductase; 5, shikimate kinase; 6, EPSP synthase; 7, chorismate synthase.

cDNAs encoding DAHP synthase have been recently isolated from *Arabidopsis thaliana* (Berlyn et al., 1989; Keith et al., 1991), tobacco (Wang et al., 1991), and tomato (Henstrand, et al., 1991). All these genes contained sequences encoding N-terminal transit peptide for targeting the protein to the chloroplast. No genes for DS-Co have been isolated thus far. The cDNA encoding DS-Mn from potato can be hybridized to a single mRNA species of about 2 kb from a number of plant species suggesting sequence homology between the genes from different species. This result is supported and extended to the protein sequence level because polyclonal antibodies raised against DS-Mn cross-react with the enzyme from carrot, maize, pigweed, squash, tomato and wheat (Dyer et al., 1990).

Production of lignin and/or phytoalexins is a common plant response to external stimulus. Phenylalanine ammonia lyase (PAL) is the first enzyme that initiates the utilization of phenylalanine for the biosynthesis of a large number of secondary metabolites. Regulation of PAL activity by wounding or fungal elicitation is very well documented (Lawton and Lamb, 1987). Recently, it has been shown that DS-Mn activity is induced by herbicide glyphosate treatment (Pinto et al., 1988), wounding (Dyer et al., 1989; Muday and Hermann, 1992), and fungal elicitors (McCue and Conn, 1989; Keith et al., 1991). This increase in enzyme activity was due to the enhanced transcription of the gene as shown by elevated levels of DS-Mn mRNA. It was further demonstrated that the time-course of stress-induced increases in the mRNAs for PAL and DS-Mn showed a similar pattern (Dyer et al., 1989; Keith et al., 1991). These results clearly demonstrate a coordinated regulation of biosynthesis of the aromatic amino acids to supply the needed amounts of aromatic compounds.

Dehydroquinate synthase catalyses the conversion of DAHP from the previous step to dehydroquinate. This reaction represents the first cyclization step in this pathway leading to the biosynthesis of aromatic compounds. The enzyme has been highly purified (4900-fold) from *Pisum sativum* (Pompliano et al., 1989) which had a native molecular weight of 66 kD. This enzyme is part of pentafunctional enzyme in yeast (Duncan et al., 1987). Similar to the induction of DAHP synthase activity in response to the environmental stimuli discussed earlier, slicing of potato tuber or sweet potato root caused an increase in DHQ synthase activity (Singh et al., 1991). Several analogs of the substrate DAHP have been found to inhibit DHQ synthase activity from plants (Pompliano et al., 1989).

Dehydroquinate dehydratase (also known as DHQ hydrolase) and shikimate oxidoreductase (also known as shikimate dehydrogenase) are the next two enzymes in this pathway. This first reaction involves dehydration of dehydroquinate to yield dehydroshikimate. Dhydroshikimate is then converted into shikimate in the second reaction by shikimate oxidoreductase. These two enzyme activities are catalyzed by a bifunctional protein which has been purified to apparent homogeneity from spinach chloroplasts (Fiedler and Schultz, 1985) and pea shoot tissue (Mousdale et al., 1987). The enzyme from both sources was monomeric with a molecular weight of 59 kD. The molecular weight estimates of the bifunctional enzyme from other higher plants were in the range of 52-73 kD (Mousdale et al., 1987).

Phosphorylation of shikimate to yield shikimate-3-phosphate is the next reaction that is catalyzed by shikimate kinase (ATP:shikimate-3-P-transferase). The enzyme has been partially purified from sorghum and mung bean seedlings (Kashiba, 1979; Bowen and Kosuge, 1979). This enzyme exhibits stimulation of activity by the thioredoxin system (Schmidt and Schultz, 1987) and light (Homeyer and Schultz, 1988). cDNA clones for shikimate kinase have recently been isolated by PCR amplification using oligonucleotide probes representing the conserved regions of the microbial enzymes (Schmid et al., 1992). The deduced protein sequence indicates a transit peptide for chloroplast uptake. As predicted, the *in vitro* synthesized shikimate kinase was imported and processed to the mature protein by tomato chloroplasts (Schmid et al., 1992).

Shikimate-3-phosphate synthesized in the previous step is condensed with phosphoenolpyruvate to form 5-enolpyruvylshikimate-3-phosphate by EPSP synthase. EPSP synthase is the most studied enzyme in this pathway because this enzyme is the site of action of a highly successful commercial herbicide, glyphosate (Steinrucken and Amrhein, 1980). Studies on the mechanism of the reaction catalyzed by EPSP synthase have shown that S-3-P is the first substrate to interact with EPSP synthase during the reaction (Bondinell et al. 1971; Wilbenmeyer et al. 1988; Anderson et al., 1988a) even though both S-3-P and PEP are capable of interacting with the free enzyme (Parr et al., 1987). Rapid chemical-quench flow kinetic methods led to the identification of a tetrahedral intermediate involved in the reaction (Anderson et al., 1988b). Inhibition kinetics with glyphosate is competitive versus PEP and uncompetitive versus S-3-P (Kishore and Shah, 1988). The enzyme reacts first with S-3-P followed by glyphosate or PEP. These results were confirmed in later studies that showed insignificant binding of glyphosate with the free enzyme (Anderson et al., 1988a).

EPSP synthase has been purified to homogeneity and the complete amino acid sequence of this protein is known (Steinrucken and Amrhein, 1984; Mousedale and Coggins, 1984; Kishore and Shah, 1988; Singh et al., 1991). Chemical modification studies have shown that Lys-22 and Arg-27, present in all EPSP synthases sequenced to date, are essential for the enzyme activity (Huynh et al., 1988; Kishore and Shah, 1988). One histidine residue has also been suggested to be important for the reaction catalysed by this enzyme (Huynh, 1987).

The plant EPSP synthase genes encode a transit peptide for the translocation of the protein in the chloroplast where it has been demonstrated to be localized (Mousdale and Coggins, 1986; Della-Cioppa et al., 1986b; Smart and Amrhein, 1987). The deduced amino acid sequence of the coding region of the protein from petunia, tomato and *Arabidopsis* show high degree of homology (84-95%), except in the region of transit peptides (23-68%). Pre-EPSP synthase synthesized *in vitro* is rapidly taken up into chloroplasts and proteolytically cleaved to the mature 48 kD protein (Della-Cioppa et al., 1986a). The protein without the transit peptide was not imported by the chloroplasts. Pre-EPSP synthase is catalytically active and this activity is inhibited by glyphosate (Della-Cioppa and Kishore, 1988). The precursor enzyme was not imported into the chloroplast in the presence of S-3-P and glyphosate.

Considerable effort has been devoted to obtain glyphosate resistant plants. Moderate to high levels of glyphosate tolerance were obtained by overproduction of EPSP synthase in plants by tissue culture selection, overexpression of the plant gene by tansformation, and expression of the mutated genes from petunia, *E. coli, Salmonella* (Smart et al., 1985; Kishore and Shah, 1988; Mazur and Falco, 1989). These mutant genes code for variants of the enzyme that are less sensitive to inhibition by glyphosate. However, none of these plants could tolerate the field use rates of glyphosate. Recently, several crop species have been transformed either with a bacterial EPSP synthase gene which produces an enzyme that has reduced sensitivity to glyphosate, or with a gene that produces an enzyme that can metabolize glyphosate (Barry et al., 1992). Both of these approaches are being claimed to provide commercial levels of glyphosate tolerance to the transformed plants.

Chorismate synthase catalyzes the final step in the pre-chorismate pathway that produces chorismic acid from EPSP. The enzyme was first detected and characterized in pea (Mousdale and Coggins, 1986). The enzyme from pea is oxygen sensitive, therefore, the enzyme activity could be detected only in the presence of reduced flavin. The enzyme has been purified to near homogeneity from a cell suspension culture of *Cordalis sempervirens* (Schmid et al., 1990). Polyclonal antibodies raised against the purified enzyme were used to isolate a cDNA clone encoding chorismate synthase (Schaller et al., 1991). The amino terminus of the protein deduced from the gene encodes a transit peptide suggesting the plastidic localization of the enzyme.

B. Biosynthesis of Tryptophan

Tryptophan is produced from chorismate through a sequence of six enzymatic reactions (Fig. 2) and the enzyme activities for each step have been detected in carrot cells, pea seeds, maize shoots and wheat (Gilchrist and Kosuge, 1980). Anthranilate synthase is the first enzyme in this pathway which removes enolpyruvyl side chain from chorismate and adds amino group donated from glutamine. This reaction requires Mg^{2+} as a cofactor. Biosynthesis of tryptophan is tightly regulated by feedback inhibition of AS.

Two genes (designated *ASAI and ASA2*) encoding AS have been isolated from *Arabidopsis* (Niyogi and Fink, 1992). Both predicted proteins contain chloroplast transit peptides at their amino termini suggesting chloroplastic localization of these proteins. Both genes appear to be expressed in all plant parts, however, expression of the two genes is differentially regulated. *ASA2* is expressed at a constitutive basal level while *ASA1* mRNA is approximately 10-times more abundant in whole plants. Interestingly, regulation of *ASA1* and *ASA2* parallels the expression of duplicated genes for DAHP synthase (Keith et al., 1991). *ASA1* responds to wounding and pathogen infection whereas *ASA2* is unaffected. Therefore, expression of the *ASA1* gene is regulated by demand for tryptophan pathway metabolites. This genetic data is supported by a 100% increase in AS activity following elicitation with culture filtrate of *Pythium aphanidermatum* and a parallel increase in the production of indole alkaloids (Poulsen et al., 1992).

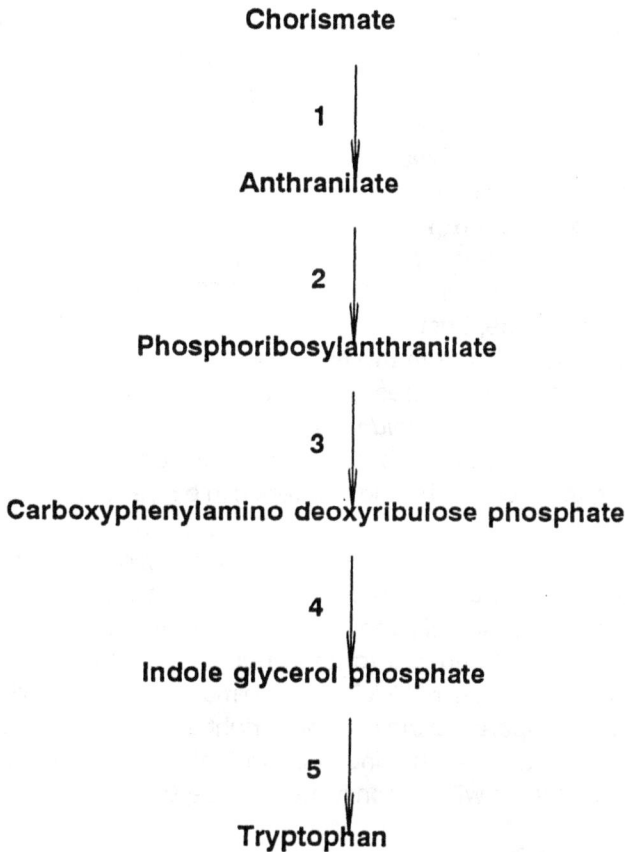

Chorismate

1

Anthranilate

2

Phosphoribosylanthranilate

3

Carboxyphenylamino deoxyribulose phosphate

4

Indole glycerol phosphate

5

Tryptophan

Fig. 2. Biosynthesis of tryptophan. Enzymes: 1, anthranilate synthase; 2, anthranilate phosphoribosyltransferase; 3, phosphoribosylanthranilate isomerase; 4, indole glycerol phosphate synthase; 5, tryptophan synthase.

Anthranilate phosphoribosyltransferase, the next enzyme in this pathway, adds a phosphoribosyl moiety to anthranilate. This enzyme gained attention because *Arabidopsis* plants with defective PAT were isolated as the first plant amino acid auxotroph in which auxotrophy shows Mendelian inheritance (Last and Fink, 1988). These tryptophan auxotrophs were selected using 5-methyl-anthranilate, a suicide substrate for this enzyme. Wild type plants containing normal PAT converted 5-MA to toxic 5-methyl-tryptophan and thereby plants did not survive. Mutant plants containing defective PAT were unable to use 5-MA and were rescued by adding a small amount of tryptophan to the medium. These mutants were identified as UV-fluorescent plants due to the accumulation of high levels of anthranilate in these plants. An *Arabidopsis* gene encoding PAT has been recenty isolated (Rose et al., 1992). All of the evidences thus far suggest a single copy gene for this enzyme in *Arabidopsis*.

In the next two reactions, as Amadori rearrangement isomerizes the phospho-ribosyl side chain into a 1-deoxy-ribulose 5-phosphate (phosphoribosyl anthrani-late isomerase), and decarboxylation with ring closure forms the completed ring system of indole (indole glycerol phosphate synthase). Tryptophan synthase cat-alyzes the last step in this pathway where indole glycerol phosphate is converted to tryptophan. The enzyme contains two sub-units. Similar to the bacterial enzyme, plant TS appears to be an $\alpha_2\beta_2$ heterotetramer. The α sub-unit converts IGP to indole and then the β sub-unit adds serine to indole to produce tryptophan.

Two unlinked genes encoding the β sub-unit of TS have been isolated from *Arab-idopsis* (Berlyn et al., 1989; Last et al., 1991) and maize (Wright et al., 1992). The deduced protein sequences from the two *Arabidopsis* TS genes contain a transit peptide for chloroplast uptake. Interestingly, expression of the two TS β genes appear to be differentially regulated to *Arabidopsis* and maize. In maize, the level of mRNAs for the two genes appeared to be similar in most tissues examined (Wright et al., 1992). In contrast, there was a 10-fold difference in expression of the two TS genes in *Arabidopsis*

TS mutants have been valuable in understanding the biosynthesis of plant growth hormone, indole acetic acid. Until recently, tryptophan was considered to be the sole precursor for IAA (Bandurski and Nonhebel, 1989). Examination of tryptophan and IAA levels in normal and *orp* (orange pericarp) mutant seedlings of maize revealed 50-fold higher levels of IAA in the *orp* mutant despite a 75% reduction in tryptophan content compared to the control (Wright et al., 1991). Additionally, stable isotopes from tryptophan were not incorporated into IAA by the mutant or normal seedlings. Further studies will demonstrate the true biosynthetic pathway for IAA.

C. Biosynthesis of Phenylalanine and Tyrosine

The proposed pathway for the biosynthesis of phenylalanine and tyrosine from chorismate is described in Fig. 3. Chorismate mutase, the first enzyme in this path-way, causes an intramolecular Claisen (oxy-Cope) rearrangement of chorismate to produce prephenate. Two major forms of CM have been isolated from a number of plants which differ in sensitivity to feedback regulation by aromatic amino acids, substrate saturation kinetics, sub-cellular localization, and immunological proper-ties (Singh et al., 1985; 1986; 1991; Singh and Conn, 1986; Jensen, 1986a, b; Kuroki and Conn, 1988a, b; Bryan, 1990). A chloroplast localized enzyme (designated CM-1) is feedback inhibited by phenylalanine and tyrosine and activated by tryptophan. This enzyme is allosteric and displays cooperativity in both substrate and effector binding. The other form of the enzyme (designated CM-2) is located in the cytosol and exhibits insensitivity to the aromatic amino acids and hyperbolic substrate satu-ration kinetics.

Chorismate

1

Prephenate

2

Arogenate

3   4

Tyrosine   Phenylalanine

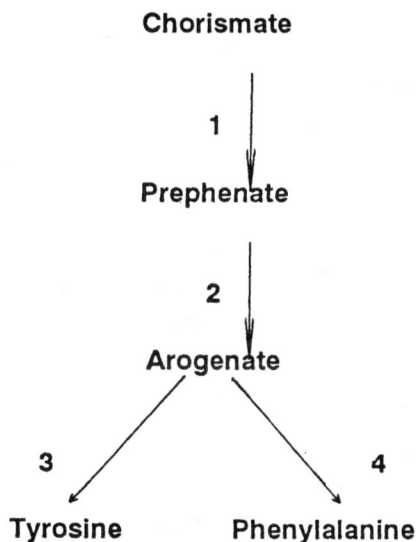

Fig. 3. Biosyntheis of tyrosine and phenylalanine. Enzymes: 1, chorismate mustase; 2, prephenate aminotransferase; 3, arogenate dehydrogenase; 4, arogenate dehydratase.

Prephenate produced in the previous step is transaminated by prephenate amino-transferase to produce arogenate. Aminotransferases with strong specificity for pre-phenate as the keto-acid amino group acceptor have been described in *Nicotiana silvestris* (Bonner and Jensen, 1985), *Sorghum bicolor* (Siehl et al., 1986a). Most, if not all, of the prephenate aminotransferase activity was localized in chloroplasts of sorghum leaf blades (Siehl et al., 1986b).

Arogenate could be used by arogenate dehydrogenase to produce tyrosine or by arogenate dehydratase to produce phenylalanine (Jensen, 1986a, b; Singh et al., 1991). Arogenate dehydrogenase from *Sorghum bicolor* is specific for both aro-genate and $NADP^+$ and does not utilize prephenate or NAD^+ (Singh et al., 1991). This activity is strongly inhibited by tyrosine. Arogenate dehydratase activity has been demonstrated in *Nicotiana sylvestris* cell cultures, spinach leaf (Jensen, 1986a, b) and *Sorghum bicolor* seedlings (Siehl and Conn, 1988). This enzyme activity is strongly inhibited by its product phenylalanine and activated by tyrosine.

D. Compartmentation

Isolated chloroplasts can synthesize phenylalanine, tyrosine, and tryptophan from NaH $^{14}CO_3$ (Bagge and Larson, 1986; Schulze-Sieber and Schultz, 1989), ^{14}C-3-phosphoglycerate (Schulze-Sieber and Schultz, 1989) and ^{14}C-phosphoenol-pyruvate (Homeyer and Schultz, 1988). This observation suggests that all of the enzymes required for the biosynthesis of aromatic amino acids, tryptophan, phen-ylalanine and tyrosine, are present in chloroplasts. This notion is supported by the biochemical and genetic data. Most of the enzymes of this pathway have been

found in the plastid fraction (Singh et al., 1991). Furthermore, all of the genes isolated thus far (DAHP synthase, EPSP synthase, chorismate synthase, anthranilate
synthase, anthranilate phosphoribosyltransferase and the β sub-unit of tryptophan
synthase) contain sequences encoding chloroplast transit peptide. Therefore, it is
reasonable to conclude that chloroplast is the primary site of aromtic amino acid
biosynthesis.

In addition to the chloroplastic form of the enzyme, biochemical studies have shown
cytosolic forms of DAHP synthase and chorismate mutase in several plant species
(Jensen, 1986a, b). Identification of these cytosolic isoenzymes have led to the speculation that the cytosol has a complete aromatic amino acid biosynthetic pathway
(Jensen, 1986a, b). Futhermore, it was suggested that each pathway has a specialized role, the chloroplastic pathway for amino acid synthesis and the cytosolic
for support of the chloroplastic pathway and for secondary products. However, lack
of any genes without chloroplast transit peptide coding sequences pose a serious
doubt on the presence of a complete pathway in the cytosol. As far as the proposed role of cytosolic isozymes is concerned, all evidence thus far shows that the
chloroplastic enzyme is induced due to various environmental stimuli (Kuroki and
Conn, 1989a; Dyer et al., 1989; McCue and Conn, 1989; Muday and Herrmann,
1992). Therefore, role of the cytosolic forms of the enzyme in the aromatic amino
acid biosynthetic pathway is not yet clear.

E. Regulation of the Pathway

Until recently, it was believed that regulation of chorismate mutase and anthranilate
synthase at the branch point were the only means of controlling the flux of carbon
through this pathway. However, it is now clear that carbon flux through this pathway is not only regulated by feedback inhibition of additional enzymes but also by
controlling the level of expression of different proteins.

High levels of tryptophan would inhibit anthranilate synthase and slow down flow
of carbon towards tryptophan biosynthesis. At the same time, elevated levels of
tryptophan will activate chorismate mutase and divert the flux of carbon to phenylalanine and tyrosine. It is clear from kinetic studies that chorismate mutase is more
sensitive to activation by tryptophan than to inhibition by tyrosine and phenylalanine.
Flux through chorismate mutase would continue until phenylalanine and tyrosine
pools become high enough to inhibit chorismate mutase. Recent results also indicate
that high levels of phenylalanine and tyrosine can inhibit their own synthesis by inhibiting the terminal enzymes, arogenate dehydratase and arogenate dehydrogenase.
Inhibition of the terminal branches would lead to the elevation of the pools of arogenate which can reduce the flow of carbon through the whole pathway by inhibiting the very first enzyme, DAHP synthase (Jensen, 1986a, b; Singh et al., 1991).

Recent studies on DAHP synthase (McCue and Conn, 1989; Keith et al., 1991),
chorismate mutase (Kuroki and Conn, 1988a), and anthranilate synthase (Niyogi
and Fink, 1992; Poulsen et al., 1992) have clearly demonstrated an elevation in
the expression of the amount of these enzymes due to wounding, exposure to fungal

elicitors, and high light intensity. Therefore, in order to meet the transient needs for the intermediates or end products of this pathway, plants up regulate the carbon flux by increasing the expression of key enzymes of this pathway. However, in order to down regulate the flow of carbon through this pathway, feedback inhibition of key enzymes of this pathway is a means adopted by the plants. Future studies will reveal if other mechanisms of regulation of this pathway are also important in plants. It would also be interesting to understand the co-ordination of up and down regulation by the ways described above.

III. Aspartate Family

Aspartate is the common precursor for the biosynthesis of lysine, threonine, isoleucine and methionine (Fig. 4-7). The isoleucine biosynthetic pathway involves enzymes that are also common to the biosynthesis of pyruvate derived amino acids, valine and leucine (Fig. 7). Therefore, isoleucine synthesis will be discussed in the section dealing with the branched chain amino acids. Biosynthesis of aspartate derived amino acids have been studied at a great length with the objective of increasing the amino acid composition and thereby improving the nutritional quality of various crops.

Aspartate kinase and aspartate semialdehyde dehydrogenase are the two common enzymes for the biosynthesis of all aspartate derived amino acids. Aspartate kinase, the first enzyme in the pathway, catalyzes the phosphorylation of aspartate to aspartyl phosphate (Fig. 4). Three different isozymes of AK which are inhibited by threonine alone, lysine alone or lysine and S-adenosylmethionine in a synergistic manner have been detected in various species (Bryan, 1990). Therefore, feedback regulation of this enzyme by the end-products has been predicted to be a major control mechanism for the regulation of carbon flow through this complex pathway (discussed later).

AK has been highly purified and characterized from several species. Interestingly, a *Daucus carota* bifunctional protein containing both AK and homoserine dehydrogenase (the third enzyme in this pathway) activities has been purified to homogeneity (Wilson et al., 1991). Oligonucleotide probes derived from the bifunctional protein sequence were used to amplify a portion of the gene from *Daucus carota* cDNA which in turn was used to isolate a full length cDNA clone which encodes the fused bifunctional protein (Matthews et al., 1992). Genes for similar bifunctional protein have also been isolated from soybean (Matthews et al., 1992) and maize (Muehlbauer et al., 1992). Protein sequence deduced from the isolated cDNAs show a chloroplast transit peptide suggesting a plastidic localization of the enzyme. Northern blot analysis revealed that this gene is equally expressed in leaf, endosperm and embryo of maize (Muehlbauer et al., 1992). Southern blot analysis results and isolation of four different AK-HSDH cDNA clones suggest that AK-HSDH exists as a small gene family in maize (Muehlbauer et al., 1992). These results do not rule out the possibility of separate AK and HSDH genes. The gene and deduced protein sequence comparison suggests that this bifunctional protein is closely related to/thrA gene in *Escherichia coli* that encodes a similar bifunctional protein AKI-HSDHI (Matthews et al., 1992)

In order to increase the aspartate derived amino acid composition of plant, a number of mutants with altered feedback sensitivity of AK have been isolated (Gengenbach et al., 1992). All of these mutants accumulated threonine indicating that threonine synthesis is directly controlled by AK. Similarly, threonine overproduction was achieved in transgenic tobacco plants by cytosolic or chloroplastic expression of a desensitized AK of *E. coli* (Shaul and Galili, 1992 b). Although threonine accumulation was higher in the transgenic plants expressing the bacterial enzyme in the chloroplast, it is possible that intermediates of this pathway are able to shuttle between cytoplasm and the plastids.

Aspartate semialdehyde dehydrogenase, the next enzyme in this pathway, has not received much attention even though the product of this enzyme is utilized in the biosynthesis of all aspartate derived amino acids. This enzyme is not inhibited by any of the intermediates or the end products of this pathway (Myers and Gengenbach, 1982). Activity of this enzyme has been detected in the extracts of pea and maize (Bryan, 1990).

A. Biosynthesis of Threonine

Aspartate semialdehyde produced in the previous step is converted to threonine by a sequence of reactions catalyzed by homoserine dehydrogenase, homoserine kinase and threonine synthase (Fig. 4). HSDH reduces aspartate semialdehyde to produce homoserine and this activity has been detected in a wide variety of plants (Bryan, 1990). This enzyme is feedback inhibited by threonine and therefore is considered to be an additional step besides AK to regulate the flow of carbon to threonine. Threonine-sensitive and threonine-insensitive forms of HSDH have been detected in several species (Bryan, 1990) and purified to homogeneity from maize suspension cultures (Walter et al., 1979). Purification of AK-HSDH bifunctional protein and isolation of cDNA clones expressing this protein has been discussed above. This discovery of the presence of both enzyme activities on the same polypeptide may have important implications in terms of feedback regulation of the pathway (discussed later).

Homoserine kinase, the next enzyme in the pathway, phosphorylates homoserine to produce phosphohomoserine. This enzyme activity has been detected and characterized in pea (Muhitch and Wilson, 1983) and barley (Aarnes, 1976). Although HK from pea was inhibited by isoleucine, threonine and S-adenosylmethionine, there is no conclusive evidence that regulation of this enzyme control the carbon flow through this pathway (discussed later).

Phosphohomoserine is diverted to either threonine by threonine synthase or by a series of enzymes (discussed later) to produce methionine. Because of this branch point, regulation of enzyme activities at this step is not surprising. Threonine synthase from several species displays an absolute requirement for SAM, a methionine derivative (Giovanelli et al., 1984). The enzyme had maximum activity in the presence of 100 μM SAM and had high affinity for phosphohomoserine (K_m = 2-7 μM). In the absence of this activator, threonine synthase is virtually inactive which should allow the flow of phosphohomoserine to methionine and SAM.

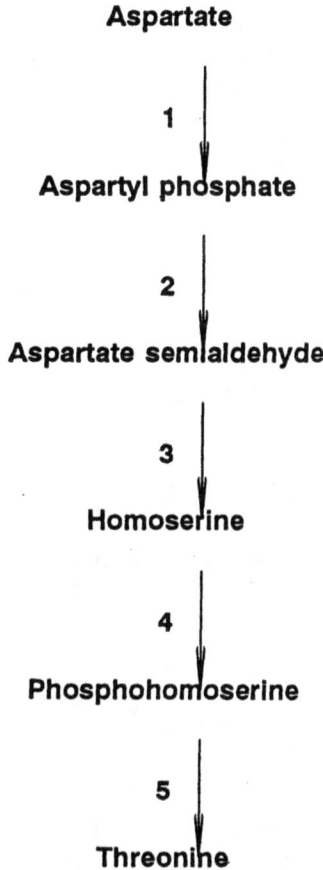

Aspartate

1

Aspartyl phosphate

2

Aspartate semialdehyde

3

Homoserine

4

Phosphohomoserine

5

Threonine

Fig. 4. Biosynthesis of threonine. Enzymes: 1, aspartate kinase; 2, aspartate semialdehyde dehydrogenase; 3, homoserine dehydrogenase; 4, homoserine kinase; 5, threonine synthase.

B. Biosynthesis of Methionine and S-adenosylmethionine

Methionine is synthesized from phosphohomoserine through a set of three enzymatic steps (Fig. 5). In the first step, cystathionine synthase produces cystathionine by condensing cysteine with phosphohomoserine. It has been shown that the specific activity of cystathionine synthase changes depending upon methionine availability *in vivo* (Thompson et al., 1982). This result suggests that regulation of the level of cystathionine synthase activity controls the carbon flow to methionine.

A three carbon skeleton derived from cysteine is removed by cystathionase to yield homocysteine. It is possible that a sulfhydrylase could directly convert phosphohomoserine to homocysteine. However, it has been demonstrated in *Lemna paucicostata* that the two enzymatic step conversion of phosphohomoserine to homocysteine is the route of homocysteine synthesis *in vivo* (Anderson, 1990). Methionine synthase transfers a methyl group from methyltetrahydrofolate to homocysteine resulting in the formation of methionine. Methionine is rapidly converted to SAM by methionine adenosyltransferase. As discussed earlier, it is SAM that regulates the activity of aspartate kinase and threonine synthase.

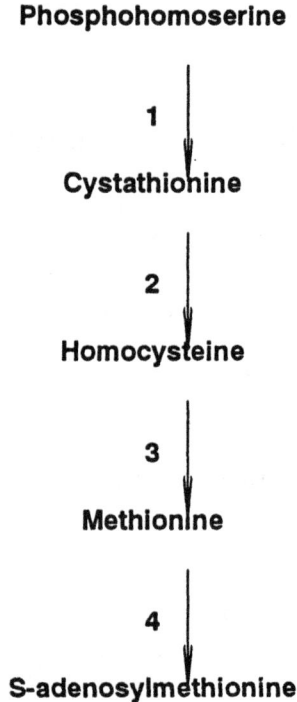

Phosphohomoserine

1

Cystathionine

2

Homocysteine

3

Methionine

4

S-adenosylmethionine

Fig. 5. Biosynthesis of methionine and S-adenosylmethionine. Phosphohomoserine, synthesized by a set of enzymes shown in Fig. 4, is the starting material for the unique pathway leading to methionine biosynthesis. Enzymes: 1, cystathionine synthase; 2, cystathionine lyase; 3, methionine synthase, 4, methionine adenosyltransferase.

The subcellular localization studies indicate involvement of both plastid and cytosolic enzymes for the biosynthesis of methionine and SAM. Fractionation of barley leaf protoplast has indicated that enzymes leading to the biosynthesis of homocysteine from aspartate are located in the chloroplast (Wallsgrove et al., 1983). Homocysteine moves out of the chloroplast to cytosol where the last two enzymes are present that convert homocysteine to methionine and SAM.

C. Biosynthesis of Lysine

Dihydrodipicolinate synthase is the first enzyme unique to the lysine biosynthetic pathway that condenses aspartate semialdehyde and pyruvate to produce dihydrodipicolinate (Fig. 6). DHPS has been detected in many species and highly purified from wheat suspension cultures (Kumpaisal et al., 1987), tobacco leaves (Ghislain et al., 1990) and maize (Frisch et al., 1991). DHPS activity isolated from plants is inhibited by low concentrations of lysine, the end product of this pathway.

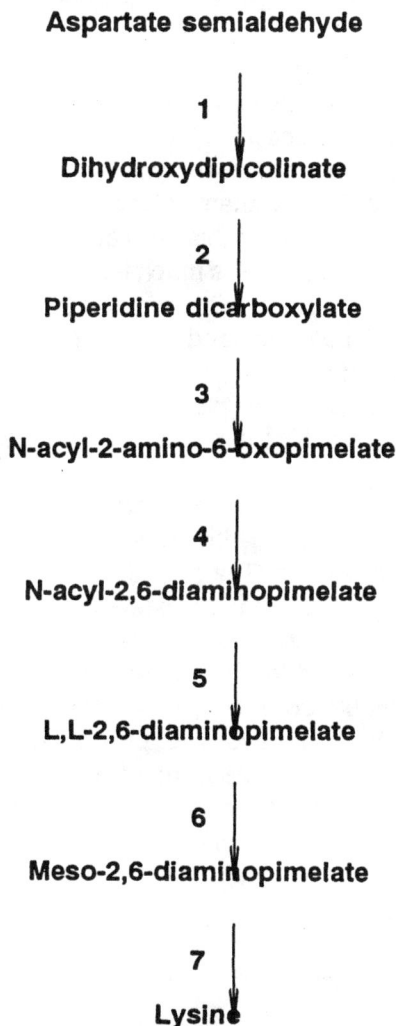

Aspartate semialdehyde

1

Dihydroxydipicolinate

2

Piperidine dicarboxylate

3

N-acyl-2-amino-6-oxopimelate

4

N-acyl-2,6-diaminopimelate

5

L,L-2,6-diaminopimelate

6

Meso-2,6-diaminopimelate

7

Lysine

Fig. 6. Biosynthesis of lysine. Aspartate semialdehyde produced by the first two enzymes shown in Fig. 4 could also be diverted to lysine. Enzymes: 1, dihydrdipicolinate synthase; 2, dihydrodipicolinate reductase; 3, piperidine dicarboxylate acylase; 4, acyldiaminopimelate aminotransferase; 5, acyldiaminopimelate deacylase; 6, diaminopimelate epimerase; 7, diaminopimelate decarboxylase.

Wheat, soybean and maize cDNA clones encoding DHPS have been isolated (Kaneko et al., 1990; Frisch et al., 1991b; Silk and Mathews, 1992). The deduced protein sequence revealed a transit peptide for chloroplast uptake of the protein which is consistent with the chloroplastic localization of the enzyme activity (Ghislain et al., 1990). The predicted molecular weight (35, 854D) of the mature DHPS poly-peptide was similar to the 38 kD molecular weight determined by SDS-polyacryla-mide gel electrophoresis of the purified enzyme from maize (Frisch et al., 1991a, b). At least two functional DHPS genes in wheat have been detected (Kaneko et al., 1990). Similarly, Southern blot analysis indicates more than one DHPS gene per haploid maize genome. Whether or not the different gene products have dif-ferent regulatory properties is unclear from the enzyme analysis data. Level of expression of DHPS from different genes is not yet known either.

Feedback inhibition of DHPS activity is considered to be the main control point for the carbon flux to lysine. The *in vitro* enzyme inhibition data is supported by the isolation of lysine-insensitive DHPS mutants of tobacco that accumulate lysine (Neg-rutiu et al., 1984). Further evidence for this regulation is apparent from plant trans-formation studies. The bacterial DHPS is much less sensitive to inhibition by lysine than the enzyme isolated from plants (Glassman, 1992; Perl et al., 1992). There-fore, the *dapA* gene from *E. coli* was used to transform tobacco (Shaul and Galili, 1992 a; Glassman, 1992) and potato (Perl et al., 1992). Expression of this bacterial DHPS in chloroplasts of the transgenic plants allowed uninterrupted flow of carbon through the lysine pathway resulting in a significant increase in level of free lysine in different plant parts.

Dihydrodipicolinic acid reductase, the next enzyme in this pathway, reduces dihy-drodipicolinic acid to piperidine-2,6-dicarboxylate. This enzyme from maize has been partially purified and characterized (Tyagi et al., 1983). In *E. coli* and most other bacteria, a set of four enzymes produce *meso*-diaminopimelate from piperidine decarboxylate (Bryan, 1990). Only the last enzyme in this set, diaminopimelate epi-merase has been demonstrated in plant (Tyagi et al., 1982). In some bacteria, piperidine decarboxylate could be directly converted to *meso*-diaminopimelate by *meso*-diaminopimelate dehydrogenase. There is one report of the presence of the enzyme in soybean (Wenko et al., 1985), however, this result has not been cor-roborated. Therefore, the biosynthetic route of *meso*-diaminopimelate in plants has not been established. Since diaminopimelate epimerase has been discovered in maize,it is likely that the biosynthetic route of lysine biosynthesis in plants is similar to the pathway in *E. coli.*

D.   Regulation of the Pathway

Aspartate amino acid biosynthesis is regulated at multiple sites because of several branch points in the biosynthetic pathway. These multiple regulation sites allow a tight control on the levels of these amino acids and also regulate the carbon flux through different pathways in such a way that no one amino acid becomes limiting.

Aspartate kinase is the primary site of regulation. Although this enzyme is feedback inhibited by lysine, threonine and S-adenosylmethionine, the recent plant transformation data suggests that regulation at this enzyme is important primarily for threonine biosynthesis. This conclusion is based on the fact that expression of a feedback inhibitor insensitive aspartate kinase caused a significant overproduction of threonine while lysine and isoleucine showed only a slight increase (Shaul and Galili, 1992 b). In this study, methionine levels were below detection levels. Despite this fact, regulation at aspartate kinase is important for other amino acids as well because threonine+lysine feeding is toxic. However, this toxicity is overcome by methionine supplementation (Hibberd and Green, 1980). Obviously, blocking carbon flow at aspartate kinase by threonine+lysine prevents methionine biosynthesis.

Additional regulation at homoserine dehydrogenase by threonine has been suggested. However, the recent discovery of homoserine dehydrogenase as part of bifunctional protein containing AK activity suggests that regulation of HSDH activity by threonine may possibly be due to a change in the protein structure brought about by the binding of threonine with AK which may affect the HSDH activity. This scenario is further complicated by the presence of separate proteins for the two enzymes. Gene isolation and their expression patterns will help resolve this issue.

Threonine biosynthesis is also regulated by activation of threonine synthase by SAM. Since this enzyme is virtually inactive in the absence of SAM, there must be adequate flow of phosphohomoserine to methionine and SAM that keeps threonine synthase activated in the transgenic plants transformed with the *E. coli* AK (Shaul and Galili, 1992 b).

In this whole complex biosynthetic pathway, only methionine biosynthesis is regulated by controlling the level of expression of an enzyme. Cystathionine synthase, the first enzyme at the branch point of phosphohomoserine partitioning, is regulated by the availability of methionine and SAM (Thompson et al., 1982). In this case, high levels of methionine repress the level of cystathionine synthase and thereby reduce carbon flux through this pathway.

Lysine biosynthesis is clearly regulated by the feedback inhibition of DHPS by lysine. The *in vitro* enzyme inhibition data is supported by the isolation of lysine-insensitive DHPS mutants that accumulate high levels of lysine (Negrutiu et al., 1984). This conclusion is further corroborated by recent studies that show that expression of DHPS from *E. coli*, an enzyme that is less sensitive to lysine inhibition than the plant counterpart, causes accumulation of free lysine.

IV. Branched Chain Amino Acids

Branched chain amino acids biosynthetic pathway has received a great deal of attention in the past 10 years because of the discovery that two classes of highly successful commercial herbicides inhibit the biosynthesis of valine, leucine and isoleucine (Ray, 1984; Shaner et al., 1984). This pathway is unique in the sense that a set of 4 enzymes carry out reactions in parallel pathways using different substrates leading to the biosynthesis of isoleucine or valine and leucine as illustrated in Fig. 7 and 8.

A. Biosynthesis of Valine and Isoleucine

Threonine dehydratase (also known as threonine deaminase) catalyzes the first step in the isoleucine biosynthetic pathway (Fig. 7). The enzyme deaminates and dehydrates L-threonine to produce α-ketobutyrate and ammonia. A form of this enzyme present primarily in the younger tissues is feed back inhibited by isoleucine (Sharma and Mazumdar, 1970). Because of its regulatory property, this form of TD is considered to be the "biosynthetic" form of the enzyme. Identification of plant mutants that are auxotrophic for isoleucine through lack of TD activity (Sidorov et al., 1981; Negrutiu et al., 1985) has demonstrated that TD is essential for isoleucine biosynthesis. Complementation of a TD-defecient *Nicotiana plumbaginifolia* mutant with ILV1 gene from *Saccharomyces cerevisiae* has confirmed the absolute requirement of TD for isoleucine biosynthesis (Colau et al., 1987).

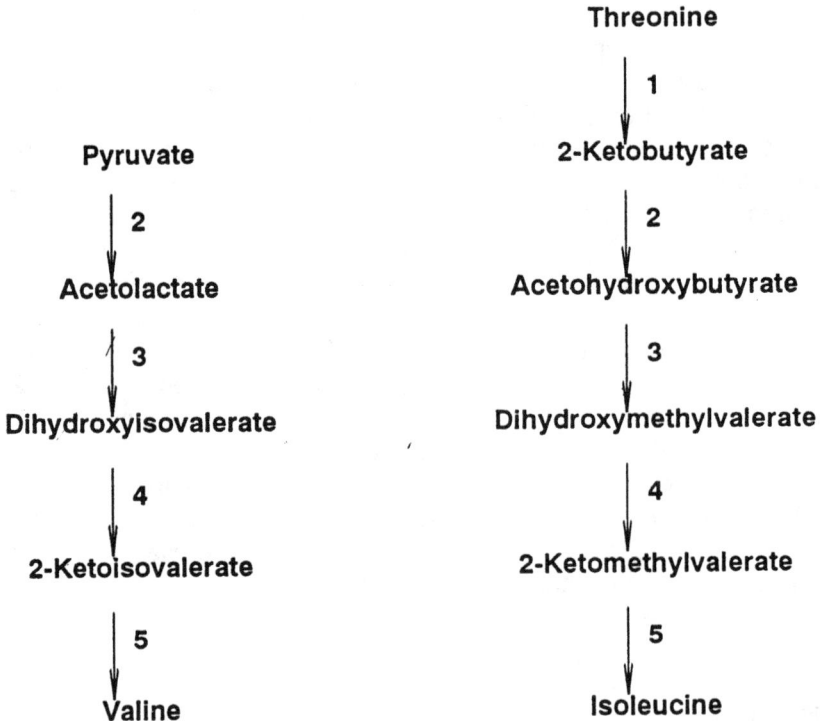

Threonine

| 1

2-Ketobutyrate

Pyruvate

| 2 | 2

Acetolactate   **Acetohydroxybutyrate**

| 3 | 3

Dihydroxyisovalerate   **Dihydroxymethylvalerate**

| 4 | 4

2-Ketoisovalerate   **2-Ketomethylvalerate**

| 5 | 5

Valine   **Isoleucine**

Fig. 7. Biosynthesis of isoleucine and valine. Threonine, the endproduct of the pathway shown in Fig. 4, is the starting precursor for isoleucine biosynthesis. Enzymes 2 to 5 carry out the two parallel reactions leading to valine and isoleucine biosynthesis. Enzymes: 1, threonine dehydratase; 2, acetohydroxyacid synthase; 3, ketoacid reductoisomerase; 4, dihydroxyacid dehydratase; 5, aminotransferase.

A biodegradative TD has been characterized in a number of parasitic and saprophytic plants (Kagan et al., 1969). High levels of the biodegradative TD were found in *Cascuta* seeds where the activity was suggested to utilize high concentrations of threonine and serine present (Madan and Nath, 1983). Similar to the enzymes from *E. coli*, both forms of the plant enzyme can use threonine and serine as substrate. Recently, a biodegradative form of TD has been identified and characterized in the senescing tomato leaves (Szamosi et al., 1993). Expression of this enzyme activity begins when the leaf has started to show the symptoms of senescence. From then on, specific activity of the enzyme increases dramatically with the highest activity in the old, yellowish senescing leaves. Appearance of the isoleucine insensitive enzyme suggests a role of this enzyme in remobilization of nitrogen from threonine and serine released from protein degradation.

A cDNA clone encoding the biosynthetic TD in tomato has been isolated (Samach et al., 1991). This gene encodes a polypeptide which contains a chloroplast transit peptide. The level of mRNA for the biosynthetic TD was >50-fold higher in sepals and >500-fold higher in the rest of the flower than in roots or leaves. However, the reason for such high levels of expression of biosynthetic TD mRNA in flowers is not understood.

The reaction product of TD, α-ketobutyrate, is condensed with pyruvate to produce acetohydroxybutyrate by acetohydroxyacid synthase. AHAS is the first common enzyme in the parallel pathways leading to valine, leucine and isoleucine biosynthesis. Therefore, AHAS also condenses 2 moles of pyruvate to yield acetolactate which is used in valine and leucine biosynthesis. AHAS is feedback inhibited by valine, leucine and isoleucine either singly or in combination (Bryan, 1990). Valine and leucine act synergistically and cause maximum inhibition of the enzyme.

AHAS has been identified as the target site for several different chemical classes of compounds (Ray, 1984; Shaner et al., 1984; Subramanian et al., 1989). A number of AHAS inhibiting compounds are currently in use as highly effective herbicides in all major crops including maize, rice, wheat, soybean, canola etc. Because of the effectiveness of these herbicides, a number of crop lines that can tolerate these herbicides have been developed through tissue culture selection, microspore selection, pollen mutagenesis, and plant transformation techniques (Haughn and Somerville, 1986; Mazur and Falco, 1989; Newhouse et al., 1991; 1992). This commercial interest has greatly aided in understanding the biosynthetic pathway of the branched chain amino acids.

AHAS is present in all plant parts and the expression of this enzyme appears to be developmentally regulated (Singh et al., 1990). Genomic and cDNA clones encoding AHAS have been isolated from several species (Mazur et al., 1987; Mazur and Falco, 1989; Bryan, 1990). Only a single copy of this gene has been identified in *Arabidopsis* (Mazur et al., 1987). However, at least two copies of this gene per haploid genome has been identified in all other species examined thus far (Mazur and Falco, 1989). The protein sequence deduced from the gene sequence shows a chloroplast transit peptide (Mazur et al., 1987) which is consistent with the chloroplastic localization of the enzyme (Miflin, 1974). A single amino acid substitution at different positions makes the enzyme insensitive to inhibition by the herbicides (Mazur and Falco, 1989).

Ketoacid reductoisomerase, the next enzyme in the pathway, reduces and iso-merizes the acetohydroxyacids from the previous step to produce dihydroxyacids. The enzyme has been purified to homogeneity from spinach (Dumas et al., 1989). A cDNA clone encoding KARI has been isolated from spinach expression library (Dumas et al., 1991). The deduced amino acid sequence shows a transit peptide. The amino acid sequence also shows the "fingerprint" region of NAD(P)H-binding site reported in several NAD(P)H-dependent oxidoreductases. There is a single KARI gene per haploid genome of spinach.

Dihydroxyacid dehydratase activity has been detected in several species (Bryan, 1990) and purified to homogeneity from spinach (Flint and Emptage, 1988). The purified enzyme has a distinct brown color because it contains a [2Fe-2 S] cluster. The Fe-S cluster is involved in the enzymatic reaction but the exact role of this cluster in the reaction is not fully understood. DHAD in its native form appears to be a hom-odimer with sub-unit molecular weights of 63.5 kD.

Ketoisovalerate or ketomethylvalerate produced in the previous step are trans-aminated to produce valine or isoleucine, respectively. Ketoisovalerate is at a branch-point and therefore it could also be diverted towards leucine biosynthesis. Aminotransferases catalyzing the terminal reaction in valine and isoleucine biosyn-thesis have been detected (Bryan, 1990), however, purification and detailed char-acterizaion of the enzyme has not been described.

B. Biosynthesis of Leucine

Ketoisovalerate, precursor for valine, is used by isopropylmalate synthase in the pathway leading to leucine (Fig. 7, 8). This enzyme condenses ketoisovalerate and acetyl-CoA to form α-isopropylmalate. Isopropylmalate synthase is feedback inhib-ited by leucine (Bryan, 1990). Isopropylmalate isomerase converts α-isopropylmalate to β-isopropylmalate which is reduced to ketoisocaproate by β-isopropylmalate dehydrogenase. The final α-ketoacid is transaminated to form leucine. O-isobutenyl oxalylhydroxamate has been identified as an inhibitor of isopropylmalate dehydro-genase (Wittenbach et al., 1992).

C. Regulation of the Pathway

There are multiple sites of regulation that control carbon flux to the biosynthesis of different branched chain amino acids. Isoleucine regulates its own biosynthesis by inhibiting the first enzyme in the pathway, threonine dehydratase. Biosynthesis of isoleucine could also be controlled by the feedback inhibition of AHAS by valine and leucine. This conclusion is based on the fact that valine and leucine feeding inhibits growth because of isoleucine starvation. This growth inhibition is reversed by isoleucine. However, this regulation may not be operating *in vivo* because it will be detrimental for plant growth. It is also possible that valine and leucine concen-tration in chloroplast may not be high enough that their combination would com-pletely shut down AHAS. This assumption is based on the fact that leucine biosynthesis is regulated by the inhibition of isopropylmalate synthase by leucine.

2-Ketoisovalerate

1

α-isopropylmalate

2

β-isopropylmalate

3

2-ketoisocaproate

4

Leucine

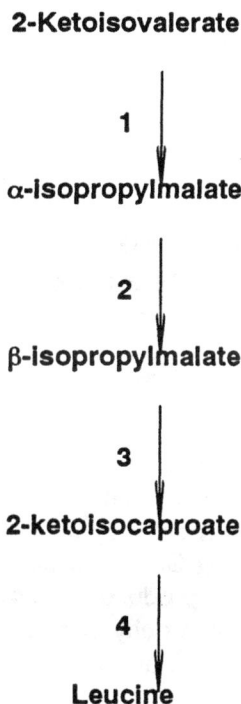

Fig. 8. Biosynthesis of leucine. Ketoisovalerate, precursor for valine (Fig. 7), could be diverted to leucine biosynthesis. Enzymes: 1, isopropylmalate synthase; 2, isopropylmalate isomerase; 3, isopropylmalate dehydrogenase; 4, aminotransferase.

Another possible mode of regulation of carbon flux through the two parallel pathways may be the affinity of the enzymes to the substrates of the two pathways since there is a set of four common enzymes in the branched chain amino acid biosynthetic pathway. This hypothesis is supported by the differential affinity of AHAS for pyruvate and α-ketobutyrate, the two substrates used in the two pathways (Barak et al., 1990). It was shown that AHAS II and III of *E. coli* have a greater affinity for 2-ketobutyrate than for pyruvate. Therefore, these two isozymes will preferentially divert carbon towards isoleucine biosynthesis. A similar situation may be present in higher plants. Additionally, similar differences in the substrate affinity may be present for the other enzymes of the pathway.

V. Histidine Biosynthesis

In 1980, Ben Miflin described histidine as the Cinderella of plant amino acids because this amino acid had been virtually ignored by plant biochemists (Miflin, 1980). Since that analogy, unfortunately, not much progress has been made in this area. It is believed that histidine biosynthetic enzymes in plants are similar to those present in bacteria (Bryan, 1990). Only histidinol dehydrogenase from *Brassica oleracea* has been purified and degenerate oligonucleotide probes deduced from the peptide sequences were used to isolate full length cDNA clone (Nagai et al., 1991). The clone was predicted to contain a transit peptide for chloroplast uptake. Several cell lines requiring histidine have been isolated that appear to lack either imidazole glycerol phosphate dehydratase or histidinol-phosphate aminotransferase (Bryan, 1990). Complementation of these cell lines with bacterial genes of the corresponding enzyme should relieve histidine auxotrophy. This approach will not only confirm that these cell lines are truly deficient in these enzymes but also confirm the involvement of these enzymes in histidine biosynthesis in plants.

VI. Proline Biosynthesis

Proline accumulation in plants in response to various types of stress has been a subject of considerable interest for a long time. Despite this importance, in addition to the obvious role of proline in protein synthesis, the biosynthetic pathway of proline biosynthesis has been only recently defined. Plants appear to synthesize proline in the same way as in *E. coli*. Glutamate kinase, the first enzyme in the pathway, phosphorylates glutamate to produce glutamyl phosphate which is reduced to glutamate semialdehyde by GSA dehydrogenase. Spontaneous cyclization of GSA gives rise to pyrroline-5-carboxylate which is reduced by P-5-C reductase to yield proline (Fig. 9).

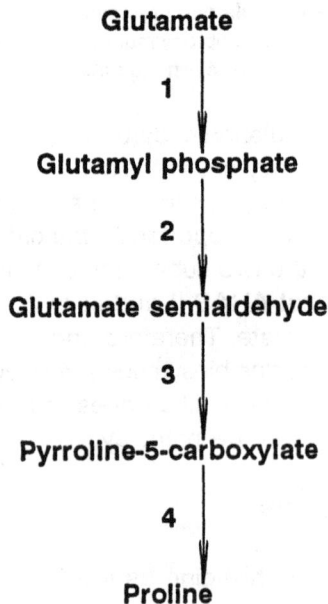

Glutamate

1

Glutamyl phosphate

2

Glutamate semialdehyde

3

Pyrroline-5-carboxylate

4

Proline

Fig. 9. Biosynthesis of proline. Enzymes: 1, glutamate kinase; 2, glutamate semialdehyde dehydrogenase; 3, pyrroline-5-carboxylate reductase.

A cDNA clone that encodes a bifunctional enzyme which exhibits both glutamate kinase and GSA dehydrogenase activity has been recently isolated from a *Vigna aconitifolia* nodule expression library by direct complementation of an *E. coli*. proline auxotroph (Hu et al., 1992). The bifunctional enzyme is feedback inhibited by proline. However, the plant enzyme is less sensitive to inhibition by proline than is the *E. coli*. glutamate kinase. The gene encoding this bifunctional enzyme is expressed in different plant parts. Salt stress induced the expression of the bifunctional enzyme in roots suggesting a key role of this enzyme in proline biosynthesis and osmoregulation in plants (Hu et al., 1992; Verma et al., 1992). Since proline is a feedback inhibitor of the bifunctional enzyme, proline synthesized under stress condition must be moved away from the compartment where these enzymes are functioning in order to allow uninterrupted synthesis of this amino acid.

P-5-C reductase has been purified to homogeneity from barley seedlings (Krueger et al., 1986). The same enzyme works in the reverse direction as proline dehydrogenase. A cDNA clone for P-5-C reductase was isolated from a soybean root nodule library using the complementation strategy (Delauney and Verma, 1990). The message for this gene was found in nodules, uninfected roots and leaves suggesting that this enzyme is expressed in all these plant parts. Two to three copies of this gene were found in the soybean genome. Osmotic stress caused an almost 6-fold increase in the expression of P-5-C reductase mRNA which indicates that this gene may respond to osmoregulation. However, up to 100-fold overexpression of this enzyme did not increase the proline content of transgenic tobacco plants suggesting that this reaction may not be the rate limiting step for proline biosynthesis (Verma et al., 1992).

Preliminary evidence shows that P-5-C can be synthesized from ornithine by ornithine aminotransferase in *Vigna aconitifolia* (Verma et al., 1992). Isolation of a cDNA clone encoding this enzyme has been briefly described (Verma et al., 1992). This result suggests that proline can be synthesized in plants from both glutamate and ornithine. However, tissue distribution and sub-cellular localization of these enzymes and of the substrates is not well understood. Therefore, the significance of the two biosynthetic pathways is not clear.

VII. Conclusions and Future Prospects

A great deal of progress has been made in the last 10 years in understanding amino acid biosynthesis in plants. If the number of papers published in this area are any indication, then it appears that research in this area is growing at an exponential rate. New intermediate (arogenate) and biosynthetic route of phenylalanine and tyrosine has been discovered. Several enzymes in different biosynthetic pathways have been identified and characterized (e.g. chorismate synthase, prephenate aminotransferase, arogenate dehydrogenase, arogenate dehydratase, enzymes in proline biosynthetic pathway etc.). With the progress in molecular techniques, genes for several enzymes have been isolated. Studies on the regulation of expression of different genes have placed serious doubts on the proposal that there are duplicate aromatic amino acid biosynthetic pathways in the cytosol and in the chloroplast. Several enzymes have been identified as the site of action of highly successful commercial herbicides. Tissue culture and genetic transformation techniques have been used to obtain herbicide resistant crops. Selection of amino acid auxotrophic mutants has validated several suggested biochemical pathways as well as challenged some proposed biosynthetic route (e.g. auxin biosynthesis). Crop transformation techniques are also being explored to alter the amino acid composition of food and feed. All the practical application of amino acid biosynthesis has fuelled the research in this area. Despite the progress made in the recent years, our understanding of amino acid biosynthesis in plants lags far behind the understanding of these pathways in *E. coli*. Continued interest and enthusiasm in this field is required to answer many remaining questions about the amino acid biosynthesis in plants.

Literature Cited

Aarnes, H. 1976. Homoserine kinase from barley seedlings. *Plant Sci. Lett.* **7**: 187-191.

Anderson, J.W. 1990. Sulfur metabolism in plants. In *The Biochemistry of Plants,* Vol. 16, (Stumpf, P.K., and Conn, E.E. eds.). Academic Press, New York. pp. 327-381.

Anderson, K.S., Sikorski, J.A., and Johnson, K.A. 1988a. Evaluation of EPSP synthase substrate and inhibitor binding by stopped flow and by equilibrium fluorescence measurements. *Biochemistry* **27**: 1604-1610.

Anderson, K.S., Sikorski, J.A., and Johnson, K.A. 1988b. A tetrahedral intermediate in the EPSP synthase reaction observed by rapid quench kinetics.. *Biochemistry* **27**: 7395-7406.

Bagge, P., and Larson, C. 1986. Biosynthesis of aromatic amino acids by highly purified spinach chloroplasts - Compartmentation and regulation of the reactions. *Physiol. Plant.* **68**: 641-647.

Bandurski, R.S., and Nonhebel, H.M. 1989. Auxins. In *Advanced Plant Physiology*, (Wilkins M.B. ed.), John Wiley & Sons, Inc., New York. pp. 1-20.

Barak, Z., Kogan, N., Gollop, N., and Chipman, D. 1990. Importance of AHAS isozymes in branched chain amino acid biosynthesis. In *Biosynthesis of Branched Chain Amino Acids*, (Barak,Z., Chipman,D.M.,and Schloss,J.V. eds.), VCH, Weinheim. pp. 91-108.

Barry, G., Kishore, G., Padgette, S., Taylor, M., Kolacz, K., Weldon, M., Re, D., Eichholtz, D., Fincher K., and Hallas, L. 1992. Inhibitors of amino acid biosynthesis: Strategies for imparting glyphosate tolerance to crop plants. In *Biosyntehsis and Molecular Reglation of Amino Acids in Plants* (Singh,B.K., Flores,H.E.,and Shannon,J.C. eds.), American Society of Plant Physiologists, Rockville. pp. 139-145.

Berlyn, M.B., Last, R.L., and Fink, G.R. 1989. A gene encoding the tryptophan synthase β subunit of *Arabidopsis thaliana*. *Proc. Natl. Acad. Sci. USA* **86**: 4604-4608.

Bondinell, W.E., Vnek, J., Knowles, P.R., Specher, M., and Sprinson, D.B. 1971. On the mechanism of 5-enolpyruvylshikimate-3-phosphate synthase. *J. Biol. Chem.* **246**: 6191-6196.

Bonner, C.A., and Jensen, R.A. 1985. Novel features of prephenate aminotransferase from cell cultures of *Nicotiana sylvestris*. *Arch. Biochem. Biophys.* **238**: 237-246.

Bowen, J.R., and Kosuge, T. 1979. *In vivo* activity, purification, and characterization of shikimate kinase from sorghum. *Plant Physiol.* **64**: 382-386.

Bryan, J. 1990. Advances in the biochemistry of amino acid biosynthesis. In *The Biochemistry of*

Plants, Vol. 16, (Stumpf P.K. and Conn E.E. eds.), Academic Press, New York. pp. 161-195.

Colau, D., Negrutiu, I., Van Montagu, M., and Hernalsteens, J-P. 1987. Complementation of a threonine dehydratase-deficient *Nicotiana plumbaginifolia* mutant after *Agrobacterium tumefaciens*-mediated transfer of *Saccharomyces cerevisiae* ILV1 gene. *Mol. Cell Biol.* **7:** 2552-2557.

Della-Cioppa, G., Bauer, S.C., Klien, B.K., Shah, D.M., and Fraley, R.T. 1986a. Translocation of the precursor of 5-enolpyruvylshikimate-3-phosphate synthase into chloroplasts of higher plants *in vitro. Proc. Natl. Acad. Sci.* USA **83:** 6873-6877.

Della-Cioppa, G., Hauptman, R.M., Fraley, R.T., and Kishore, G.M. 1986b. Overproduction of 5-enolpyruvylshikimate-3-phosphate (EPSP) synthase in plastids of *Petunia hybrida* suspension culture cells confers resistance to the herbicide glyphosate. *Curr. Top. Plant Biochem. Physiol.* **5:** 194.

Della-Cioppa, G., and Kishore, G.M. 1988. Import of a precursor protein into chloroplasts is inhibited by the herbicide glyphosate. *EMBO. J.* **7:** 1299-1305.

Delauney, A., and Verma, D.P.S. 1990. A soybean pyrroline-5-carboxylate reductase cDNA was isolated by functional complementation in *Escherichia coli* and was found to be osmoregulated. *Mol. Gen. Genet.* **221:** 299-305.

Dumas, R., Joyard, J., and Douce, R. 1989. Purification and characterization of acetohydroxyacid reductoisomerase from spinach chloroplasts. *Biochem.J.* **262:** 971-976.

Dumas, R., Lebrun, M., and Douce, R. 1991. Isolation, characterization and sequence analysis of a full-length cDNA clone encoding aceohydroxyacid reductoisomerase from spinach chloroplasts. *Biochem. J.* **277:** 469-475.

Duncan, K., Edwards, M.R., and Coggins, J.R. 1987. The pentafunctional *arom* enzyme of *Saccharomyces cerevisiae* is a mosaic of monofunctional domains. *Biochem. J.* **246:** 375-386.

Dyer, W.E., Henstraand, J.M., Handa, A.K., and Herrmann, K.M. 1989. Wounding induces the first enzyme of the shikimate pathway in solanaceae. *Proc. Natl. Acad. Sci.USA.* **86:** 7370-7373.

Dyer, W.E., Weaver, L.M., Zhao, J., Kuhn, D.N., Weller, S.C., and Herrmann, K.M. 1990. A cDNA encoding 3-deoxy-D-arabino-heptulosonate 7-phosphate synthase from *Solanum tuberosum* L. *J. Biol. Chem.* **265:** 1608-1614.

Fiedler, E., and Schultz, G. 1985. Localization, purification, and characterization of shikimate oxidoreductase-dehydroquinate hydrolyase from stroma of spinach chloroplasts. *Plant Physiol.* **79:** 212-218.

Flint, D., and Emptage, M. 1988. Dihydroxyacid dehydratase from spinach contained a [2Fe-2S] cluster. *J. Biol. Chem.* **263:** 3558-3564.

Frisch, D.A., Gengenbach, B.G., Tommey, A.M., Sellner, J.M., Somers, D.A., and Myers, D.E. 1991. Isolation and characterization of dihydrodipicolinate synthase from maize. *Plant Physiol.* **96:** 444-452.

Frisch, D.A., Tommey, A.M., Gengenbach, B.G., and Somers, D.A. 1991. Direct genetic selection of a maize cDNA for dihydrodipicolinate synthase in an *Escherichia coli dapA* auxotroph. *Mol. Gen. Genet.* **228:** 287-293.

Ganson, R.J., D'Amato, T.A., and Jensen, R.A. 1986. The two-isozyme system of 3-deoxy-D-arabino-heptulosonate 7-phosphate synthase in *Nicotiana sylvestris* and other higher plants. *Plant Physiol.* **82:** 203-210.

Gengenbach, B., Somers, D., Keith, R., Muehlbauer, G., Sellner, J., Bittel, D., and Shaver, J. 1992. Cellular and molecular genetic regulation of the synthesis of lysine, methionine and threonine-the aspartate family amino acids. In *Biosynthesis and Molecular Regulation of Amino Acids in Plants*, (Singh,B.K., Flores H.E.,and Shannon,J.C. eds.), American Society of Plant Pysiologists, Rockville. pp. 195-207.

Ghislain, M., Frankard, V., and Jacobs, M. 1990. Dihydrodipicolinate synthase of *Nicotiana sylvestris*, a chloroplast-localized enzyme of the lysine pathway. *Planta* **180**: 480-486.

Gilchrist, D., and Kosuge, T. 1980. Aromatic amino acid biosynthesis and its regulation. In *The Biochemistry of Plants*, Vol. 5, (Stumpf, P.K.,and Conn, E.E. eds.), Academic Press, New York. pp. 507-531.

Giovanelli, J., Veluthambi, K., Thompson, G.A., Mudd, S.H.,and Datko, A.H. 1984. Threonine synthase of *Lemna paucicostata*. *Plant Physiol*. **76**: 285-292.

Glassman, K. 1992. A molecular approach to elevating free lysine in plants. In *Biosynthesis and Molecular Regulation of Amino Acids in Plants,* (Singh, B.K., Flores, H.E.,and Shannon, J.C. eds.), American Society of Plant Physiologists, Rockville. pp. 217-228.

Haughn, G.W., and Somerville, C.R. 1986. Sulfonylurea-resistant mutants of *Arabidopsis thaliana*. *Mol. Gen. Genet*. **204**: 430-434.

Henstrand, J.M.,Herrmann, K.M., and Handa, A.K. 1991. Molecular cloning, characterization and wound induction of tomato 3-deoxy-D-arabino-heptulosonate 7-phosphate synthase. *Plant Physiol*. **93**: s-31.

Hibberrd, K.A., and Green, C.E. 1980. Inheritance and expression of lysine plus threonine resistance selected in maize tissue culture. *Proc. Natl. Acad. Sci*. USA **79**: 559-563.

Homeyer, U., and Schultz, G. 1988. Activation by light of plastidic shikimate pathway in spinach. *Plant Physiol. Biochem*. **26**: 365-370.

Hu, A.C., Delauney, A.J., and Verma, D.P.S. 1992. A bifunctional enzyme pyrroline-5-carboxylate synthetase catalyzes the first two steps in proline biosynthesis in plants. *Proc. Natl. Acad. Sci.*USA **89**: 9354-9358.

Huynh, Q.K. 1987. Reaction of 5-enolpyruvylshikimate-3-phosphate synthase with diethyl pyrocarbonate: Evidence for an essential histidine residue. *Arch. Biochem. Biophys*. **258**: 233-239.

Huynh, Q.K., Kishore, G.M., and Bild, G.S. 1988. 5-enolpyruvylshikimate-3-phosphate synthase from *Escherichia coli*: Identification of lys-22 as a potential active site residue. *J. Biol. Chem*. **263**: 735-739.

Jensen, R.A. 1986a. Tyrosine and phenylalanine biosynthesis; relationship between alternative pathways, regulation and subcellular location. In *Recent Advances in Phytochemistry*, Vol. 20. (Conn,E.E. ed.), Plenum, New York. pp. 57-81:

Jensen, R.A. 1986b. The shikimate/arogenate pathway; link between carbohydrate metabolism and secondary metabolism. *Physiol. Plant*. **66**: 164-168.

Kagan, Z.S., Sinelnikova, E.M., and Kretovich, W.L. 1969. L-threonine dehydratases of flowering parasitic and saprophytic plants. *Enzymologia* **36**: 335-352.

Kaneko, T., Hashimito, T., Kumpaisal, R., and Yamada, Y. 1990. Molecular cloning of wheat dihydrodipicolinate synthase. *J. Biol. Chem*. **265**: 17451-17455.

Keith, B., Dong, X., Ausubel, F.M., and Fink, G.R. 1991. Differential induction of 33-deoxy-D-arabinoheptulosonate 7-phosphate synthase genes in *Arabidopsis thaliana* by wounding and pathogenic attack. *Proc. Natl. Acad. Sci.*USA **88**: 8821-8825.

Kishore, G., and Shah, D. 1988. Amino acid biosynthesis inhibitors as herbicides. *Annu. Rev. Biochem*. **57**: 627-663.

Koshiba, T. 1979. Alicyclic acid metabolism in plants XII. Partial purification and some properties of shikimate kinase from *Phaseolus mungo* seedlings. *Plant Cell Physiol*. **20**: 803-809.

Krueger, R., Jage, H-J., Hintz, M., and Pahlich, E. 1986. Purification to homogeneity of pyrroline-5-carboxylate reductase of barley. *Plant Physiol*. **80**: 142-144.

Kumpaisal, R., Hashimoto, T., and Yamada, Y. 1987. Purification and characterization of dihydrodipicolinate synthase from wheat suspension cultures. *Plant Physiol*. **85**: 145-151.

Kuroki, G., and Conn. E.E. 1988a. Increased chorismate mutase levels as a response to wounding in *Solanum tuberosum* tubers. *Plant Physiol*. **86**: 895-898.

Kuroki, G., and Conn, E.E. 1988b. Purification and characterization of an inducible aromatic amino acid-sensitive form of chorismate mutase from *Solanum tuberosum* tubers. *Arch. Biochem. Biophys.* **260**: 616-621.

Last, R.L., Bissinger, P.H., Mahoney, D.J., Radwanski, E.R., and Fink, G.R. 1991. Tryptophan mutants in *Arabidopsis:* The consequences of duplicated tryptophan synthase β genes. *Plant Cell* **3**: 345-358.

Last, R., and Fink, G. 1988. Tryptophan-requiring mutants of the plant *Arabidopsis thaliana. Science* **240**: 305-310.

Lawton, M.A., and Lamb, C.J. 1987 . Transcriptional activation of plant defence by fungal elicitor, wounding and infection. *Mol. Cell. Biol.* **7**: 335-341.

Madan, V.K., and Nath, M. 1983. Threonine (serine) dehydratase from *Cuscuta campestris* Yunck. *Biochem. Physiol. Pflanzen.* **178**: 43-51.

Matthews, B., Weisemann, J., Lewin, K., Wadsworth, G., and Gebhardt, J. 1992. Cloning and analysis of cDNAs encoding bifunctional aspartate kinase-homoserine dehydrogenase activities in carrot and soybean. In *Biosynthesis and Molecular Regulation of Amino Acids in Plants* (Singh,B.K., Flores, H.E., and Shannon, J.C. eds.), American Society of Plant Physiologists, Rockville. pp. 294-295.

Mazur, B.J., Chui, C.F., and Smith, J.K. 1987. Isolation and characterization of plant genes coding for a acetolactate synthase, the target enzyme for two classes of herbicides. *Plant Physiol.* **85**: 1110-1117.

Mazur, B.J., and Falco, S.C. 1989. The development of herbicide resistant crops. *Annu. Rev. Plant Physiol.* **40**: 441-470.

McCue, K.F., and Conn, E.E. 1989. Induction of 3-deoxy-D-arabino-heptulosonate 7-phosphate synthase activity by fungal elicitor in cultures of *Petroselinum crispum. Proc. Natl. Acad. Sci. USA* **86**: 7374-7377.

Meister, A. 1992. Biochemistry of the amino acids. In *Frontiers and New Horizons in Amino Acid Research* (Takai, K. ed.), Elsevier, Weinheim. pp. 357-372.

Miflin, B.J. 1974. The location of nitrite reductase and other enzymes related to amino acid biosynthesis in the plastids of root and leaves, *Plant Physiol.* **54**: 550-555.

Miflin, B.J. 1980. Histidine biosynthesis. In *The Biochemistry of Plants*, Vol. 5, (Stumpf, P.K. and Conn, E.E. eds.), Academic Press, New York. pp. 533-541.

Mousdale, D.M., Campbele, M.S., and Coggins, J.R. 1987. Purification and characterization of bifunctional dehydroquinase-shikimate:NADP oxidoreductase from pea seedlings. *Phytochemistry.* **26**: 2665-2670.

Mousedale, D.M., and Coggins, J.R. 1984. Purification and properties of 5-enolpyruvylshikimate-3-phosphate synthase from seedlings of *Pisum sativum* L. *Planta* **160**: 78-83.

Mousdale, D.M., and Coggins, J.R. 1985. Subcellular localization of the common shikimate pathway enzymes in *Pisum sativum* L. *Planta* **163**: 241-249.

Mousdale, D.M., and Coggins, J.R. 1986. Detection and subcellular localization of a higher plant chorismate synthase. *FEBS Lett.* **205**: 328-332.

Muday, G.K., and Herrmann, K.M. 1992. Wounding induces one of two isozymes of 3-deoxy-D-arabino-heptulosonate 7-phosphate synthase in *Solanum tuberosum. Plant Physiol.* **98**: 496-500.

Muehlbauer, G., Somers, D., Matthews, B., and Gengenbach, B. 1992. Isolation and characterization of a maize aspartate kinase - homoserine dehydrogenase cDNA clone. In *Biosynthesis and Molecular Regulation of Amino Acids in Plants*, (Singh, B.K., Flores, H.E.,and Shannon, J.C. eds.), American Society of Plant Physiologists, Rockville. pp. 324-325.

Muhitch, M.J., and Wilson, K.G. 1983. Chloroplasts are the subcellular location of both soluble and membrane-associated homoserine kinases in pea leaves. *Z. Pflanzenphysiol.* **110**: 39-46.

Myers, D.E., and Gengenbach, B.G. 1982. The isolation and purification of enzymes involves in the metabolism of the aspartate family of amino acids in maize. *Plant Physiol.* **71**:s-714.

Nagai, A., Ward, E., Beck, J., Tada, S., Chang, J-Y., Scheideger, A., and Ryals, J. 1991. Structural and functional conservation of histidinol dehydrogenase between plants and microbes. *Proc. Natl. Acad. Sci. USA* **88**: 4133-4137.

Negruitiu, I., Brouwer, D. De, Dirks, R., and Jacobs, M. 1985. Amino acid auxotrophs from protoplast cultures of *Nicotiana plumbaginifolia* Viviani. Mol. Gen. Genet. **199**: 330-337.

Negrutiu, I., Cattoir-Reynearts, A., Verbruggen, I., and Jacobs, M. 1984. Lysine overproducer mutants with an altered dihydrodipicolinate synthase from protoplast culture of *Nicotiana sylvestris*. *Theor. Appl. Genet.* **68**: 11-20.

Newhouse, K.E., Singh, B.K., Shaner, D.L. and Stidham, M.A. 1991. Mutations in corn (*Zea mays* L.) conferring resistance to imidazolinone herbicides. *Theor. Appl. Genet.* **83**: 65-70.

Newhouse, K.E., Smith, W.A., Starrett, M.A., Schaefer, T.J., and Singh, B.K. 1992. Tolerance to imidazolinone herbicides in wheat. *Plant Physiol.* **100**: 882-886.

Niyogi, K.K., and Fink, G.R. 1992. Two anthranilate synthase genes in *Arabidopsis:* Defense-related regulation of the trypotophan pathway. *Plant Cell* **4**: 721-733.

Parr, G., Padgette, S.R., Brundage, L., and Kishore, G.M. 1987. Elucidation of the stoichiometry and ligand dissociation constants for EPSP synthase by fluorescence studies. *Fed. Proc.* **46**: 1977.

Perl, A., Shaul, O., and Galili, G. 1992. Regulation of lysine synthesis in transgenic potato plants expressing a bacterial dihydrodipicolinate synthase in their chloroplasts. *Plant Mol Biol.* **19**: 815-823.

Pinto, J.E.B., Dyer, W.E., Wellerr, S.C., and Herrmann, K.M. 1988. Glyphosate induces 3-Deoxy-D-arabino heptulosonate 7-phosphate synthase in potato (*Solanum tuberosum*) cells grown in suspension culture. *Plant Physiol.* **87**: 891-893.

Pompliano, D.L., Reimer, L.M., Myrvold, S., and Frost, J.W. 1989. Probing lethal metabolic perturbations in plants with chemical inhibition of dehydroquinate synthase. *J. Am. Chem. Soc.* **111**: 1866-1871.

Poulsen, C., Bongaerts, R.J.M., and Verpoorte, R. 1992. Purification and characterization of anthranilate synthase from *Catharanthus roseus*. In *Biosynthesis and Molecular Regulation of Amino Acids in Plants* (Singh, B.K., Flores, H.E.,and Shannon, J.C. eds.), American Society of Plant Physiologists, Rockville. pp. 302-304.

Ray, T.B. 1984. Site of action of chlorsulfuron. *Plant Physiol.* **75**: 827-831.

Rose, A.B., Casselman, A.L., and Last, R.L. 1992. A Phosphoribosylanthranilate transferase gene is defective in blue fluorescent *Arabidopsis thaliana* tryptophan mutants. *Plant Physiol.* **100**: 582-592.

Samach, A., Harevent, D., Gutfinger, T., Ken-Dror, S., and Lifschitz, E. 1991. Biosynthetic threonine deaminase gene of tomato: Isolation, structure and up regulation in floral organs. *Proc. Natl. Acad. Sci. USA* **88**: 2678-2682.

Schaller, A., Schmid, J., Leibinger, U., and Amrhein, N. 1991. Molecular cloning and analysis of a cDNA coding for chorismate synthase from higher plant *Corydalis sempervirens*. *J. Biol. Chem.* **266**: 21434-21438.

Schmid, J., Schaller, A., Leibinger, U., Boll, W., and Amrhein, N. 1992. The *in vitro* synthesized tomato shikimate kinase precursor is enzymatically active and is imported to the mature enzyme by chloroplasts. *Plant J.* **2**: 375-383

Schmid, J., Windhofer, V., and Amrhein, N. 1990. Purification of chorismate synthase from a cell culture of the higher plant *Corydalis sempervirens*. *Arch. Biochem. Biophys.* **282**: 437-442.

Schmidt, C.L., and Schultz, G. 1987. Stimulation by thioredoxin of shikimate kinase from spinach chloroplasts. *Physiol. Plant.* **70**: 65-67.

Schulze-Siebert, D., and Schultz, G. 1989. Formation of aromatic amino acids and valine from $^{14}CO_2$ or 3-[U-^{14}C] phosphoglycerate by isolated intact spinach chloroplasts. Evidence for a chloroplastic 3-phosphoglycerate→2-phosphoglycerate→phosphoenolpyruvate→pyruvate pathway. *Plant Sci.* **59**: 167-174.

Shaner, D.L., Anderson, P.A., and Stidham, M.A. 1984. Imidazolinone: Potent inhibitors of acetohydroxyacid synthase. *Plant Physiol.* **76**: 545-546.

Shaul, O., and Galili, G. 1992. Increased lysine biosynthesis in tobacco plants that express high levels of bacterial dihydrodipicolinate synthase in their chloroplasts. *Plant. J.* **2**: 203-209.

Shaul, O., and Galili, G. 1992. Threonine overproduction in transgenic tobacco plants expressing a mutant desensitized aspartate kinase of *Escherichia coli. Plant Physiol.* **100**: 1157-1163.

Sharma, R.K., and Mazumdar, R. 1970. Purification, properties and feedback control of threonine dehydratase from spinach. *J. Biol. Chem.* **245**: 3008-3014.

Sidorov, V., Menczel, L., and Maliga, P. 1981. Isoleucine-requiring *Nictoiana* plant deficient in threonine deaminases. *Nature* **294**: 87-88.

Siehl, D.L., and Conn, E.E. 1988. Kinetic and regulatory properties of arogenate dehydratase in seedlings of *Sorghum bicolor. Arch. Biochem. Biophys.* **260**: 822-829.

Siehl, D.L., Connelly, J.A., and Conn, E.E. 1986a. Tyrosine biosynthesis in *Sorghum bicolor.* Characteristics of prephenate aminotransferase. *Z. Naturforsch.* **41**: 79-86.

Siehl, D.L., Singh, B.K., and Conn, E.E. 1986b. Tissue distribution and subcellular localization of prephenate aminotransferase in leaves of *Sorghum bicolor. Plant Physiol.* **81**: 711-713.

Silk, G., and Matthews, B. 1992. Molecular cloning of the soybean gene encoding dihydrodipicolinate synthase. In *Biosynthesis and Molecular Regulation of Amino Acids in Plants,* (Singh, B.K., Flores, H.E., and Shannon, J.C. eds.), American Society of Plant Physiologists, Rockville. pp. 341-342.

Singh, B.K., and Conn, E.E. 1986. Chorismate mutase isoenzymes from *Sorghum bicolor.* Immunological characterization. *Arch. Biochem. Biophys.* **246**: 617-621.

Singh, B.K., Connelly, J.A., and Conn, E.E. 1985. Chorismate mutase isoenzymes from *Sorghum bicolor.* Purification and properties. *Arch. Biochem. Biophys.* **243**: 374-384.

Singh, B.K., Connelly, J.A., and Siehl, D.L. 1991. Shikimate pathway: Why does it mean so much to so many? In *Oxford Surveys of Plant Molecular and Cell Biology. Vol. VII,* (Miflin, B.J. ed.), Oxford University Press, Oxford. pp. 143-186.

Singh, B.K., Lonergan, S.G., and Conn, E.E. 1986. Chorismate mutase isoenzymes from selected plants and their immunological comparison with the isoenzymes from *Sorghum bicolor. Plant Physiol.* **81**: 717-722.

Singh, B.K., Newhouse, K.E., Stidham, M.A., and Shaner, D.L. 1990. Acetohydroxyacid synthase-Imidazolinone interaction. In *Biosynthesis of Branched Chain Amino Acids,* (Barak, Z., Chipman, D.M., and Schloss, J.V. eds.). VCH, Weinheim. pp. 357-372.

Smart, C.C., and Amrhein, N. 1987. Ultrastructural localization by protein A gold immunocytochemistry of 5-enolpyruvylshikimate-3-phosphate synthase in a plant cell culture which overproduces the enzyme. *Planta* **170**: 1-6.

Smart, C.C., Johanning, D., Muller, G., and Amrhein, N. 1985. Selective overporduction of EPSP synthase in a plant cell culture which tolerates high doses of the herbicide glyphosate. *J. Biol. Chem.* **260**: 16338-16346.

Steinrucken, H.C., and Merhein, N. 1980. The herbicide glyphosate is a potent inhibitor of 5-enolpyruvylshikimate-3 phosphate synthase. *Biochem. Biophys. Res. Commun.* **94**: 1207-1212.

Steinrucken, H.C., and Amrhein, N. 1984. 5-enolpyruvylshikimate-3 phosphate synthase of *Klebsiella pneumoniae:* purification and properties. *Eur. J. Biochem.* **143**: 341-349.

Subramanian, M.V., Loney, V., and Pao, L. 1989. Mechanism of action of 1,2,4-triazolo [1,5-a] pyrimidine sulfonamide herbicides. In *Prospects for Amino Acid Biosynthesis Inhibitors in Crop Protection and Pharmceuticl Chemistry, Monograph No. 42,* (Copping, L.G., Dliziel, J., and Dodge, A.D. eds). British Crop Protection Council, Surrey. pp. 97-100.

Szamosi, I.T., Shaner, D.L., and Singh, B.K. 1993. Identification and characterization of biodegradative form of threonine dehydratase in senescing tomato leaf. *Plant Physiol.* **101**: 999-1004.

Thompson, G.A., Datko, A.H., and Mudd, S.H. 1982. Methionine synthesis in *Lemna.* Studies on the regulation of cystathionine synthase, O-phosphohomoserine sulfhydrylase, and O-acetylserine sulfhydrylase. *Plant Physiol.* **69**: 1077-1083.

Tyagi, V.V.S., Henke, R.R., and Farkas, W.R. 1982. Occurrence of diaminopimelic epimerase in maize. *Biochim. Biophys. Acta.* **719**: 363-369.

Tyagi, V.V.S., Henke, R.R., and Farkas, W.R., 1983. Partial purification and characterization of dihydrodipicolinic acid reductase from maize. *Plant Physiol.* **73**: 687-691.

Verma, D.P.S., Hu, C., Deluney, A., Miao, G., and Hong, Z. 1992. Deciphering proline biosynthesis pathways in plants by direct, trans-, and co-complementation in bacteria. In *Biosynthesis and Molecular Regulation of Amino Acids in Plants* (Singh, B.K., Flores, H.E., and Shannon, J.C. eds.), American Society of Plant Physiologists, Rockville. pp. 128-138.

Wallsgrove, R.M., Lea, P.J., and Miflin, B.J. 1983. Intracellular localization of aspartate kinase and the enzymes of threonine and methionine biosynthesis in green leaves. *Plant Physiol.* **71**: 780-784.

Walter, T.J., Connelly, J.A., Gengenbach, B.G., and Wold, F. 1979. Isolation and characterization of two homoserine dehydrogenases from maize suspension cultures. *J. Biol. Chem.* **254**: 1349-1355.

Wang, Y., Herrmann, K.M., Weller, S.C., and Goldsbrough, P.B. 1991. Cloning and nucleotide sequence of a cDNA encoding 3-deoxy-D-arabino-heptulosonate 7-phosphate synthase from tobacco. *Plant Physiol.* **97**: 847-848.

Wenko, L.K., Treick, R.W., and Wilson, K.G. 1985. Isolation and characterization of a gene encoding meso-diaminopimelate dehydrogenase from *Glycine max. Plant Mol. Biol.* **4**: 197-204.

Wibbenmeyer, J., Brundage, L., Padgette, S.R., Likos, J.J., and Kishore, G.M. 1988. Mechanism of the EPSP synthase catalyzed reaction: Evidence for the lack of the covalent carboxyvinyl intermediate in catalysis. *Biochem. Biophys. Res. Commun.* **153**: 760-766.

Wilson, B.J., Gray, A.C., and Matthews, B.F. 1991. Bifunctional protein in carrot contains both aspartokinase and homoserine dehydrogenase activities. *Plant Physiol.* **97**: 1323-1328.

Wittenbach, V., Rayner, D., and Schloss, J. 1992. Pressurre points in the biosynthetic pathway for branched-chain amino acids. In *Biosynthesis and Molecular Regulation of Amino Acids in Plants,* (Singh, B. K., Flores, H.E., and Shannon, J.C., eds.), American Society of Plant Physiologists, Rockville. pp. 69-88.

Wright, A.D., Moehlenkamp, C.A., Perrot, G.H., Neuffer, M.G., and Cone, K.C. 1992. The maize auxotrophic mutant *orange pericarp* is defective in duplicate genes for tryptophan synthase β. *Plant Cell* **4**: 711-719.

Wright, A.D., Sampson, M.B., Neuffer, M.G., Michalczuk, L., Slovin. J.P., and Cohen, J.D. 1991. Indole-3-acetic acid biosynthesis: *De novo* synthesis in the maize mutant *orange pericarp,* a tryptophan auxotroph. *Science* **254**: 998-1000.

Nitrogen Nutrition in Higher Plants, 1995
Editors : H.S. Srivastava & R.P. Singh
Associated Publishing Co., New Delhi, India
pp. 273-309

Biosynthesis and Utilization of Ureides in Tropical Legumes

RANDHIR SINGH

I. Introduction

Ammonia is the first stable product of N_2 fixation in legume nodules (Bergersen, 1965; Kennedy, 1966). Produced in bacteroids, it is excreted into the host cell cytoplasm where it is assimilated and used in the synthesis of organic nitrogen for transport. Based on the product used for transport, nitrogen fixing legumes can be classified as amide exporters or ureide exporters (Fig. 1). The amide exporters transport asparagine (ASN), glutamine (GLN) or 4-methylene glutamine (MeGLN) and generally belong to the tribes *Vicieae, Genisteae* and *Trifolieae,* including legumes of mostly temperate origin such as pea, lupin, broad bean, alfalfa and clover. 4-methylene glutamine accounts for 90% of the nitrogen in the root bleeding sap of nodulated peanut (*Arachis hypogaea*), belonging to the tribe *Aeschynomenaeae* (Done and Fowden, 1951; Fowden, 1954; Winter et al., 1981). The ureide exporters transport either allantoin (ALN) and allantoic acid (ALA) or citrulline and are members of the tribe *Phaseoleae*. Allantoin and allantoic acid account for about 60 to 90% of the total N in the xylem sap of many tropical legumes including soybeans (McClure and Israel, 1979; Streeter, 1979; Schubert, 1981), cowpea (Herridge et al., 1978; Pate et al., 1980), gardenbeans (Pate, 1973; Thomas et al., 1979; Cookson et al., 1980), pigeonpea (Luthra et al., 1981; Sheoran et al., 1981) and other legumes grown symbiotically (Pate et al., 1980; Pate and Atkins, 1983). Though the above-mentioned products are the major nitrogenous solutes in the xylem stream of N_2-fixing plants, lesser amounts of many other protein and non-protein amino acids are also present (Herridge et al., 1978; Streeter, 1972, 1979; Cookson et al., 1980; Rawsthorne et al., 1980; Coker and Schubert, 1981; Pate and Atkins, 1983; Rainbird, 1983). Similarly, low levels of ALN and ALA occur in xylem sap of amide exporting legumes and vice-versa. The absolute amounts and relative proportion of these primary and secondary nitrogenous solutes vary depending upon the stage of development and environmental factors (Matsumoto et al., 1977a, 1977b; Israel and McClure, 1980; Pate et al., 1980; Patterson and LaRue, 1983).

Synthesis of ureides in nodules is closely associated with the process of N_2-fixation, as they are rapidly labelled after exposure of nodulated roots to labelled N_2 (Herridge et al., 1978; Ohyama and Kumazawa, 1978; Schubert and Coker, 1982; Schubert et al., 1981). After 2 h of exposure to labeled N_2, ureides account for about 90% of the ^{15}N in the xylem exudate. Moreover, nodulated legumes contain higher concentration of ureides than non-nodulated (Kushizaki et al., 1964; Matsumoto et al., 1976; 1977a, 1977b; Fujihara et al., 1977; Fujihara and Yamaguchi, 1978a; Thomas et al., 1980; Thomas and Schrader, 1981). Similarly, presence of nitrate or ammonium in the rooting medium decreases the content of ureides in xylem sap, which further correlates with the decrease in nodule mass and N_2-fixing activity (Matsumoto et al., 1976; 1977a, 1977b; Israel and McClure, 1980; McNeil and LaRue, 1984), indicating that ureides syntheized in nodules are the primary products of recent N_2 fixation in legumes of tropical origin. Hence, a clear understanding of the pathways of ureide biosynthesis and their assimilation into major

NODULE CELL N_2

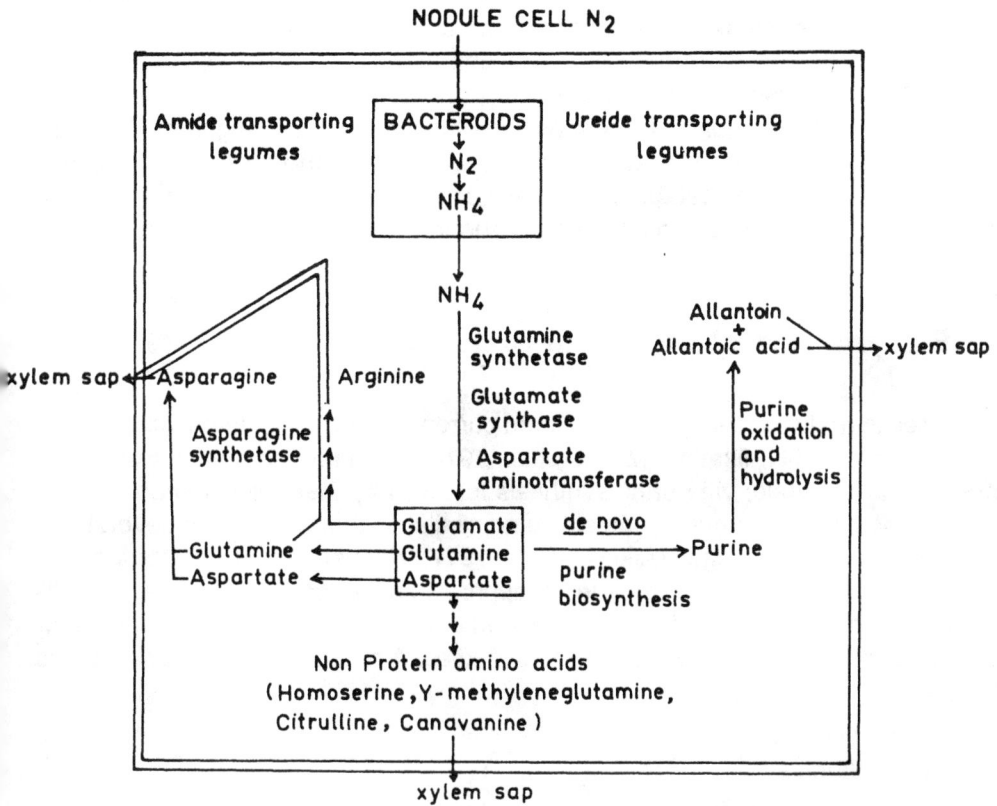

Fig. 1. Forms of fixed nitrogen transported in legume plants.

N constituents of the cell, is a pre-requisite for improving the C and N economy of ureide producing legumes and to reap economic yields with higher harvest index for N and dry matter. Ureide metabolism in general and with reference to nodulated legumes has been reviewed in the past (Thomas and Schrader 1981; Atkins, 1982; Schubert, 1986; Rao and Singh, 1988; Atkins and Beevers, 1990; Schuber and Boland, 1990). However, since then many reports have appeared which have broadened our knowledge of ureide biogenesis in nodules of tropical legumes. The purpose of the present review, therefore, is to highlight only the recent findings, with particular emphasis on metabolic pathways and regulation of ureide metabolism. Possible significance and consequences of ureide based nitrogen metabolism has also been discussed.

II. Biosynthesis of Ureides in Nodules

Ureides in nodules are synthesized from currently delivered photosynthates and products of recent N_2 fixation (Meeks et al., 1978; Ohyama and Kumazawa, 1979, 1980a, 1980b, 1980c) via purine synthesis followed by their oxidation and hydrolysis; the pathway of purine synthesis being similar to that operates in several animals and micro-organisms (Atkins et al., 1977; Nguyen, 1979; Reinbothe and Mothes, 1962). The various experimental approaches followed to delineate the pathway of ureide biosynthesis could be grouped under; tracer studies, precursor studies, inhibitor studies, and developmental studies. A brief account of these studies is given below.

A. Tracer Studies

In conventional pathway of purine biosynthesis, CO_2, glycine, formate, aspartate, glutamine are the precursors of purines. Hence, if the conventional pathway operates in nodules, by feeding any of the above precursors in a radioactive form, one must find the incorporation of the label into purine intermediates, purines and ureides. When (^{14}C) glycine was fed to nodule slices of cowpea, there was much greater incorporation of label into ureides (ALN and ALA) as compared to that from (^{14}C) glucose and (^{14}C) acetate (Atkins et al., 1980b). Similar results were obtained when cell-free extracts of cowpea and soybean (Atkins et al., 1982a) and organelle preparations from soybean (Schubert and Boland, 1982; Boland and Schubert, 1983a) and cowpea nodules (Shelp et al., 1983) were incubated with labelled glycine. Subsequently, cell-free preparations from both cowpea and soybean nodules were shown to form purine nucleotides from glutamine, aspartic acid, glycine, CO_2, a source of activated 'C_1' units (such as methanyl - and formyl-THFA), Mg^{2+}-ATP and Mg^{2+}-PRPP at rates commensurate with those of N_2 fixation (Boland et al., 1982; Atkins et al., 1982b; Shelp and Atkins, 1983; Boland and Schubert, 1983a).

Carbon-6 of the pruine ring is derived from CO_2 in a reaction catalysed by phosphoribosylaminoimidazole carboxylase (AIR carboxylase) and is lost specifically when uric acid is oxidized to ALN. This fact was exploited by Boland and Schubert (1982a) in their *in vivo* studies carried out with soybean, where they exposed nodulated roots to $^{14}CO_2$ in presence of allopurinol (a hypoxanthine analogue which inhibits xanthine dehydrogenase) to block uric acid production. Labelled xanthine recovered from nodules was then degraded by xanthine oxidase and uricase to produce ALN. This led to a loss of 67-80% of the labelled carbon. Labelled xanthine was also detected in the absence of allopurinol. In nodulated cowpea roots incubated with $^{15}N_2$ and allopurinol, labelled xanthine accumulated at rates equivalent to the rate of ureide labelling in plants not treated with allopurinol (Atkins et al., 1988a). These results confirm that purines with IMP as the primary product are synthesized *de novo* in nodules and subsequently oxidized as part of the pathway of ureide biogenesis.

The fate of IMP and ureide synthesis from various other purine derivatives was investigated by employing labelled IMP, hypoxanthine, guanine and adenine (Woo et al., 1980a, 1980b). Cell-free extracts of cowpea nodules converted labelled guanine or hypoxanthine into a mixture of uric acid and ALN. The time course of labelling pattern indicated a possible order of ^{14}C transfer from (8-^{14}C) guanine, to uric acid, ALN and ALA. The cell free system, however, could not synthesize ureides from (8-^{14}C) adenine. In the same cell free system, formation of ^{14}C ALN and ALA (Atkins, 1981) and other intermediates of purine catabolism viz., IMP, xanthosine, inosine, hypoxanthine and xanthine from (8-^{14}C) IMP (Atkins et al., 1982b; Shelp and Atkins, 1983) was also shown. Over the period of assay (upto 1 h), the decrease in label of IMP was accompanied by an increase of the label in ureides and other intermediates. Proplastid fraction of soybean nodules also metabolized labelled IMP to ureide intermediates as discussed above (Boland and Schubert, 1983a).

B. Precursor Studies

Glutamine is the donor of N-3 and N-9 of purines according to the conventional pathway. The incorporation of glutamine N occurs in the reaction 2 and 5 catalyzed by phosphoribosyl amidotransferase (PRAT) and phosphoribosyl formyl glycinamide synthetase (FGAR synthetase) resulting in the synthesis of PRA and FGAR, respectively. Effect of glutamine, NH_3 and asparagine as amide group donor in purine synthesis, as measured by (^{14}C) glycine incorporation, was studied in nodule extracts of cowpea and soybean (Atkins et al., 1982b). In these experiments, glutamine was found to be the efficient amido group donor, as judged by the amount of label in IMP+AICAR.

Ureide biosynthesis in nodule fraction increases by the addition of different purines and purine derivatives (Triplett et al., 1980). In soybean, lima bean and pigeonpea nodules, maximum ALA production was observed with ALN, followed by hypoxanthine, inosine, uric acid, xanthine, xanthosine, IMP and XMP in decreasing order (Amarjit and Singh, 1984; Triplett et al., 1980). The production of ALA required NAD^+ and was accompanied by an increase in NADH formation as determined by absorbance at 340 nm. NADH production was maximum with hypoxanthine, followed by xanthine, xanthosine, inosine, IMP, uric acid and XMP in decreasing order. However, neither NAD was required in the production of ALA from uric acid and ALN nor there was production of ALA and NADH in the presence of allopurinol (Amarjit and Singh, 1984; Atkins, 1981; Triplett et al., 1980), indicating that NAD is presumably utilized in the xanthine dehydrogenase (XDH) reaction and the ALA is produced via the metabolic route xanthine-urate-allantoin-allantoic acid.

The capacity of all the four purine nucleotides, XMP, IMP, GMP and AMP to be metabolized to ureides was assessed by Atkins (1981) in cell free extracts of cowpea nodules. Ureides were readily synthesized from XMP and IMP but not from AMP and very slowly from GMP, though GMP and AMP were metabolized to their respective nucleosides and free bases. These conversions required NAD^+ and were inhibited by allopurinol. Furthermore, the rate of conversion of IMP and XMP into ureides were commensurate with known rates of N_2 fixation in cowpea (Atkins, 1981; Atkins et al., 1988a, 1988b) and soybean (Triplett et al., 1980). In the same system, in presence of allopurinol, label from $(1-^{14}C)$ glycine was mainly incorporated into xanthine accompanied by NAD^+ reduction and very little label appeared in inosine or hypoxanthine (Shelp and Atkins, 1983). Similarly, in soybean root nodules, allopurinol treatment resulted in accumulation of $^{14}CO_2$ into xanthine rather than in hypoxanthine (Boland and Schubert, 1982a). These findings indicated that IMP oxidation is the major metabolic route in the formation of xanthine rather than dephosphorylation of IMP to inosine. The possibility that xanthine can arise from GMP (GMP-guanosine-guanine-xanthine) formed from IMP was ruled out, as there was no synthesis of either GMP, guanosine or guanine from IMP. This is also supported by the fact that metabolism of GMP was extremely slow compared to ureide formation from IMP or XMP (Atkins et al., 1982b).

C. Inhibitor Studies

The most important inhibitor utilized in the study of ureide biogenesis has been allopurinol (4-hydroxypyrazolo (3,4-d) pyrimidine), a structural analog of hypoxanthine and an irreversible inhibitor of xanthine dehydrogenase. Allopurinol applied to cowpea roots rapidly inhibited ureide export and caused accumulation of xanthine in the nodules (Atkins et al., 1980b, 1988a, 1988b). Similarly, treatment of nodulated soybean plants with allopurinol resulted in significant decrease in ureide concentration in the stems and a concomitant accumulation of xanthine in the nodules (Fujihara and Yamaguchi, 1978a, 1978b). Nodulated roots of soybean exposed to $^{14}CO_2$ in presence of allopurinol, had much higher levels of xanthine (Boland and Schubert, 1982a). However, the specific activity of xanthine was similar in both control and allopurinol treated plants, indicating that xanthine is being accumulated not as a result of breakdown of stored purine precursors but due to *de novo* purine synthesis. This has been confirmed in a recent research, where Atkins et al. (1988a) demonstrated intact attached nodules of cowpea to form ^{15}N-xanthine from [15]N_2 at rates equivalent to those of ureide synthesis, in presence of 0.5 mM allopurinol. Xanthine accumulated in nodules and was exported in increasing amounts in xylem of allopurinol treated plants. Other intermediates of purine oxidation, *de novo* purine synthesis, and ammonia assimilation did not increase, confirming the direct assimilation of fixed nitrogen into purines.

D. Developmental Studies

The activities of the enzymes of initial ammonia assimilation viz. glutamine synthetase (GS) and aspartate aminotransferase (AAT), of *de novo* purine synthesis viz. phosphoribosyl pyrophosphate synthetase (PRPP synthetase), PRAT, phosphoglycerate dehydrogenase (PGDH), SHM and methylene FH_4 dehydrogenase, and of purine catabolism viz. XDH, uricase and allantoinase, increased in soybean nodules in parallel with the increase in nodule mass, nitrogenase activity, leghemoglobin content and apparent rate of ureide export from the nodules (Schubert, 1981; Reynolds et al., 1982a, 1982b). In general, the activities of all these enzymes were equal to or several times greater than that of nitrogenase activity and rate of ureide export. Similar was the case for nucleotidase and nucleosidase of pigeonpea nodules (Amarjit and Singh, 1985a). Furthermore, the levels of enzymes involved in biosynthesis and degradation of purines were much higher in nodules of ureide exporting soybean and pintobean than in nodules of amide exporting lupin and pea (Reynolds et al., 1982a, 1982b; Christensen and Jochimsen, 1983). However, the enzymes involved in the initial assimilation of ammonia except AS viz. GS, GOGAT and AAT were present at comparable levels in nodules of amide and ureide producing plants. Higher activities of uricase and allantoinase (Woo et al., 1980a; 1980b; Singh et al., 1985) and XDH (Woo et al., 1980a, 1980b) have also been reported in nodules of ureide producing *Vigna* species. The induction of enzymes of *de novo* purine biosynthesis has also been correlated with the onset of N_2 fixation, the induction of ammonium assimilatory enzymes and the increase in ureide levels in the xylem sap of uriede producing tropical legumes (Atkins et al., 1980b, 1984b; Reynolds et al., 1982b; Schubert, 1981; Sheoran et al., 1981).

The above studies discussed under the four different heads have given convincing evidences that ureides in nodules of tropical legumes arise from the oxidation of *de novo* synthesized purines. The pathway of purine biosynthesis is outlined here in Fig. 2, which is quite similar to that operative in animal and microbial systems.

E. Sources of Precursors for Purine Biosynthesis

Purine biosynthesis as outlined in Fig. 2 requires an input of nitrogen in the form of glycine, the amide nitrogen of glutamine, and the amino nitrogen of aspartate. The sources of carbon are glycine, CO_2 and two single carbon atoms added as derivatives of tetrahydrofolic acid (FH_4). The source of ribose 5-P as usual is the pentose phosphate pathway, which is quite active in the nodules of ureide producing legumes (Copeland et al., 1989; Hong and Copeland, 1990; Gupta et al., 1991). The regulatory enzymes (glucose-6-P and 6-phosphogluconate dehydrogenases) have recently been partially purified and characterized from nodules of pigeonpea (Gupta and Singh, 1991). Kohl et al. (1988) in a recent study suggested that proline biosynthesis by pyrroline-5-carboxylate reduction (P5CR) generates $NADP^+$ required for the synthesis of purine precursor, ribose-5-P through pentose phosphate pathway. However, comparison of pentose phosphate pathway and pyrroline-5-carboxylate reductase activities of ureide versus amide exporting nodules did not support the above hypothesis (Kohl et al., 1990). Ribose released during the hydrolysis of nucleotides/nucleosides by nucleotidases/nucleosidases could also be recycled with the help of the enzyme, ribosephosphotransferase (ribokinase). However, the relative contribution of these two routes as a source of ribose-5-P is not known.

Ammonia assimilation via the combined activities of GS and GOGAT in cytosol of nodules serves as the source of glutamine and aspartate. The two enzymes have been isolated and purified from legume nodules (McParland et al., 1976; Boland and Benny, 1977; Groat and Schrader, 1982; Cullimore et al., 1983). GS is present mainly in cytosol, while GOGAT exists both in cytosol and in the plastid (Awonaike et al., 1981; Shelp et al., 1983; Shelp and Atkins, 1984). These enzymes get induced during nodule development (Robertson et al., 1975a, 1975b; Boland et al., 1978; Groat and Vance, 1981; Schubert, 1981; Reynolds et al., 1982a, 1982b; Atkins et al., 1984a, 1984b). Aspartate is synthesized from glutamate and oxaloacetate via AAT (Ryan et al., 1972; Reynolds and Farrden, 1979). Serine via the action of the enzymes serine hydroxymethyl transferase (SHM), and methylene FH_4 dehydrogenase serves as the source of glycine and the single carbon moiety. Both these enzymes are now known to be present in the nodules of ureide producing legumes (Boland et al., 1982; Boland and Schubert, 1982b; Reynolds et al., 1982b; Shelp et al., 1983; Mitchell et al., 1986). Serine in turn is synthesized by oxidation of phosphoglycerate (glycolytic intermediate) by a dehydrogenase to form phosphohydroxy pyruvate, which is transaminated to give phosphoserine. Hydrolysis of phosphate ultimately yields serine. Evidences in favour of this sequence of reactions have come from the existence of significant levels of phosphoserine in plant soluble fractions from nodules of ureide producing legumes (Reynolds et al., 1982a). But this amino acid could not be detected in amide producing legume, lupin. Similarly, the levels of phosphoglycerate dehydrogenase were several fold higher in the nodules of ureide producing legumes compared to amide producing legumes. The activity of phosphoglycerate dehydrogenase, as well as that of glutamate dependent aminotransferase, SHM and methylene FH_4 dehydrogenase, increased during nodule development and the onset of ureide production (Reynolds and Blevins, 1986;

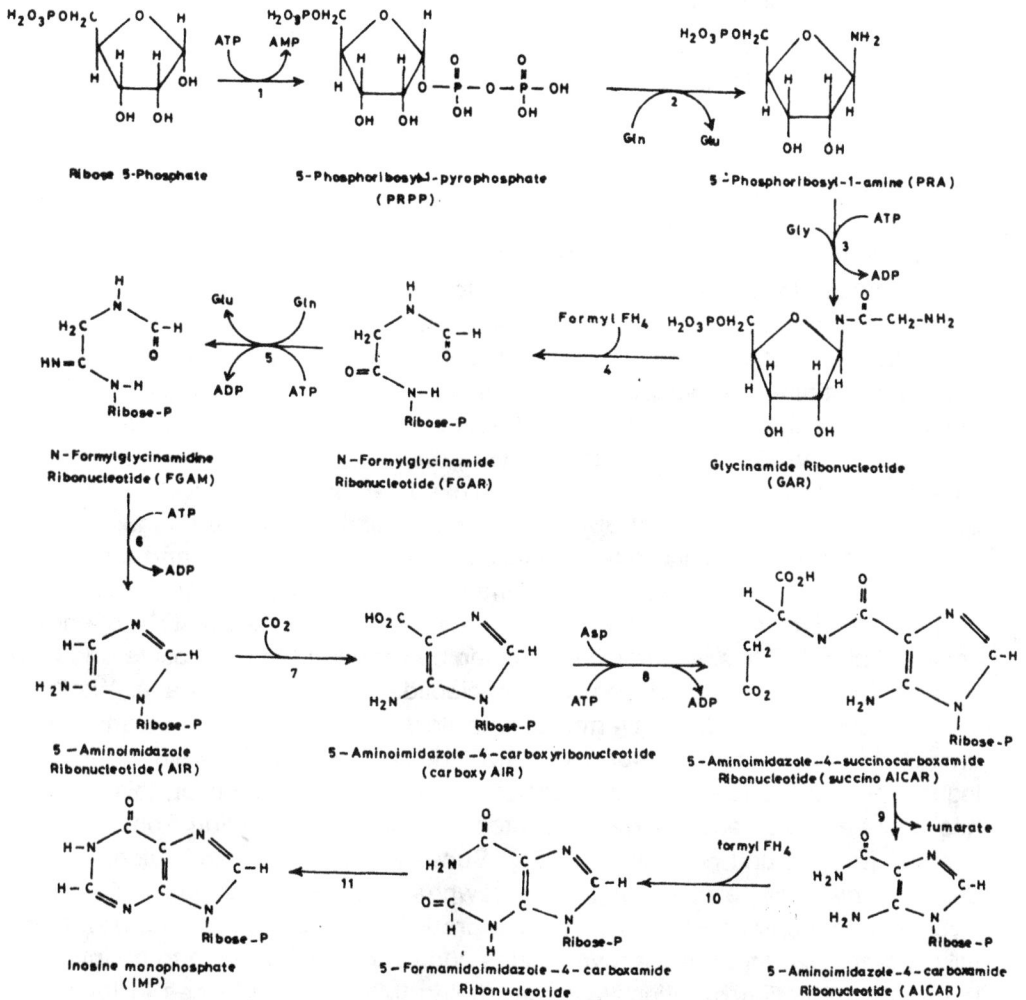

Fig. 2. The pathway of *de novo* purine ribonucleotide biosynthesis. The enzymes catalysing the numbered reactions are (1) PRPP synthetase, (2) PRPP amidotransferase, (3) GAR synthetase, (4) GAR transformylase, (5) FGAR amidotransferase, (6) AIR synthetase, (7) AIR carboxylase, (8) Succino-AICAR synthetase, (9) Adenylosuccinase, (10) AICAR transformylase, and (ii) IMP cyclohydrolase.

Reynolds et al., 1982b). Dark CO_2 fixation by PEP carboxylase seems to contribute very little carbon for incorporation into ureides (Coker and Schubert, 1980;
Schubert and Coker, 1985).

F. Conversion of Purines into Ureides

The conversion of IMP, the immediate product of purine biosynthesis, to ALN
and thence to ALA has been the subject of considerable interest. In earlier studies, it was presumed that IMP is converted to hypoxanthine through the actions
of 5'-nucleotidase and nucleosidase, which is then oxidized to xanthine via xanthine oxidase. Xanthine is then converted to uric acid, ALN and ALA through the
actions of uricase, allantoinase and allantoicase, respectively. This sequence of
reactions was not consistent with the observed accumulation of xanthine (and
not hypo-xanthine) in extracts of nodules treated with allopurinol (Atkins et al.,
1980, 1988a; Boland and Schubert, 1982a; Fuzihara and Yamaguchi, 1978a;
Triplett et al., 1980). Secondly, the enzyme responsible for the oxidation of xanthine is not xanthine oxidase, but a NAD-linked xanthine dehydrogenase (Atkins
et al., 1980b; Triplett et al., 1980). These observations led to the suggestion that
alternative pathways might to responsible for IMP catabolism (Boland and Schubert, 1982a; Reynolds et al., 1982b). One of the alternatives suggested was the
involvement of the enzyme IMP dehydrogenase, which catalyses NAD^+-dependent
conversion of IMP to XMP, which could then be converted to xanthosine and xanthine by the reactions catalyzed by 5'-nucleotidase and nucleosidase. Xanthine
in the sequence is then converted to uric acid by NAD^+-linked xanthine dehydrogenase. Early attempts to detect IMP dehydrogenase in ureide-producing
legume nodule tissue were unsuccessful (Farnden, 1981). However, this enzyme
has now been successfully demonstrated in a proplastid fraction from soybean
nodules (Boland and Schubert, 1983b). Subsequently, Shelp and Atkins (1983)
detected the enzyme in extracts from cowpea nodules and suggested it to be
involved in the primary route for xanthine production. The enzyme has now been
fully characterized from the above tissues and conclusively shown to be involved
in ureide production by effectively scavenging IMP. The evidences in favour of
the above suggested catabolic route leading to ureide biogenesis include: (i) high
levels of xanthine dehydrogenase, uricase, and allantoinase in nodules of ureide
producing legumes such as cowpea (Atkins, 1981; Atkins et al., 1980, 1984 a),
soybean (Tajima and Yamamoto, 1975; Fujihara and Yamaguchi, 1978a, 1978b;
Schubert, 1981), gardenbean (Thomas et al., 1980; Thomas and Schrader, 1981;
Reynolds et al., 1982a, 1982b), pigeonpea (Amarjit and Singh, 1984) and mungbean (Singh et al., 1985), (ii) low levels of these enzymes in nodules of amide exporting legumes (Reynolds et al., 1982a, 1982b; Christensen and Jochimsen, 1983),
(iii) increase in the levels of these enzymes in response to the onset of N_2 fixation,
NH_4 assimilation and ureide export within the nodule (Atkins et al., 1980b; 1984a;
Reynolds et al., 1982b; Schubert, 1981; Atkins, 1981; Amarjit and Singh, 1984)

and (iv) decrease in the levels of ureides in nodules and ureide export in the xylem and an increase in xanthine in nodules on addition of allopurinol (Fujihara and Yama-guchi, 1978a, 1978b; Atkins et al., 1980b; Triplett et al., 1980; Atkins, 1981).

III. Subcellular Compartmentalization of the Pathway of Ureide Biosynthesis

The pathway of ureide biogenesis is now known to be highly compartmenta-lized both at cellular and subcellular levels (Rao and Singh, 1988). The locali-zation studies have shown GS to be located in the cytoplasmic fraction of legume nodules, while glutamate synthase is associated, at least in part, with the plastid fraction (Awonaike et al., 1981; Boland and Schubert, 1982a; Shelp et al., 1983; Shelp and Atkins, 1984; Verma et al., 1986). Of the two isoenzymes of AAT in nodules, one with higher mobility is present in the plastid fraction (Boland and Schubert, 1982a).

Out of the various enzymes of *de novo* purine biosynthesis, PRAT from soy-bean nodules has been clearly shown to be located in proplastid fraction (Boland and Schubert, 1982a). This led Boland and his associates to suggest that the plastid might be the site of purine biosynthesis in nodules. This was confirmed when purines could be synthesized *in vitro* by a proplastid fraction from.soybean (Boland and Schubert, 1982a; 1983a) and cowpea (Atkins and Shelp, 1983; Shelp et al., 1983) nodules. In these studies, proplastid was shown to incorporate (U-^{14}C) glycine into purines, primarily IMP in the presence of added PRPP, glutamine, aspartate, ATP, bicarbonate, methenyl FH_4, $MgCl_2$and KCl. Label from (U-^{14}C) serine and (3-^{14}C) serine were also incorporated into purines by proplastid frac-tion when FH_4 and $NADP^+$ were substituted for methenyl tetrahydrofolate, indi-cating the presence of a functional pathway for conversion of serine to glycine plus N^5, N^{10}-methenyl tetrahydrofolate and N^{10}-formyltetrahydrofolate in this frac-tion. Labelled bicarbonate could also be incorporated into IMP and inosine by this fraction. However, labelled formate failed to get in, indicating the absence of a formyltetrahydrofolate synthetase from this fraction. The occurrence of IMP dehydrogenase in this fraction was examined directly by incubating the fraction with labelled IMP with or without pyridine nucleotide. The products identified con-sisted of a mixture of inosine, XMP and xanthosine in the presence of added NAD^+ or $NADP^+$, while inosine was the main product in the absence of pyridine nucleotide. These results indicated the presence of IMP dehydrogenase and nucleo-tidase in the plastid fraction from soybean nodules. However, Shelp and Atkins (1983) reported the presence of IMP dehydrogenase in the cytoplasm of cowpea nodules.

Enzymes needed for the synthesis of purine precursors, i.e. from the pathway from phosphoglycerate to glycine plus formyltetrahydrofolate have also been shown to be located in the plastid fraction. Specific enzymes identified include triose phos-phate isomerase, phosphoglycerate dehydrogenase, serine hydroxymethylase, methylene tetrahydrofolate dehydrogenase, phosphoserine aminotransferase and phosphoserine phosphatase (Boland and Schubert, 1982a; Shelp et al., 1983; Reynolds and Blevins, 1986). The enzymes of purine oxidation i.e. XDH, uricase and allantoinase are now known to be localized in the cytoplasm, peroxisome and endoplasmic reticulum fractions of soybean nodules, respectively (Hanks et al., 1981). The presence of uricase in peroxisomes has now been confirmed by immun-ocytochemical techniques (Bergmann et al., 1983; Triplett and Hao, 1988; Vanden Bosch and Newcomb, 1986; Webb and Newcomb, 1987).

Based on the localization studies described here, Schubert and Boland (1990) have proposed a model for the sub-cellular localization of reactions of purine synthesis and ureide biogenesis in nodules of ureide-exporting tropical legumes. According to this model, ammonium produced by the bacteroids is incorporated into glutamine in the cytosol surrounding the peribacteroid sac. Further meta-bolism leading to the synthesis of IMP and its conversion to xanthine occurs in the proplastid fraction. Xanthine appears to be translocated to cytosol where it is oxidized to uric acid. Uric acid is oxidized to allantoin in the peroxisomes by uricase, producing H_2O_2, which is degraded by catalase, also located in the per-oxisomes. Allantoin then appears in the xylem sap or is hydrolysed to allantoic acid by allantoinase in the endoplasmic reticulum and allantoic acid secreted into the xylem. However, species difference in minor details cannot be ruled out at this stage.

IV. Cellular Organization of Ureide Biosynthesis

The central zone or tissue of several legume nodules consists of a large number of both infected and uninfected cells (Bergersen and Goodchild, 1973; Newcomb, 1981; Newcomb and Tandon, 1981; Shelp et al., 1983; Newcomb et al., 1985). The ratio of these two cell types may vary from 1:1 to 1:3. In soybean nodules, the number of uninfected cells was 1.5 times more than infected cells. However, the volume of the uninfected cells was much less (one-third) as compared to that of infected cells (Selker and Newcomb, 1985). The properties of these cells and of the pro-toplasts derived from (Davey et al., 1973; Woo and Broughton, 1979; Hanks et al., 1983) are summarized in Table 1. Plastids are present in both cell types. Accumu-lation of starch during development is characteristic of plastids in uninfected cells while plastids in infected cells exhibit a pronouced increase in internal structure (Webb and Newcomb, 1987) suggesting that plastids undergo developmentally programmed, cell specific differentiation.

Hanks et al. (1983) fractionated protoplasts of infected and uninfected cells on sucrose step gradients and determined the levels of enzymatic activity for enzymes involved in ureide biogenesis for each fraction. Uricase and allantoinase were present at much higher levels in the uninfected protoplast fraction. However, the enzymes involved in production of precursors of purine biosynthesis (AAT, PGA dehydrogenase etc.) were equally distributed in the two cell types. Based on these results, it was difficult to determine conclusively the primary location of plastids responsible for the synthesis of purines to be used for ureide production. Shelp et al. (1983) conducted a parallel study with cowpea nodules and obtained similar results as discussed above. They observed that the majority (about 85%) of the activity for enzymes associated with NH_4^+ assimilation, synthesis of purine precursors and *de novo* purine synthesis was in the infected cell fraction. GS was exclusively present in infected cells while glutamate synthase followed the distribution of enzymes of *de novo* purine synthesis. Based on the above distribution, they concluded that infected cell is the primary site of ammonium assimilation and purine synthesis, although maintenance levels of these enzymes may be present in all cells (Schubert and Boland, 1984; Schubert, 1986). Kouchi et al. (1988) also examined the distribution of enzymes of ammonium assimilation, carbon metabolism and ureide biogenesis in soybean nodule protoplasts. Their results for the distribution of uricase and allantoinase were similar to those reported by Hanks et al. (1983). Immunogold labelling experiments confirmed the location of GS in cytoplasm of infected cells of soybean nodules (Verma et al., 1986). Similarly, immunofluorescent labelling studies proved uricase to be primarily present in the uninfected cells of soybean nodules (Bergmann et al., 1983). Immunogold-labelled antibodies directed against uricase conclusively substantiated the above conclusion (Vanden Bosch and Newcomb, 1986). Similar results were obtained for cowpea nodules (Webb and Newcomb, 1987). Triplett (1985) showed XDH to be present only in infected cells, which was not confirmed by immunofluorescent probes (Nguyen et al., 1986). Additional studies are required to establish with certainty the location of enzymes and organelles involved in purine synthesis and catabolism.

Table 1. A comparison of the characteristics of infected and uninfected cells.

Infected cell	*Uninfected cell*
1. Large	Small
2. Increased number of mitochondria and plastids	Lesser number of mitochondria and plastids.
3. Microbodies, rarely, present as indicated by electron microscopy and diaminobenzidine test	Enlarged and increased number of microbodies
4. Little or no smooth endoplasmic reticulum	Smooth endoplasmic reticulum well developed and abundant
5. Larger protoplasts, irregular in shape and granular appearance	Smaller protoplast, spherical
6. Less fragile protoplasts	More fragile protoplasts
7. Higher density protoplasts	Protoplasts are of lower density
8. Starch granules absent	Starch granules present

The exact nature of the metabolite (i.e. IMP, XMP, xanthosine, xanthine, uric acid) transported between the two cell types is not known since plasma membranes are generally impermeable to nucleotides (Muller et al., 1982). Transport commonly requires conversion of the nucleotide to the nucleoside or free base plus an active transport system. Hence xanthosine and/or xanthine are probable metabolites transported from plastids to cytosol and then from infected to uninfected cells. However, IMP, being freely soluble seems to be better choice because of the relatively low solubility of xanthosine, xanthine and uric acid. The localization of XDH in uninfected cells is consistent with the transport of XMP, xanthosine. The above problems associated with transport may be partially circumvented if glutamine is translocated from infected cells to uninfected cells where purine synthesis and catabolism take place. The clustering of plastids, peroxisomes and ER around the periphery of infected and uninfected cells as observed by Newcomb and his colleagues in their electron micrographs, may be physiologically important in facilitating the transfer of metabolites between cells. All these probabilities have been indicated in the model proposed by Schubert and Boland (1990) for the cellular and subcellular distribution of enzymes of ammonium assimilation and ureide biogenesis in nodules of ureide exporting legumes (Fig. 3). According to this model, ammonium assimilation, *de novo* purine synthesis and associated reactions, and conversion of IMP to xanthine seem to occur in infected cells. Xanthine on transport to uninfected cells is converted to uric acid in cytosol and further conversion of uric acid to ALN and ALA occurs in peroxisomes and ER, respectively. Places of uncertainty in the model have been indicated with broken lines and question marks. The model suggested can be rationalized on the basis of known properties of the enzymes involved. A number of enzymes of purine synthesis as well as some intermediates in the pathway are oxygen labile (Rowe et al., 1978). Hence, their compartmentalization in infected cells would protect them as these cells contain low oxygen concentration because of the presence of leghemoglobin. On the other hand, uricase has a K_m for O_2 of approx. 30 μM (Rainbird and Atkins, 1981; Lucas et al., 1983) requiring high levels of O_2 for activity and would be ineffective at low pO_2 present in infected cells. Localization of uricase in the uninfected cells circumvents this apparent O_2 limitation. However, the model is still speculative and requires confirmation.

V. Properties of Enzymes of Ureide Biosynthesis

The pathway of conventional purine biosynthesis and the reaction sequence leading to the formation of ureides, ALN and ALA, from IMP, the end product of purine synthesis, is shown here in Fig. 2. Several enzymes involved in the above pathways have been purifed from nodules and their properties investigated. These enzymes have been grouped here under the heads:

A. Enzymes involved in purine synthesis
B. Enzymes involved in glycine and methenyl tetrahydrofolate synthesis and
C. Enzymes of ureide formation and IMP

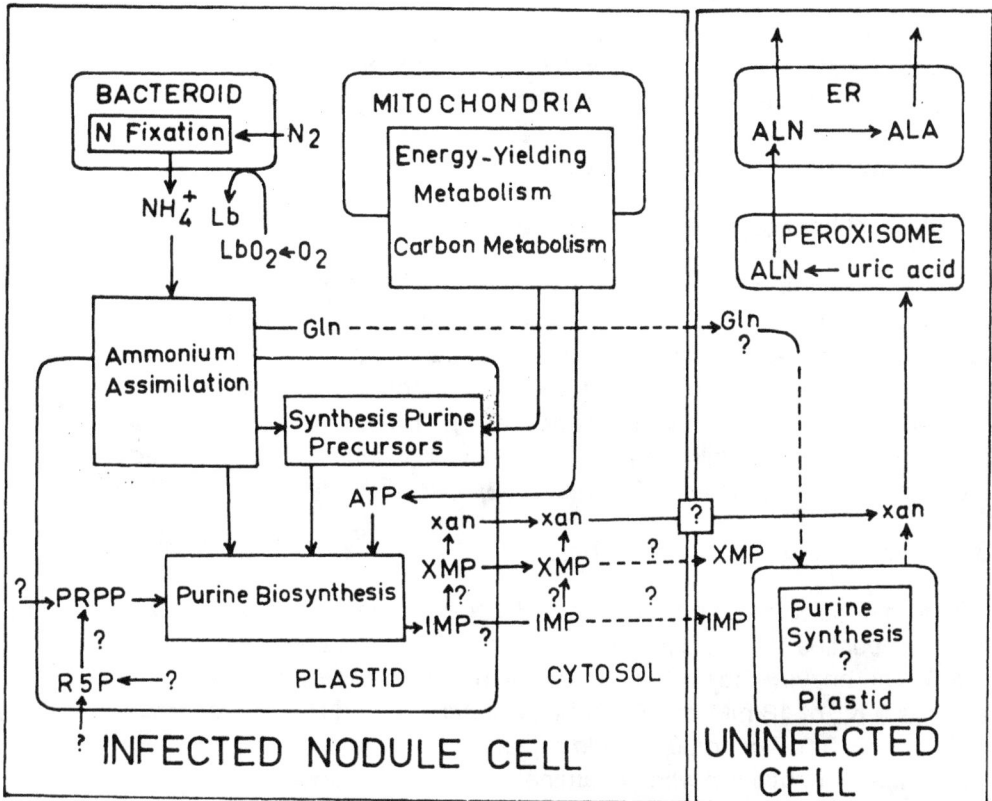

Fig. 3.   Model for the cellular compartmentalization of the reactions of nitrogen fixation, ammonium assimilation, purine synthesis, and ureide biogenesis in infected and uninfected cells of soybean root nodules. Question marks and/or dashed lines indicate uncertainties.

A. Enzymes Involved in Purine Synthesis

1. Ribokinase (EC 2.7.1.11)

Ribokinase was studied in dialysed extracts of soybean nodules by Christensen and Jochimsen (1983). This enzyme can serve to recycle ribose formed in purine nucleotidase reaction, for purine synthesis. The enzyme requires Mg^{2+} and ATP and has an optimum pH at 8.6. It is specific for ribose with K_m of 300 µM. Deoxyglucose could not inhibit the enzyme.

2. 5'-Phosphoribosyl-1-pyrophosphate synthetase (EC 2.7.6.1)

The enzyme partially purified from soybean nodules (Schubert and Boland, 1990) has two pH optima, one at pH 7.5 and the other above pH 9.0. The enzyme is extremely labile and is difficult to purify. It is inhibited by inorganic phosphate, 50% inhibition brought about by 20 mM KPi. The mungbean enzyme has K_m values of 0.18 mM for ATP and 0.014 mM for ribose-5-P. The enzyme has absolute requirement for Mg^{2+} as a divalent cation. The enzyme is inhibited by nucleotides and is regulated by energy charge (Atkins, 1989).

3. Phosphoribosyl-1-pyrophosphate amidotransferase (EC 2.4.2.14)

The enzyme catalysing the committed step reaction of *de novo* purine synthesis, has been purified 1500-fold from soybean nodules and characterized (Reynolds et al., 1984). The enzyme has a high Mr of 8×10^6 indicative of a multicomplex nature, and has a K_m of 18 mM and 0.4 mM for glutamine and PRPP respectively, at the optimum pH 8.0. Ammonium could replace glutamine as the nitrogen donor, with V_{max} twice that obtained with glutamine. The enzyme requires Mg^{2+} and the reactive molecular species is Mg $PRPP^{3-}$ complex. Glutamate inhibits the enzyme competitively with respect to glutamine with a Ki of 30 mM, whereas PPi inhibits uncompetitively with respect to PRPP with a Ki of 0.65 mM. This inhibition pattern is consistent with ordered sequential mechanism of enzyme action. The enzyme is also inhibited by the end products of purine biosynthesis, XMP and IMP. The inhibition of the enzyme by the products of the reaction and end products of purine biosynthesis suggests that the first committed step is likely to be an important regulatory step in *de novo* purine synthesis. Besides the above form, low molecular weight forms of PRAT have also been purified to virtual homogeneity from soybean nodules (Schubert, 1987).

The other enzymes of purine biosynthetic pathway have not yet been isolated and purified from nodule sources.

B. Enzymes Involved in Glycine and Methenyl Tetrahydrofolate Synthesis

1. Phosphoglycerate dehydrogenase (EC 1.1.1.95)

The enzyme has been purified 200 fold from soybean nodules and the kinetic properties investigated (Boland and Schubert, 1983b). The enzyme catalyses a reversible reaction, with a pH optimum of 9.4 for the forward reaction and 6.1 for the reverse reaction. At optimum pH, K_m value for the forward reaction are 0.25 mM for NAD^+ and 0.29 mM for 3-phosphogylcerate, and at pH 7.5, K_m values for reverse reaction are 0.012 mM for NADH and 0.15 mM for 3-phosphohydroxy-pyruvate. The enzyme can also utilize NADPH in the reverse reaction with K_m of 0.15 mM. In reverse direction, NAD^+ inhibits the enzyme competitively with respect to NADP, with a Ki of 0.35 mM. However, 3-phosphoglycerate is a weak inhibitor, inhibiting the enzyme noncompetitively with respect to phosphohydroxypyruvate. Serine does not inhibit the nodule enzyme.

2. Phosphoserine aminotransferase (EC 2.6.1.52)

The enzyme purified 250-fold from soybean nodules, has an Mr of 85 kD with a subunit size of 42 kD (Reynolds and Blevins, 1986; Reynolds et al., 1988). At its optimum pH of 8.0, the enzyme has K_m values of 0.5 mM for glutamate and 0.067 mM for phosphohydroxypyruvate. Neither aspartate nor alanine can act as alternate amino group donors and phosphohydroxypyruvate cannot be replaced by B-hydroxypyruvate, pyruvate or oxalacetate. Nodule enzyme is again not inhibited by serine.

3. Serine hydroxymethylase (EC 2.1.2.1.)

The importance of this enzyme stems from the likely possibility that the products of its reaction viz., glycine and methylene FH_4 serve as carbon donor for the purine ring. The enzyme has been purified 1550-fold from soybean nodules by Mitchell et al. (1986). The enzyme has an Mr of 230 kD with four identical subunits. Serine has a biphasic binding pattern, with limiting K_m values of 1.5 and 40 mM. The Km for FH_4 is 0.25 mM. The reaction is catalysed by a sequential mechanism, with serine the first substrate to bind, and glycine the last product to leave. The enzyme is insensitive to a wide variety of metabolites which have been reported to affect the activity in other species. Unlike the enzyme from mung bean (Rao and Appaji Rao, 1982), nodule enzyme is not allosterically regulated.

4. Methylene tetrahydrofolate dehydrogenase (EC 1.5.1.15)

The enzyme detected in soybean (Boland and Schubert, 1982; Boland et al., 1982) and cowpea (Schubert and Boland, 1984) nodule extracts, has been partially puri-fied from soybean nodules by ammonium sulfate fractionation but the activity was lost during gel filtration (Boland and Schubert, unpublished). The K_m value for methylene tetrahydrofolate was 8 µM and for $NADP^+$ 4 µM.

5. Methynyltetrahydrofolate cyclohydrolase (EC 3.5.4.9)

This enzyme involved in the interconversion of methenyltetrahydrofolate and formyltetrahydrofolate was detected in soybean nodules (Reynolds et al., 1982a, 1982b) but has not been purified from this source. The enzyme partially purified from pea seedlings by Suzuku and Iwai (1973) has a pH optimum of 7.7. The reaction catalysed is reversible with K_m values of 0.04 mM for the methenyl derivative and 0.2 mM for the formyl derivative.

C. Enzymes of Ureide Formation and IMP

1. IMP dehydrogenase (EC 1.1.1.205)

The enzyme purified to apparent homogeneity from nodules of cowpea (Atkins et al., 1985) has Mr of 200 kD and is composed of four identical subunits. The enzyme is highly specific for IMP and has a K_m of 0.009 mM for this substrate. At saturating concentrations of both the substrates, the enzyme has a pH optimum between 9 and 10. $NADP^+$ could not replace NAD^+ in the reaction and molecular oxygen did not act as an alternative oxidant. Depending on IMP concentration, K_m for NAD^+ may vary from 0.018 to 0.035 mM. None of the intermediates of ureide metabolism inhibit the enzyme. The enzyme is inhibited by AMP, GMP and NADH; inhibition by GMP being competitive with respect to IMP. The enzyme is activated by K^+ (Kd = 1.6 mM). The activation is offset by Mg^{2+} in a pattern of concentration dependence similar to that of a competitive inhibitor.

2. 5-Nucleotidase (EC 3.1.3.5)

The enzyme from nodules of pigeonpea has been resolved into two forms, N-I and N-II, having Mr of 52 and 119 kD, respectively (Amarjit and Singh, 1986). Both forms have pH optima in the acidic range (between pH 5.2 and 5.7) with either CMP, GMP, XMP, IMP or AMP as the substrate. Upto pH 6.6, both forms show higher activity with CMP followed by GMP, XMP, IMP, and AMP, respectively. However, the activity changes with pH in the alkaline range making the enzyme relatively more active with purine nucleotides. Neither of the forms has a requirement for any of the metal ions. Fe^{2+} inhibits the enzyme activity; the inhibition at 5, 10 and 15 mM concentration being 11, 43 and 47%, respectively with N-I and 14, 47 and 52%, respectively with N-II. K_m values for AMP, IMP, GMP, CMP and XMP are 0.10, 0.18, 0.40, 0.40 and 0.77 mM, respectively with N-I and 0.12, 0.20, 0.40, 0.40 and 0.99 mM, respectively with N-II. The enzyme is inhibited non-competitively by adenosine and inosine; Ki values being 1.78, 0.25 and 0.30; 3.50, 2.12 and 0.75 mM, respectively with AMP, IMP and XMP as the substrate. In soybean nodules, the enzyme activity did not correlate well with ureide production, suggesting that the enzyme is primarily involved in some other metabolic functions (Doremus and Blevins, 1988a). Nucleoside diphosphatase purified from soybean root nodules (Doremus and Blevins, 1988b) is similarly involved in other metabolic functions rather than in ureide production.

3. Nucleosidase (EC 3.2.2.1)

This enzyme has been purified to near homogeneity from cowpea nodules by following ammonium sulfate precipitation, chromatography on DEAE-sephacel, guanosine-affinity sepharose, and AH-sepharose-4B and preparative polyacrylamide gel electrophoresis (Atkins et al., 1989). The enzyme has a broad pH optimum around pH 8.0. The native enzyme has Mr of 160 kD with subunit Mr of about 31 kD. The purified enzyme hydrolyses both purine (xanthosine, inosine, guanosine, and adenosine) as well as pyrimidine (uridine, thymidine and cytidine) nucleosides to their bases. Among the purine nucleosides, the V_{max} values are in the ratio 28:7:1:0:4 for xanthosine, inosine, adenosine and guanosine, respectively. K_m values are similar for xanthosine and inosine (0.80 mM). Inosine, adenosine, guanosine and uridine competitively inhibit xanthosine hydrolysis with Ki values of 0.78, 1.10, 0.36 and 1.20 mM, respectively. Xanthosine and adenosine inhibit inosine hydrolysis competitively with Ki values of 0.80 and 1.05 mM, respectively, but guanosine inhibits noncompetitively (Ki= 0.98 mM).

4. Xanthine dehydrogenase (EC 1.1.1.204)

The enzyme has been purified from nodules of *Phaseolus vulgaris* (Boland, 1981) and soybean (Triplett et al., 1982; Boland and Schubert, 1983a). The soybean enzyme is a molybdo flavoprotein with two subunits of Mr 141 kD containing two molybdenum and eight iron atoms per molecule. The enzyme contains FMN as the cofactor and has pH optimum between 8 and 9 for hypoxanthine and broadly from pH 8 to 10 for xanthine. The *Phaseolus* enzyme catalyses NAD^+ dependent oxidation of both hypoxanthine and xanthine, with K_m values of 0.009 and 0.0048 mM, respectively. At pH 7.5, K_m for NAD^+ was 0.03 mM. $NADP^+$ or phenazine methosulfate could not act as an alternative oxidant, and rates with O_2 were only 4% of the control. The enzyme catalyses the reaction by a ping-pong type of mechanism. Product inhibition patterns with NADH and uric acid are mixed linear for both inhibitors with respect to both substrates, suggesting that product inhibition is not simple competition, but also involves dead end inhibition, when the product binds to the wrong form of the enzyme. Regulation of its activity seems to be mediated by the end products urate and NADH and cosubstrate NAD^+. The activity of the enzyme should depend on the rate at which urate is further metabolised to ureides and is transported away from the site of location of XDH. Similarly, the rate at which NAD^+ is regenerated is also important for the enzyme activity.

5. Uricase (EC 1.7.3.3)

The enzyme has been purified from cowpea (Rainbird and Atkins, 1981; Lucas et al., 1983) soybean (Bergmann et al., 1983; Lucas et al., 1983) nodules. Cowpea enzyme has Mr of 100 kD a pH optimum between 9 and 10, and an isoelectric point of 6.8. K_m values are 0.018 and 0.29 mM, respectively for uric acid and O_2. Metal chelating reagents inhibit the enzyme activity. The enzyme was also inhibited by some divalent cations, and low concentrations of ammonia, glutamine and xanthine, and stimulated by Fe^+.

The enzyme purified to apparent homogeneity from soybean nodules has Mr of 124 kD with subunit Mr of 33 kD suggesting the enzyme to be a tetramer (Bergmann et al., 1983). However, the native Mr as determined by gel filtration was found to be 100 kD. The differences in apparent molecular weight may be the result of the unusual shape of the enzyme. The soybean enzyme has a PI equal to approximately 9 and contains about 1.0 mol Cu^{2+}/mol. On the other hand, Lucas et al. (1983) reported soybean nodule enzyme to exist as monomer at pH 7.5 with Mr of 32, kD and dimer at pH 8.8 with Mr of 70 kD. At a pH optimum of 9.5, the soybean enzyme has K_m values of 0.010 and 0.031 mM, respectively for uric acid and oxygen. Xanthine inhibits the enzyme competitively with respect to uric acid (Ki = 0.010 mM). The enzyme was not inhibited significantly by a variety of amino acids, ammonia, adenine or allopurinol.

6.   Allantoinase (EC 3.5.2.5)

This enzyme detected in nodules of pigeonpea (Amarjit and Singh, 1984), soybean (Schubert, 1981) and cowpea (Atkins et al., 1980b) has been purified by Amarjit and Singh (1985b). Like the enzyme from other sources, it has a neutral pH optimum of 7.5-7.7 and K_m value of 13.3 mM. However, the K_m for partially purified enzyme from peanut nodules was 8.7 mM (Venkateswara et al., 1988). The enzyme was unaffected by metal ions and neither inhibited nor stimulated by a wide variety of nucleotides, nucleosides, purines, or amino acids. The enzyme was inhibited substantially by sulfhydryl reducing agents, which could be eliminated by dialysis, suggesting that the active enzyme requires both disulfide linkages and free sulfhydryl groups for activity. The soybean nodule enzyme has a broad pH optimum between 7 and 8.5 and a K_m value of 20 mM. The enzyme has recently been purified to apparent homogeneity from soybean nodules by using a combination of chromatographic techniques. The enzyme activity was stable in the presence of a variety of ionic and nonionic detergents and 5 M urea and was not inhibited by 10 mM DTT. The enzyme did not require any metal ion for activity or stability and appears to be a glycosylated protein.

VI.   Regulation of Ureide Biosynthesis

The regulation of ureide biogenesis in nodules is not well understood. However, as usual with any metabolic pathway, the potential points of regulation again could be at the level of fine and coarse control of enzyme activity. The fine control involves the alteration of enzyme activity through changing levels of metabolites, whereas the coarse control is determined at the gene level. Besides these, isozymes and compartmentation could be other elements of regulation. Since ureide biogenesis in N_2-fixing nodules of ureide-transporting plants is the main source of nitrogen nutrition for the plant, the pathway might be expected to be very active and perhaps only limited by the availability of precursors. The various elements of regulation have been discussed below:

A. Fine Control of Enzymes

5-Phosphoribosyl-1-pyrophosphate amidotransferase, which catalyses the committed step reaction of *de novo* purine synthesis is the regulatory enzyme in the pathway of purine synthesis (Reynolds et al., 1984). IMP and XMP, the products of purine biosynthesis and the first step towards ureides, inhibit the enzyme. The enzyme is also inhibited by GMP; and AMP, XMP and AMP are linear competitive inhibitors with respect to PRPP, with Ki values of 1.2 and 2.0 mM, respectively. IMP and GMP inhibitions of the enzyme promote co-operativity in the binding of PRPP, with Hill coefficients of 1.7 in both cases. IMP with a Ki of 0.3 mM has only one binding site per enzyme molecule. These kinetic data are consistent with PRAT occupying a central control point, responding rapidly and effectively to small changes in the product of its metabolic pathway (IMP).

IMP dehydrogenase, which catalyses the first step following purine biosynthesis seems to be another regulatory enzyme in the pathway of ureide production. The enzyme has high affinity for its substrate (IMP), and is inhibited by purine nucleotides other than IMP. None of the intermediates of ureide biogenesis downstream from IMP inhibits the enzyme (Atkins et al., 1985). These properties of the enzyme are consistent with the ability of this enzyme to effectively divert IMP, without restriction into the ureide pathway. When purine pool levels are high, excess purines will be funnelled into the ureide pathway and IMP dehydrogenase will be inhibited. This will result in a build up of IMP, which would inhibit *de novo* purine synthesis.

Metabolism of purines downstream of XMP is unlikely to be subject to regulation (Schubert and Boland, 1990). Many of the steps in this portion of the pathway are irreversible, and none of the enzymes exhibits regulatory properties. Since there are no products in this part of the pathway of metabolic significance other than in ureide production, there is little need for regulation.

B. Coarse Control of Enzymes

Control of gene expression during development also regulates the biogenesis of ureides in nodules of ureide producing plants. Nodule specific host proteins (nodulins), which are products of the plant genome, are expressed in nodules (Legocki and Verma, 1980). Though the enzymatic function of all these proteins has not yet been elucidated, nodulin-35 from soybean is a nodule specific uricase located in peroxisomes of uninfected nodule cells (Nguyen et al., 1985; Vanden Bosch and Newcomb, 1986), a glutamine synthetase (Lara et al., 1983; Cullimore and Miflin, 1984; Cullimore et al., 1984; Sen Gupta et al., 1986) or a PEP carboxylase (Deroche et al., 1983). From the pattern of appearance of enzymes in developing soybean nodules, it seems that the enzymes of ammonium assimilation are induced in two phases (Schubert, 1981; Reynolds et al., 1982b). Specific activities of the enzymes of ammonium assimilation increase in parallel with the appearance of nitrogenase. However, in nodules of *Phaseolus vulgaris*, GS in nodules is induced prior to the onset of active nitrogen fixation (Sen Gupta et al., 1986). Enzymes associated with purine synthesis and oxidation are similarly induced, but increase more slowly, in parallel with a decrease in the activity of asparagine synthetase (Reynolds et al., 1982b). This changeover is reflected in a dramatic decrease in asparagine transport and an increase in ureide transport in the xylem.

The nature of the inducer molecule(s) for these enzymes is not known. However, it appears that NH_4^+ excreted from the bacteroids triggers the induction of enzymes (Robertson et al., 1975a, 1975b). This has been confirmed by feeding ammonium to plants with split root systems. In these experiments, levels of nodule specific form of aspartate amino transferase as well as ALA were 10 times higher in soybean roots fed ammonium than in roots supplied with nitrogen-free nutrient solution. Growth of cowpea plants in an atmosphere of 80% Ar : 20% O_2 (conditions under which NH_4^+ would not be produced) had little effect on the induction of nitrogenase activity and leghemoglobin but prevented the induction of enzymes of NH_4^+ assimilation, *de novo* purine synthesis, and purine oxidation (Atkins et al., 1984a, 1984b), suggesting that NH_4^+ produced as a result of N_2 fixation (or a product of NH_4^+ assimilation) specifically induces these activities. When the production of NH_4^+ was interrupted by short term (3 day) exposure of nodulated roots of cowpea to Ar:O_2, nitrogenase activity, leghemoglobin content, GC activity and allantoinase activity were relatively unaffected while activities of glutamate synthase, *de novo* purine synthesis, XDH and uricase were severely inhibited.

Nitrate inhibits ureide synthesis in nodules, replacing the system by synthesis of asparagine (Matsumoto et al., 1977a; Cookson et al., 1980; Israel and McClure, 1980; Pate et al., 1980; McNeil and LaRue, 1984). The molecular and biochemical nature of this switching mechanism is, however, not known. Schuller et al. (1986) conducted detailed experiments on the effect of nitrate treatment on nodule metabolism and reported that *in situ* C_2H_2 reduction by nodulated soybean plants was inhibited by 50 to 70% within 2 days after treatment of plants with 10 mM nitrate. Acetylene reduction activity of isolated bacteriods as well as *in vitro* activities of GS, glutamate synthase, XDH, uricase and allantoinase in the host plant cytoplasm were not affected during this 2 days period. After 7 days of treatment, C_2H_2 reducing activity of bacteroids was inhibited by 100% and glutamate synthase by 43%. Activities of the other enzymes were unaffected. Ureide levels in the xylem sap declined by 23 and 70% after 2 and 7 days of exposure to nitrate. In the experiments of Schubert and DeShone (unpublished), PRPP synthetase activity declined in parallel with the reduction in nitrogenase activity after 2 days of exposure to 10 mM nitrate. These results suggest that the regulatory molecules and kinetics of induction for enzymes of NH_4^+ assimilation, purine synthesis and purine oxidation may be different.

The effect of rhizobial strain on the induction and levels of enzymes of ureide biogenesis and concentration of ureides in the xylem has not been investigated in detail. Except one report, where ureide level in the sap of soybean inoculated with one strain were 10-20 fold higher than those of the plants inoculated with another strain, all other studies carried to this date on this aspect have indicated that rhizobial strain has no effect on ureide production.

A thorough understanding of the regulation of the pathways for purine synthesis and ureide biogenesis would require isolation of genes for key enzymes and subsequent examination of transcriptional and translational control mechanisms. Work in this direction has already been initiated in a number of laboratories to isolate genes encoding the enzymes of purine synthesis. Genes for uricase have already been isolated. However, we need to understand the temporal regulation of coordinated induction and repression of enzymes of NH_4^+ assimilation and ureide biogenesis.

C. Other Means of Regulation

Besides the above discussed control mechanisms, occurrence of isoenzymes of enzymes involved in ureide biogenesis, may have regulatory functions in the process of ureide formation. However, studies have not yet been carried out and we still do not know the biochemical and molecular mechanisms which control the levels and activity of these enzymes.

The localization of the overall pathway of ureide biogenesis in different organelles and cell types is also of significance in the control of the pathway. This compartmentation may provide conducive environment for certain parts of the pathway (e.g microaerophilic environment for N_2 fixation and purine synthesis, high pO_2 for uricase activity), and may facilitate metabolite channelling and ureide export by maintaining pool levels of key intermediates and regulating concentrations of effectors (Schubert and Boland, 1990).

VII. Metabolism of Ureides

In tropical legumes dependent on symbiotically fixed nitrogen, ureides are metabolized mainly in the leaves (Pate and Atkins, 1983). However, about 33% of the nitrogen delivered to the developing fruit particularly in the first half of their growth, are still accounted by ureides, which are metabolized mainly in the pod and seedcoat (Rainbird et al., 1984). Allantoinase, the enzyme that catalyses the hydrolysis of ALN to ALA is abundant in stems, leaves, nodules and fruits of legumes (Thomas and Schrader, 1981; Amarjit and Singh, 1985a). The further breakdown of ALA to give the necessary products can occur by any of the two sequences shown in Fig. 4. The distinguishing feature of these two sequences is the production of urea in the allantoicase pathway, in which allantoicase (allantoate amidinohydrolase, EC 3.5.3.4) catalyses the hydrolysis of ALA to give ureidogylcolate and urea. Ureidoglycolate is further hydrolysed to a second molecule of urea and glyoxylate either by allantoicase itself or ureide-glycolase (ureidoglycolate : urea lyase). In the allantoate amide-hydrolase pathway, ALA is hydrolysed first to ureidoglycine plus CO_2 and NH_4^+. Ureidoglycine on further metabolism either enzymatically or spontaneously is converted to ureidoglycolate. The latter is cleaved to CO_2 and $2NH_4^+$ plus glyoxylate in a reaction catalysed by ureidoglycolate amidohydrolase, an enzyme which has recently been purified from developing french bean fruits (Wells and Lees, 1991).

Recent researches conducted in this area have established that ALN in leaf tissue is degraded to $4NH_4^+$, $2CO_2$ and glyoxylate without the involvement of urea (Winkler et al., 1987b; Schubert and Boland, 1990; Singh, 1990). This has been proved by use of labelled allantoin (2, 7-^{14}C-allantoin), which released $^{14}CO_2$ at linear rates (Winkler et al., 1987a). The $^{14}CO_2$ release was not affected by the irreversible urease inhibitor, phenylphosphordiamidate at concentrations eliminating all urease activity. Also, neither ^{14}C-urea nor any other ^{14}C-metabolite other than ALA could be detected in leaf tissue. On the contrary, when the leaf discs were incubated with (4, 5- ^{14}C) allantoin, $^{14}CO_2$, ^{14}C-glyoxylate, ^{14}C-glycine and ^{14}C-serine were released. These observations are consistent with the action of allantoate amidohydrolase and a second amidohydrolase activity (ureidoglycolate amidohydrolase), that degrades the intermediate, ureidoglycolate as shown in Fig. 4. Studies of allantoin catabolism carried out in soybean tissue culture have also indicated allantoin to be completely catabolized without the involvement of a urea intermediate. Phenylphosphordiamidate eliminated urease activity and growth on urea but did not affect the release of $^{14}CO_2$ nor caused an accumulation of (^{14}C) urea when the cells were fed (2-^{14}C) allantoin (Winkler et al., 1987b). The availability of nitrogen for growth in the absence of urease activity argues for the amidohydrolase pathway. Winkler et al. (1985, 1988) have further shown soybean seed coat extracts to contain an enzyme complex that could degrade allantoin to CO_2 and glyoxylate in a ratio of 2.3 to 1 and phenylphosphordiamidate neither inhibited $^{14}CO_2$ release from (2, 7-^{14}C) allantoate, nor caused an accumulation of (^{14}C) urea. This extract was thus shown to contain both allantoate amidohydrolase and ureidoglycolate amidohydrolase activities (Winkler et al., 1988). The enzyme ureidoglycolate amidohydrolase has now been purified from developing french bean fruits to get a preparation which was free of allantoinase and urease activities (Wells and Lees, 1991). The location of the enzyme in peroxisomes further ensures the utilization of the products of ureidoglycolate degradation in this organelle. Thus there is now very strong evidence that allantoate can be degraded directly by the enzymes allantoate amidohydrolase and ureidoglycolate amidohydrolase (Fig. 4), rather than by enzymes which release urea.

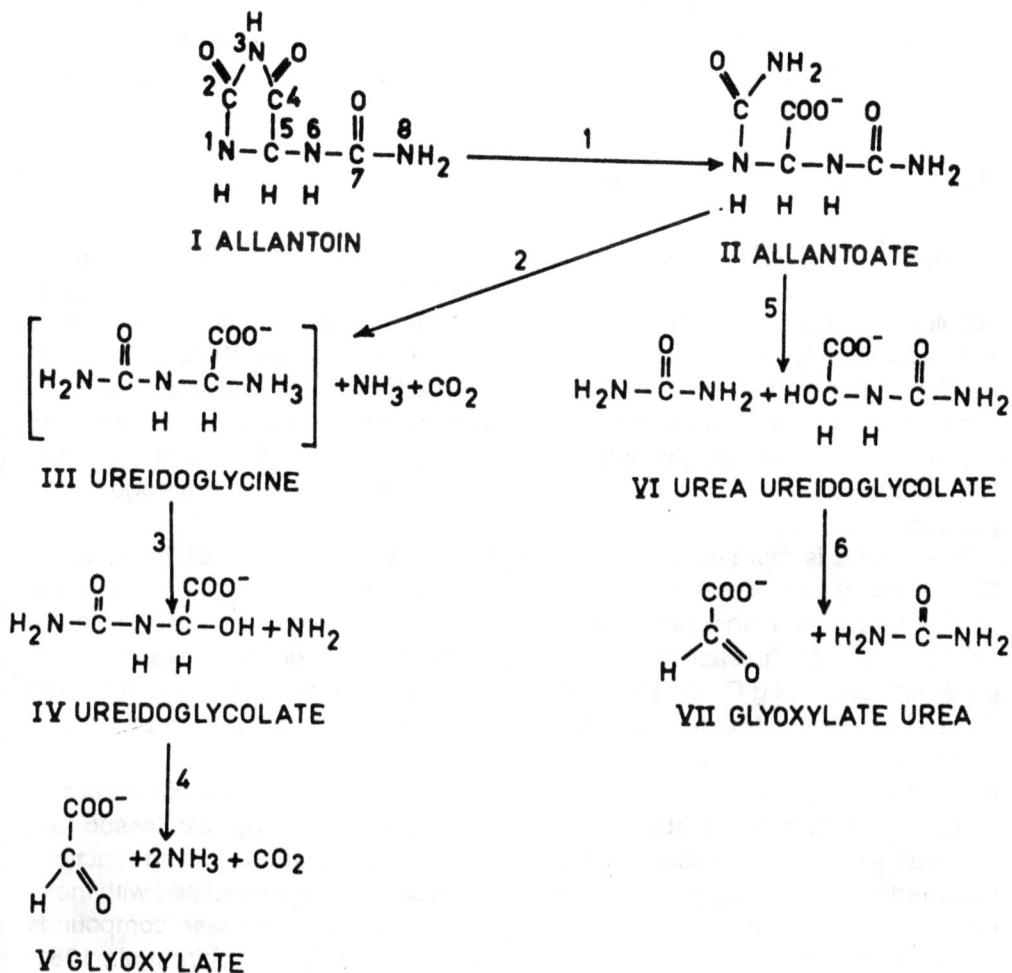

Fig. 4. Catabolic pathways of allantoin assimilation. (1) Allantoin amidohydrolase, (2) Allantoate amidohydrolase, (3) Ureidoglycine amidohydrolase, (4) Ureidoglycolate amidohydrolase, (5) Allantoate amidinohydrolase, (6) Ureidoglycolate urea-lyase.

The reports of urea formation during catabolism of ureides (Atkins et al., 1982; Shelp and Ireland, 1985) can be criticised on the following basis: (a) the urea found could have been a product of aphid metabolism; (b) urea could have been released nonenzymically from ALA by acid hydrolysis during the isolation procedure; (c) urea could have been released nonenzymically from ALA by the extraction procedure (boiling 80% ethanol); (d) the unlabelled urea reported could have resulted from sources other than ALA, such as arginine.

VIII. Physiological Significance of Ureides

Why the plant had to adopt the strategy of ureide biosynthesis and their transport? Why the products of initial ammonium assimilation are not exported out of nodules but retained in the nodules and diverted towards purine biosynthesis and subsequently ureide formation? What physiological and metabolic advantages ureides confer on the plan? Why ureides are predominant in nodulated plants as compared to non-nodulated counterparts? We seem to be far away from providing a clear answer to all these questions. In the following discussion, an attempt has been made to analyse the possible significance of ureide based nitrogen assimilation.

The ureides is more efficient form of N transport both in terms of the moles of C used for carbon skeletons and the energetics of synthesis, as in ureides, the ratio of C:N is one whereas, in amides it is 2. The ureides produced are more efficient in use of photoassimilates compared to amide exporters. Cowpea (ureide producer) uses 5.5 g C per g N fixed while lupin consumes 6.9 g C per g N fixed (Pate et al., 1982). Allantoin and ALA are also the least expensive to produce in terms of ATP consumption ($\sim$ 6.0 ATPs per N compared to $\sim$ 9 ATPs per N for arginine, $\sim$ 13 ATP per N for citrulline). This is mainly because of the fact that the ureide producing plants have the ability to recapture much of the energy costs associated with purine synthesis by coupling purine oxidation and other associated reactions to the reduction of $NAD(P)^+$. In addition, the carbon skeleton associated with these ureides is already highly oxidized and, therefore, the costs of these compounds are quite low (Layzell et al., 1979; Rawsthorne et al., 1980; Minchin et al., 1981; Pate et al., 1981; Pate and Atkins, 1983). Hence ALN and ALA are by far the most efficient species for transporting and storing nitrogen. Ultimately, this may be the primary reason why plants use these ureides to transfer and store nitrogen.

Nitrogen assimilation in the form of ureides may also have the advantage of storability. Ureides being relatively less soluble, compared to amides (amides are approx 5 times more soluble than ureides), are better suited for storage. Potassium salt of ALA may be equally or even more soluble than asparagine (Winkler et al., 1988). The lower solubility of ALN/ALA might be advantageous to plants in sequestering nitrogen in storage tissues and organs. This viewpoint gains support from the fact that ureides do accumulate to varying extent in different plant parts, particularly during the reproductive phase, and these ureides may (Matsumoto et al., 1977a) or may not (Luthra et al., 1983a, 1983b; Amarjit et al., 1986) be retained till maturity. Moreover, ureides storage may have advantage in special situations where photosynthetic and/or the sink activity is affected. Under these conditions, N can be stored in the form of ureides till the photosynthetic activity and/or sink activity is regained. Hence, storing N in the form of ureides may help in co-ordinating the rates of protein synthesis in relation to the availability of carbon skeletons from photosynthesis.

Sprent (1980) has suggested that the reduced solubility of ALN and ALA compared to asparagine may be one reason why high levels of these ureides are found in tropical legumes. The growth pattern of nodules in these legumes is essentially determinate with vascular strands fused apically to form a series of closed loops connected to the main vascular network at the base of the nodule. Because of the reduced resistance to water flow, this closed loop system provides for a higher flux of transpirational water through the system to flush the products of fixation from the nodule. On the contrary, growth pattern of nodules in amide exporting legumes is indeterminate with an open branched vascular system with reduced water flow. Hence, these species require special transfer cells to actively secrete nitrogenous solutes into the xylem (Pate et al., 1969; Cunning et al., 1974). As a consequence, the molar concentration of amides in the xylem is several fold higher than the corresponding concentration of ALN and ALA in the tropical legumes (Wong and Evans, 1971; Pate et al., 1979), which implies that the water use efficiency for nitrogen export is higher for amide exporters. This may be the reason that the use of ALN and ALA to transport N is restricted to tropical species (Sheoran et al., 1981; Atkins et al., 1982a). The water requirement (i.e. ml water per mmol compound transported or ml water required to transport 1 mg of N) of ALA and ALN is the highest, those of glutamine and asparagine intermediate and of amino acids like glutamate etc. the lowest. However, solubility may vary depending on the pH and form of compound (i.e. salt) present. Accordingly, the transport of ALN and ALA in species from more temperate or arid regions is disadvantageous, especially under conditions of environmental stress (low temp., drought).

Tropical plants have to bear high temperature and water stress and CO_2 limiting conditions due to closure of stomata under these conditions. Ureides, before assimilation, are degraded to NH_4^+ and CO_2. The CO_2 released may be refixed by RuBP carboxylase. It may thus be speculated that assimilation in the form of ureides may help raising the internal levels of CO_2 which in turn, may optimise photosynthetic rates.

Uriede producing legumes export much greater amounts of malate from the nodules and amount of malate exported seems to increase with increase in the extent of ureide being transported. This is all the more paradoxical in view of the fact that PEP carboxylase activity (CO_2 fixing activity) is apparently much higher in clusterbean (amide producer) than in pigeonpea (Sawhney et al., 1987). From where does this malate originate in the nodules? Malate may arise from the combined reactions of glycolysis, TCA and anaplerotic reaction of PEP carboxylase, and may serve one or more of the following functions: (a) accumulation of malate may be an osmotic adjustment, which is particularly relevant under conditions of high temperature and water stress, (b) ureides having lower C:N ratio required additional carbon for the assimilation of nitrogen. Malate exported alongwith the ureides may serve as the carbon source for the assimilation of nitrogen in the shoot and (c) allantoin can be hydrolysed non-enzymically at alkaline pH while allantoic acid in solution is labile at neutral or slightly acidic pH. Hence, pH of the xylem sap can be a major factor determining the proportion of different ureides being transported in the xylem stream. Malate may contribute to the maintenance of pH in the xylem sap.

IX. Conclusions and Future Prospects

Ureides, ALN and ALA, are key organic molecules for transporting and storing reduced nitrogen in legumes of tropical origin. These compounds in nodules are synthesized from currently delivered photosynthates and products of recent nitrogen fixation via purine synthesis followed by their oxidation and hydrolysis; the pathway of purine synthesis being similar to that operative in several animals and micro-organisms. The elevated levels of enzymes involved in purine biosynthesis and production of purine precursors in ureide producing nodules and the concurrent induction of these enzymes with the commencement of ureide production support a direct link between ureide production and purine biosynthesis. Several of the enzymes involved in purine biosynthesis and their subsequent conversion into ureides have been purified and the effects of potential regulatory molecules on these proteins have been examined. However, a number of enzymes in these pathways are yet to be purified and characterized. It has been suggested that ammonium or some related metabolite induces the synthesis of GS which in turn is needed for the subsequent expression of other enzymes in the pathway. Nodule-specific forms of some of these proteins, particularly those of uricase, GS, PEP carboxylase etc., have been identified. However, the knowledge regarding the factors controlling developmental and tissue specific expression of these nodulins is still rudimentary. Metabolic reactions leading to the formation of ureides are compartmentalized both at cellular and subcellular levels in nodules. However, uncertainty still exists with respect to the nature of the intermediate transported from the infected cell to the uninfected cell as well as the site of purine synthesis. In addition, the site(s) of PRPP synthesis (plastid and/or cytosolic) and the path and site of synthesis of ribose-5-P are still not defined. Similarly, factors controlling cellular and subcellular differentiation in this tissue have not been defined. A concerted effort using biochemical, molecular, and immunological approaches will be required to fill the above discussed gaps.

Literature Cited

Amarjit, and Singh, R. 1984. Ureide biogenesis and the enzymes of ammonia assimilation and ureide biosynthesis in nitrogen fixing pigeonpea (*Cajanus cajan*) nodules. *J. Bio Sci.* **6**: 185-192.

Amarjit, and Singh, R. 1985a. Enzymes of ammonia assimilation and ureide biogenesis in developing pigeonpea (*Cajanus cajan* L.) nodules, *J. Bio Sci.* **7**: 375-385.

Amarjit, and Singh, R. 1985b. Allantoinase from nodules of pigeonpea (*Cajanus cajan* L.). *Phytochemistry* **24**: 415-418.

Amarjit, and Singh, R. 1986. Properties of 5'-nucleotidase from nodules of pigeonpea. *Phytochemistry* **25**: 2267-2270.

Amarjit, Sheoran, I.S., and Singh, R. 1985. Effect of water stress on the enzymes of nitrogen metabolism in mungbean (*Vigna radiata* L. Wilczek) nodules. *Plant Cell Environ.* **8**: 195-200.

Amarjit, Saharan, M.R., and Singh, R. 1986. Ureide metabolism in developing pods of pigeonpea (*Cajanus cajan* L.) *Proc. Indian Natn. Sci. Acad.* B **52**: 673-678.

Atkins, C.A. 1988. Synthesis, transport and utilization of translocate and solutes of nitrogen. In *Nitrogen in Higher plants*, (Abrol,Y.P. ed.) Research Studies Press Ltd., Taunton. pp. 223-295.

Atkins, C.A. 1981. Metabolism of purine nucleotides to form ureides in nitrogen fixing nodules of cowpea (*Vigna unguiculata* L. Walp.) *FEBS Lett.* **125**: 89-93.

Atkins, C.A. 1982. Ureide metabolism and the significance of ureides in legumes. In *Advances in Agricultural Microbiology*, (Subba Rao,N.S. ed.) Oxford and IBH Publishing Co., New Delhi. pp. 53-88.

Atkins, C.A., and Shelp, B.J. 1983. Cellular and subcellular organization of ureide biosynthesis in cowpea nodules. *Plant Physiol.* **72S**: 113.

Atkins, C.A., Herridge, D.F., and Pate, J.S. 1977. The economy of carbon and nitrogen in nitrogen fixing annual legumes: experimental observations and theoretical consideration. In *Isotopes in Biological Dinitrogen Fixation*, (Welsh,C.N. ed.) FAO-IAEA Advisory Conference Vienna. pp. 211-242.

Atkins, C.A., Pate, J.S., Griffiths, G.J., and White, S.T. 1980a. Economy of carbon and nitrogen in nodulated and non-nodulated cowpea (*Vigna unguiculata* L. Walp). *Plant Physiol.* **66**: 978-983.

Atkins, C.A., Rainbird, R.M., and Pate, J.S. 1980b. Evidence for a purine pathway of ureide synthesis in N_2-fixing nodules of cowpea. *Z. Pflanzenphysiol.* **97**: 249-260.

Atkins, C.A., Pate, J.S., Ritchie, A., and Peoples, M.B. 1982a. Metabolsim and translocation of allantoin in ureide producing grain legumes. *Plant Physiol.* **70**: 476-482.

Atkins, C.A., Ritchie, A., Rowe, P.B., Mc Cairns, E., and Sauer, D. 1982b. *De novo* purine synthesis in nitrogen fixing nodules of cowpea (*Vigna unguiculata* L. Walp) and soybean (*Glycine max* L. Merr). *Plant Physiol.* **70**: 55-60.

Atkins, C.A., Pate, J.S., and Shelp, B.J. 1984a. Effects of short-term N_2 deficiency on nitrogen metabolism of legume nodules. *Plant Physiol.* **76**: 705-710.

Atkins, C.A., Shelp, B.J., Storer, P.J., and Pate, J.S. 1984b. Nitrogen nutrition and the development of biochemical functions associated with nitrogen fixation and ammonia assimilation of nodules in cowpea seedlings. *Planta* **162**: 327-333.

Atkins, C.A., Shelp, B.J., and Storer, P.J. 1985. Purification and properties of inosine monophosphate oxidoreductase from nitrogen fixing nodules of cowpea (*Vigna unguiculata* L. Walp). *Arch. Biochem. Biophys.* **236**: 807-814.

Atkins, C.A., Storer, P.J., and Pate, J.S. 1988 a. Pathways of nitrogen assimilation in cowpea nodules studied using $^{15}N_2$ and allopurinol. *Plant Physiol.* **86**: 204-207.

Atkins, C.A., Sanford, P.J., Storer, P.J., and Pate, J.S. 1988b. Inhibition of nodule functioning in cowpea by a xanthine oxidoreductase inhibitor, allopurinol. *Plant Physiol.* **88**: 1229-1234.

Atkins, C.A., Storer, P.J., and Shelp, B.J. 1989. Purification and properties of purine nucleosidase from N_2 fixing nodules of cowpea (*Vigna unguiculata*). *Plant Physiol.* **134**: 447-452.

Awonaike, K.O., Lea, P.J., and Miflin, B.J. 1981. The location of the enzymes of ammonia assimilation in root nodules of *Phaseolus vulgaris* L. *Plant Sci. Lett.* **23**: 189-195.

Bergersen, F.J. 1965. Ammonia—an early stable product of nitrogen fixation by soybean root nodules. *Aust. J. Biol. Sci.* **18**: 1-9.

Bergersen, F.J., and Goodchild, D.J. 1973. Aeration pathways in soybean root nodules. *Aust. J. Biol. Sci.* **26**: 729-740.

Bergmann, H., Preddie, E., and Verma, D.P.S. 1983. Nodulin-35: A subunit of specific uricase (uricase II) induced and localized in the uninfected cells of soybean nodules. *EMBO Journal.* **2**: 2333-2339.

Boland, M.J. 1981. NAD: xanthine dehydrogenase from nodules of navy beans: Partial purification and properties. *Biochem. Int.* **2**: 567-574.

Boland, M.J., and Benny, A.C. 1977. Enzymes of nitrogen metabolism in legume nodules. Purification and properties of NADH-dependent glutamate synthase from lupin nodules. *Eur. J. Biochem.* **79**: 355-362.

Boland, M.J., and Schubert, K.R. 1982a. Purine biosynthesis and catabolism in soybean root nodules: Incorporation of $^{14}CO_2$ in to xanthine. *Arch. Biochem. Biophys.* **213**: 486-491.

Boland, M.J., and Schubert, K.R. 1982b. The biosynthesis of glycine and methenyl tetrahydrofolate. precursors for ureide synthesis in soybean nodules. *Plant Physiol.* **69S**: 112.

Boland, M.J., and Schubert, K.R. 1983a. Biosynthesis of purines by a proplastid fraction from soybean nodules. *Arch. Biochem. Biophys.* **220**: 179-187.

Boland, M.J., and Schubert, K.R. 1983b. Phosphoglycerate dehydrogenase from soybean nodules. Partial purification and some kinetic properties. *Plant Physiol.* **71**: 658-661

Boland, M.J., Fordyce, A.M., and Greenwood, R.M. 1978. Enzymes of nitrogen metabolism in legume nodules; A comparative study. *Aust. J. Plant Physiol.* **5**: 553-559.

Boland, M.J., Hanks, J.F., Reynolds, P.H.S., Blevins, D.G., Tolbert, N.E., and Schubert, K.R. 1982. Subcellular organization of ureide biogenesis from glycolytic intermediates and ammonium in nitrogen fixing soybean nodules. *Planta* **155**: 45-51.

Boland, M.J., Blevins, D.G., and Randall, D.D. 1983. Soybean nodule xanthine dehydrogenase. A kinetic study. *Arch. Biochem. Biophys.* **222**: 435-441.

Christensen, T.M.I.E., and Jochimsen, B.U. (1983). Enzymes of ureide synthesis in pea and soybean. *Plant Physiol.* **72**: 56-59.

Coker, G.T. III, and Schubert, K.R. 1980. Asparagine synthesis in soybean roots and nodules. *Plant Physiol.* **65S**: 111.

Coker, G.T. III, and Schubert, K.R. 1981. Carbondioxide fixation in soybean roots and nodules. I. Characterization and comparison with N_2 fixation and composition of xylem exudate during early nodule development. *Plant Physiol.* **67**: 691-696.

Cookson, C., Hughes, H., and Coombs, J. 1980. Effects of combined nitrogen on anaplerotic carbon assimilation and bleeding sap composition in *Phaseolus vulgaris* L. Planta **148**: 338-345.

Copeland, L., Vella, J., and Hong, Z. 1989. Enzymes of carbohydrate metabolism in soybean nodules. *Phytochemistry.* **28**: 57-61.

Cullimore, J.V., and Miflin, B.J. 1984. Immunological studies on glutamine synthetase using antisera raised to the two plant forms of the enzyme from *Phaseolus* root nodules. *J. Exp. Bot.* **35**: 581-587.

Cullimore, J.V., Lara, M., Lea, P.J., and Miflin, B.J. 1983. Purification and properties of two forms of glutamine synthetase from the plant fraction of *Phaseolus* root nodules. *Planta* **157**: 245-253.

Cullimore, J.V., Gebhardt, C., Saarelainen, R., Miflin, B.J., Idler, K.B., and Barker, R.F. 1984. Glutamine synthetase of *Phaseolus vulgaris* L. Organ specific expression of a multigen family. *J. Mol. Appl. Genet.* **2**: 589-599.

Davey, M.R., Cocking, E.C., and Bush, E. 1973. Isolation of legume root nodule protoplasts. *Nature* **244**: 400-401.

Deroche, M.E., Carroyl, E., and Jolivet, E. 1983. Purification and properties of pyruvate kinase from legume nodules. *Physiol. Veg.* 21: 1075-1081.

Done, J., and Fowden, L. 1951. A new amino acid in the groundnut plant (*Arachis hypogaea*): Evidence of the occurrence of r-methyleneglutamine and r-methylenegumatic acid. *Biochem. J.* 51: 451-458.

Doremus, H.D., and Blevins, D.G. 1988a. Nucleoside diphosphatase and 5'-nucleotidase activities of soybean root nodules and other tissues. *Plant Physiol.* 87: 36-40.

Doremus, H.D., and Blevins, D.G. 1988b. Purification and characterization of a specific nucleoside diphosphatase from soybean root nodules. *Plant Physiol.* 87: 41-45.

Farnden, K.J.F. 1981. Aminotransferases in legume nodules. *In Current Perspectives in Nitrogen Fixation*, (Gibson, A.H., and Newton,W.E. eds.). Australian Acad. Sci. Canberra. pp. 271-273.

Fowden, L. 1954. The nitrogen metabolism of groundnut plants. The role of r-methyleneglutamine and r-methyleneglutamic acid. *Ann. Bot.* 18: 417-440.

Fujihara, S., and Yamaguchi, M. 1978a. Effects of allopurinol [4-hydroxypyrazolo- (3, 4-d)-pyrimidine] on the metabolism of allantoin in soybean plants. *Plant Physiol.* 62: 134-138.

Fujihara, S., and Yamaguchi, M. 1978b. Probable site of allantoin formation in nodulating soybean plants. *Phytochemistry.* 17: 1239-1243.

Fujihara, S., Yamamoto, K., and Yamaguchi, M. 1977. A probable role of allantoin and the influence of nodulation on its production in soybean plants. *Plant and Soil* 48: 233-242.

Groat, R.G., and Vance, C.P. 1981. Root nodule enzymes of ammonia assimilation in alfalfa (*Medicago sativa* L.). Developmental pattern and response to applied nitrogen. *Plant Physiol.* 67: 1198-1203.

Groat, R.G., and Schrader, L.E. 1982. Isolation and immunochemical characterization of plant glutamine synthetase in alfalfa (*Medicago sativa* L.) nodules. *Plant Physiol.* 70: 1759-1761.

Gunning, B.E.S., Pate, J.S., Minchin, F.R., and Marks, I. 1974. Quantitiative aspects of transfer cell structure in relation to vein loading in leaves and solute transport in legume nodules. *In Transport at the Cellular Level. Symp. Soc. Exp. Biol.* 28: 87-125.

Gupta, A.K., and Singh, R. 1991. Characterization of glucose-6-phosphate and 6-phosphogluconate dehydrogenases from pigeonpea (*Cajanus cajan* L.) nodules. *Proc. Indian Natn. Sci. Acad.* B57: 329-338.

Gupta, A.K., Sheoran, I.S., and Singh, R. 1991. Carbohydrate metabolism in relation to ammonia assimilation in nodules of pigeonpea (*Cajanus cajan* L.). *Proc. Indian Natn. Sci. Acad.* B57: 295-302.

Hanks, J.F., Tolbert, N.E., and Schubert, K.R. 1981. Localization of enzymes of ureide biosynthesis in peroxisomes and microsomes of nodules. *Plant Physiol.* 68: 65-69.

Hanks, J.F., Schubert, K.R., and Tolbert, N.E. 1983. Isolation and characterization of infected and uninfected cells from soybean nodules. Role of uninfected cells in ureide synthesis. *Plant Physiol.* 71: 869-873.

Herridge, D.F., Atkins, C.A., Pate, J.S., and Rainbird, R.M. 1978. Allantoin and allantoic acid in the nitrogen economy of the cowpea (*Vigna unguiculata* L. Walp.). *Plant Physiol.* 62: 495-498.

Hong, Z., and Copeland, L. 1990. Pentose phosphate pathway enzymes in nitrogen fixing leguminous root nodules. *Phytochemistry.* 29: 2437-2440.

Israel, D.W., and McClure, P.R. 1980. Nitrogen translocation in the xylem of soybeans. *In World Soybean Research Conference II: Proceedings*, (Corbin,F.T. eds.), Boulder, Westview. pp. 111-127,

Kennedy, I.R. 1966. Primary products of symbiotic nitrogen fixation. I. Short-term exposures of serrodella nodules to $^{15}N_2$. *Biochim. Biophys. Acta* 130: 285-294.

Kohl, D.H., Schubert, K.R., Carter, M.B., Hagedorn, C.H., and Shearer, C. 1988. Proline metabolism in N_2-fixing root nodules: Energy transfer and regulation of purine synthesis. *Proc. Natl. Acad. Sci. USA* 85: 2036-2040.

Kohl, D.H., Lin, J.J., Shearer, G., and Schubert, K.R. 1990. Activities of the pentose phosphate pathway and enzymes of proline metabolism in legume root nodules. *Plant Physiol.* 94: 1258-1264.

Kouchi, H., Fukai, K., Katagisi, H., Manamisawa, K., and Tajima, S. 1988. Isolation and enzymological characterization of infected and uninfected cell protoplasts from root nodules of *Glycine max. Physiol. Plant.* 73: 327-334.

Kushizaki, M., Ishizuka, J., and Akamatsu, F. 1964. Physiological studies on nutrition of soybean plants. 2. Effects of nodulation on the nitrogen constituents of soybean plants. *J. Sci. Soil Manure* (Japan) 35: 323-327.

Lara, M., Cullimore, J.V., Lea, P.J., Miflin, B.J., Johnston, A.W.B., and Lamb, J.W. 1983. Appearance of a novel form of plant glutamine synthetase during nodule development in *Phaseolus vulgaris* L. *Planta* 157: 254-258.

Layzell, D.B., Rainbird, R.M., Atkins, C.A., and Pate, J.S. 1979. Economy of photosynthate use in nitrogen fixing legume nodules. *Plant Physiol.* 64: 888-891.

Legocki, R.P., and Verma, D.P.S. 1980. Identification of nodule specific host proteins (nodulins) involved in the development of Rhizobium-legume symbiosis. *Cell* 20: 153-163.

Lucas, K., Boland, M.J., and Schubert, K.R. 1983. Uricase from soybean root nodules. Purification, properties and comparison with the enzyme from cowpea. *Arch. Biochem. Biophys.* 226: 190-197.

Luthra, Y.P., Sheoran, I.S., and Singh, R, 1981. Pigeonpea (*Cajanus cajan* L.) A ureide producing legume. *Curr. Sci.* 50: 270-271.

Luthra, Y.P., Sheoran, I.S., Rao, A.S., and Singh, R. 1983a. Ontogenetic changes in the level of ureides and enzymes of their metabolism in various plant parts of pigeonpea (*Cajanus cajan* L.). *J. Exp. Bot.* 34: 1358-1370.

Luthra, Y.P., Sheoran, I.S., Rao, A.S., and Singh, R. 1983b. A probable pathway of ureide assimilation in pigeonpea pods. *Curr. Sci.* 52: 957-960.

Matsumoto, T., Yamamoto, Y., and Yatazawa, M. 1976. Role of root nodules in the nitrogen nutrition of soybean. II. Fluctuation in allantoin concentration of the bleeding sap. *J. Sci. Soil Manure (Japan)* 47: 436-469.

Matsumoto, T., Yatazawa, M., and Yamamoto, Y. 1977a. Effects of exogenous nitrogen compounds on the concentrations of allantoin and various constituents in several organs of soybean plants. *Plant Cell Physiol.* 18: 613-624.

Matsumoto, T., Yatazawa, M., and Yamamoto, Y. 1977b. Incorporation of ^{15}N into allantoin in nodulated soybean plants supplied with $^{15}N_2$. *Plant Cell Physiol.* 18: 459-462.

McClure, P.R., and Israel, D.W. 1979. Transport of nitrogen in the xylem of soybean plants. *Plant Physiol.* 644: 411-416.

McNeil, D.L., and LaRue, T.A. 1984. Effect of nitrogen source on ureides in soybean. *Plant Physiol.* 66: 720-725.

McParland, R.H., Guevara, J.G., Becker, R.R., and Evans, H.J. 1976. The purification and properties of the glutamine synthetase from the cytosol of soybean root nodules. *Biochem. J.* 153: 597-606.

Meeks, J.C., Wolk, C.P., Schilling, N., Shaffer, P.W., Avissar, Y., and Chien, W.S. 1978. Initial organic products of fixation of ^{13}N-dinitrogen by root nodules of soybean (*Glycine max*). *Plant Physiol.* 61: 980-983.

Minchin, F.R., Summerfield, R.J., Hadley, P., Roberts, E.H., and Rawsthorne, S. 1981. Carbon and nitrogen nutrition of nodulated roots of grain legumes. *Plant Cell Environ.* 4: 5-26.

Mitchell, M.K., Reynolds, P.H.S., and Blevins, D.G. 1986. Serine hydroxymethylase from soybean root nodules: Purification and kinetic properties. *Plant Physiol.* 81: 553-557.

Muller, M., Kraupp, M., Chiba, P., and Rumpold, H. 1982. Regulation of purine uptake in normal and neoplastic cells. *Adv. Enzyme Regul.* 21: 239-256.

Newcomb, W. 1981. Biology of the Rhizobiaceae, (Giles, K.L.,and Atherly, A.G. eds.), Academic Press, New York. pp. 247-298

Newcomb, E.H., and Tandon, S.K. 1981. Uninfected cells of soybean root nodules: Ultrastructure suggests key role in ureide production. *Science* 212: 1394-1396.

Newcomb, E.H., Tandon, S.K., and Kowal, R.R. 1985. Ultrastructural specialization for ureide production in uninfected cells of soybean root nodules. *Protoplasma* 125, 1-12.

Nguyen, J. 1979. Effect of light on deamination and oxidation of adenylic compounds in cotyledons of *Pharbitis nil. Physiol. Plant,* 46: 225-229.

Nguyen, J., Zelechowska, M., Foster, V., Bergmann, H., and Verma, D.P.S. 1985. Primary structure of the soybean nodulin-35 gene encoding nodule specific uricase localized in peroxisomes of uninfected cells of soybean. *Proc. Natl. Acad. Sci. USA* 82: 5040-5044.

Nguyen, J., Machal, L., Vidal, J., Perrot-Rechenmann, C., and Gadal, P. 1986. Immunochemical studies on XDH of soybean root nodules. Ontogenic changes in the level of enzyme and immunocytochemical localization. *Planta* 167: 190-195.

Ohyama, T., and Kumazawa, K. 1978. Incorporation of N into various nitrogenous compounds in intact soybean nodules after exposure to $^{15}N_2$ gas. Soil Sci. *Plant Nutr.* 24: 525-533.

Ohyama, T., and Kumazawa, K. 1979. Assimilation and transport of nitrogenous compounds originated from $^{15}N_2$-fixation and $^{15}NO_3$ absorption. *Soil Sci. Plant Nutr.* 25: 9-19.

Ohyama, T., and Kumazawa, K. 1980a. Nitrogen assimilation in soybean nodules. I. The role of GS-GOGAT system in the assimilation of ammonia produced by N_2 fixation. *Soil Sci. Plant Nutr.* 26: 109-115.

Ohyama, T., and Kumazawa, K. 1980b. Nitrogen assimilation in soybean nodules. II. $^{15}N_2$ assimilation in bacteroid and cytosol fractions of soybean nodules. *Soil Sci. Plant Nutr.* 26: 205-213.

Ohyama, T., and Kumazawa, K. 1980c. Nitrogen assimilation in soybean nodules. III. Effects of rhizosphere pO_2 on the assimilation of $^{15}N_2$ in nodules attached to intat plants. *Soil Sci. Plant Nutr.* 26: 321-324.

Pate, J.S. 1975. Uptake, assimilation and transport of nitrogen compounds of plants. *Soil Biol. Biochem.* 5: 109-119.

Pate, J.S., and Atkins, C.A. 1983. Nitrogen uptake, transport and utilization. In *Nitrogen Fixation,* (Broughton,W.J. ed.), Oxford University Press. 3: 245-298.

Pate, J.S., Gunning, B.E.S., and Briarty, L.G. 1969. Ultrastructure and functioning of the transport system of the leguminous root nodule. *Planta* 85: 11-34.

Pate, J.S., Atkins, C.A., Hamel, K., McNeil, D.L., and Layzell, D.B. 1979. Transport of organic solutes in phloem and xylem of a nodulated legume. *Plant Physiol.* 63: 1082-1088.

Pate, J.S., Atkins, C.A., White, S.T., Rainbird, R.M., and Woo, K.C. 1980. Nitrogen nutrition and xylem transport of itrogen in ureide-producing grain legumes. *Plant Physiol.* 65: 961-965.

Pate, J.S., Atkins, C.A., Herridge, D.F., and Layzell, D.B. 1981. Synthesis, storage and utilization of amino compounds in white lupin (*Lupinus albus* L.). *Plant Physiol.* 67: 37-42.

Pate, J.S., Atkins, C.A., and Rainbird, R.M. 1982. Theoretical and experimental costing of nitrogen fixation and related process in nodules of legumes. In *Current Perspectives in Nitrogen Fixation,* (Gibson,A.H.,and Newton,W.E. ed.), Australian Academy of Science, Canberra. pp. 105-116.

Patterson, T.G., and LaRue, T.A. 1983. N_2 fixation and ureide content of soybeans. Environmental effects and source sink manipulations. *Crop Sci.* 23: 819-824.

Rainbird, R.M. 1983. Elements in the cost of nitrogen fixation with special reference to the legume cowpea (*Vigna unguiculata* L. Walp). *Ph.D. Thesis,* Univ. West. Australia, Nedlands.

Rainbird, R.M., and Atkins, C.A. 1981. Purification and some properties of urate oxidase from nitro-gen fixing nodules of cowpea. *Biochim. Biophys. Acta.* **659**: 132-140.

Rainbird, R.M., Thorne, J.H., and Hardy, R.W.F. 1984. Role of amides, amino acids and ureides in the metabolism of developing soybean seeds. *Plant Physiol.* **74**: 329-334.

Rao, D.N., and Appaji Rao, N. 1982. Purification and regulatory properties of mungbean (*Vigna radiata* L.) serine hydroxymethyl transferase. *Plant Physiol.* **69**: 11-18.

Rao, A.S., and Singh, R. 1988. Ureide metabolism in nodulated legumes. *In Advances in Frontier Areas of Plant Biochemistry* (Singh,R.,and Sawhney, S.K. eds.), Prentice Hall of India, New Delhi. pp. 281-316.

Rawsthorne, S., Minchin, F.R., Summerfield, R.J., Cookson, C., and Coombs, J. 1980. Carbon and nitrogen metabolism in legume root nodules. *Phytochemistry.* **19**: 341-355.

Reinbothe, H., and Mothes, K. 1962. Urea, ureides and guanidines in plants. *Annu. Rev. Plant Physiol.* **13**: 129-150.

Reynolds, P.H.S., and Farnden, K.J.F. 1979. The involvement of aspartate aminotransferases in ammonium assimilation in lupin nodules. *Phytochemistry.* **18**: 1625-1630.

Reynolds, P.H.S., and Blevins, D.G. 1986. Phosphoserine aminotransferase in soybean nodules: Demonstration and localization. *Plant Physiol.* **81**: 293-296.

Reynolds, P.H.S., Boland, M.J., and Farnden, K.J.F. 1981. Enzymes of nitrogen metabolism in legume nodules: Partial purification and properties of the aspartate aminotransferase from lupin nodules. *Arch. Biochem. Biophys.* **209**: 524-533.

Reynolds. P.H.S., Blevins, D.G., Boland, M.J., Schubert, K.R., and Randall, D.D. 1982a. Enzymes. of ammonia assimilation in legume nodules. A comparison between ureide and amide-transporting plants. *Physiol. Plant.* **55**: 255-260.

Reynolds, P.H.S., Boland, M.J., Blevins, D.G., Schubert, K.R., and Randall, D.D. 1982b. Enzymes of amide and ureide biogenesis in developing soybean nodules. *Plant Physiol.* **69**: 1334-1338.

Reynolds, P.H.S., Blevins, D.G., and Randall, D.D. 1984. 5'-phosphoribosylpyrophosphate amido-transferase from soybean root nodules. Kinetic and regulatory properties. *Arch. Biochem. Biophys.* **229**: 623-631.

Reynolds, P.H.S., Hine, A., and Rodber, K. 1988. Serine metabolism in legume nodules. Purifi-cation and properties off phosphoserine aminotransferase. *Physiol. Plant.* **74**: 194-199.

Robertson, J.G., Farnden, K.J.F., Warburton, M., and Banks, J.B. 1975a. Induction of glutamine synthetase during nodule development in lupin. *Aust. J. Plant Physiol.* **2**: 265-272.

Robertson, J.G., Warburton, M., and Farnden, K.J.F. 1975b. Induction of glutamate synthase during nodule development in lupin. *FEBS Lett.* **55**: 33-37.

Rowe, P.B., Mc Cairns, E., Madsen, G., Sauer, D., and Elliott, H. 1978. *De novo* purine synthesis in avian. Copurification of the enzymes and properties of the pathway. *J. Biol. Chem.* **253**: 7711-7721.

Ryan, E., Bodley, F., and Fotrel, P.F. 1972. Purification and characterization of aspartate amino-transferases from soybean root nodules and *Rhizobium japonicum*. *Phytochemistry.* **11**: 957-963.

Sawhney, V., Saharan, M.R., and Singh, R. 1987. Nitrogen fixing efficiency and enzymes of CO_2 assimilation in nodules of ureide and amide producing legumes. *J. Plant Physiol.* **129**: 201-210.

Schubert, K.R. 1981. Enzymes of purine biosynthesis and catabolism in *Glycine max*. Comparison of activities with N_2 fixation and composition of xylem exudate during nodule development. *Plant Physiol.* **68**: 1115-1122.

Schubert, K.R. 1982. The Energetics of Biological Nitrogen Fixation. *Workshop Summaries-I. Am. Soc. Plant Physiol.* pp. 1-30.

Schubert, K.R., 1986. Products of biological nitrogen fixation in higher plant: Synthesis, transport and metabolism. *Annu. Rev. Plant Physiol.* **37**: 539-574.

Schubert, K.R. 1987. Purification of glutamine dependent phosphoribosylpyrophosphae amidotransferase from soybean root nodules. *Plant Physiol.* 83S: 20.

Schubert, K.R., and Coker, G.T. III 1982. Studies of nitrogen and carbon assimilation in N₂ fixing plants. Short-term studies using ¹³N and ¹¹C. *In Recent Developments in Biological and Chemical Research with Short-lived Radio-isotopes. Washington, DC: Am. Chem. Soc.* pp. 317-339.

Schubert, K.R.; and Boland, M.J. 1982. Purine biosynthesis by proplastids in soybean nodules. *Plant Physiol.* 69:S-112.

Schubert, K.R., and Boland, M.J. 1984. The cellular and sub-cellular organization of the reactions of ureide biosynthesis in nodules of tropical legumes. *In Advances in Nitrogen Fixation Research* (Veeger,C., and Newton,W.E. eds.), The Hague, Nijhoff/Junk. pp. 445-451.

Schubert, K.R., and Coker, G.T. III 1985. Carbon metabolism in soybean roots and nodules. Role of dark CO₂ fixation. *World Soybean Res. Conf. III Proc.* (Shibles, R. ed.), Boulder: Westview. pp. 815-823.

Schubert, K.R., and Boland, M.J. 1990. The ureides. *In The Biochemistry of Plants: A comprehensive Treatise*, (Stumpf,P.K.,and Conn,E.E. eds.), Academic Press, New York. 16: 197-282.

Schubert, K.R., Coker, G.T. III, and Firestrone, R.B. 1981. Ammonia assimilation in *Alnus glutinosa* and *Glycine max.* Short term studies using ¹³NH₄. *Plant Physiol.* 67: 662-665.

Schuller, K.A., Day, D.A., Gibson, A.H., and Gresshoff, P.M. 1986. Enzymes of ammonia assimilation and ureide biosynthesis in soybean nodules. Effect of nitrate. *Plant Physiol.* 80: 646-650.

Selker, J.M.L., and Newcomb, E.H. 1985. Spatial relationship between uninfected and infected cells in root nodules of soybean. *Planta* 165: 446-454.

Sen Gupta-Gopalan, C., and Pitas, J.W. 1986. Expression of nodule specific glutamine synthetase genes during nodule development in soybean. *Plant Mol. Biol.* 7: 189-199.

Shelp, B.J., and Atkins, C.A. 1983. Role of inosine monophosphate oxidoreductase in the formation of ureides in nitrogen fixing nodules of cowpea (*Vigna unguiculata* L. Walp). *Plant Physiol.* 72: 1029-1034.

Shelp, B.J., and Atkins, C.A. 1984. Subcellular location of enzymes of ammonia assimilation and asparagine synthesis in root nodules of *Lupinus albus* L. *Plant Sci. Lett.* 36: 225-230.

Shelp, B.J., and Ireland, R.J. 1985. Ureide metabolism in leaves of nitrogen fixing soybean plants. *Plant Physiol.* 77: 779-783.

Shelp, B.J., Atkins, C.A., Storer, P.J., and Canvin, D.T. 1983. Cellular and subcellular organization of pathways of ammonia assimilation and ureide synthesis in nodules of cowpea (*Vigna unguiculata* L. Walp). *Arch. Biochem. Biophys.* 224: 429-441.

Sheoran, I.S., Luthra, Y.P., Kuhad, M.S., and Singh, R. 1981. Effect of water stress on some enzymes of nitrogen metabolism in pigeonpea. *Phytochemistry.* 20: 2675-2677.

Sheoran, I.S., Luthra, Y.P., Kuhad, M.S., and Singh, R. 1982. Clusterbean—A ureide or amide producing legume. *Plant Physiol.* 70: 917-918.

Singh, R. 1990. Synthesis, transport and metabolism of products of biological nitrogen fixation in nodulated legumes. *In Plant Biochemistry Research in India* (Singh, R. ed.), Society for Plant Physiology and Biochemistry, New Delhi. pp. 81-88.

Singh, D., Rao, A.S., Nainawatee, H.S., and Singh, R. 1985. Ureide and enzymes of their metabolism in nodules of mungbean as influenced by plant genotype and rhizobial strain. *H.A.U. J. Res.* 15: 148-152.

Sprent, J.I. 1980. Root nodule anatomy, type of export product and evolutionary origin in some leguminosae. *Plant Cell Environ.* 3: 35-43.

Streeter, J.G. 1972. Nitrogen nutrition of field grown soybean plants. I. Seasonal variations in soil nitrogen and nitrogen composition of stem exudate. *Agron. J.* 64: 311-314.

Streeter, J.G. 1979. Allantoin and allantoic acid in tissues and stem exudate from field grown soybean plants. *Plant Physiol.* **63**: 478-480.

Suzuki, N., and Iwai, K. 1973. The occurrence and properties of methenyl tetrahydrofolate cyclohydrolase in plants. *Plant Cell Physiol.* **14**: 319-327.

Tajima, S., and Yamamoto, Y. 1975. Enzymes of purine catabolism in soybean plants. *Plant Cell Physiol.* **16**: 271-282.

Thomas, R.J., and Schrader, L.E. 1981. Ureide metabolism in higher plants. *Phytochemistry.* **20**: 361-371.

Thomas, R.J., Feller, V., and Erisman, E.H. 1979. Allantoic acid metabolism in non-nodulated bush beans. *Plant Physiol.* **63S**: 50.

Thomas, R.J., Feller, V., and Erisman, E.H. 1980. Ureide metabolism in non-nodulated *Phaseolus vulgaris* L. *J. Exp. Bot.* **31**: 409-418.

Triplett, E.W. 1985. Intercellular nodule localization and nodule specificity of xanthine dehydrogenase in soybean. *Plant Physiol.* **17**: 1004-1009.

Triplett, E.W., and Hao, A.C. 1988. Molecular and serological comparisons of purified xanthine dehydrogenase from root nodules of three ureide producing legumes. *Physiol Plant.* **74**: 164-169.

Triplett, E.W., Blevins, D.G., and Randall, D.D. 1980. Allantoic acid synthesis in soybean root nodule cytosol via xanthine dehydrogenase. *Plant Physiol.* **65**: 1203-1206.

Triplett, E.W., Blevins, D.G., and Randall, D.D. 1982. Purification and properties of soybean nodule xanthine dehydrogenase. *Arch. Biochem. Biophys.* **219**: 39-46.

Vanden Bosch, K.A., and Newcomb, E.H. 1986. Immunogold localization of nodule-specific uricase in developing soybean root nodules. *Planta* **167**: 425-436.

Venkateswara Rao, N., Subhash Reddy, R., and Sivarama Sastry, K. 1988. Allantoinases of nodulated *Arachis hypogaea*. *Phytochemistry.* **27**: 693-695.

Verma, D.P.S., Fortin, M.G., Stanley, J., Mauro, V.P., Purohit, S., and Morrison, N. 1986. Nodulins and nodulin genes of *Glycine max*. *Plant Mol. Biol.* **7**: 51-61.

Webb, M.A., and Newcomb, E.H. 1987. Cellular compartmentation of ureide biogenesis in root nodules of cowpea (*Vigna unguiculata* L. Walp). *Planta* **172**: 162-175.

Wells, X.E., and Lees, E.M. 1991. Ureidoglycolate amidohydrolase from developing french bean fruits (*Phaseolus vulgaris* L.). *Arch. Biochem. Biophys.* **287**: 151-159.

Winkler, R.G., Polacco, J.C., Blevins, D.G., and Randall, D.D. 1985. Enzymic degradation of allantoate in developing soybeans. *Plant Physiol.* **79**: 787-792.

Winkler, R.G., Blevins, D.G., Polacco, J.C., and Randall, D.D. 1987a. Ureide catabolism of soybeans. II. Pathway of catabolism in intact leaf tissue. *Plant Physiol.* **83**: 585-591.

Winkler, R.G., Blevins, D.G., Polacco, J.C., and Randall, D.D. 1987b. Ureide catabolism in nitrogen fixing legumes. *Trends Biochem. Sci.* **13**: 97-100.

Winkler, R.G., Blevins, D.G., and Randall, D.D. 1988. Ureide catabolism in soybeans. III. Ureidoglycolate amidohydrolase and allantoate amidohydrolase are activities of an allantoate degrading enzyme complex. *Plant Physiol.* **86**: 1084-1088.

Winkler, R.G., Blevins, D.G., Polacco, J.C., and Randall, D.D. 1988. Ureide catabolism in N_2-fixing legumes. *Trends Biochem. Sci.* **13**: 97-100.

Winter, H.C., Powell, G.K., and Dekker, E.E. 1981. 4-methylene glutamine in peanut plants: Dynamics of formation levels, turnover in relation to other free amino acids. *Plant Physiol.* **68**: 588-593.

Wong, P.P., and Evans, H.J. 1971. Poly-B-hydroxybutyrate utilization by soyban (*Glycine max* Merr.) nodules and assessment of its role in maintenance of nitrogenase activity. *Plant Physiol.* **47**: 750-755.

Woo, K.C., and Broughton, W.J. 1979. Isolation and metabolism of *Vigna unguiculata* root nodule protoplasts. *Planta* **145**: 487-495.

Woo, K.C., Atkins, C.A., and Pate, J.S. 1980a. Ureide synthesis in cell free extracts of cowpea (*Vigna unguiculata* L.) nodules. *Plant Physiol.* **66**: 655-657.

Woo, K.C., Atkins, C.A., and Pate, J.S. 1980b. Biosynthesis of ureides from purines in a cell free system from nodule extracts of cowpea (*Vigna unguiculata* L. Walp). *Plant Physiol.* **66**: 735-739.

Woo, K.C., Atkins, C.A., and Pate, J.S. 1981. Ureide synthesis in a cell free system from cowpea (*Vigna unguiculata* L. Walp) nodules. *Plant Physiol.* **67**: 1156-1160.

Nitrogen Nutrition in Higher Plants, 1995
Editors : H.S. Srivastava & R.P. Singh
Associated Publishing Co., New Delhi, India
pp. 311-322.

Polyamines and Plant Senescence

CHING HUEI KAO

Abbreviations: ACC, 1-aminocyclopropane-1-carboxylic acid; ADC, arginine decarboxylase; DAO, diamine oxidase; DCH, dicyclohexylamine; DFMA, difluoromethylarginine; DFMO, difluoromethylornithine; dSAM, decarboxylated SAM; MTA, 5'-methylthioadenosine, MGBG, methylglyoxal bis (guanylhydrazone); ODC, ornithine decarboxylase; SAM, S-adenosylmethionine; SAMDC, SAM decarboxylase; PAO, polyamine oxidase.

I. Introduction

Polyamines are ubiquitous nitrogen compounds classified as plant growth substances (Evans and Malmberg, 1989). They are widespread in higher plants. The most common polyamines in higher plants include putrescine, spermidine and spermine (Fig. 1). Polyamines have been studied in animals and bacteria for more than 50 years. However, it is only relatively recently that their presence and significance

$$H_2N\text{-}(CH_2)_3\text{-}NH_2$$

Diaminopropane

$$H_2N\text{-}(CH_2)_4\text{-}NH_2$$

Putrescine

$$H_2N\text{-}(CH_2)_4\text{-}NH(CH_2)_3NH_2$$

Spermidine

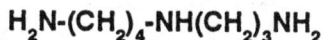

$$H_2N\text{-}(CH_2)_3\text{-}HN\text{-}(CH_2)_4\text{-}NH\text{-}(CH_2)_3\text{-}NH_2$$

Spermine

Fig. 1. Structure of polyamines found in higher plants.

has been recognized in higher plants. Interest in plant polyamines arose mainly as a results of the pioneering work of Bagni and Serafini-Fracassini (1974), who showed that polyamines stimulated growth in *Helianthus tuberosus* explants. Experimental evidence now indicates that plant polyamines may be involved in growth, differentiation or morphogenesis, stress and senescence. This review will be limited to a discussion of the involvement of polyamines in plant senescence. To initiate our discussion, some aspects of polyamine biochemistry are also briefly reviewed. For more general discussion on plant polyamines, the reader is referred to several recent reviews (Evans and Malmberg, 1989; Flores, 1990; Galston and Sawhney, 1990; Smith, 1984; 1990; Kakkar and Rai, 1993).

II. Biosynthesis of Polyamines

Figure 2 shows the biosynthetic pathways of polyamines. Putrescine is probably derived from ornithine in all plants, as it is in animals and micro-organisms, utilizing the enzyme ornithine decarboxylase. Unlike animals, putrescine in higher plants also arises from arginine by the action of the enzyme arginine decarboxylase. Arginine can be converted to agmatine by ADC; agmatine then loses urea, forming putrescine. Both ADC and ODC are pyridoxal phosphate-dependent enzymes. It has been shown that ADC and ODC pathways have different tissue distributions (Smith, 1985) and different regulation (Hiatt et al., 1986). Ornithine decarboxylase is primarily linked to rapid cell division (Cohen et al., 1982; Heimer and Mizrahi, 1982), whereas ADC is usually linked to various stress responses (Flores, 1990).

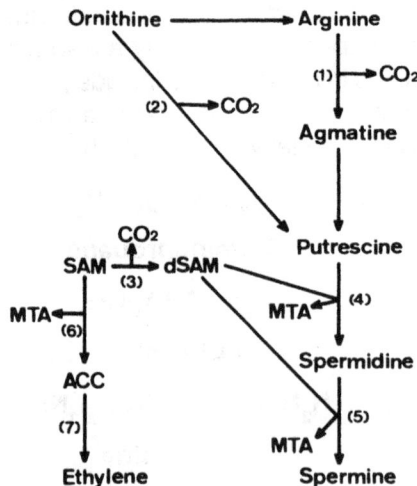

Fig. 2.   Pathways for biosynthesis of polyamines and ethylene. The enzymes involved: 1, ADC; 2, ODC; 3, SAMDC; 4, spermidine synthase; 5, spermine synthase; 6, ACC synthase; 7, ethylene-forming enzyme.

Putrescine is converted into spermidine by the action of an aminopropyltransferase called spermidine synthase. A second aminopropyltransferase termed spermine synthase adds an additional propylamine moiety to spermidine, forming spermine. The other product of the aminopropyltransferase reactions is 5'-methylthioadenosine. The source of these propylamine moiety is decarboxylated S-adenosylmethionine, which is formed by the action of S-adenosylmethionine decarboxylase. By the action of 1-aminocyclopropane-1-carboxylic acid synthase and ethylene-forming enzyme, SAM can also be converted into ethylene, an important hormone in higher plants.

The availability of specific or potent inhibitors for enzymes in the biosynthetic pathways of polyamines has provided an enormous stimulus to most biochemical and physiological studies of plant polyamines. Difluoromethylarginine and difluorornithine have been developed to function as irreversible inhibitors for ADC and ODC, respectively (Metcalf et al., 1978; Kallio and McCann, 1981). Methylglyoxal bis (guanylhydrazone) is a potent inhibitor of SAMDC (Williams-Ashman and Schenone, 1972). Spermidine synthase is inhibited by DCH (Hibasami et al., 1980). These inhibitors are useful and have been widely used to change levels of polyamines in plant tissues. However, not all inhibitors will result in the changes of polyamine levels as expected. There is at least one report indicating that DCH is not an inhibitor of spermidine synthase (Batchelor et al., 1986). Rice leaves treated with DCH resulted in a reduction in level of spermine rather than spermidine (Chen and Kao, 1991b). A review by Slocum and Galston (1987) discussed the application of these inhibitors in higher plants in considerable detail.

III. Metabolism of Polyamines

Polyamines can be oxidized by diamine oxidases (DAOs) and polyamine oxidases (PAOs). Diamine oxidases are widespread in higher plants (Rinaldi et al., 1985; Smith, 1985) and are particularly active in leguminous plants. All the plant DAOs reported so far have been dimers, several of which have been shown to contain two copper atoms per mole of the enzyme (Smith, 1985). The following reactions are typical of those catalyzed by DAO:

Putrescine + $O_2 \rightarrow$ Pyrroline + NH_3 + H_2O_2

Spermidine + $O_2 \rightarrow$ Aminopropylpyrroline + NH_3 + H_2O_2

In higher plants PAOs have been found only in the Gramineae family (Smith, 1985) and in water hyacinth (Yanagisawa et al., 1987). PAOs have been purified and characterized from maize and barley roots and oat leaves (Smith, 1972, 1977; Suzuki and Yanagisawa, 1980). Polyamine oxidase, a flavoprotein, oxidizes spermidine and spermine to give, respectively, pyrroline and aminopropylpyrroline plus the coproduct diaminopropane and hydrogen peroxide:

Spermidine + $O_2 \rightarrow$ Pyrroline + Diaminopropane + H_2O_2

Spermine + $O_2 \rightarrow$ Aminopropylpyrroline + Diaminopropane + H_2O_2

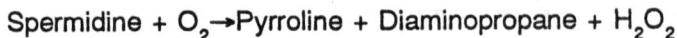

Polyamines can be conjugated with several compounds in higher plants. Conjugates of putrescine, spermidine and spermine with cinnamic acid and its derivatives are widespread in higher plants (Smith, 1981). Accumulating evidence suggests that hydroxycinnamoyl acid amides may play important role in flowering (Maritin-Tanguay, 1985). Putrescine hydroxylcinnamoyl transferase has been isolated and purified from tobacco callus (Meurer-Grimes et al., 1989; Negrel, 1989). It has been shown that polyamines can be bound to RNA, DNA and proteins (Bagni et al., 1981; Serafini-Fracassini et al., 1984; Serafini-Fracassini and Mossetti, 1985; Apelbaum et al., 1988; Mizrahi et al., 1989). Serafini-Fracassini et al. (1988, 1989) provided the first evidence for the occurrence of transglutaminase, an enzyme capable of binding polyamines covalently to proteins in plants. Using thin-layer tobacco tissue culture, Apelbaum et al. (1988) demonstrated that the post-translational modification of an unique protein by attachment of spermidine may be causally connected to the appearance of flower buds.

IV. Polyamines and Senescence

Senescence can be defined as endogenously controlled deterioative changes, which are natural causes of death of cells, tissues, organs, or organisms (Nooden, 1988). Thus, senescence is a natural developmental process and it is completely endogenous. Several lines of evidence indicate that polyamines may play an important role in the regulation of senescence. The evidence consists mainly of (1) delay of senescence through polyamine treatments, and (2) correlation of senescence with a decline in endogenous levels of polyamines.

A.   Influence of Exogenous Polyamines

Interest in the effects of polyamines on senescence arose mainly as a result of the pioneering work by Galston and his colleagues of Yale university. During attemps to regenerate a cereal plant from a single oat leaf protoplast, Altman et al. (1977) found that oat protoplasts underwent progressive senescence. This senescence is manifested by morphological deterioation and ultimate lysis of protoplasts, by a decrease in incorporation of [³H] uridine and [³H] leucine into macromolecules, and by a sharp increase in ribonuclease activity (Altman et al., 1977). Experiments conducted by Altman et al. (1977) and Galston et al. (1978) provided evidence that polyamines may prevent senescence-linked decrease of RNA and protein synthesis and rise in ribonuclease activity in oat protoplasts, thereby stabilizing them against lysis. When polyamines were applied to detached oat leaves in darkness, polyamines also retarded chlorophyll degradation and inhibited the rise in ribonuclease activity and α-amino nitrogen (Kaur-Sawhney and Galston, 1979). As little as 10 min exposure to 1 mM spermine at the beginning of the dark period induces a marked retardation of chlorophyll degradation over 2 days. The antisenescent action of exogenously applied polyamines has been confirmed in various leaf tissues, such as barley (Popovic et al., 1979), radish (Altman, 1982), soybean (Cheng and Kao, 1983), rice (Cheng and Kao, 1983; Cheng et al., 1984). *Hydrilla* (Kar and Choudhuri, 1986), and *Brassica* (Pjon et al., 1990). Polyamines retard chlorophyll degradation in detached leaves in darkness, but accelerate it under light condition (Kaur-Sawhney and Galston, 1979; Cheng et al., 1984; Kar and Choudhuri, 1986; Pjon et al., 1990). This effect has been considered as a photo-chemical reaction (Kar and Choudhuri, 1986). However, polyamines do retard pro-tein degradation in detached rice leaves under light (Cheng et al., 1984). Diami-nopropane, an oxidation product of polyamines, has also been shown to retard senescence of detached leaves (Fuhrer et al., 1982; Shih et al., 1982; Cheng et al., 1984).

To determine the role of polyamines in fruit ripening of tomatoes, fruit of normal-ripening line and cultivar (Alcobaca-red and Rutgers) were vacuum infiltrated through the stem scar with solutions of polyamines and some of their precursors and meta-bolites. Polyamines, diaminopropane and methionine were effective in extending the shelf life of tomatoes (Davies et al., 1990). Polyamines also retarded softening of apple fruit tissue during postharvest storage (Wang and Kramer, 1990). Contrary to the results observed in leaves and fruits, addition of putrescine and spermidine in the culture solution resulted in shorten longevity of cut carnation flowers (Downs and Lovell, 1986).

Muhitch et al., (1983) found that putrescine and cadaverine at 1mM and sper-midine and spermine at 0.1 μM retarded senescence of Paul's scarlet rose cell sus-pension cultures. Life of suspension-cultured pear fruit cells was also found to be retarded by 0.5 to 1.0 mM spermidine (Ke and Romani, 1988).

B. Correlation with Endogenous Polyamines

If polyamines are indeed to play an important role in the regulation of senes-cence, the decrease in the levels of polyamines and activities of the polyamine bio-synthetic enzymes would be expected. Kaur-Sawhney et al. (1982) demonstrated that the levels of spermidine and activity of ADC decreased during senescence of both attached and excised oat leaves. They concluded that these decreases are critical events in a complex induction system, and not simply results of leaf sen-escence. The decreases in levels of polyamines and activities of polyamine bio-synthetic enzymes have also been demonstrated during senescence of attached rice and pea leaves (Srivastava et al., 1981; Chen and Kao, 1991a).

Tomato fruits also show evidence of inverse correlation of polyamine levels with ripening. Fruits of Liberty tomato ripen slowly and have a prolonged keeping qual-ity. An unusual increase in putrescine during ripening is correlated with an increased storage life (Saftner and Baldi, 1990).

To demonstrate the physiological role of polyamines in tomato ripening, Dibble et al. (1988) reported on the levels of polyamines in the pericarp tissues of Alcobaca *(alc)*, its spontaneous revertant Alcobaca-red, and the standard variety Rutgers (both *Alc*) during the process of ripening. Fruits of the Alcobaca, which ripen slowly and have prolonged keeping qualities, contain three times as much putrescine as the normal variety at the ripe stage. They suggested that the enhanced putrescine level in this line may be responsible for the ripening and storage features. However, in a subsequent report they demonstrated that the increase in Alcobaca fruit appeared to be an age related phenomenon as the increase in putrescine levels took place in Alcobaca whether or not the fruits ripen, a process in which Alcobaca is light dependent (Rastogi and Davies, 1991).

Several other examples have been reported of systems where polyamine levels did not decrease during senescence. Working with a photoperiod-inducible senescence in the apical bud of peas, Smith and Davies (1985) found that there was no decrease of polyamines prior to the appearance of early symptoms of senescence. Birecka et al. (1984) reported that Heliotropium leaves, with a very low polyamine levels, exhibited only a weak senescence syndrome, whereas in detached darkened leaves of *Avena sativa* and *Nicotiana alata* having high polyamine levels, senescence was very pronounced. These results led them to conclude that exogenous polyamines could have a nonspecific action on senescence, an action apparently not exercised by endogenous polyamines in spite of their high accumulation.

Another way to approach the question of whether or not polyamines have functional role in the control of senescence involves the use of inhibitors of polyamine synthesis. Using this approach, Birecka et al. (1990) showed that dark-induced leaf senescence of barley and oats was not correlated with the levels of endogenous polyamines. Chen and Kao (1991b) found that levels of putrescine in detached rice leaves decreased with incubating duration of dark incubation. To characterize further the role of endogenous putrescine in dark-induced senescence, inhibitors of biosynthesis of polyamines were used to change levels of intracellular putrescine in detached rice leaves (Chen and Kao, 1991b). Difluoromethylarginine and DFMO significantly decreased levels of putrescine without affecting senescence of detached rice leaves. By contrast, treatments with DCH and MGBG resulted in elevated levels of putrescine in detached rice leaves but also promoted senescence. These observations suggest that a lowering of the level of putrescine is unlikely to be the factor responsible for the senescence of detached rice leaves in darkness. This conclusion is further supported by the observations that the effect of hormone (abscisic acid or benzyladenine) on senescence of detached rice leaves are separated from that on polyamine levels or polyamine biosynthesis (Chen and Kao, 1991b; 1992).

LT-8 is a chlorophyll-deficient mutant of rice derived from Norin no. 8. Chen et al. (1991a) reported on the levels of polyamines in intact rice leaves of LT-8 and Norin no. 8. Chlorophyll level in the normal leaves decreased with increasing age. However, the chlorophyll level in the mutant leaves began to decrease only when more than 60% of the initial protein had been degraded. If chlorophyll is used as the prime indicator of senescence, the leaves of the mutant would be regarded as nonsenescent or slowly senescent. However, both spermidine and spermine levels in the mutant leaves decreased at the similar rate with increasing age as in the normal leaves (Chen et al., 1991a).

C. Mechanism of Polyamine Action

It has been postulated that the senescence signal itself acts by decreasing polyamines as one of the critical control steps. Since no close inverse correlation between polyamines and senescence was observed in several systems, endogenous polyamine levels do not seem to be the part of the chain of events leading the initiation of senescence. This suggestion is further supported by the observation that endogenous putrescine levels accumulated by several defined stresses (such as potassium deficiency, acid, osmotic and water stress, cadmium toxicity), several of which accelerate the senescence of leaves (Flores, 1990). An important question now is how exogenous polyamines retard senescence.

Cheng et al. (1984) have studied senescence of detached rice leaves and found that exogenous diaminopropane, spermidine and spermine all appeared to retard senescence in the dark and all promoted chlorophyll degradation in the light. The presence of β-hydroxyethylhydrazine, an inhibitor of conversion of polyamines to diaminopropane by PAO, reversed the effects on spermine and spermidine in the light, indicating a possible requirement for conversion to diaminopropane. However, neither intact nor detached rice leaves contained detectable diaminopropane throughout senescence (Chen et al., 1991a, Chen and Kao, 1991a, b).

Ca^{2+} was shown to inhibit competitively the observed effects of exogenous polyamines and diaminopropane (Kaur-Sawhney and Glaston, 1979; Shih et al., 1982; Cheng et al., 1984), suggesting that initial attachment to a membrane site may be required. Thus, the hypothesis that polyamines stabilize membranes must be involved in the control of senescence was postulated. Roberts et al. (1986) presented evidence to show that exogenously applied polyamines simply reflects membrane rigidification rather than true physiological responses. Recent experiments conducted by DiTomaso et al. (1989) showed that putrescine did not replace calcium in maintaining membrane stability. Thus, the postulated role for polyamines in stabilizing cell membranes is far from clear.

Ethylene is a senescence-promoting hormone and accelerates fruit ripening (Mattoo and Aharoni, 1988). Ethylene and polyamines are known to have opposite effects in relation to fruit ripening and senescence. Both ethylene and polyamines share SAM as a common precursor (Fig. 2). Thus, the mechanism by which exogenous polyamines retard senescence may be related to the possibility that they inhibit the biosynthesis of ehtylene (Fuhrer et al., 1982; Shih et al., 1982). Indeed, polyamines have been reported to inhibit ethylene production in a number of plant tissues, including fruits, leaves, petals and hypocotyls (Apelbaum, 1981; Suttle, 1981; Fuhrer et al., 1982; Ke and Romani, 1988). In order to understand whether the mechanism of the retardation of senescence of detached rice leaves by polyamines is related to the inhibition of the biosynthesis of ethylene by these compounds, Chen et al. (1991b) studied the effects of polyamines, specifically putrescine, on the biosynthesis of ethylene. Unexpectedly, they found that polyamines effectively promoted the production of ethylene in detached rice leaves under both light and dark condition. Putrescine stimulated ethylene production via the enhancement of the synthesis of ACC and the conversion of ACC to ethylene. Exogenous polyamines also stimulate ethylene biosynthesis by detached tobacco and soybean leaf tissues (Pennazio and Roggero, 1989; 1990). In some instances, polyamines have been shown to have no detectable effect on the production of ethylene (Downs and Lovell, 1986; Kramer and Wang, 1990). Thus, the mechanism that polyamines may retard senescence by inhibiting ethylene production is not clear.

V. Conclusions and Further Prospects

There is no doubt that exogenous polyamines retard senescence. However, we do not know the exact mechanism by which polyamines bring about retardation, nor do we know the subcellular sites where polyamines act. In the past few years, much attention has been paid to the changes of free polyamine levels during senescence. It would be interesting to know whether polyamines covalently bound with proteins or with hydroxycinnamic acids are causally related to senescence. Considerable attention has been given to polyamine biosynthesis during senescence. Polyamine degradation, however, has been investigated only in a few cases. To determine the functional role of polyamines, it is important to understand the metabolic fate of these compounds during senescence. We also need to know whether polyamine effects on senescence are artifacts or nonspecifically toxic as suggested by some workers. More work is required to further clarify the involvement of polyamines in senescence. There is potential for major breakthroughs. The prospects are promising.

Acknowledgements

The author would like to acknowledge with gratitude the contribution of colleagues and students, past and present, to the personal research reported. This research has been supported by grants from the National Science Council of the Republic of China.

Literature Cited

Altman, A. 1982. Retardation of radish leaf senescence by polyamines. *Physiol. Plant.* **54**: 189-193.
Altman, A., Kaur-Sawhney, R., and Galston, A.W. 1977. Stabilization of oat leaf protoplasts through polyamine-mediated inhibition of senescene. *Plant Physiol.* **60**: 570-574.

Apelbaum, A., Burgoon, A.C., Anderson, J.D., Lieberman M., Ben-Arie, R., and Mattoo A.D. 1981. Polyamines inhibit biosynthesis of ethylene in higher plant tissue and fruit protoplasts. *Plant Physiol.* **68**: 453-456.

Apelbaum, A., Canellakis, Z.N., Applewhite, P.B., Kaur-Sawhney, R., and Galston, A.W. 1988. Binding of spermidine to a unique protein in thin-layer tobacco tissue culture. *Plant Physiol.* **88**: 996-998.

Bagni, N., and Serafini-Fracassini, D. 1974. The role of polyamines as growth factors in higher plants and their mechanism of action. *In: Plant Growth Substances* 1973. Hirokawa Publishing Co., Tokyo. pp. 1205-1217.

Bagni, N., Adams, P., and Serafini-Fracassini, D. 1981. RNA, proteins and polyamines during growth in germinating apple pollen. *Plant Physiol.* **68**: 727-730.

Batchelor, K.W., Smith, R.A., and Watson, N.S. 1986. Dicyclohexylamine is not an inhibitor of spermidine synthase. *Biochem. J.* **233**: 307-308.

Birecka, H., DiNolfo, T.E., Martin, W.B., and Frohlich, M.W., 1984. Polyamines and leaf senescence in pyrrolizidine alkaloid-bearing *Heliotropium* plants. *Phytochemistry* **23**: 991-997.

Birecka, H. Birecki, M., and Ireton, K. 1990. Endogenous polyamine levels and dark-induced leaf senescence. *Plant Physiol.* **93**: S-757.

Chen, C.T., and Kao, C.H. 1991a. Senescence of rice leaves XXIX. Ethylene production, polyamine level and polyamine biosynthetic enzyme activity during senescence. *Plant Sci.* **78**: 193-198.

Chen, C.T., and Kao, C.H. 1991b. Senescence of rice leaves XXX. Levels of endogenous polyamines and dark-induced senescence of rice leaves. *Plant Cell Physiol.* **32**: 935-941.

Chen, C.T., and Kao, C.H. 1992. Senescence of rice leaves XXXII. Effects of abscisic acid and benzyladenine on polyamines and ethylene production during senescence. *J. Plant Physiol.* **139**: 617-620.

Chen, C.T., Li, C.C., and Kao, C.H. 1991a. Senescence of rice leaves XXXI. Changes of chlorophyll, protein, and polyamine contents and ethylene production during senescence of a chlorophyll-deficient mutant. *J. Plant Growth Regul.* **10**: 201-205.

Chen, S.L., Chen, C.T., and Kao, C.H. 1991b. Polyamines promote the biosynthesis of ethylene in detached rice leaves. *Plant Cell Physiol.* **32**: 813-817.

Cheng, S.H., and Kao, C.H. 1983. Localized effect of polyamines on chlorophyll loss. *Plant Cell Physiol.* **24**: 1463-1467.

Cheng, S.H., Shyr, Y.Y., and Kao, C.H. 1984. Senescence of rice leaves XII. Effects of 1, 3-diaminopropane, spermidine and spermine. *Bot. Bull. Acad. Sin.* **25**: 191-196.

Cohen, E., Arad, S., Heimer, Y.M., and Mizrahi, Y. 1982. Participation of ornithine decarboxylase in early stages of tomato fruit development. *Plant Physiol.* **70**: 540-543.

Davies, P.J., Rastogi, R., and Law, D.M. 1990. Polyamines and their metabolism in ripening fruit. In *Polyamines and Ethylene: Biochemistry, Physiology, and Interactions.* (Flores, H.E., Arteca, R.N., and Shannon, J.C. eds.) American Society of Plant Physiologists, Rockville, MD. pp. 112-125.

Dibble, A.R.G., Davies, P.J., and Mutschlr, M.A., 1988. Polyamine content of long-keeping alcobaca tomato fruit. *Plant Physiol.* **86**: 338-340.

DiTomaso, J.M., Shaff, J.E., and Kochian, L.V. 1989. Membrane-mediated putrescine transport and its role in stress-induced phytotoxicity. *Plant Physiol.* **89**: S-147.

Downs, C.G., and Lovell, P.H. 1986. The effect of spermidine and putrescine on the senescence of cut carnations. *Physiol. Plant.* **66**: 679-684.

Evans, R.T., and Malmberg, R.L., 1989. Do polyamines have roles in plant development? *Annu. Rev. Plant Physiol. Plant Mol. Biol.* **40**: 235-269.

Flores, H.E. 1990. Polyamines and plant stress. In *Stress Responses in Plants: Adaptation and Acclimation Mechanism*. (Alscher, R.G., and Cumming, J.R. eds.) Wiley-Liss, New York. pp. 217-239.

Fuhrer, J., Kaur-Sawhney, R., Shih, L.M., and Galston, A.W. 1982. Effects of exogenous 1,3-diaminopropane and spermidine on senescence of oat leaves II. Inhibition of ethylene biosynthesis and possible mode of action. *Plant Physiol.* **70**: 1597 -1600.

Galston, A.W., and Sawhney, R.K. 1990. Polyamines in plant physiology. *Plant Physiol.* **94**: 406-410.

Galston, A.W., Altman, A., and Kaur-Sawhney, R. 1978. Polyamines, ribonucleases and the improvement of oat leaf protoplasts. *Plant Sci. Lett.* **11**: 67-79.

Heimer, Y.M., and Mizrahi, Y. 1982. Characterization of ornithine decarboxylase of tobacco cells and tomato ovaries. *Biochem J.* **201**: 373-376.

Hiatt, A.C., Mc Indoo, J., and Malmberg, R.L. 1986 Regulation of polyamine synthesis in tobacco. *J. Biol. Chem.* **261**: 1293-1298.

Hibasami, H., Tanaka, M., Nagai, J., and Ikeda, T. 1980. Dicyclohexylamine, a potent inhibitor of spermidine synthetase in mammalian cells. *FEBS Lett.* **116**: 99-101.

Kallio, A., and Mc Cann, P.P. 1981. Difluoromethylornithine irreversibly inactivates ornithine decarboxylase of *Pseudomonas aeruginosa*, but does not inhibit the enzymes of *E. coli*. *Biochem. J.* **20**: 3163-3166.

Kar, R.K., and Choudhuri, M.A. 1986. Effects of light and spermine on senescence of *Hydrilla* and spinach leaves. *Plant Physiol.* **80**: 1030-1033.

Kakkar, R.K.,and Rai, U.K. 1993. Plant Polyamines in flowering and fruit ripening. *Phytochemistry* **33** : 1281-1288.

Kaur-Sawhney, R., and Galston, A.W. 1979. Interaction of polyamines and light on biochemical process involved in leaf senescence. *Plant Cell Environ.* **2**: 189-196.

Kaur-Sawhney, R; Shih, L.M., Flores. H.E., and Galston, A.W. 1982. Relation of polyamine synthesis and titer to aging and senescence in oat leaves. *Plant Physiol.* **69**: 405-410.

Ke, D., and Romani, R.J. 1988. Effects of spermidine on ethylene production and the senescence of suspension-cultured pear fruit cells. *Plant Physiol. Biochem.* **26**: 109-116.

Kramer, G.F. and Wang, C.Y. 1990. Inhibition of softening of apples by postharvest polyamine infiltration. *Plant Physiol.* **93**: S-442.

Martin-Tanguy, J. 1985. The occurrence and possible function of hydroxycinnamoyl acid amides in plants. *Plant Growth Regul.* **3**: 381-399.

Mattoo, A.K., and Aharoni, N. 1988. Ethylene and plant senescence. In *Senescence and Aging in Plants. (Nooden, L.D., and Leopold, A.C. eds)*. Academic Press, San Diego. pp. 241-280.

Metcalf, B., Gey, P., Danzin, C., Jung, M., and Vevert, J. 1978. Catalytic irreversible inhibition of mammalian ornithine decarboxylase (E.C.4.1.1.17) by substrate and product analogues. *J. Am. Chem. Soc.* **100**: 2551-2553.

Meurer-Grimes, B., Berlin, J., and Strack, D. 1989. Hydroxycinnamoyl-CoA: putrescine hydroxylcinnamoyltransferase in tobacco cell cultures with high and low levels of caffeoylputrescine. *Plant Physiol.* **89**: 488-492.

Mizrahi, Y., Applewhite, P.B., and Galston, A.W. 1989. Polyamine binding to proteins in oat and *Petunia* protoplasts. *Plant Physiol.* **91**: 738-743.

Muhitch, M.J., Edwards, L.A., and Fletcher, J.S. 1983. Influence of diamines and polyamines on the senescence of plant suspension cultures. *Plant Cell Rep.* **2**: 82-84.

Negrel, J. 1989. The biosynthesis of cinnamoylputrescines in callus tissue cultures of *Nicotiana tabacum. Phytochemistry* **28**: 477-481.

Nooden, L.D. 1988. The phenomena of senescence and aging. In *Senescence and Aging in Plants*, (Nooden, C.D., Leopold, A.C., eds.) Academic Press, San Diego. pp. 1-50.

Pennazio, S., and Roggero, P. 1989. Stimulation of ethylene production by exogenous spermidine in detached tobacco leaves in the light. *Biol. Plant.* **31**: 58-66.

Pennazio, S., and Rogero, P. 1990. Exogenous polyamines stimulate ethylene synthesis by soybean leaf tissues. *Ann. Bot.* **65**: 45-50.

Pjon, C.J., Kim, S.D., and Pak, J.Y. 1990. Effect of spermidine on chlorophyll content, photosynthetic activity and chloroplast ultrastructure in the dark and under light. *Bot. Mag. Tokyo* **103**: 43-48.

Popovic, R.B., Kyle, D.J., Cohen, A.S., and Zalik, S. 1979. Stabilization of thylakoid membranes by spermine during stress-induced senescence of barley leaf discs. *Plant Physiol.* **64**: 721-726.

Rastogi, R., and Davies, P.J. 1991. Polyamine metabolism in ripening tomato fruit II. Polyamine metabolism and synthesis in relation to enhanced putrescine content and storage life of *alc* tomato fruit. *Plant Physiol.* **15**: 41-45.

Rinaldi, A., Floris, G., and Giartosio, A 1985. Plant amine oxidases. In *Structure and Function of Amine Oxidases*. (Mondovi, B., ed.) CRC Press, Boca Raton, FL. pp. 51-62.

Roberts, D.R., Dumbroff, E.B., and Thompson, J.E. 1986. Exogenous polyamines alter membrane fluidity in bean leaves—a basis for potential ministerpretation of their true physiological role. *Planta* **167**: 395-401.

Saftner, R.A., and Baldi, B.G. 1990. Polyamine levels and tomato fruit development: possible interaction with ethylene. *Plant Physiol.* **92**: 547-550.

Serafini-Fracassini, D., and Mossett, U. 1985. Free and bound polyamines in different physiological stages of *Helianthus tuberosus* tubers. In *Recent Progress in Polyamine Research*. (Selmeci L, Brosnan, M.E., Seiler, N. eds.) Akademiai Kiado, Budapest. pp. 551-560.

Serafini-Fracassini, D., Torrigiani, P., and Branca, C. 1984. Polyamines bound to nucleic acids during dormancy and activation of tuber cells of *Helianthus tuberosus*. *Physiol. Plant.* **6**: 351-357.

Serafini-Fracassini, D., Del Duca, S., and Orzi, D. 1988. First evidence for polyamine conjugation mediated by an enzymic activity in plants. *Plant Physiol.* **87**: 757-761.

Serafini-Fracassini, D., Del Duca, S., and Torrigiani, F. 1989. Polyamine conjugation during the cell cycle of *Helianthus tuberosus:* non-enzymatic and transglutaminase-like binding activity. *Plant Physiol. Biochem.* **27**: 659-668.

Shih, L.M., Kaur-Sawhney, R., Fuhrer, J., Samanta, S., and Galston, A.W. 1982. Effects of exogenous 1,3-diaminopropane an spermidine on senescence of oat leaves I. Inhibition of portease activity, ethylene production, and chlorophyll loss as related to polyamine content. *Plant Physiol.* **70**: 1592-1596.

Slocum, R.D., and Galston, A.W., 1987. Inhibition of polyamine biosynthesis in plants and plant pathogenic fungi. In *Inhibition of Polyamine Metabolism*. (McCann, P.P., Pegg, A.E., and Sjoerdsma, A. eds.) Academic Press, San Diego. pp. 305-316.

Smith, M.A., and Davies, P.J. 1985. Effect of photoperiod on polyamine metabolism in apical buds of g-2 peas in relation to the induction of apical senescence. *Plant Physiol.* **79**: 400-405.

Smith, T.A. 1972. Purification and properties of the polyamine oxidase of barley plants. *Phytochemistry* **11**: 895-910.

Smith, T.A. 1977 Further properties of polyamine oxidase from oat seedlings. *Phytochemistry* **16**: 1647-1649.

Smith, T.A., 1981. Amines, In *The Biochemistry of Plants, Vol 7*. (Conn, E.E. ed.) Academic Press. NY. pp. 249-268.

Smith, T.A. 1984. Putrescine and inorganic ions. *Adv. Phytochemistry.* **18**: 7-54.

Smith, T.A. 1985. Di-and polyamine oxidases of higher plants. *Biochem. Soc. Trans.* **13**: 319-322.

Smith, T.A. 1990. Plant polyamines metabolism and function. In *Polyamines and Ethylene: Biochemistry, Physiology, and Interactions*. (Flores, H.E., Arteca, R.N., Shannon, F.C. eds.) American Society of Plant Physiologists, Rockville, MD. pp. 1-23.

Srivastava, S.K., Raj, A.D.S., and Naik, B.I. 1981. Polyamine metabolism during ageing and senescence of pea leaves. *Indian J. Exp. Biol.* **19**: 437-440.

Suttle, J.C. 1981. Effect of polyamines on ethylene production. *Phytochemistry* **20**: 1477-1480.

Suzuki, Y., and Yanagisawa, H. 1980. Purification and properties of maize polyamine oxidase: a flavoprotein. *Plant Cell Physiol.* **21**: 1985-1094.

Wang, C.Y., and Kramer, G.F. 1990. Effect of polyamine treatment on ethylene production of apples. In *Polyamines and Ethylene: Biochemistry, Physiology, and interactions.* (Flores, H.E., Arteca, R.N., Shannon, J.C., eds.) American Society of Plant Physiologists, Rockville, MD. pp. 411-413.

Williams-Ashman, H., and Schenone, H. 1972. Methylglyoxal bis(guanylhydrazone) as a potent inhibitor of S-adenosylmethionine decarboxylases. *Biochem. Biophys. Res. Commun.* **46**: 288-295.

Yanagisawa, H., Kato, A., Hoshiai, S., Kamiya, A., and Torii, N. 1987. Polyamine oxidases from water hyacinth. Purification and properties. *Plant Physiol.* **85**: 906-909.

Nitrogen Nutrition in Higher Plants, 1995
Editors : H.S. Srivastava & R.P. Singh
Associated Publishing Co., New Delhi, India
pp. 323-336.

Interactions Between Nitrogen and Carbon Assimilation in Green Cells

CATALINA LARA

Abbreviations: GS-GOGAT, Glutamine synthetase-glutamate synthase, NiR, nitrite reductase; NR, nitrate reductase.

I. Introduction

Nitrate is the more abundant form of combined inorganic nitrogen in soils and the primary nitrogen source for photosynthetic organisms, although ammonium can also be used as a nitrogen source and some plants, such as legumes, can benefit from symbiotic N_2 fixation.

Nitrate assimilation can be considered as the conversion of nitrate-N into the α-amino group of amino acids. This pathway includes at least three basic steps: (i) nitrate transport into the cell, mediated by specific carriers and requiring an energy supply, (ii) reduction of nitrate to ammonium which occurs in two sequential exergonic reactions, the two-electron reduction of nitrate to nitrite, catalyzed by NR and the six-electron reduction of nitrite to ammonium catalyzed by NiR and, (iii) incorporation of ammonium to carbon skeletons through the glutamine synthetase-glutamate synthase cycle, which consumes two more electrons, one energy-rich phosphate bond and one α-ketoglutarate molecule per glutamate produced. From glutamate, amino-N can be transferred to a variety of α-ketoacids by transamination reactions, forming the different amino acids. This basic scheme should be modified, depending on the organism under consideration, to include additional intermediate steps, such as storage in vacuoles and/or translocation from the roots to the leaves, which are the major site of nitrate assimilation in herbaceous plants.

In this chapter, the various and complex interactions between nitrogen and carbon assimilation in green cells and tissues are analyzed. Special emphasis is placed in the more general situation in which nitrate is the available nitrogen source in the environment. Evidence is discussed supporting the contention that nitrate assimilation is a genuine photosynthetic process. An attempt is made to update and summarize the present status of knowledge and to provide an overview of the mechanisms by which primary carbon and nitrogen assimilation are integrated and balanced in green cells.

In order to facilitate their analysis, two levels of interaction will be distinguished: (i) incidence of nitrate (and ammonium) assimilation on CO_2 fixation; and (ii) regulatory role of CO_2 fixation on nitrate assimilation. The reader must, however, keep in mind that these two types of interactions coexist and their effects are manifested jointly under any physiological conditions.

II. Photosynthetic Nature of Nitrate Assimilation

Nitrate assimilation is a major anabolic pathway in plant metabolism. It has been estimated that the plant kingdom assimilates about 10^{10} tons of nitrate-N per year, and that the proportion of electrons being used for nitrate assimilation is, on an average and net basis, about one fourth of that utilized for CO_2 fixation (Guerrero et al., 1981). Four electrons are required to assimilate CO_2 to the level of carbohydrate and 10 electrons are consumed in the assimilation of nitrate-N into amino-N. This implies that in particular cases, such as in microalgae with an C/N ratio of 5, or young plant leaves exhibiting C/N ratios of 7-8, as much as one half or one third, respectively, of the reducing equivalents used for carbon assimilation are used for nitrate assimilation.

Although all assimilatory power utilized by plant metabolism originates in photosynthesis, a matter of controversy has been the nature of the energetic connection between nitrate assimilation and the photochemical reactions of photosynthesis (see Beevers and Hageman, 1980; Losada et al., 1981; and references therein.)

Quantitative photochemical conversion of nitrate to ammonium has been demonstrated by measuring nitrate-dependent O_2 evolution in the absence of CO_2 in whole cells of green algae (Warburg and Negelein, 1920; Syrett and Morris, 1963; Thacker and Syrett, 1972; Syrett, 1981; Larsson et al., 1982) and cyanobacteria (Flores et al., 1983a), as well as in reconstituted systems of thylakoid membranes (Candau et al., 1976; Ortega et al., 1976; Losada and Guerrero, 1979) or isolated chloroplasts supplemented with NR (House and Anderson, 1980). In all cases 2 mol of O_2 were evolved per mol of NO_3^- reduced to NH_4^+, according to the equation:

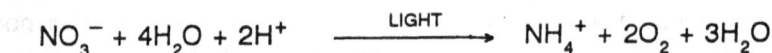

$$NO_3^- + 4H_2O + 2H^+ \xrightarrow{\text{LIGHT}} NH_4^+ + 2O_2 + 3H_2O$$

Evidence for the direct coupling of nitrate assimilation to water photolysis has also been obtained in microalgae under physiological conditions with simultaneous CO_2 and nitrate assimilation. In intact cells of the green algae *Chlorella* and *Scenedesmus* at saturating light and CO_2, nitrate enhances O_2 evolution while the rate of CO_2 fixation is practically unchanged. Consequently, the photosynthetic CO_2/O_2 quotient becomes lower than one when nitrate assimilation is operative (Cramer and Myers, 1949, Van Niel et al., 1953; Larsson et al., 1982). As CO_2 fixation is operating at full capacity under these light and CO_2 saturating conditions, the extra O_2 evolution induced by nitrate cannot be due to the refixation of CO_2 produced from carbohydrate oxidation but to photosynthetic nitrate assimilation, nitrate acting as an additional electron acceptor in photosynthesis (Van Niel et al., 1953). In the cyanobacterium *Anacystis nidulans* at saturating CO_2, nitrate increases the rate of O_2 evolution in a wide range of light intensities, the stimulation being maximal at saturating light (Romero and Lara, 1987). This indicates that when CO_2 is the only electron acceptor available, the rate of non-cyclic electron flow is not limited by light but by the rate of CO_2 fixation through the reductive pentose phosphate cycle. Addition of nitrate as a second electron acceptor releases non-cyclic electron flow from this limitation, allowing higher rates of electron transport from water to ferredoxin. Nitrate, therefore, increases the maximum capacity of the photosynthetic apparatus (Romero and Lara, 1987), although the apparent quantum yield of O_2 evolution remains unchanged (Romero et al., 1989). Interestingly, the stoichiometry between the extra O_2 evolved and the NO_3^- taken up under these conditions in *Anacystis* is very low at low light intensity, but it increases with photon flux becoming close to the theoretical value of 2.5 mol of O_2 per mol of NO_3^- at saturating light. This indicates that the stimulation of non-cyclic electron flow induced by nitrate at saturating photon flux density suffices to support the reductant requirements of nitrate assimilation, while at low light it does not. Accordingly, nitrate depresses CO_2 fixation at limiting but not at saturating light, this depression reflecting the competition between both processes for photosynthetically generated assimilatory power (Romero and Lara, 1987; Lara and Guerrero, 1989).

Another series of studies performed mainly in plant leaves have shown that high endogenous carbohydrate levels, as well as exogenously supplied sucrose, increase nitrate reduction in the light (Klepper et al., 1971; Beevers and Hageman, 1980; Pace et al., 1990), apparently favouring the alternative view that CO_2 fixation and further oxidation of the resulting carbohydrates is an obligate energetic link between the photochemical reactions of photosynthesis and nitrate assimilation. Recent reports have shown, however, that the photosynthetic CO_2/O_2 quotient in barley shoots is lower in nitrate-grown than in ammonium-grown plants or in NR-deficient mutants, O_2 evolution in nitrate-grown shoots exceeding CO_2 consumption by about a 25% (Bloom et al., 1989). Moreover, illuminated intact leaves from barley and pea evolve O_2 in a CO_2 saturating atmosphere at a higher rate when nitrate had previously been supplied through the transpiration stream. No stimulation of O_2 evolution occurs otherwise after feeding ammonium. The nitrate-induced extra

O_2 evolution is maximal at saturating photon flux densities and results in no changes in the apparent quantum yield of O_2 evolution (De la Torre et al., 1991). These results clearly indicate that also in plant leaves, nitrate assimilation uses photosynthetically generated reductant and stimulates the rate of non-cyclic electron flow, by acting as a second electron-accepting process in addition to CO_2 fixation. This is by no means in contradiction with the situation in roots and other non-photosynthetic tissues in which necessarily nitrate assimilation is energetically dependent on the oxidation of carbohydrates (Jackson et al. 1980; Beevers and Hageman 1980; Oaks and Hirel, 1985).

In summary, available evidence indicate that in microalgae and plant leaves nitrate assimilation is a direct photosynthetic process that increases the capacity of the photosynthetic apparatus for noncyclic electron flow, overcoming the limitation imposed by CO_2 fixation through the reductive pentose phosphate cycle. This does not exclude that, if NR is in fact located in the cytoplasm, export of reducing power from the chloroplast via the malate-oxoloacetate shuttle or the phosphate translocator (Heineke et al., 1991) should be involved in nitrate reduction.

III. Incidence of Nitrate and Ammonium Assimilation on CO_2 Fixation

Nitrogen assimilation affects both the rate of CO_2 fixation and the distribution of newly fixed carbon between metabolite fractions. These effects manifest depending on the experimental conditions and the N-nutrition of the cells (Turpin, 1991; Lara, 1992). Three types of metabolic events may be distinguished:

A. Competition for Assimilatory Power

The photosynthetic nature of nitrate assimilation brings about a first point of interaction, i.e., the competition for reducing equivalents between CO_2 and nitrate assimilation at limiting light intensity. In this respect, nitrate assimilation depresses CO_2 fixation at low, but not at high, light intensities in either N-sufficient spinach protoplasts (Rathnam, 1978), or *Scenedesmus* (Larsson et al., 1985) and *Anacystis* (Romero and Lara, 1987) cells, supplied with saturating CO_2 concentrations. As discussed for *Anacystis* in the previous section, no competition for assimilatory power should exist at saturating light, the photosynthetic apparatus being able to generate enough reductant to sustain the simultaneous assimilation of CO_2 and nitrate (Lara and Guerrero, 1989).

When the available nitrogen source is ammonium there is no depression of CO_2 fixation at limiting light intensity, indicating that no competition for assimilatory power exists when a reduced form of inorganic nitrogen is assimilated. Interestingly, a positive effect of ammonium on the rate of CO_2 fixation has been observed in blue-green (Lawrie et al., 1976; Romero and Lara, 1987; Coronil et al., 1993) and green (Kanazawa et al., 1972) algae, and in isolated higher plant cells (Rehfield and Jensen, 1973; Paul et al., 1978; Woo and Canvin, 1980a,b). Several mechanisms have been proposed to explain this phenomenon. Most notable is the proposal of activation of phosphoenol pyruvate carboxylase and pyruvate kinase by ammonium ions (Bassham et al., 1981; Kanazawa et al., 1983). In spinach (Woo and Canvin, 1980b) and *Anacystis* (Romero and Lara, 1987; Coronil et al., 1993) cells, however, the positive effect of ammonium on CO_2 fixation is abolished when ammonium assimilation is specifically inhibited, indicating that the stimulation is not produced by the cation (NH_4^+) itself but it is a consequence of its incorporation to carbon skeletons. The stimulation by ammonium of CO_2 fixation is specially evident under conditions in which carbon assimilation is not limited by light or CO_2 availability (Woo and Canvin, 1980a; Romero and Lara, 1987; Coronil et al., 1993).

It would appear, therefore, that when nitrate is the nitrogen source assimilated by the cells, the actual effect on the rate of CO_2 fixation should result from the balance between the negative effect of nitrate reduction (reflecting competition for assimilatory power) and the positive effect of ammonium incorporation to carbon skeletons (which will be discussed below), the preponderance of each effect being dependent on the light intensity. This is consistent with the observation that when ammonium assimilation is inhibited, the nitrate-induced depression of CO_2 fixation at limiting light is potentiated (Romero and Lara, 1987).

B. Diversion of Carbon Flow from Carbohydrate to Amino Acid Biosynthesis

Nitrogen assimilation produces changes in the distribution of newly fixed carbon, diverting carbon flow to organic and amino acid biosynthesis, at the expense of reserve carbohydrate biosynthesis (Kanazawa et al., 1972; Paul et al., 1978; Woo and Canvin, 1980a,b Romero and Lara, 1987; Lara et al., 1987b). In the cyanobacterium *Anacystis nidulans*, under light and CO_2 saturating conditions, the percentage of recently fixed carbon incorporated into acid soluble metabolites increases 2-fold in the presence of nitrate and 3-fold in that of ammonium. Within this fraction, the most dramatic changes are observed in the percentage of C incorporated into amino acids, which increases about 6-fold under concomitant nitrate assimilation and about 11-fold under ammonium assimilation. In contrast, the pool size of total amino acids increases less than 2-fold, indicating that it is the turnover of amino acids what is dramatically accelerated. This is particularly evident for the cases of glutamate and aspartate, the main α-amino donors in transamination reaction (Coronil et al., 1993). Thus, the operativity of the GS-GOGAT cycle and transamination reactions under conditions of N assimilation would accelerate the carbon flow through the glycolytic pathway and the tricarboxylic acid pathway, which in the cyanobacteria lacks both 2-oxoglutarate dehydrogenase and succinyl-CoA synthetase and function with biosynthetic purposes to yield 2-oxoglutarate on one end, and succinate, on the other end. This acceleration of carbon flow would entail most probably, stimulation of phosphoenolpyruvate carboxylation and citrate synthesis (Coronil et al., 1993). This diversion of carbon flow in N-sufficient cells has not negative effects on the operativity of the Calvin cycle and, in fact, it is produced at the expense of carbon flow towards reserve carbohydrate biosynthesis. Under conditions of nitrate, and specially ammonium, assimilation, the flux of newly fixed carbon to glycogen biosynthesis in *A. nidulans* decreases and even, at low light intensity, ammonium stimulates glycogen degradation, increasing the carbon skeletons available for amino acid biosynthesis (Garcia-Gonzalez et al., 1992).

C. Competition for Carbon Skeletons

Whereas the above described stituation seems to apply in general to N-sufficient cells, substantial differences are, however, apparent in N-starved or N-limited cells and tissues.

Exposure of N-limited eukaryotic microalgae to a N-source results in a transient situation in which nitrate or ammonium are assimilated at rates 2 to 4 fold higher than in N-sufficient cells and, simultaneously, CO_2 fixation is severely depressed to about one third of its original value (Cresswell and Syrett, 1981; Hipkin et al., 1983, Elrifi and Turpin, 1986). In this situation, studies carried out on N-limited *Selenastrum minutum* cells have shown a dramatic drop in the levels of ribulose bisphosphate upon N-supply and a very drastic diversion of carbon flow to amino acid biosynthesis. This phenomenon, which has been called N-induced photosynthetic suppression, is currently explained as a result of competition for carbon skeletons between the Calvin cycle and nitrogen assimilation (Elrifi and Turpin, 1986; 1987; Elrifi et al., 1988; Turpin, 1991). Under conditions of very rapid nitrate or ammonium assimilation, the GS-GOGAT pathway and the transmination reactions would create a sink for carbon skeletons, much stronger than under N-sufficient conditions, depleting the Calvin cycle intermediates and restricting CO_2 fixation through the availability of CO_2 acceptors. If the supplied N-source is nitrate, competition for photogenerated reductant may also contribute to restrict CO_2 fixation (Thomas et al., 1976). In addition to this, it has also been proposed that some key enzymes, like sedoheptulose bisphosphatase and phosphoribulokinase, would be inactivated in the short-term after exposure to N (Smith et al., 1989) while phosphoenolpyruvate carboxylase activity would be stimulated (Guy et al., 1989; Vanlerberghe et al., 1990).

Work performed with N-depleted wheat, barley and spinach leaves has shown that the supply of nitrate or of ammonium to the leaf produces, in a time scale of 2-3 hours, a depression of net sucrose synthesis, which correlated with an increase in the rate of nitrogen uptake and assimilation in amino acid (Van Quy et al., 1991a; Champigny et al., 1992). Being N-depleted leaves, some inhibition of CO_2 fixation was also evident in these studies. Interestingly, nitrogen supply increased about two-fold the levels of fructose-2,6-bisphosphate, a potent inhibitor of fructose-1, 6-bisphosphatase, the enzyme catalyzing the first committed reaction in sucrose biosynthesis, and led to a decrease in extractable sucrose-phosphate synthase activity. Also, the increased carbon flux to amino acids was associated with a decrease in the level of phosphoenolpyruvate, and an increase in extractable phosphoenolpyruvate carboxylase activity (Van Quy et al., 1991b, Champigny et al., 1992). On the basis that both sucrose-phosphate synthase and phosphoenolpyruvate carboxylase are modulated by covalent modification by a phosphorylation-dephosphorylation mechanism, it has been suggested that nitrate, or a product of its assimilation, would act as a signal metabolite activating cytosolic protein kinase activity. The activated kinase(s) would phosphorylate sucrose-phosphate synthase, turning it into its low activity form, and phosphoenolpyruvate carboxylase, turning it into the high activity form (Van Quy and Champigny, 1992; Champigny and Foyer, 1992). This proposal, which would account for the carbon partition observed under nitrate or ammonium assimilation conditions, is an attractive possibility which deserves experimental confirmation.

In summary, these three main metabolic effects of nitrogen assimilation on CO_2 fixation may be operative depending on the conditions. The diversion of carbon flow to amino acids elicited by ammonium (or nitrate) assimilation should always occur, facilitating CO_2 fixation. When the light intensity is low, nitrate (but not ammonium) assimilation depresses CO_2 fixation as a result of competition for assimilatory power. Finally, when the rates of nitrate (or ammonium) assimilation are higher than normal, as it occurs after N-starvation, competition for carbon skeletons may result in severe depression of CO_2 fixation.

IV. Regulatory Role of CO_2 Fixation on Nitrate Assimilation

A general phenomenon observed in microalgae and higher plants is that nitrate cannot be assimilated unless CO_2 fixation is operative or a carbohydrate source is available. It is known that nitrate assimilation in leaves is restricted at subambient CO_2 levels (Aslam et al., 1979; Kaiser and Foster, 1989; Pace et al. 1990). Also, high endogenous carbohydrate levels, as well as exogenously supplied sucrose increase nitrate reduction in the light (Klepper et al., 1971; Beevers and Hageman, 1980; Pace et al., 1990). Nitrate assimilation in the unicellular cyanobacterium *Anacystis nidulans* also exhibits an strict requirement for CO_2 fixation, and, in fact, a correlation has been shown to exist between the rate of nitrate uptake and that of carbon assimilation (Flores et al., 1983b; Romero et al., 1985). These features had been formerly considered as a proof of the energy dependence of nitrate assimilation upon carbohydrate oxidation. As discussed above however, evidence is now available indicating that nitrate assimilation in green cells uses electrons directly derived from the photosynthetic apparatus. We will discuss now results showing that the carbon dependence exhibited by nitrate assimilation is regulatory in nature and determines the operativity of rate-limiting steps of the process and, hence, its overall control. Thus, in cyanobacteria and other microalgae, such as diatoms, active CO_2 fixation determines nitrate transport activity (Lara et al., 1987a; Syrett, 1988), whereas in green algae and plant leaves, CO_2 fixation determines the control of NR activity (Pistorius et al., 1976; Kaiser and Foster, 1989; Kaiser and Brendle-Behnish, 1991; De Cires et al., 1993).

A. Control of Nitrate Transport Activity

Although it is evident that nitrogen assimilation requires carbon compounds as amino acceptors, the dependence of nitrate assimilation upon the provision of fixed carbon in *Anacystis nidulans* is much more stringent than that exhibited by ammonium assimilation (Lara and Romero, 1986). This strict carbon dependence of nitrate assimilation seems to be related to its feed-back inhibition by some organic nitrogen products(s), which manifests specially after ammonium addition. Studies with different inhibitors have shown that prevention of ammonium assimilation protects nitrate uptake from the negative effects of ammonium and eliminate the requirement for CO_2 fixation (Flores et al., 1983b, Romero et al., 1985). A common link appears, therefore, to underlay the carbon requirement and the ammonium-dependent inhibition of nitrate assimilation, and, in fact, there is evidence indicating that either carbon reserves or high CO_2 fixation rates counteract the negative effects of ammonium assimilation (Flores et al., 1983b; Romero et al., 1987).

The target of this regulatory mechanism in *A. nidulans* is nitrate transport across the plasma membrane. The nitrate transport activity was shown to be sensitive to a rapid and reversible modulation exerted by products of both ammonium and CO_2 assimilation, thus providing evidence that photosynthetic nitrate assimilation in cyanobacteria is regulated at the level of substrate supply to the cell (Lara et al., 1987a; Guerrero et al., 1990). This provides the basis to understand the carbon dependence of nitrate assimilation in these organisms.

Similar situation appears to be operative in some eukaryotic microalgae, such as the diatom *Phaeodactylum tricornutum*, in which intracellular nitrate accumulation is inhibited by ammonium addition or CO_2 deprivation (Cresswell and Syrett, 1979; Syrett 1988), and the green alga *Ankistrodesmus braunii,* in which nitrate uptake, but not nitrate reduction, is hampered in the absence of carbon assimilation (Ullrich and Eisele, 1978).

B. Control of Nitrate Reductase Activity

Early work with green algae provided evidence that light, nitrate supply, CO_2 and O_2 were involved in the control of NR activity. Thus, in *Chlamydomonas* and *Chlorella* cells, NR is reversibly inactivated upon exposure of the cells to a CO_2-free atomosphere (Losada, 1973; Pistorius et al., 1976). A redox interconversion mechanism has been established for the algal enzyme with an oxidized-active form, and a reduced-inactive form able to bind cyanide and stabilize the inactive state (Solomonson, 1974; Pistorius et al., 1976; Guerrero et al., 1981; Solomonson and Barber, 1990).

Very recently, it has been shown that NR from spinach and barley leaves is reversibly modulated, being active in the light and inactive in the dark (Kaiser et al., 1992; Riens and Heldt, 1992; Huber et al., 1992; De Cires et al., 1993). Spinach NR is also reversibly inactivated upon exposure of the leaves to CO_2-free air (Kaiser and

Foster, 1989; Kaiser and Brendle-Behnisch, 1991). Inhibition of extractable NR activity by CO_2 removal led to a significant decrease in V_{max}, not affecting otherwise the K_m values for nitrate and NADH (Kaiser and Brendle-Behnisch, 1991). These changes in NR activity appear to be the result of covalent modification of the enzyme protein by a phosphorylation/dephosphorylation mechanism (Kaiser and Spill, 1991; Kaiser et al.,1992; Huber et al., 1992; Mackintosh, 1992).

The nature of the signal(s) mediating the light-activation of leaf NR and its CO_2 dependence is still an open question. On the basis of opposing effects of ATP and AMP on *in vitro* NR modulation and of an inverse correlation between the ATP/AMP ratio and the extractable NR activity from spinach leaves, Kaiser and coworkers have suggested that darkness and CO_2 deprivation would modify the cytosolic ATP and AMP levels, which would, in turn, determine the activation state of NR (Kaiser and Spill, 1991; Kaiser et al., 1992). Riens and Heldt (1992) have questioned that a light-dark transition could bring about such a dramatic increase in the cytosolic ATP/AMP ratio as required for *in vitro* NR inactivation, and have proposed that another signal of unknown nature may coordinate the activation state of NR with photosynthesis. On the basis of a hyperbolic correlation between extractable NR activity from barley leaves and *in vivo* CO_2 fixation rate as well as of the protective effect against dark-inactivation exerted by hexose and hexose-phosphate feeding, we proposed that CO_2 fixation products are regulatory factors of NR activity in barley leaves, mediating both the light-dark modulation of NR and its CO_2 dependence (De Cires et al., 1993). In the dark, low levels of CO_2 fixation products would determine NR inactivation. Under illumination, increased levels of CO_2 fixation products would stimulate the activation system acting on NR. This response would be mimicked by feeding carbohydrates via the transpiration stream, resulting in protection against the dark-inactivation of NR (De Cires et al., 1993). Interestingly, it has been reported that sucrose mimics the light-induction of NR gene transcription in *Arabidopsis*, suggesting that carbon assimilation products are also involved in the light response of NR synthesis (Cheng et al., 1992).

It is interesting to mention that, in contrast to the situation described in green algae (Pistorius et al., 1976), neither nitrate nor ammonium supply have any effect on the activation state of NR in leaves (Huber et al., 1992; De Cires et al., 1993). Although it has been proposed that the most important factor in the regulation of NR activity is the flux of nitrate into the leaves (Shaner and Boyer, 1976; Schrader and Thomas, 1981; Barneix et al., 1984, Gojon et al., 1991), nitrate seems to play a different role in this respect. It is clearly required for expression of NR, increasing NR activity by enhancing NR synthesis (Remmler and Campbell, 1986; Oaks et al., 1988; Melzer et al., 1989, Solomonson and Barber, 1990, Gowri et al., 1992). Also, increased substrate availability would produce higher rates of *in vivo* nitrate reduction, observed as nitrate-dependent O_2 evolution (De la Torre et al., 1991), without affecting the activation state of NR.

In summary, this evidence strongly support the contention that nitrate assimilation in green cells is closely coupled to and regulated by CO_2 fixation under physiological conditions. Thus, by acting at the level of nitrate transport to the cell-determining substrate availability, in the cyanobacteria, or by altering the activation state of NR in plant leaves, CO_2 fixation would control nitrate reduction, balancing the assimilation of the two primordial bioelements, carbon and nitrogen.

Acknowledgements

Support from spanish DGICYT (grant PB 91-0611) and PAI (grant no. 3101) is gratefully acknowledged. Thanks are also due to P. Perez de Leon for skilful secretarial assistance.

Literature Cited

Aslam, M., Huffaker, R.C., Rains, D.W., and Rao, K.P. 1979. Influence of ambient carbon dioxide concentration on nitrate assimilation by intact barley seedlings. *Plant Physiol.* **63**: 1205-1209.

Barneix, A.J., James, D.M., Watson, E.F., and Hewitt, E.J. 1984. Some effects of nitrate abundance and starvation on metabolism and accumulation of nitrogen in barley *(Hordeum vulgare* cv. Sonja). *Planta* **162**: 469-476.

Bassham, J.A., Larsen, P.O., Lawyer, A.L., and Cornwell, K.L. 1981. Relationship between nitrogen metabolism and photosynthesis. In *Nitrogen and Carbon Metabolism*, (Bewley, J.D. ed.) Nijhoff/Junk, The Hague. pp. 135-163.

Beevers, L., and Hageman, R.H. 1980. Nitrate and nitrite reduction. In *The Biochemistry of Plants*, Vol. 5. (Miflin, B.J. ed.) Academic Press, New York. pp. 115-168.

Bloom, A.J., Caldwell, R.M., Finazzo, J., Warner,R.L., and Weisbart, J. 1989. Oxygen and carbon dioxide fluxes from barley shoots depend on nitrate assimilation. *Plant Physiol.* **91**: 352-356.

Candau, P., Manzano, C., and Losada, M. 1976. Bioconversion of light energy into chemical energy through reduction with water of nitrate to ammonia. *Nature* **262**: 715-717.

Champigny, M.-L., and Foyer, C. 1992. Nitrate activation of cytosolic protein kinase diverts photosynthetic carbon from sucrose to amino acid biosynthesis. *Plant Physiol.* **100**: 7-12.

Champigny, M.L., Brauer, M., Bismuth, E., Thi Manh, C., Siegl, G., Van Quy, L., and Stitt, M. 1992. The short-term effect of NO_3^- and NH_3 assimilation on sucrose synthesis in leaves. *J. Plant Physiol.* **139**: 361-368.

Cheng, C.L., Acedo, G.N., Cristinsin, M., and Conkling, M.A. 1992. Sucrose mimicks the light induction of *Arabidopsis* nitrate reductase gene transcription. *Proc. Natl. Acad. Sci. USA*: 1861-1864.

Coronil, T., Lara, C., and Guerrero, M.G. 1993. Shift in carbon flow and stimulation of amino-acid turnover induced by nitrate and ammonium assimilation in *Anacystis nidulans*. *Planta* **189**: 461-467.

Cramer, M., and Myers, J. 1949. Nitrate reduction and assimilation in *Chlorella*. *J. Gen. Physiol.* **32**: 93-102.

Cresswell, R.C., and Syrett, P.J. 1979. Ammonium inhibition of nitrate uptake by the diatom *Phaeodactylum tricornutum*. *Plant Sci. Lett.* **14**: 321-325.

Cresswell, R.C., and Syrett, P.J. 1981. Uptake of nitrate by the diatom *Phaeodactylum tricornutum*. *J. Exp. Bot.* **32**: 19-26.

De Cires, A., De La Torre, A., Delgado, B., and Lara, C. 1993. Role of light and CO_2 fixation in the control of nitrate reductase activity in barley leaves. *Planta* **190**: 277-283.

De la Torre, A., Delgado, B., and Lara, C. 1991. Nitrate-dependent O_2 evolution in intact leaves. *Plant Physiol.* **96**: 898-901.

Elrifi, I.R., and Turpin, D.H. 1986. Nitrate and ammonium induced photosynthetic suppression in N-limited *Selenastrum minitum*. *Plant Physiol.* **81**: 273-279.

Elrifi, I.R., and Turpin, D.H. 1987. The path of carbon flow during NO_3^- induced photosynthetic suppression in N-limited *Selenastrum minutum*. *Plant Physiol.* **83**: 97-107.

Elrifi, I.R., Holmes, J.N., Weger, H.G., Mayo, W.P., and Turpin, D.H. 1988. RuBP limitation of photosynthetic carbon fixation during NH_3 assimilation: interactions between photosynthesis, respiration and ammonium assimilation in N-limited green algae. *Plant Physiol.* **87**: 395-401.

Flores, E., Guerrero, M.G., and Losada, M. 1983a. Photosynthetic nature of nitrate uptake and reduction in the cyanobacterium *Anacystis nidulans*. *Biochim. Biophys. Acta* **722**: 408-416.

Flores, E., Romero, J.M., Guerrero, M.G., and Losada, M. 1983b. Regulatory interaction of photosynthetic nitrate utilization and carbon dioxide fixation in the cyanobacterium *Anacystis nidulans*. *Biochim. Biophys. Acta* **725**: 529-532.

Garcia-Gonzalez, M., Sivak, M.N., Guerrero, M.G., Preiss, J., and Lara, C. 1992. Depression of carbon flow to the glycogen pool induced by nitrogen assimilation in intact cells of *Anacystis nidulans*. *Physiol. Plant* **86**: 360-364.

Gojon, A., Wakrim, R., Passama, L., and Robin, P. 1991. Regulation of NO_3^- assimilation by anion availability in excised soybean leaves. *Plant Physiol.* **96**: 398-405.

Gowri, G., Kenis, J.D., Ingemarsson, B., Redinbaugh, M.G., and Campbell, W.H. 1992. Nitrate reductase transcript is expressed in the primary response of maize to environmental nitrate. *Plant Mol. Biol.* **18**: 55-64.

Guerrero, M.G., Vega, J.M., and Losada, M. 1981. The assimilatory nitrate-reducing system and its regulation. *Annu. Rev. Plant Physiol.* **32**: 169-204.

Guerrero, M.G., Romero, J.M., Rodriguez, R., and Lara, C. 1990. Nitrate transport in cyanobacteria. In *Inorganic Nitrogen in Plants and Microorganisms*, (Ullrich, W.R., Rigano, C., Fuggi, A., and Aparicio, P.J. eds.), Springer, Berlin. pp. 79-85.

Guy, R.D., Vanlerberghe, G.C., and Turpin, D.H. 1989. Significance of phosphoenolpyruvate carboxylase during ammonium assimilation. Carbon isotope discrimination in photosynthesis and respiration by the N-limited green alga *Selenastrum minutum*. *Plant Physiol.* **89**: 1150-1157.

Heineke, D., Riens, BB., Grosse, H., Hoferichter, P., Peter, U., Flugge, U.I., and Heldt, H.W. 1991. Redox transfer across the inner chloroplast envelope membrane. *Plant Physiol.* **95**: 1131-1137.

Hipkin, C.R., Thomas, R.J., and Syrett, P.J. 1983. Effects of nitrogen deficiency on nitrate reductase, nitrate assimilation and photosynthesis in unicellular marine algae. *Mar. Biol.* **77**: 101-105.

House, C.H., and Anderson, J.M., 1980. Light-dependent reduction of nitrate by pea chloroplasts in the presence of nitrate reductase and C_4- dicarboxylic acids. *Phytochemistry* **19**: 1925-1930.

Huber, J.L., Huber, S.C., Campbell, W.M., and Redinbaugh, M.G. 1992. Reversible light-dark modulation of spinach leaf nitrate reductase activity involves protein phosphorylation. *Arch. Biochem. Biophys.* **296**: 58-65.

Jackson, W.A., Volk, R.J., and Israel, D.W. 1980. Energy supply and nitrate assimilation in root systems. In *Carbon-Nitrogen Interactions in Crop Production*, (Tanaka, A. ed.). The Japan Society for Promotion of Science, Tokyo. pp. 25-40.

Kaiser, W.M., and Foster, J. 1989. Low CO_2 prevents nitrate reduction in leaves. *Plant Physiol* **91**: 970-974.

Kaiser, W.M., and Brendle-Behnish, E. 1991. Rapid modulation of spinach leaf nitrate reductase activity by photosynthesis. I. Modulation *in vivo* by CO_2 availability. *Plant. Physiol.* **96**: 363-367.

Kaiser, W.M., and Spill, D. 1991. Rapid modulation of spinach nitrate reductase by photosynthesis. II. *In vitro* modulation by ATP and AMP. *Plant Physiol* **96**: 368-375.

Kaiser, W.M., Spill, D., and Brendle-Behnish, E. 1992. Adenine nucleotides are apparently involved in the light-dark modulation of spinach-leaf nitrate reductase. *Planta* **186**: 236-240.

Kanazawa, T.K., Kanazawa, M.R., Kirk, M.R., and Bassham, J.A., 1972. Regulatory effects of ammonia on carbon metabolism in *Chlorrella pyrenoidosa* during photosynthesis and respiration. *Biochim. Biophys. Acta.* **265**: 656-669.

Kanazawa, T., Distefano, M., and Bassham, J.A. 1983. Ammonia regulation of intermediary metabolism in photosynthesizing and respiring *Chlorella pyrenoidosa*: Comparative effects of methylamine. *Plant Cell Physiol.* **24**: 979-986.

Klepper, L., Flesher, D., and Hageman, R.H. 1971. Generation of reduced NAD for nitrate reduction in green leaves. *Plant Physiol.* **45**: 580-590.

Lara, C. 1992. Photosynthetic nitrate assimilation: interactions with CO_2 fixation. In *Trends in Photosynthesis Research* (Barber, J., Guerrero, M.G., and Medrano, H. eds.) Intercept, Andover. pp. 195-208.

Lara, C., and Guerrero, M.G. 1989. The photosynthetic assimilation of nitrate and its interactions with CO_2 fixation. In *Techniques and New Developments in Photosynthesis Research*, (Barber, J., and Malkin, R. eds.) Plenum, New York. pp. 393-341.

Lara, C., and Romero, J.M. 1986. Distinctive light and CO_2-fixation requirements of nitrate and ammonium utilization by the cyanobacterium *Anacystis nidulans. Plant Physiol.* **81**: 686-688.

Lara, C., Romero, J.M., and Guerrero, M.G. 1987a. Regulated nitrate transport in the cyanobacterium *Anacystis nidulans J. Bacteriol.* **169**: 4376-4378.

Lara, C., Romero, J.M., Coronil. T., and Guerrero, M.G. 1987b. Interactions between photosynthetic nitrate assimilation and CO_2 fixation in cyanobacteria. In *Inorganic Nitrogen Metabolism*, (Ullrich, W.R., Aparicio, P.J., Syrett, P.J., and Castillo, F. eds.), Springer, Berlin. pp. 45-52.

Larsson, M., Ingermarssor, B., and Larsson, C.-M. 1982. Photosynthetic energy supply for NO_3^- assimilation in *Scenedesmus.*. Physiol. Plant. **55**: 301-308.

Larsson, M., Olsson, T., and Larsson, C.-M. 1985. Distribution of reducing power between photosynthetic carbon and nitrogen assimilation in *Scenedesmus. Planta* **1164**: 246-253.

Lawrie, A.C., Codd, G.A., and Stewart, W.D.P. 1976. The incorporation of nitrogen into products of recent photosynthesis in *Anabaena cylindrica Lemm. Arch. Microbiol.* **107**: 155-24.

Losada, M. 1973. Interconversion of nitrate and nitrite reductase of the assimilatory type. In *Metabolic Interconversion of Enzymes,* (Fisher, E.H., Krebs, E.G., Neurath, H., and Stadtman, E.R. eds.) Springer-Verlag, Berlin. pp. 257-270.

Losada, M., and Guerrero, M.G. 1979. The photosynthetic reduction of nitrate and its regulation. In *Photosynthesis in Relation to Model, System* (Barber, J. ed.) Elsevier-North Holland, Amsterdam. pp. 365-408.

Losada, M., Guerrero, M.G., and Vega, J.M. 1981. The assimilatory reduction of nitrate. In *Biology of Inorganic Nitrogen and Sulfur,* (Bothe, H., and Trebst, A. eds.) Springer, Berlin. pp. 30-63.

Mackintosh, C. 1992. Regulation of spinach leaf nitrate reductase by reversible phosphorylation. *Biochim. Biophys. Acta* **1137**: 21-126.

Melzer, J.M., Kleinhofs, A., and Warner, R.L. 1989. Nitrate reductase regulation: effects of nitrate and light on nitrate reductase mRNA accumulation. *Mol. Gen. Genet.* **217**: 341-346.

Oaks, A., and Hirel, B. 1985. Nitrogen metabolism in roots. *Ann. Rev. Plant Physiol.* **36**: 345-365.

Oaks, A., Poulle, M., Goodfellow, V.J., Class, A., and Deising, H. 1988. The role of nitrate and ammonium ions and light on the induction of nitrate reductase in maize leaves. *Plant Physiol.* **88**: 1067-1072.

Ortega, T., Castillo, F., and Cardenas, J. 1976. Photolysis of water coupled to nitrate reduction by *Nostoc muscorum* subcellular particles. *Biochem. Biophys. Res. Commun.* **71**: 885-891.

Pace, G.M., Volk, R.J., and Jackson, W.A. 1990. Nitrate reduction in response to CO_2-limited photosynthesis. Relationship to carbohydrate supply and nitrate reductase activity in maize seedlings. *Plant Physiol.* **92**: 286-292.

Paul, J.S., Cornwell, K.L., and Bassham, J.A. 1978. Effects of ammonia on carbon metabolism in photosynthesizing isolated cells from *Papaver somniferum* L. *Planta* **142**: 49-54.

Pistorius, E.K., Gewitz, H.S., Voss, H., and Vennesland, B. 1976. Reversible inactivation of nitrate reductase in *Chlorella vulgaris* in vivo. *Planta* **128**: 73-80.

Rathman, C.K.M. 1978. Malate and dihydroxyacetone phosphate-dependent nitrate reduction in spinach leaf protoplasts. *Plant Physiol.* **62**: 220-223.

Rehfield, D.W., and Jenseen, R.G. 1973. Metabolism of separated leaf cells. III. Effect of calcium and ammonium on product distribution during photosynthesis with cotton cells. *Plant Physiol.* **52**: 17-22.

Remmler, J.L., and Campbell, W.H. 1986. Regulation of corn leaf nitrate reductase. II. Synthesis and turnover of the enzyme's activity and protein. *Plant Physiol.* **80**: 442-447.

Riens, B., and Heldt, H.W. 1992. Decrease in nitrate reductase activity in spinach leaves during a light-dark transition. *Plant Physiol.* **98**: 573-577.

Romero, J.M., and Lara, C. 1987. Photosynthetic assimilation of NO_3^- by intact cells of the cyanobacterium *Anacystis nidulans*. Influence of NO_3^- and NH_4^+ assimilation on CO_2 fixation. *Plant Physiol.* **83**: 208-212.

Romero, J.M., Lara, C., and Guerrero, M.G. 1985. Dependence of nitrate utilization upon active CO_2 fixation in *Anacystis nidulans*: A regulatory aspect of the interaction between photosynthetic carbon and nitrogen metabolism. *Arch. Biochem. Biophys.* **237**: 396-401.

Romero, J.M., Coronil, T., Lara, C., and Guerrero, M.G. 1987. Modulation of nitrate uptake in *Anacystis nidulans* by the balance between ammonium assimilation and CO_2 fixation. *Arch. Biochim. Biophys.* **256**: 578-584.

Romero, J.M., Lara, C., and Sivak, M.N. 1989. Changes in net O_2 exchange induced by inorganic nitrogen in the blue-green alga *Anacystis nidulans*. *Plant Physiol.* **91**: 28-30.

Schrader, L.E., and Thomas, R.J. 1981. Nitrte uptake, reduction and transport in the whole plant. In *Nitrogen and Carbon Metabolism,* (Bewley, J.D. ed.) Nijhoff/Junk, The Hague. pp. 49-93.

Shaner, D.L., and Boyer,J.S. 1976. Nitrate reductase activity in maize (*Zea mays* L.) leaves. I. Regulation by nitrate flux. *Plant Physiol.* **58**: 499-504.

Smith, R.G., Vanlerberghe, G.C., Stitt, M., and Turpin, D.H. 1989. Short-term metabolite changes during transient ammonium assimilation by the N-limited green alga *Selenastrum minutum*. *Plant Physiol.* **91**: 749-755.

Solomonson, L.P. 1974. Regulation of nitrate reductase activity by NADH and cyanide. *Biochim. Biophys. Acta* **334**: 2967-308.

Solomonson, L.P., and Barber, M.J. 1990. Assimilatory nitrate reductase: functional properties and regulation. *Annu. Rev. Plant Physiol. Plant Mol. Biol.* **41**: 225-253.

Syrett, P.J. 1981. Nitrogen metabolism in microalgae. *Can. Bull. Fish. Aquat. Sci.* **210**: 182-210.

Syrett, P.J. 1988. Uptake and utilization of nitrogen compounds. In *Biochemistry of the Algae and Cyanobacteria,* (Rogers, L.J., and Gallon, J.R. eds.) Clarendon Press, Oxford. pp. 23-29.

Syrett, P.J., and Morris, I. 1963. The inhibition of nitrate assimilation by ammonium in *Chlorella*. *Biochim. Biophys. Acta* **67**: 566-575.

Thacker, A., and Syrett, P.J. 1972. The assimilation of nitrate and ammonium by *Chlamydomonas reinhardii*. *New Phytol.* **71**: 423-433.

Thomas, R.J., Hipkin, C.R., and Syrett, P.J. 1976. The interaction of nitrogen assimilation with photosynthesis in nitrogen deficient cells of *Chlorella*. *Planta* **133**: 9-13.

Turpin. D.H. 1991. Effects of inorganic nitrogen availability on algal photosynthesis and carbon metabolism. *J. Phycol.* **27**: 14-20.

Ullrich, W.R., and Eisele, R. 1978. Relations between nitrate uptake and nitrate reduction in *Ankistrodemus braunii*. In *Transmembrane Ionic Exchanges in Plants,* (Ducet, G., Heller, R., and Tellier, M., eds.) CNRS, Paris. pp. 307-315.

Vanlerberghe, G.C., Schuller, K.A., Smith, R.G., Feil, R., Plaxton, W.P., and Turpin, D.H. 1990. Relationship between NH_4^+ assimilation rate and *in vivo* phosphoenolpyruvate carboxylase activity. Regulation of anaplerotic carbon flow in the green alga *Selenastrum minutum*. *Plant Physiol.* **94**: 284-290.

Van Niel, C.B., Allen, M.B., and Wright, B.E. 1953. On the photochemical reduction of nitrate by algae. *Biochim. Biophys. Acta* **12**: 67-74.

Van Quy, L., Lamaze, T., and Champigny, M.-L. 1991a. Short-term effect of nitrate on sucrose synthesis in wheat leaves. *Planta* **185**: 53-57.

Van Quy, L., Foyer, C., and Champigny, M.-L. 1991b. Effect of light and NO_3^- on wheat leaf phosphoenolpyruvate carboxylase activity. *Plant Physiol.* **97**: 1476-1482.

Van Quy, L., and Champigny, M.-L. 1992. NO_3^- enhances the kinase activity for phosphorylation of phosphoenolpyruvate carboxylase and sucrose phosphate synthase proteins in wheat leaves. Evidence from the effects of mannose and okadaic acid. *Plant Physiol.* **99**: 344-347.

Warburg, O., and Negelein, E. 1920. Ube die reduktion der salppetersaure in grunen zellen. *Bio-Chem. Z.* **110**: 66-116.

Woo, K.C., and Canvin, D.T. 1980a. Effect of ammonia on photosynthetic carbon fixation in isolated spinach leaves cells. *Can. J. Bot.* **58**: 505-510.

Woo, K.C., and Canvin, D.T. 1980b. Effect of ammonia, nitrite, glutamate, and inhibitors of N metabolism on photosynthetic carbon fixation in isolated spinach leaves cells. *Can. J. Bot.* **58**: 511-516.

Nitrogen Nutrition in Higher Plants, 1995
Editors : H.S. Srivastava & R.P. Singh
Associated Publishing Co., New Delhi, India
pp. 337-365.

Nitrogen Mobilization During Senescence

SIBDAS GHOSH, GOPINADHAN PALIYATH, DAVID PEIRSON,
and RONALD A. FLETCHER

I. Introduction

Availability of nitrogen is a major factor that influences plant growth and productivity. Healthy plants constitute 1-4% of their dry weight as reduced nitrogen. Nitrate (NO_3^-) is the primary source of nitrogen for growth and development of field crops even though NH_4^+ or urea fertilizers may have been applied. This is due to the fact that soil microorganisms oxidize the NH_4^+ and/or urea-N to NO_3^- by the process of nitrification under normal conditions. When the supply of inorganic nitrogen is inadequate in the soil, certain leguminous and actinorhizal plants are able to obtain nitrogen through a symbiotic association with nitrogen-fixing microorganisms.

In return, plants provide photosynthate to support growth of nodules, the symbiotic organ of the plant containing nitrogen-fixing bacteria. During nitrogen fixation, atmospheric nitrogen (dinitrogen: N_2) is reduced to ammonium (NH_4^+) and made available for amination reactions, by which ammonium ions are transferred to carbon containing compounds for the production of amino acids and other nitrogen containing organic compounds. Of the total nitrogen fixed biologically (140 million tonnes per annum), 80% is contributed by symbiotic associations and 20% by free living organisms (Vance, 1978, Gutschik, 1980). This chapter focusses on the significance of nitrogen mobilization during senescence.

II. Forms of Mobilizable Nitrogen

Organic form of nitrogen, primarily transported to the growing regions of vegetative and reproductive organs of the plant, contributes to the soluble nitrogen pool. As nitrogen is needed, these compounds are withdrawn from the pool and catabolized to release NH_4^+ and other byproducts. The products of catabolism are subsequently reassimilated, producing the specific components of amino acids required for the synthesis of other metabolites such as aromatic amino acids, chlorophylls, nucleotides, nucleic acids, coenzymes, polyamines, alkaloids, and macromolecules that include DNA, RNA, structural proteins and enzymes. Turnover of macromolecules is prominent during reproductive growth and also contributes to the pool of soluble nitrogen. Of all the storage forms of nitrogen, proteins are the best form suitable for recycling nitrogen, especially during adverse conditions such as stress or senescence. According to Millard (1988), the advantages of storing nitrogen in the protein form rather than as NO_3^- are: (1) it optimizes the potential for carbon fixation, (ii) it avoids complications with leaf turgor and (iii) it permits continued reduction of NO_3^-. A considerable amount of nitrogen may be stored as secondary products and unusual metablolites that generally have no apparent turnover and are end products with no significant metabolic functions in plants. It has been suggested that 1-4% of the total nitrogen of vegetative parts of a mature plant is immobilized into glycosides, 1-8% is immobilized into nonprotein amino acids (e.g., homoserine and djenkolic acid), 0.5% is immobilized into glycosinolates and only 0.08-0.6% is immobilized into components such as alkaloids (Pate, 1986).

Seeds are the major storage organs for nitrogen. Protein forms of nitrogen are characteristic to the plant family and even species. Seed storage proteins contain adequate levels of most of the essential amino acids. In dicots, the proteins are stored in the cotyledons. In leguminosae (fabaceae), up to 80% of the total proteins in seed appear to be reserve proteins and they are rapidly degraded and mobilized during seedling establishment. The major storage proteins of legumes are globulins. Legumin, a type of storage globulin, possess high molecular weight (330 kD) and a pI of 4.8. Tricilin, another type of globulin with a molecular weight 180 kD is a more soluble protein with a pI of 5.5. Vicilin is a complex multimeric globulin. Convicilin in *Pisum sativum* comprises of 71 kD molecular weight subunits that complexes together to form a 290 kD molecular weight protein (Miege, 1982; Feller, 1990). In monocots, the proteins are stored in the endosperm. Protein content varies from 8% in *Oryza* to 12% in *Avena*. Seeds of *Triticum, Hordeum* and *Secale* have an average of 10-15% of proteins by dry weight. Outer layer of the endosperm (aleurone) is particularly rich in proteins and phosphorus reserves. The monocot protein reserves are primarily prolamins and glutelins. In addition, globulins are located both in embryo and endosperm. During very early stages of germination, globulins serve as the protein reserve for growing regions of the seedling even before the proteins stored in endosperm are degraded. In *Zea mays* and *Hordeum*, the main reserve of endosperm is usually prolamins. In *Oryza sativa*, glutelins constitute 80% of the total protein content.

In addition to proteins, a large part of reduced nitrogen in plant tissues is made up of amino acids. Over 200 non-protein amino acids have been discovered in plants which occur free in the amino acid pool. Non-protein amino acids are generally considered as secondary plant products and their role in plant growth and development is rather controversial. Two pools of free amino acids are believed to exist, one which is metabolically active contributing to protein synthesis and the other for storage, comprising of amino acids derived from protein breakdown as well as by biosynthesis. In cotyledons of germinating pea, homoserine which was absent in the seed, accounted for 70% of the free amino acid pool. Glutamic and aspartic acids are the major precursors for homoserine (Larson and Beevers, 1965).

Relatively little is known about the metabolism and role of most of the uncommon amino acids in plants. Plants which synthesize uncommon amino acids frequently accumulate them in very high concentrations. 5-Hydroxy-L-tryptophan can account for 14% of the seed weight of *Griffonia simplicifolia*, canavanine, 7-10% of the seed weight of *Diocloea megacarpa* and L-3,4-dihydroxyphenylalanine, more than 8% of seed weight of *Mucuna mutissiana* (Bell, 1976).

Though the varieties of nitrogenous compounds are enormous, only a limited number of them are used for storage and transport. The amino groups of the amino acids appear to be transferred to glutamate to form the amido nitrogen of glutamine and this is the principal nitrogen containing compound exported from the cotyledons to the developing axis in pumpkin seedlings. Although the reserve proteins are enriched in asparagine and arginine, and cotyledons may contain soluble nitrogen-rich amino acids such as canavanine, it appears that these compounds are not extensively translocated to the embryonic axis. Instead, they are metabolized in the reserve tissue. Initially, arginine and canavanine are hydrolyzed by arginase which accumulates in the cotyledons during germination. The products of arginine hydrolysis are ornithine and urea. Occurrence of ornithine in the exudate of pumpkin cotyledons suggest that it may be transported to the growing axis. In addition, it can be metabolized to glutamic acid, gamma amino butyric acid (GABA) and proline. Urea produced is hydrolyzed by urease to ammonia and carbon dioxide. Ammonia may be incorporated as amido nitrogen of glutamine by glutamine synthetase. Glutamine could then be translocated to the growing axis. High levels of glutamine in the cotyledonary exudate is consistent with this suggestion (Chou and Splittstoesser, 1972).

In legumes, asparagine is the principal transport amino acid and arginine the prevalent amino acid in proteins (Atkins et al., 1975). These two amino acids also serve similar functions in trees such as apple and spruce. Asparagine is used for short distance transport. Arginine is the predominant transport component, but glutamine, glutamic acid and aspartic acid also make considerable contributions. The relative importance of the different compounds depends to some extent on the source of inorganic nitrogen. Thus, maize plants grown on NH_4^+ salts transport their nitrogen chiefly as amides, but when grown on nitrate, glutamic acid is the predominant component. In lupins and other legumes, asparagine forms 60-70% of the nitrogen in the xylem. Phloem contains 10-20 times as much nitrogen as found in xylem, though the relative amino acid composition does not differ to a great extent. Glutamine, along with glutamate and serine are the primary transport compounds from leaves to the developing fruit of *Datura* (Paliyath, 1979).

It is interesting to note the storage and utilization of hydroxyarginine, a non-protein amino acid in the angiospermous plant parasite *Cuscuta* (Cuscutaceae). Being an obligate plant parasite, *Cuscuta* obtains bulk of its reduced nitrogen from phloem of the host plant. It appears that, whatever the phloem components of the hosts are, *Cuscuta* invariably converts most of the absorbed reduced nitrogen into hydroxyarginine. This amino acid could accumulate in the range of 0.5-1% by dry weight in the growing tip regions of the vine. In addition to hydroxyarginine, arginine and histidine accumulate to similar levels in the growing regions. However, in the phloem exudate of *Cuscuta* hydroxyarginine constitute 90% of the amino acids transported. The rest of the amino acid fraction in the phloem exudate comprises of arginine and histidine. It appears that after excision of the vine, the basal regions undergo gradual senescence and during this process the stored amino acids are mobilized to the tip region for growth and development (Paliyath and Mahadevan, 1988).

III. Nitrogen Mobilization During Senescence

The process of recycling nitrogen frequently involves catabolism, reassimilation and resynthesis of nitrogen-containing metabolites. It has been estimated that about 50% of the total nitrogen available in leaf and stem of wheat is catabolized and transported to the developing grains where it is converted and used for the synthesis of storage proteins (Harper et al., 1987). These proteins normally constitute 80-90% of the total proteins present in seeds which in turn are utilized by the young developing seedlings as a source of nitrogen and carbon, and in some cases sulphur (Okita et al., 1989). In annual crops, seed formation is accompanied by the mobilization and transport of nitrogen from vegetative organs to the developing grains (cereals) or pods (legumes). Peoples and Dalling (1988) estimated that the leaves, stem, pod walls and glumes can mobilize nitrogen at a higher rate (> 65%) than by the root (≤ 30%). This observation is consistent with the fact that the root is the last organ to senesce. In mature rice, the major source of nitrogen (64%) for developing leaves and ears is the nitrogen released from older senescing leaves (Mae et al., 1983; 1985; Kamachi et al., 1991). Leaf proteins constitute 75-85% of the reduced nitrogen while nucleic acids only account for 10-20% of the reduced nitrogen. Of the total leaf proteins, soluble and insoluble (membrane bound) fractions constitute 62% and 38% respectively. Ribulose-1,5-bisphosphate carboxylase oxygenase (Rubisco), the major soluble leaf protein accounting for 35% of total reduced nitrogen in mesophyll cells, is the single most abundant source of nitrogen for reallocation. Conventional wisdom would suggest that apart from its key role in CO_2 fixation through Calvin cycle, Rubisco is a major storage form of nitrogen especially for annual seed crops. Moreover, the half life of Rubisco is higher in young expanding leaves compared to that in fully expanded mature leaves (Brady, 1988). Thus, Rubisco may be a perfect candidate for short term storage. Indeed, this may be the case since Rubisco decreases and becomes the principal source of transported nitrogen during senescence (Huffaker and Paterson, 1974; Mae et al., 1985). In perennial crops, protein breakdown in shoot bark (autumnal senescence) is of prime importance in supplying nitrogen for early spring growth (Kang and Titus, 1980a; 1980b; Kang et al., 1982).

Senescence is associated with the degradation of proteins and nucleic acids and loss of polyamines and chlorophyll along with disintegration of chloroplast and nodule structures. All these processes lead to the supply of mobilizable N_2. Although senescence involves a series of highly ordered degradative events that leads to the death of an organ or of the whole plant, it is necessary not only for the development of seeds but also for the survival and establishment of new progeny. Studies using ^{15}N have demonstrated a substantial decrease in ^{15}N-labelled proteins with no significant loss in total reduced nitrogen in vegetative organs during growth (Brady, 1988). During early stage of growth, vegetative organs, still assimilating nitrogen, may turn over 1-2% of the ^{15}N assimilated per day. During later stage of growth this may rise to 2-3% per day. By comparison, contribution of nitrogen by senescing tissue to the nitrogen economy of plant can be three-fold higher. Proteolysis plays an important role in the nitrogen economy during senescence by increasing reduced nitrogen pool for subsequent translocation to other developing parts of plants.

IV. Proteolysis During Senescence

During growth, the cellular proteins are in a dynamic state of flux involving both synthesis and degradation. This phenomenon, protein turnover, is of critical importance to the cell. The purpose of protein turnover is not only to maintain a constant supply of metabolic enzymes but also to remove proteins which are misfolded, denatured or otherwise abnormal. Indeed, during senescence, a rise in the activities of proteases is well documented. However, proteolysis is not restricted to the phase of senescence. For example, peptide bonds are hydrolysed throughout leaf development and proteolytic processing occurs as a part of protein transport from endoplasmic reticulum into chloroplasts, mitochondria or the apoplast. These processing enzymes are highly specific for the cleavage of a defined peptide from the precursor protein. This chapter discusses protein degradation during.

senescence and does not cover protein processing and protein turnover in non-senescing tissues.

Based on the position of the peptide bond hydrolyzed, proteases are grouped into two main classes (Dalling, 1986): (i) endopeptidases and (ii) exopeptidases. Endopeptidases hydrolyze peptide bonds releasing residues of more than three amino acids and are grouped into four categories depending on the amino acids present at the site of hydrolysis: (a) serine endopeptidases characterized by diverse substrate specificities and location of serine and histidine residues at the catalytic site, (b) cysteine endopeptidases with cysteine at the active site, (c) aspartate endopeptidases with aspartic acid at the active site and (d) metallo endopeptidases with zinc as an essential metal at the catalytic site. Plant endopeptidases characterized so far are sensitive to sulfhydryl group inhibitors, but whether they can be correctly categorized as cysteine-proteases is not clear. The pH optima for the proteolysis of haemoglobin, casein, or Rubisco are often found in the range 4 to 7. The pH optima for proteolytic enzymes in extracts of senescing leaves depended on the substrate used. The maximum activity against Rubisco or cytochrome c was observed at pH 5. Plant endopeptidases, not belonging to the cysteine-proteinases, were found to have pH optima in both acidic and alkaline range.

Exopeptidases hydrolyze proteins to release single amino acids, dipeptides or tripeptides and further division of this class of enzymes is based on whether the release of amino acids occurs from N-terminus (i.e., catalyzed by aminopeptidases) or C-terminus (i.e., catalyzed by carboxypeptidases). Different forms of aminopeptidases have been detected in plants (Scandalios and Espirita, 1969; Ott and Scandalios, 1976; Collier and Murray, 1977; Wittenbach, 1978; Waters and Dalling, 1979; Tazaki and Ishikura, 1983). Various forms of aminopeptidases have been separated by gel electrophoresis and ion exchange chromatography. They differ in their specificities for the N-terminal amino acids of the substrate and in their susceptibility to inhibitors. Optimal pH for aminopeptidase activity is in the neutral or slightly alkaline region. Carboxypeptidases have been isolated and characterized from leaves of citrus (Zuber and Matile, 1968; Sprossler et al., 1971; Zuber, 1976), french bean (Wells, 1965, Carey and Wells, 1972) and tomato plants (Walker-Simmons and Ryan, 1977a; 1977a; 1980). While using synthetic peptides as substrates (e.g., N-carbobenzoxy dipeptides) maximal activities were observed in the pH range between 5 and 6.5. Leaf carboxypeptidases liberate a wide range of C-terminal amino acids from proteins or oligopeptides. Plant carboxypeptidases are sensitive to diisopropylfluorophosphate or to PMSF, indicating that a serine residue may be in or near the active site.

Based on the mechanisms and location of proteolysis, there are two major categories of protein degradation in plants (i) vacuolar and (ii) ubiquitin-dependent, although these are speculative to some extent. Localization of hydrolytic enzymes such as acid phosphatase, nuclease phosphodiesterase, α-mannosidase, N-acetyl-β-D-glucosaminidase, phosphodiesterase, and β-fructosidase in isolated vacuoles supports the hypothesis that vacuoles of higher plants function in a manner similar to lysosomes in animal cells. Activities of both endopeptidases and exopeptidases have been detected in the vacuole. These enzymes do not seem to discriminate specific substrates among various proteins, hence any control of proteolysis must be exerted at the step of entry into the vacuoles.

A. Vacuolar Proteolysis

There are two hypotheses on how the selective degradation of proteins is achieved in the vacuoles (Dalling and Nettleton, 1986): (i) pinocytosis and (ii) change in transfer properties of tonoplasts. During pinocytosis, vacuole engulfs whole or fragmented organelles and their constituents such as proteins, lipids, pigments and nucleic acids (Lamppa et al., 1980; Wittenbach et al., 1982). However, Thomas and Stoddart (1975) working with the nonyellowing mutant of *Festuca pratensis* concluded that the senescence of chloroplasts involved mechanisms other than pinocytosis. They found that the degradation of the extrinsic proteins of thylakoids and proteins of the stroma in both the mutant and the wild type follow a similar pattern. However, intrinsic proteins of thylakoids, in particular LHCP (proteins associated with the light harvesting complex) and other proteins associated with chlorophyll (chlorophyll a, b-binding proteins) are retained resulting in the nonyellowing appearance of the mutant. Thus, such a discrimination would seem to be totally against the pinocytosis hypothesis. Moreover, ultrastructural studies of senescing tissue failed to depict the existence of any chloroplasts, or chloroplast fragments within the vacuole (Thomas, 1977) and there are no significant changes in the number of chloroplasts during senescence. According to the second hypothesis, specificity of degradation is achieved by selective movement of particular or targeted compounds into the vacuole which to some extent is an extended idea of pinocytosis or autophagy. In *Lemna minor*, stress-induced changes in permeability of the tonoplasts to amino acids has been reported by Cooke et al. (1980a; 1980b). They have also correlated increased protein degradation induced by nitrate starvation with increased protein transfer across the tonoplasts.

B. Ubiquitin-Dependent Proteolysis

The role of ubiquitin in proteolysis was discovered in a study on the mode of action of ATP-dependent protein breakdown in reticulocytes (Ciechanover et al., 1978; Hershko et al., 1978). As the name implies, ubiquitin, a highly conserved protein of 76 amino acids, is present in eukaryotes in both free and covalently bound forms, the latter conjugated to various intracellular proteins (Hershko and Ciechanover, 1986). The covalent conjugation of ubiquitin is an obligatory intermediary step in proteolysis and there are specific proteases that hydrolyse only conjugated proteins (Hershko et al., 1980). Ubiquitin-protein conjugates are linked by isopeptide bonds between the C-terminal glycine residue of ubiquitin molecule and ε-NH_2 groups of lysine residues in proteins. It is a widely accepted fact that ubiquitin is a component in the ATP-dependent nonlysosomal protein degradation pathway and the targeted proteins must be ubiquitinated before they are recognized by the proteases. However, the existence of specific ubiquitinated protein conjugates which are not rapidly degraded indicates that ubiquitin may play other physiological roles yet to be elaborated. Ubiquitination of proteins appear to possess multiple functions: (i) in tagging proteins for degradation via nonlysosomal ATP-dependent pathway, (ii) in chromatin structure and (iii) possibly in mediating cell-cell interactions by modification of cell surface molecules (Hershko et al., 1981). Proteolytic degradation could well be correlated with high levels of ubiquitin in senescing tissues, although no direct evidence is available.

Initial reaction in the ligation process is the activation of the polypeptide by a specific ubiquitin-activating enzyme (E_1) that adenylates the C-terminus of ubiquitin using ATP. Activated ubiquitin molecule is then attached through a thiol ester linkage to another site on E_1 with the concomitant release of AMP. Next, the ubiquitin moiety is transferred by a trans-esterification reaction to ubiquitin carrier proteins (E_2) which comprise of a family of related proteins. Finally, ubiquitin is covalently attached to target proteins with or without the participation of a family of ubiquitin-protein ligases (E_3).

Of all E_2 proteins, only one participates in E_3-dependent ligation of ubiquitin to substrate protein. Enzymes that belong to the family of ubiquitin-protein-ligases appear to have a central role in the selection of subtrate proteins for subsequent proteolysis. The protein-binding sites of E_3, show remarkable specificities, and to date there are three different types of substrates that bind to specific sites of the E_3: (i) Type I substrate proteins (lysozyme or ribonuclease) that are represented by proteins having basic N-terminus moieties, (ii) Type II substrate proteins (lactoglobulin or pepsinogen) that have a Leu residue at the N-terminus and (iii) Type III substrate proteins which are extremely suitable for the ubiquitin ligation, but have neither basic nor bulky hydrophobic N-terminus residues (Hershko, 1988).

One of the structural determinants in protein recognition by ubiquitin ligase was found to be a free amino group at the N-terminus. Substrate proteins with Arg, Lys, Asp, Asn, Trp, Leu, Phe, His, Glu, Tyr, Gln, and Ile had half lives of 2-30 min whereas those with Met, Ser, Ala, Thr, Val, Gly, and Cys at the N-terminus were essentially stable (Hershko, 1988). Even though the N-terminus is undoubtedly an important signal this may not be the only recognition feature for substrate proteins. For example, RNAase A, having Lys at the N-terminus is not found to be a good substrate for ubiquitin-ligase system from reticulocytes, but it can be converted to an excellent substrate following the oxidation of its methionine residues to sulfoxide derivatives (Hershko et al., 1986). Most proteins that have stabilizing residues at their N-terminus are not intracellular, but are secreted proteins in which the cleavage of the signal peptide created the final N-terminus. Therefore the main function of N-terminus recognition mechanism could be to eliminate all foreign or secreted proteins.

In addition to the role of ligation in ATP-dependent nonlysosomal proteolysis, it has been demonstrated that purified ubiquitin has an intrinsic endoprotease activity which is comparable to other endopeptidases (Fried et al., 1987). Activity of ubiquitin as an endopeptidase is optimal at pH 8.0, highly selective, stimulated by Ca^{2+} or other cofactors, and inhibited by PMSF and diisopropyl fluorophosphate. Ubiquitin may also act as a functional protease when conjugated to other proteins and thus be regulated, in part, by this specific immobilization. Finally, the ubiquitination of a protein can convert that protein into an *adhoc* protease that can play a role in a variety of cellular events. Rapidly degraded proteins contain regions enriched by the four amino acids (Pro, Glu, Ser and Thr) and referred to as PEST regions (Rogers et al., 1986). PEST sequences have been hypothesized as signals for rapid turnover. With one exception (Shanklin et al., 1987), the degradation of proteins containing PEST by the ubiquitin pathway has not been unequivocally determined.

C. Regulation by Cellular Factors

Though one of the early events in senescing systems is the well documented increase in protease activities, little is known about the regulation of proteolysis itself during senescence. Degradation of any given protein is influenced by various parameters such as the presence of peptide hydrolases in the same compartment, the susceptibility of the protein to proteolytic attack and the biochemical conditions in this compartment including pH, ions, and concentrations of metabolites. Increase in the levels of proteolytic enzymes is due to several factors such as *de novo* synthesis of the enzyme (Martin and Thimann, 1972; Patterson and Huffaker, 1975), activation of a zymogen (Hochkeppel, 1973), disappearance of inhibitors (Watanbe and Kondo, 1983) or changes in the compartmentation (Cooke et al., 1980a, 1980b; Wittenback et al., 1982). Reduction in protease activity after application of protein synthesis inhibitors to senescing tissues suggests that *de novo* synthesis of protease is involved in the rapid breakdown of proteins during senescence. Senescing leaves contain proteinaceous protease-inactivating factors which are heat sensitive, precipitated by ammonium sulfate and excluded by Sephadex G-25. (Streit and Feller, 1982). Inactivation can be delayed by the addition of (i) casein, (ii) heat-denatured leaf extracts, (iii) papain, (iv) trypsin or (v) chymotrypsin to leaf extracts (Streit and Feller, 1983a; 1983b).

Substrate specificity is a major factor controlling proteolysis. Nitrate reductase is known to be a highly susceptible enzyme whereas glutamate dehydrogenase is very stable under the same *in vitro* conditions (Batt and Wallace, 1983; Streit and Feller, 1983b); Also chloroplastic glutamine synthetase was more rapidly inactivated than cytosolic glutamine synthetase in an extract from senescing leaves of rice containing both forms (Streit and Feller, 1983a; Kamachi et al., 1991). Other proteins present in the same compartment may play an active role in protecting a particular enzyme, since they may act as competitive substrates to proteolytic attack. Thus, enzymes in leaf extracts can be protected *in vitro* by the addition of bovine serum albumin (BSA) or casein. Susceptibility of a protein can be altered by covalent modifications (Bond and Offermann, 1981) or by interactions with low-molecular-weight compounds such as polyamines (Kaur-Sawhney et al., 1982). In some cases, modifications of substrate proteins could accelerate proteolytic breakdown (Feller, 1986).

Spatial separation of proteolytic enzymes and their substrates is another important mechanism for controlling degradation of proteins in senescing cells. For example, carboxypeptidases and endopeptidases with pH optima ranging from 4 to 7 were detected mainly in the vacuoles, whereas alkaline endopeptidase activity with a pH optimum around 9 was found to be located in the apoplastic compartment of bean leaves (Van der Wilden, 1983). Bulk of the leaf proteins are located in the cytosol and outside the vacuole, and a high percentage of these proteins are present in the chloroplasts. Most of the leaf proteins are, therefore, separated by membranes from the major proteolytic enzymes in the vacuole. It has been suggested that the properties of the tonoplast may be altered during senescence allowing the release of vacuolar proteases into the cytoplasm, or, alternatively, facilitating the entry of selected proteins into the vacuole (Cooke et al., 1980a; 1980b). Most studies suggest that constituents of the chloroplasts are degraded within the chloroplasts themselves as proteolytic activities were detected in isolated intact chloroplasts (Ragster and Chrispeels, 1981; Lin and Wittenback, 1981; Dalling et al., 1983) Thus, the activity of a particular protease *in vitro* should be interpreted with caution because *in vivo* these enzmes may be separated from their potential substrates by membranes.

The role of concerted activities of different proteases in the catabolism of proteins is yet to be elucidated fully. It is likely that proteolytic enzymes located in different compartments may be involved in the complete hydrolysis of a particular protein. Peptides produced by endopeptidases may be further hydrolysed to free amino acids in a different cellular compartment. Furthermore, pH value, inorganic ions, or metabolites may influence proteolysis by interacting with either the proteolytic enzymes or the substrate proteins. The pH optimum for the degradation of a protein depends not only on the properties of the protease but also that of substrate proteins. Thus, changes in the pH value of a compartment are likely to alter the rate of proteolysis. Degradation of various proteins may be affected differently by a shift in the pH. The pH profiles for degradation of Rubisco by endogenous proteases of purified chloroplasts from barley differed in the presence and absence of SDS (Dalling et al., 1983). Free fatty acid content in the membrane can change *in vivo* during senescence. Degradation of membrane lipids and increased concentrations of free fatty acids could also be involved in the regulation of proteolysis in chloroplasts (Thomas, 1982). Moreover, the susceptibility of enzymes to proteolytic attack can be altered by substrates, cofactors, stimulators or inhibitors (Holzer and Heinrich, 1980; Streit and Feller, 1982; 1983b). For example, stabilization of glutamine synthetase can be achieved in the presence of ATP but not in the presence of ADP or AMP (Streit and Feller, 1982). By contrast, inactivation of nitrate reductase was not affected by ATP, ADP or AMP but was influenced by pyridine nucleotides. In some cases ubiquitin was essential for the ATP-dependent proteolysis

(Hershko et al., 1983; 1984; Ciechanover et al., 1984) whereas in other systems ubiquitin had no effect (Waxman and Goldberg, 1982; Desautels and Goldberg, 1982; Hershko et al., 1983; Tanaka et al., 1984).

D. Degradation of Membrane Proteins during Senescence

Deterioration of membranes is an early event that occur during senescence (Paliyath and Droillard, 1992). The membrane fraction comprising of plasma membrane, endoplasmic reticulum and tonoplast constitute the major components of membranes, and their deterioration is associated with the loss of compartmentation. Catabolism of membrane lipids has a deleterious influence on membrane protein conformation. These changes could result in the loss of enzyme functions as observed in the inactivation of calcium pump during senescence (Paliyath and Thompson, 1988). Damaged proteins are subject to proteolytic degradation. Also, free radicals generated during membrane lipid degradation cause the denaturation of membrane proteins (Duxbury et al., 1991a). These proteins in turn are subjected to proteolytic degradation (Duxbury et al., 1991b). Recent observations suggest that proteolytic activity in membranes of senescing bean cotyledons could be localized in a fraction of non-sedimentable membrane vesicles referred to as deteriosomes (Yao et al., 1991). The fate of the released peptides and their further degradation has yet to be established.

V. Mobilization of Nitrogen During Leaf Senescence

By comparison to the senescence of organs such as root and stem, senescence of leaves has been studied widely (Thimann, 1980; Thomas and Stoddart, 1980; Woolhouse, 1982). Senescing leaves lose their ability to assimilate nitrate. Along with this, increased degradation of nitrogen containing macromolecules are observed. In general, the amino acids released are exported rapidly to other parts of the plant. An active long-distance transport is a prerequisite for the redistribution of nitrogen and other nutrients during natural senescence. Depending upon the species, the senescence of leaves may be synchronous with either the end of the growing season, the reproductive phase of plants, or may occur as a sequential dieback of the lower (oldest) leaves throughout development. Whatever may be the case, the senescence of individual leaves is likely to be similar at the molecular level. The earliest physiological and biochemical changes observed during foliar senescence are typically a decline in the rate of photosynthesis and progressive deterioration in the functions of chloroplasts. Associated with the loss of chloroplast function, is the decline in chlorophyll and soluble proteins (e.g., Rubisco). In cereals, seed formation is accompanied by the mobilization of nitrogen from senescing tissues (e.g., stems and leaves) to developing grains. Nitrogen enters the seed primarily as amino acids, of which glutamate appears to be the predominant constituent. It has been estimated that about 50% of the total nitrogen available in leaf and stem of wheat is catabolized and transported to the developing grains where it is converted mainly into storage proteins.

A. Breakdown of Chlorophyll

In general, chlorophyll breakdown is one of the first visible symptoms of leaf senescence. There are two enzymes that have been shown to be capable of chlorophyll degradation: (i) chlorophyllase (chlorophyll-chlorophyllido-hydrolase, EC 3.1.1.14) that catalyzes the removal of the phytol chain from the porphyrin head group, which is the first step in the catabolism of chlorophyll (Purvis and Barmore, 1981; Kuroki et al., 1981; Mazumdar et al., 1991) and (ii) peroxidase (Huff, 1982). Although, both chlorophyll and chlorophyllase are components of the thylakoid membrane, the role of chlorophyllase in normal chlorophyll turnover appears to be low in healthy and non-stressed plants. This is perhaps due to effective compartmentation of chlorophyll from chlorophyllase within the membrane or because of metabolic constraints on the activation of chlorophyllase. Large increase in chlorophyllase activity has been observed during senescene of excised leaves, during ripening of fruits and exposure of fruits to ethylene. However, it is difficult to assign a role for the peroxidase *in situ*. Peroxidative and oxidative enzymes which catalyze the bleaching of chlorophyll become active upon disorganization of the thylakoids and breakdown of the membrane during senescence (Martinoia et al., 1982).

B. Degradation of Nucleic Acids

Loss of nucleic acids during leaf aging has been demonstrated by Kessler and Engelberg (1962). A number of subsequent studies have demonstrated degradation of both DNA and RNA during aging of organelles. The highest RNAse activity of the subcellular particles was found to be localized in the chloroplasts, while ribosomes and mitochondria had very little activity associated with them. Contrary to expectations, RNAse and chlorophyllase activities in senescing bean and radish leaves declined. When senescence was delayed by cytokinin treatment the activities of these enzymes increased (Phillips et al., 1969). These results suggest that, in addition to their degradative role, RNAse and chlorophyllase may take part in the turnover of their substrates as well.

C. Polyamine Catabolism

Putrescine, spermidine and spermine are the most common polyamines known to occur in high levels in young, actively growing tissues. Polyamine content declines as organs age and senesce (Galston and Kaur-Sawhney, 1987). The decline in polyamines coincides with the appearance of senescence-related symptoms. Polyamines undergo oxidative deamination by the action of amine oxidases. To date, two major classes of oxidases are well characterized: copper containing diamine oxidases (EC 1.4.3.6) and flavin containing polyamine oxidases (EC 1.4.3.4). The 4-aminobutyraldehyde formed by the oxidation of putrescine or spermidine is further oxidized by a NAD-dependent pyrroline dehydrogenase (Flores and Filner, 1985). The product γ-aminobutric acid, can be transaminated and the resulting succinic acid may be incorporated into the TCA cycle. The recyling of nitrogen and carbon from polyamine, perhaps, play an important role during senescence and hence can be used as a source of organic nitrogen. For instance, a tobacco cell line capable of utilizing putrescine as a source of nitrogen has been reported (Flores and Filner, 1985).

Polyamines have also been suggested to affect protease activity in senescing tissues. Polyamines applied exogenously are potent inhibitors of senescence of oat leaf protoplasts (Kaur-Sawhney et al., 1980) and also of leaves and storage tissue from several plants (Altman and Bachrach, 1981). Consistent with this observation, the application of exogenous polyamines to leaf segments delayed senescence by decreasing protease activity and chlorophyll loss (Kaur-Sawhney et al., 1982). Reduced proteolytic activity was attributed to the binding of polyamines applied exogenously to the enzyme. In senescing systems, relatively low rates of decline in the levels of spermidine and spermine, combined with a concurrent rise in putrescine may serve to stabilize membranes (Dumbroff, 1990). This helps to moderate the rise in ethylene by suppressing possible free-radical mediated conversion of 1-aminocyclopropane-1-carboxylic acid to ethylene. This might also help to modulate the onset and course of senescence by affecting competitive demands for S-adenosyl methionine by the ethylene and polyamine biosynthetic pathways (Dumbroff, 1990). Absencce of competition between the two pathways in a specific plant system (Kushad et al., 1988) does not prevent polyamines from moderating the onset of senescence through their functions as free radical scavengers and as stabilizers of plant membranes (Dumbroff, 1990). The role of polyamines is discussed in detail in another chapter of this book.

D. Proteolysis

Degradation of proteins in senescing leaves is likely to depend upon the integrated action of many groups of enzymes whose activities are modulated differentially during senescence. It is possible that some of these enzymes may be involved in the turnover of functional proteins while others may contribute to nitrogen mobilization. Nonetheless, the degradation of proteins can be envisaged as a chain reaction, initiated by the endopeptidases and sustained by a series of exopeptidases, each with varying substrate specificities.

There has been considerable interest in the decline of chloroplast proteins because of their dominant role in carbon assimilation. In addition, Rubisco serves as the largest nitrogen reserve for redistribution to developing sinks. The level of Rubisco protein declines shortly after full leaf expansion and continues to fall more rapidly than most other leaf proteins throughout senescence (People and Dalling, 1988). This characteristic and preferential degradation of Rubisco during senescence led to the *in vitro* characterizaion of proteolytic enzymes involved in hydrolyzing Rubsico. Six proteases have been partially purified from crude extracts of wheat leaves (People et al., 1979). In soybean (Ragster and Chrispels, 1981) and barley leaves (Miller and Huffaker, 1981) three proteases that degrade Rubsico have been identified. During flag leaf senescence, decline in Rubisco was correlated with two peak activities of proteolysis. One of these appeared during the early period of protein mobilization when Rubisco was lost at a rate equivalent to that of other soluble proteins. By the time 30% of the total soluble protein had been mobilized, the second rise in proteolytic activity occurred after which the *in vivo* loss of Rubisco proceeded at a relatively faster rate than other proteins (Waters et al., 1980). This biphasic pattern of change in activity was perplexing since it did not appear to be due to an activation of preexisting proteases. The *in vitro* degradation of Rubisco in cowpea leaves was characterized by a single peak of activity at the beginning of seed filling, however, this resembled the second phase of proteolysis in wheat by being closely coincident with the commencement of preferential loss of leaflet Rubisco (Peoples et al., 1980; 1983). Much of the *in vitro* proteolytic activity against Rubisco was identified as vacuolar (Wittenbach et al., 1982), although the existence of proteases involved in the degradation of Rubisco have been localized within the chloroplasts (Nettleton et al., 1985). There are significant correlations between the rates of crude proteolytic activity *in vitro* at acidic pH and rates of *in vivo* loss of Rubisco-nitrogen in senescing leaves (Dalling et al., 1976).

Another group of proteolytic enzymes involved in proteolysis comprises of the aminopeptidases. The activities of aminopeptidases were found to be high in developing and mature parts of the plants. A large proportion of aminopeptidase activity was localized in chloroplasts and only traces of the activity were detected in the vacuoles (Waters et al., 1982; Heck et al., 1987). In senescing leaves of both annual and perennial plants, aminopeptidase activities decreased in a similar manner to leaf proteins content. The high aminopeptidase activities in the developing and mature tissues suggest that they may be involved in protein turnover and in post-translational modification, rather than in senescence.

By contrast to aminopeptidase activity, carboxypeptidase activity was localized to a large extent in the vacuoles of higher plant cells. Carboxypeptidase activity was relatively low in expanding leaves of wheat, while the highest activities were detected in fully expanded mature leaves. However, during senescence, carboxypeptidases were found to remain active much longer than aminopeptidases. It is possible that carboxypeptidase contributes to the rapid degradation of leaf proteins by further hydrolysis of peptides produced by endopeptidases.

Levels of endopeptidase activities that are observed during development and senescence, differ considerably between various groups of annual and perennial plants (Keist, 1980; Feller 1990). Monocarpic annuals are of special interest because of their rapid mobilization of nutrients from senescing leaves. Also, a close relationship between leaf senescence and seed filling exists in annual plants. On the other hand, nutrient mobilization from leaves of perennial plants is not closely related to seed maturation and may be controlled in a different manner. Endopeptidase activities are low in expanding and mature leaves of annual crops and a marked increase in the activities of these enzymes occurred during senescence. Endopeptidase activity attained maximal levels when aminopeptidase and carboxypeptidase activities were already declining. The rise in endopeptidase activity begins at the tip of the individual leaves and progresses through the basal part and finally reaches the leaf sheaths. Often, the maximal levels of endopeptidase activity were observed when bulk of leaf nitrogen had already been exported to other parts of the plant. During development and maturation of cereal grains, endopeptidase activities were highest in outer pericarp, in the cross cells, and in intact glume, lemma and palea. During this period, activities remained low in the embryo and in the endosperm. Nitrogen was mobilized sequentially from different senescing layers of wheat kernels. High endopeptidase acitivity along with decreasing nitrogen content was first detected in the outer pericarp and later in the chlorophyll containing cross cells. Thus, endopeptidases play an important part in proteolysis, and thereby in the mobilization of nitrogen from senescing plant parts. In senescing leaves of legumes, an increase in total proteolytic activity was observed only in few cases. The activities of these proteolytic enzymes were enhanced by thiol reagents and inhibited by φ-hydroxymercuribenzoate or N-ethylmaleimide. It has previously been shown that the sulfhydryl-dependent enzymes from soybean leaves were endopeptidases. These endopeptidases are most likely carboxypeptidases since their activities are inhibited by phenylmethylsulfonylfluoride. In pea plants, increased proteolytic activities were observed in senescing pods but not in senescing leaves (Storey and Beevers, 1977). In some cases, activities of these enzymes declined during senescence while increased activities were observed during leaf expansion and maturation. The variability in the proteolytic activities of senescing legume leaves suggest that different factors such as the occurrence of different peptide hydrolases, nutritional status of the plants and environmental stress may influence the expression of these enzymes.

Endopeptidase activities were also detected in leaves of other plants including tobacco, pineapple, papaya, *Ficus* and apple, and needles of larch and pine. Although a large proportion of these enzymes are located in the vacuoles, there are indications that relatively non-specific endopeptidases are present in other cell compartments (Peoples and Dalling, 1978; Ragster and ChrispeIts, 1981; Thomas and Huffaker 1981; Wanger et al., 1981). Changes in activity patterns of endopeptidases were observed in cereals, legumes and other dictots, however, senescence-specific endopeptidases have not been unequivocally identified so far.

The properties of proteases observed in senescing leaves differed from those in extracts obtained from germinating seeds of the same species (Feller et al., 1977; 1978; Feller, 1979; 1981; Ryan and Walker-Simmons, 1981). The pH optimum, temperature optimum and stability of azocaseine-hydrolyzing activities were different in crude extracts obtained either from bean leaves, or cotyledons of germinating bean seeds. The pH profiles and the inhibition properties of protease activity in endo-sperm of germinating maize seeds differed from those of leaves. These observa-tions suggest that differences exist in the endopeptidase protein-forms or endopeptidase activity patterns between senescing leaves and germinating seeds.

E.   Mobilization of Leaf Nitrogen

The level of Rubisco, the major soluble protein in leaves, decreases with sen-escence and serves as the principal source of transported nitrogen. Glutamine is a major free amino acid in mature leaves of rice. It has recently been shown that glutamine and asparagine accounted for 42% and 12% respectively of the total amino acids in the phloem sap of rice plants (Hayashi and Chino, 1990). These amides are derived from glutamic and aspartic acids and ammonia released by the hydro-lysis of Rubisco and other leaf proteins. Glutamine synthetase is a major enzyme involved in the conversion of glutamate and NH_4^+ to glutamine in senescing leaves, and the resultant glutamine could well be the source for amino groups for the trans-amination of aspartic acid to asparagine (Miflin and Lea, 1977; Oaks and Hirel, 1985). However, glutamine synthetase activity is known to decrease rapidly during both natural and dark-induced senescence. In many plants, there are two isoforms of glutamine synthetase, one located in the cytosol and the other located in the cho-loroplast stroma. The physiological function of chloroplastic glutamine synthetase is considered to be the reassimilation of NH_4^+ released during photorespiration. Indeed, in cereals, leaf cytosolic glutamine synthetase remains constant or increase slightly during senescence, whereas chloroplastic glutamine synthetase protein and its message decline during senescence (Mann et al., 1980; Streit and Feller, 1983a; 1983b; Kamachi et al., 1991). Cytosolic glutamine synthetase is, therefore, impor-tant in the synthesis of glutamine for export. Similar results were obtained for ami-notransferase activity in leaves of *Lolium temulentum* during senescence (Thomas, 1975). Activities of the various forms of aminotransferases present in the cholor-oplasts decreased earlier than the cytosolic forms. In general, the pattern of activ-ities of the enzymes involved in nitrogen metabolism changes during senescence. For example, the activities of nitrate reductase, glutamine synthetase and gluta-mate synthase decrease while glutamate dehydrogenase and endopeptidases (pro-teases) remain active longer, and often reach maximal activities during senescence. It has been proposed that glutamate dehydrogenase could be involved in the cat-abolism of glutamate rather than in its synthesis (Miflin and Lea, 1977, Thomas, 1978).

VI.   Mobilization of Nitrogen During Germination

During germination and initial stages of seedling establishment, the plant is depen-dent upon the nutrient reserves laid down in the seed during its development on the mother plant. Carbohydrates, proteins, lipids and phytins are the most commonly occurring reserves in the seeds. In monocots, the reserves are located mainly in the endosperm, whereas in dicots, the reserves are generally found in the cotyledons. In germinating seeds, catabolism of storage proteins releases amino acids which provide the nitrogen and carbon skeletons for growing plants.

In germinating cereals, proteolytic activity and the rate of storage protein hydrolysis are low during the early stages of germination, after which both these activities increase rapidly. In contrast to exopeptidases, minimal increase in the activities of endopeptidases was observed during early stages of germination of wheat. A new set of endopeptidases are formed during late stages of germination. Increase in endopeptidase activity was attributed to the *de novo* synthesis of these enzymes promoted by hormones (Preston and Kruger, 1979). In wheat, at the advanced stages of germination, degradation of storage proteins in endosperm proceeds concurrently with a fourfold to sixfold increase in endopeptidase activity. Occurrence of high levels of amino acids and low levels of peptides suggest that the initial products of endopeptidase activity must be degraded extremely fast. The enzymes responsible for this task are likely to be the carboxypeptidases that are present in abundance and have broad specificities against proteins such as glutelins (Preston and Kruger, 1976a; 1976b; 1976c). In barley, rye and oats, the activities of the proteolytic enzymes appear to be similar to those in wheat (Preston and Kruger, 1986).

By comparison to the cereals such as wheat, barley and oat grown in temperate climates, in cereals such as maize, sorghum and rice that are grown in tropical and subtropical areas, a much higher ratio of endopeptidase to carboxypeptidase activity was detected. Thus, a different pattern of nitrogen mobilization from the endosperm reserves can be expected in tropical cereals. Indeed, endopeptidase activity increases in maize endosperm during germination and this coincides with an increase in protein breakdown and a decrease in total nitrogen. For example, degradation of zein and glutelin in maize endosperm began after 20 hours of germination while major loss of protein occurred between 3 and 8 days after germination, coinciding with an increase in endopeptidase activity (Harvey and Oaks, 1974). Similarly, Feller (1978) found that the endopeptidase and carboxypeptidase activities increased rapidly from day 2 onwards, reaching a plateau between day 4 to day 6 of germination. This was accompanied by a simultaneous decrease in endosperm nitrogen. In general, an increase in endopeptidase activity is, in part, due to gibberellic acid-regulated *de novo* synthesis of the enzyme (Horiguchi and Kitagishi, 1976; Oaks et al., 1983).

VII. Mobilization of Nitrogen During Nodule Senescence

In the early reproductive phase of annual plants, root growth is often slow and this feature has been correlated with their reduced rates of mineral uptake and production of hormones (Nooden, 1980). Actively growing roots contain relatively low concentration of nitrogen of which 25% are in soluble proteins, 25% are represented by amino acids and ureides, and the remainder is represented by an insoluble component (not readily released from root cellular debris by buffer extraction). In senescing root systems, a gradual decline in soluble protein coincides with mobilization of 20-30% of total root nitrogen. Depending on the species, redistribution of nitrogen from the roots including the nodules contributes between 0.4-16% of nitrogen required by developing seeds (Peoples and Dalling, 1988).

In annual legumes, maximal rates of nitrogen fixation are attained by nodules at the beginning of the flowering stage, followed by a peak period of photosynthesis when nodule senescence begins (Marschner, 1986; Pate, 1989). During seed filling period, amino compounds and carbohydrates become limiting. With perennial and forage legumes, senescence of nodules tends to be more dependent on vegetative growth patterns or grazing and defoliation than to reproductive growth (Vance et al., 1979). Exposure of leguminous plants to inorganic nitrogen, drought or low temperature induce senescence of nodules (Groat, 1981). This is associated with lower levels of leghaemoglobin and protein content, and a decline in their nitrogen fixation capability (Becana et al., 1985). However, the mechanism (s) by which these factors initiate and influence the senescence and functioning of nodules is not clear.

Natural senescence of nodules is marked by a decrease in measurable amount of nitrogenase activity and a reduction in soluble protein from the cortical and the cytosolic compartments of the nodules. However, nitrogen fixation by the bacteroids may not be affected until late stages of senescence. Nitrogenase catalyzes symbiotic nitrogen fixation and this requires coupling to an efficient ATP-generating system (oxidative phosphorylation). The component I (MoFe-protein dinitrogenase) and II (Fe-protein, dinitrogen reductase) of nitrogenase are extremely oxygen-labile (Becana and Rodriguez-Barrueco, 1989). It has been demonstrated that the oxygen supply to bacteroids significantly increases after exposure of nodules to high doses of NO_3^-, drought or low temperature, leading to the nitrogenase inactivation by O_2 during senescence (Monk et al., 1989). On the other hand, protection against oxygen toxicity leads to the biochemical modifications of nitrogenase complex and increased rates of low-efficiency respiration (Robson and Postage, 1980). The enzymes superoxide dismutase, catalase and peroxidase have been identified in nodules of leguminous plants (Robson and Postage, 1980). During development of soybean nodules, total glutathione, ascorbate peroxidase and dehydroascorbate reductase activities increased significantly and were directly correlated with nitrogenase activity and leghaemoglobin content (Dalton et al., 1986). A higher antioxidant status, including a higher activity of superoxide dismutase contributes to lower levels of lipid peroxidation, that has been correlated with greater tolerance to environmental stresses (Monk et al., 1989).

A. Proteolysis in Roots and Nodules

Complete senescence of nodules and roots is of significant interest because these organs are the sources of newly assimilated nitrogen after nitrogen fixation and soil mineral nitrogen uptake. Delayed senescence of these tissues is critical for continued maintenance of nitrogen mobilization to the remainder of the plant. In crude extracts obtained from roots, the presence of aminopeptidase, carboxypeptidase and endopeptidase activities have been well characterized (Murray et al., 1979; Wallace and Shannon, 1981; Feller, 1990). Root tip appears to be enriched in protein, carboxypeptidases and aminopeptidases relative to the remainder of the root. The ability of crude extract from roots to degrade endogenous root proteins or haemoglobin at pH 5.0 has been demonstrated throughout the periods of fruit development and root senescence (Peoples et al., 1983). It has been shown that in soybean total root protein reached a maximum level coinciding with flowering and decreased to a low level during pod browning and seed desiccation (Peoples et al., 1983).

The half lives of root-and nodule-proteins have been estimated to be between 2 and 14 days (Coventry and Dilworth, 1976; Bisseling et al., 1980). During senescence of alfalfa nodules induced by defoliation (Vance et al., 1979) or by high nitrate (Becana et al., 1985), loss of protein coincides with the rise in proteolysis in host cell cytosol and bacteroid fraction. For example, a 400% increase in cytosolic protease activity was observed in senescing alfalfa nodules on day 7 after defoliation which coincided with a 50% loss in leghaemoglobin and soluble proteins. High levels of proteolytic activity were detected in extracts from nodule cortex and cytosol prior to the initial decline in nitrogen fixation. This coincided with a loss in total soluble protein, preferential decrease in leghaemoglobin and rapid loss in symbiotic capacity (Peoples et al., 1985; Pladys and Rigaud, 1985). Functional nodules primarily degraded proteins by metallo-and serine-proteases (endopeptidase and aminopeptidase) while during seed growth, protein degradation mediated by proteases sensitive to sulfhydryl group inhibitors became predominant. Nodule proteases formed during flowering are distinct from proteases formed during senescence of nodules (Peoples and Dalling, 1988). *In vitro* studies showed that proteases with acidic pH optima, strongly depressed symbiotic capability of bacteroids, whereas proteases with alkaline pH optima induced a lower O_2 requirement for optimal nitrogenase activity (Peoples and Dalling, 1988). Similar decline in bacteroid nitrogenase activity and optimal O_2 tensions were also observed during natural nodule aging. Perhaps, the acidic proteases are responsible for the degradation of most of the host cell proteins because the alkaline proteases only become apparent very late, during senescence as in french bean and cowpea nodules (Pladys and Rigaud, 1985; Peoples and Dalling, 1988). However, this may not be applicable in all senescing nodule systems, as extensive studies failed to detect proteases with acidic pH optima in soybean nodules (Peiffer et al., 1983).

Bacteroid proteases in cowpea and soybean nodules were found to have neutral and alkaline pH optima. Although, a high proportion of total nodule proteins are localized in bacteroids, bacteroid proteases have lower activities compared to their counterparts localized in nodule cytosol or cortex (Peoples and Dalling, 1988). Also, very little changes were observed in the activity of bacteroid proteases throughout the life of nodules. In general, degradation appeared to start primarily in the plant components during nodule senescence, such that the cytosolic and cortical proteins declined while bacteroid proteins were maintained until the late stage of nodule senescence.

VIII. Transport of Reduced Nitrogen

In cereals, grain filling accompanies the mobilization and transport of nitrogen from leaves and other source tissues to the developing seeds. Approximately 50% of the total available nitrogen present in stem and leaf tissues are catabolized, transported to the developing grains and converted into storage proteins (Okita et al., 1989). Nitrogen-fixing plants can be grouped into two major classes (Schubert, 1986): (i) amide exporters (legumes of mostly temperate origin and majority of the actinorhizal plants) and (ii) ureide exporters (legumes of mostly tropical origin). The amide exporters usually transport asparagine, glutamine or 4-methyleneglutamine while ureide exporters transport either allantoin and allantoic acid or citrulline. The rationale for exporting amides or ureides is under continued research. The transport of nitrogen through ureides may be more efficient, both in terms of the moles of carbon used for carbon skeletons and the cost of synthesis (Schubert, 1986). In many ureide exporting legumes, a shift from ureide synthesis to amide synthesis occurs when NO_3^- or NH_4^+ are present in the rooting media (Tajima and Yamamoto, 1975; Matsumoto et al., 1977; Pate et. al., 1981; Paterson and LaRue, 1983).

In senescing leaves of legumes and cereals, degradation of proteins to provide the complete amino acid complement for phloem export also caused synthesis of these amino acids prior to their loading onto the phloem (Alkins et al., 1983). For example, senescing leaves of lupin synthesize amino acids such as glutamate, serine, valine, isoleucine, tyrosine and phenylalanine, or to a lesser degree another group of amino acids that includes glutamine, threonine and GABA. Again, catabolism of asparagine and aspartic acid, combined with excess of glycine and lysine released by proteolysis, provided the primary nitrogen for synthesis of other amino acids needed for phloem loading (Atkin et al., 1983). In the case of cereals, surplus glutamine was found to be the principal nitrogen available for the synthesis of required amino acids (Simpson and Dalling, 1981). Among these, glutamic acid serves as a dual substrate and can be utilized by both aspartate and alanine aminotransferases to form aspartic acid and alanine respectively. These amino acids can then be used by transaminases and other biosynthetic enzymes to produce a series of different amino acids (Peoples and Dalling, 1988). This is supported by the observations that glutamate, aspartate and alanine represent between 50-80% of all amino acids present in soluble nitrogen pool in senescing leaves of both legumes and cereals (Atkin et al., 1980; Simpson and Dalling, 1981).

In nitrogen and carbon recycling steps of tree seeds which contain 10-40% of the nitrogen as arginine, hydrolysis of arginine, ornithine and urea by arginase and urease are the first events to occur (Thompson, 1980). In fruit trees (e.g., apple), much of the leaf nitrogen is translocated to cells under the bark during senescence for winter storage (Kang and Titus, 1980b). This stored nitrogen is then mobilized to leaves and flower buds for their spring growth. There was a 30-fold increase in the total nitrogen content of the phloem sap during flowering of apple trees in early spring, which sharply declined as extension growth proceeded. In the bark of dormant apple trees, arginine was the predominant transport component in autumn and winter, while asparagine, aspartate and glutamine were the major transport compounds in summer (O'Kennedy et al., 1975; O'Kennedy and Titus, 1979).

At the onset of flowering, nodulated root of white lupin consumes over 50% of net photosynthate synthesized by shoots and thereby monopolizes the translocate from the lower leaves (Pate, 1980; 1986). Net amount of nitrogen redistributed between nodules and roots is relatively small when compared with the contribution from leaves. However, roots play an important role in the nitrogen economy of plants especially during senescence, by cycling phloem-born nitrogen back to the shoot via the xylem stream (Pate et al., 1981; Simpson et al., 1983). The compositional differences between phloem and xylem saps indicate involvement of major enzymatic processes occurring in roots and nodules as discussed above.

It has been estimated that close to 90% of the nitrogen required by the developing fruits of legumes enters through the phloem (Atkins et al., 1975) reflecting a 10-20 fold higher concentration of nitrogen in phloem sap than in xylem sap (Pate, 1984). However, during early stages of fruit growth, a high proportion of nitrogen enters the fruit through xylem when soluble nitrogen pools are being established for the development of fruits and seeds (Atkins et al., 1975; Pate, 1984). Similar observations were also made in cowpea where phloem supplied respectively 97% and 72% of carbon and nitrogen required for fruit growth, and the small contribution from xylem occurred primarily during early stages of growth (Pate, 1984).

IX. Conclusion and Future Prospects

Mobilization of nitrogen from senescing tissue to regions of growth or storage alleviates to a certain extent the necessity for continued fixation of nitrogen by the plant and eliminates the unnecesary and wasteful utilzation of chemical energy. This is especially evident in monocarpic plants where fixed nitrogen is remobilized and stored in developing seeds for the future use of growing embryo. Most angiosperms of the temperate zone have coupled the autumnal senescence of leaves to the mobilization of nitrogen into phloem and roots, where they are stored for future utilization during spring growth. The storage could be in the form of proteins, amino acids or other nitrogen-rich compounds such as amides and guanidino compounds. Various types of proteolytic activity, expressed and regulated by hormones and environmental factors releases the nitrogen stored in proteins as amino acids. Moreover, interconversion between amino acids results in the formation of the ideal transport amino acids that are characteristic to the family and even individual species. Thus, mobilization of nitrogen during senescence serves an essential physiological function, integrating two temporally and spatially separated events, senescence and growth.

Literature Cited

Altman, A., and Bachrach, U. 1981. Involvement of polyamines in plant growth and senescence. *In Advances in Polyamine Research. Vol. 3.* (Caldarera, C.M., Zappia, V.,and Bachrach, U. eds.) Raven Press, New York. pp. 365-375.

Atkins, C.A., Pate, J.S., and McNeil, D.L. 1980. Phloem loading and metabolism of xylem-born amino compounds in fruiting shoots of a legume. *J. Exp. Bot.* **31**: 1509-1520.

Atkins, C.A., Pate, J.S., Peoples, M.B., and Joy, K.W. 1983. Amino acid transport and metabolism in relation to the nitrogen economy of a legume leaf. *Plant Physiol.* **71**: 841-848.

Atkins, C.A., Pate, J.S., and Sharkey, P.J. 1975. Asparagine metabolism-key to the nitrogen nutrition of developing legume seeds. *Plant Physiol.* **56**: 807-812.

Batt, R.G., and Walace, W. 1983. A comparison of the effect of trypsin and a maize root proteinase on nitrate reductase and other enzymes from maize. *Biochim. Biophys. Acta* **744**: 205-211.

Becana, M., Aparicio-Tejo, P.M., and Sanchez-Diaz, M. 1985. Levels of ammonia, nitrite and nitrate in alfalfa root nodules supplied with nitrate. *J. Plant Physiol.* **119**: 359-367

Becana, M., and Rodriguez-Barrueco, C. 1989. Protective mechanisms of nitrogenase against oxygen excess and partially-reduced oxygen intermediates. *Physiol. Plant.* **75**: 429-438.

Bell, E.A. 1976. "Uncommon" amino acids in plants. *FEBS Lett.* **64**: 29-35.

Bisseling, T., Van Straten, J., Houwaard, F. 1980. Turnover of nitrogenase and leghaemoglobin in root nodules of *Pisum sativum. Biochim. Biophys. Acta* **610**: 360-370.

Bond, J.S., and Offermann, M.K. 1981; Initial events in the degradation of soluble cellular enzymes: Factors affecting the stability and proteolytic susceptibility of fructose- 1,6- bisphosphate aldolase. *Arch. Biol. Med. Germ.* **40**: 1365-1374.

Brady, C.J. 1988. Nucleic acid and protein synthesis. *In Senescence and Aging in Plants.* (Nooden, L.D.,and Leopold, A.C. eds.) Academic Press Inc. New York. pp. 147-179.

Carey, W.F., and Wells, J.R.F. 1972. Phaseolain: a plant carboxypepetidase of unique specificity, *J. Biol. Chem.* **247**: 5573-5579.

Ciechanover, A., Finley, D., and Varshavasky, A. 1984. Ubiquitin dependence of selective protein degradation demonstrated in the mammalian cell cycle mutant ts85. *Cell* **37**: 57-66.

Ciechanover, A., Hod, Y., and Hershko, A. 1978. A heat-stable polypeptide component of an ATP-dependent proteolytic system from reticulocytes. *Biochem. Biophys. Res. Commun.* **81**: 1100-1105.

Chou, K-H., and Splittstoesser, W.E. 1972. Glutamate dehydrogenase from pumpkin cotyledons. *Plant Physiol.* **49**: 550-554.

Cooke, R.J., Roberts, K., and Davies, D.D. 1980a. Model for stress-induced protein degradation in *Lemna minor. Plant Physiol.* **66**: 1119-1122.

Cooke, R.J., Grego, S., Roberts, K., and Davies, D.D. 1980b. The mechanism of deuterium oxide-induced protein degradation in *Lemna minor. Planta* **148**: 374-380.

Collier, M.D., and Murray, D.R. 1977. Leucyl β-napthylamidase activities in developing seeds and seedlings of *Pisum sativum* L. *Aust. J. Plant Physiol.* **4**: 571-582.

Coventry, D.R., and Dilworth, M. J. 1976. Synthesis and turnover of leghaemoglobin in lupin root nodules. *Biochim. Biophys. Acta* **447**: 1-10.

Dalling, M.J. 1986. *Plant Proteolytic Enzymes,* Vol. II. Boca Raton, Florida, USA: CRC Press.

Dalling M.J., Boland, G., and Wilson, J. H. 1976. Relation between acid proteinase activity and redistribution of nitrogen during grain development in wheat. *Aust. J. Plant Physiol.* **3**: 721-730.

Dalling, M.J., and Nettleton, A.M. 1986. Chloroplast senescence and proteolytic enzymes. *In Plant Proteolytic Enzymes. Vol. II* (Dalling, M.J. ed.) CRC Press. Orlando Florida. pp. 125-153.

Dalling M.J., Tang, A., and Huffaker, R.C. 1983. Evidence for the existence of peptide hydrolase activity associated with chloroplasts isolated from barley mesophyll protoplasts. *Z. Pflanzenphysiol.* **111**: 311-318.

Dalton, D.A., Russell, S.A., Hanus, F.J., Pascoe, G.A., and Evans, H.J. 1986. Enzymatic reactions of ascorbate and glutathione that prevent peroxide damage in soybean root nodules. *Proc. Natl. Acad. Sci. USA* **83**: 3811-3815.

Desautels, M., and Goldberg, A.L. 1982. Liver mitochondria contain an ATP-dependent, vanadate-sensitive pathway for the degradation of proteins. *Proc. Natl. Acad. Sci. USA* **79**: 1869-1873.

Dumbroff, E.B. 1990. Polyamine-functions and relationships with ethylene and cytokining. In *Biochemistry and Physiology of Polyamines in Plants.* (Slocum, R.D., and Flores, H.E., eds.) CRC Press. Boca Raton, Florida, pp. 256-266.

Duxbury, C.L., Legge, R.L, Paliyath, G., Barber, R.J., and Thompson, J.E. 1991a. Alteration in membrane protein conformation in response to senescence-related changes in membrane fluidity and sterol concentration. *Phytochemistry* **30**: 63-68.

Duxbury, C.L., Legge, R.L., Paliyath, G., and Thompson, J.E. 1991b. Lipid breakdown in smooth microsomal membranes from bean cotyledons alters membrane proteins and induces proteolysis. *J. Exp. Bot.* **42**: 103-112.

Feller, U. 1978. Changes in nitrogen contents and in proteolytic activities in different parts of field-grown wheat ears (*Triticum aestivum* L.) during maturation, *Plant Cell Physiol.* **19**: 1489-1496.

Feller, U. 1979. Nitrogen mobilization and proteolytic activities in germinating and maturing bush beans (*Phaseolus vulgaris* L.). *Z. Pflanzenphysiol.* **95**: 413-422.

Feller, U. 1981. In vitro stability and inactivation of peptide hydrolases extracted from *Phaseolus vulgaris* L. *Plant Cell Physiol.* **22**: 1095-1104.

Feller, U. 1986. Proteolytic enzymes in relation to leaf senescence. In *Plant Proteolytic Enzymes Vol II.* (Dalling, M.J. ed.) CRC Press: Orlando, Florida, pp. 49-68.

Feller, U. 1990. Nitrogen remobilization and protein degradation during senescence. In *Nitrogen in Higher Plants.* (Abrol, Y.P. eds.) New York: Research Studies Press Limited, Sommerset, England and John Wiley and Sons Inc. pp. 195-222.

Feller, U., Soong, T-S. T., and Hageman, R.H. 1977. Leaf proteolytic activities and senescence during grain development of field grown corn (*Zea mays* L.). *Plant Physiol.* **59**: 290-294.

Feller, U., Soong, T-S. T., and Hageman, R.H. 1978. Patterns of proteolytic enzyme activities in different tissues of germinating corn (*Zea mays* L.). *Planta* **140**: 155-162.

Flores, H.E., and Filner, P. 1985. Polyamine catabolism in higher plants: characterization of pyrroline dehydrogenase. *Plant Growth Regul.* **3**: 277-291.

Fried, V.A., Smith, H.T., Hildebrandt, E., and Weiner, K. 1987. Ubiquitin has intrinsic proteolytic activity: Implications for cellular regulation. *Proc. Natl. Acad. Sci. USA* **84**: 3685-3689.

Galston, A.W., and Kaur-Sawhney, R. 1987. Polyamines and senescence in plants. In *Plant Senescence: It Biochemistry and Physiology.* (Thompson, W.W., Nothnagel, E.A., and Huffaker, R.C. eds.) American Society of Plant Physiologists. Maryland, Rockville, pp. 167-181.

Groat, R.G. 1981. Root nodule enzymes of ammonia assimilation in alfalfa (*Medicago sativa* L.). *Plant Physiol.* **67**: 1198-1203.

Gutschick, V.P. 1980. Energy flow in the nitrogen cycle, especially in fixation. In *Nitrogen Fixation, Vol. I.* (Newton, W.E., and Orme-Johnson, W.H. eds.) University Park Press. Baltimore, pp. 17-27.

Harper, L.A., Sharpe, P.P., Langdale, G.W., and Giddens, J.E. 1987. Nitrogen cycling in a wheat crop: Soil, plant, and aerial nitrogen transport. *Agron. J.* **79**: 965-973.

Harvey, B.M. R., and Oaks, A. 1974. The hydrolysis of endosperm protein in *Zea mays.* The role of gibberelic acid in the hydrolysis of endosperm reserves in *Zea mays. Planta* **121**: 67-74.

Hayashi, H., and Chino, M. 1990. Chemical composition of phloem sap from the uppermost internode of the rice plant. *Plant Physiol.* **31**: 247-251.

Heck, U., Martinoia, E., and Matile. Ph. 1981. Subcellular localization of acid proteinase in barley mesophyll protoplasts. *Planta* **151**: 198-200.

Hershko, A. 1988. Ubiquitin-mediated protein degradation. *J. Biol. Chem.* **263**: 15237-15240.

Hershko, A., and Ciechanover, A. 1986. The ubiquitin pathway for the degradation of intracellular proteins. *Prog. Nucleic Acid Res. Mol. Biol.* **33**: 19-56.

Hershko, A., Ciechanover, A., Haas, A.L., and Rose, I.A. 1980. Role of α-amino group of protein in ubiquitin-mediated protein breakdown. *Proc. Natl. Acad. Sci. USA* **81**: 7021-7025.

Hershko, A., Ciechanover, A., and Rose, I.A. 1981. Identification of the active amino acids reserve of the polypeptide of ATP-dependent protein breakdown. *J. Biol. Chem.* **256**: 1525-1528.

Hershko, A., Eytan, E., and Ciechanover, A. 1978. In Protein Turnover and Lysosome Function. (Segal, H.L., and Doyle, D.J. eds.) Academic Press, New York. pp. 149-169.

Hershko, A., Heller, H., Elias, S., and Reiss, Y. 1983. Components of ubiquitin-protein ligase system. Resolution, affinity purification, and role in protein breakdown. *J. Biol. Chem.* **258**: 8206-8214.

Hershko, A., Heller, H., Eytan, E., and Reiss, Y. 1986. The protein membrane-binding site of the ubiquitin-protein ligase system. *J. Biol. Chem.* **261**: 11992-11999.

Hershko, A., Leshinskly E., Ganoth, D., and Heller, H. 1984. ATP-dependent degradation of proteins in reticulocytes without affecting ubiquitin conjugation. *J. Biol. Chem.* **259**: 2803-2809.

Hochkeppel, H-K. 1973. Isolerung einer endopeptidase aus alternden tabakblattern und ihre beziehung zum vergilben. *Z. Pflanzenphysiol.* **69**: 329-343.

Holzer, H., and Heinrich, P.C. 1980. Control of Proteolysis. *Annu. Rev. Biochem.* **49**: 63-91.

Horiguchi, T., and Kitagishi, K. 1976. Protein metabolism in rice seedling. I. Effect of inhibitors of protein synthesis on the degradation of seed protein during germination. *Soil sci. Plant Nutr.* **22**: 327-343.

Huff. A. 1982. Peroxidase-catalyzed oxidation of chlorophyll by hydrogen peroxide. *Phytochemistry* **21**: 261-265.

Huffaker, R.C., and Paterson, L. W. 1974. Protein turnover in plants and possible means of its regulation. *Annu. Rev. Plant Physiol.* **25**: 363-392.

Kamachi, K., Yamaya, T., Mae, T., and Ojima, K. 1991. A role for glutamine synthetase in the remobilization of leaf nitrogen during natural senescence in rice leaves. *Plant Physiol.* **96**: 411-417.

Kang,SM., Matsui, H., and Titus, J. S. 1982. Characteristics and activity changes of proteolytic enzymes in apple leaves during autumnal senescence. *Plant Physiol.* **70**: 1367-1372.

Kang,SM., and Titus, J.S. 1980a. Isolation and partial characterization of an acid endopeptidase present in dormant apple shoot bark. *Plant Physiol.* **66**: 984-989.

Kang,SM., and Titus, J.S. 1980b. Qualitative and quantitative changes in nitrogenous compounds in senescing leaf and bark tissues of the apple. *Physiol. Plant.* **50**: 285-290.

Kaur-Sawhney, R., Flores, H.E., and Galston, A.W. 1980. Polyamine-induced DNA synthesis and mitosis in oat leaf protoplasts. *Plant Physiol.* **65**: 368-371.

Kaur-Sawhney, R., Shih, L., Cegielska, T., and Galston, A. W. 1982. Inhibition of protease activity by polyamines. Relevance for control of leaf senescence. *FEBS Lett.* **145**: 345-349.

Keist, M. 1980. Vergleich der extrahierbaren Aktivitaeten von Peptidhydrolasen in Blaettern Verschiedener Pflanzengruppen vor und waehrend der Senezenz, Lizentiatsarbeit, University of Berne, Switzerland.

Kessler, B., and Engelberg, N. 1962. Ribonucleic acid and ribonuclease activity in developing leaves. *Biochim. Biophys. Acta* **55**: 70-82.

Kuroki, M., Shioi, Y., and Sasa, T. 1981. Purification and properties of soluble chlorophyllase from tea leaf sprouts. *Plant Cell Physiol.* **22**: 717-725.

Kushad, M.M., Yelenosky, G., and Knight, R. 1988. Interrelatioship of polyamine and ethylene biosynthesis during avocado fruit development and ripening. *Plant Physiol.* **87**: 463-467.

Lamppa, G.K., Elliot, L.V., and Bendich, A.J. 1980. Changes in chloroplast number during pea leaf development. An analysis of a protoplast population. *Planta* **148**: 437-443.

Larson, L.A., and Beevers, H. 1965. Amino acid metabolism in young pea seedlings. *Plant Physiol.* **40**: 424-431.

Lin, W., and Witterbach, V.A. 1981. Subcellular localization of proteases in wheat and corn mesophyll protoplasts. *Plant Physiol.* **67**: 969-972.

Mae, T., Hoshino, T., and Ohira, K. 1985. Protease activities and loss of nitrogen in senescing leaves
of field grown rice (*Oryza sativa* L.) *Soil Sci. Plant Nutr.* **31**: 589-600.

Mae, T., Makino, A., and Ohira, K. 1983. Changes in the amounts of ribulose bisphosphate carboxylase synthesized and degraded during the life span of rice leaf (*Oryza sativa* L.). *Plant
Cell Physiol.* **24**: 1079-1086.

Mann, A. F., Fentem, P. A., and Stewart, G.R. 1980. Tissue localization of barley (*Hordeum vulgare*) glutamine synthetase isozymes. *FEBS Lett.* **110**: 265-267.

Marschner, H. 1986. *Mineral Nutrition of Higher Plants.* Academic Press, New York. pp. 181.

Martin, C., and Thimann, K.V. 1972. The role of protein synthesis in the senescence of leaves. I.
The formation of protease. *Plant Physiol.* **49**: 64-71.

Martinoia, E., Dalling, M.J., and Matil, Ph. 1982. Catabolism of chlorophyll: Demonstration of
chloroplast-localized peroxidase activities. *Z. Pflanzenphysiol.* **107**: 269-273.

Matsumoto, T., Yatazawa, M., and Yamamoto, Y. 1977. Distribution and change in the contents
of allantoin and allantoic acid in developing, nodulating and non-nodulating soybean plants.
Plant Cell Physiol. **18**: 353-359.

Mazumdar, S., Ghosh, S., Glick, B.R., and Dumbroff, E.B. 1991. Activities of chlorophyllase, phosphoenolpyruvate carboxylase and ribulose- 1,5-bisphosphate carboxylase in the primary
leaves of soybean during senescence and drought. *Physiol. Plant.* **81**: 473-480.

Miege, M.-N. 1982. Protein types and distribution. *In Encyclopaedia of Plant Physiology, Vol. 14A.
Nucleic Acids and Proteins in Plants I.* (Boulter, D., and Parthier, B. eds.) Springer Verlag
New York. pp. 291-345.

Miflin, B.J., and Lea, P.J. 1977. Amino acid metabolism. *Ann. Rev. Plant Physiol.* **28**: 299-329.

Millard, P. 1988. The accumulation and storage of nitrogen by herbaceous plants. *Plant Cell Environ.* **11**: 1-8.

Miller, B.L., and Huffaker, R.C. 1981. Partial purification and characterization of endoproteinases
from senescing barley leaves. *Plant Physiol.* **68**: 930-936.

Monk, I.S., Fagerstedt, K.V., and Crawford, R.M.M. 1989. Oxygen toxicity and superoxide dismutase as an antioxidant in physiological stress. *Physiol. Plant.* **76**: 456-459.

Murray, D.R., People, M.M., and Waters, S.P. 1979. Proteolysis in the axis of the germinating pea
seed. I. Changes in protein degrading enzyme activities of the radicle and primary root. *Planta*
147: 111-116.

Nettleton, A.M., Bhalla, P.L., and Dalling M.J., 1985. Characterization of peptide hydrolase activity
associated with thylakoids of the primary leaves of wheat. *J. Plant. Physiol.* **119**: 35-43.

Nooden, L.D. 1980. Senescence in the whole plant. In *Senescence in Plants.* (Thimann, K.V. ed.)
CRC Press. Boca Raton. Florida, pp. 219-258.

Oaks, A., Winspear, M.J., and Misra, S. 1983. Hydrolysis of endosperm protein in *Zea mays* (W_{64A}
X W_{182E}). 3rd Int. Symp. Pre-Harvest Sprouting Cereals. (Kruger, J.E., and LaBerge, D.E.
eds.) Westview Press. Boulder, Col. pp. 204-210.

Oaks, A., and Hirel, B. 1985. Nitrogen metabolism in roots. *Annu. Rev. Plant Physiol.* **36**: 345-365.

O'Kennedy, B.T., Hennerty, M.J., and Titus, 1975. Changes in the nitrogen reserves of apple shoots
during the dormant season. *J. Hort. Sci.* **50**: 321-329.

O'Kennedy, B.T., and Titus, J.S. 1979. Isolation and mobilization of storage proteins from apple
shoot bark. *Physiol. Plant.* **45**: 419-424.

Okita, T., Aryan, A., Reeves, C., Kim, W.T., Leisy, D., Hnilo, J., and Morrow, D. 1989. Molecular
aspects of storage protein and starch synthesis in wheat and rice seeds. *In Recent Advances in Phytochemistry, Vol. 23: Plant Nitrogen Metabolism.* (Poulton, J. E., Tomeo, J. T., and
Conn, E.E. eds.) Academic Press, London. pp. 289-327.

Ott, L.A., and Scandalios. J. G. 1976. Genetically defined peptidases of maize. I. Biochemical characterization of allelic and non-allelic forms, *Biochem. Genet.* **14**: 619-634.

Paliyath, G. 1979. Investigations on growth, auxin transport, haustoria formation and free amino acids in the angiospermic plant parasite *Cuscuta chinensis*, Lamk. *Ph. D. Thesis,* Indian Institute of Science, Bangalore, India.

Paliyath, G., and Droillard, M.J. 1992. Mechanisms of membrane deterioration and disassembly during senescence. *Plant Physiol. Biochem.* **30**: 789-812.

Paliyath, G., and Mahadevan, S. 1988. Changes in the levels of free amino acids in various regions of *Cuscuta* during growth. *Plant Cell Physiol.* **29**: 945-950.

Paliyath, G., and Thompson, J.E. 1988. Senescence-related changes in ATP-dependent uptake of Ca^{2+} into microsomal vesicles from carnation petals. *Plant Physiol.* **88**: 295-302.

Pate, J.S. 1980. Transport and partitioning of nitrogenous solutes. *Ann. Rev. Plant Physiol.* **31**: 313-340.

Pate, J.S. 1984. The carbon and nitrogen nutrition of fruit and seed: Case studies of selected grain legumes. In Seed Physiology-Development. (Murray, E.R. ed.) Vol. 1. Academic Press. New York. pp. 41-81.

Pate, J.S. 1986. Xylem-to-phloem transfer: A vital component of the nitrogen-partitioning system of a nodulated legume. *In Phloem Transport.* (Cronshaw, J., Lucas, W.J.,and Giaquinta, R.T. eds.) A.R. Liss, Inc. New York. pp. 445-462.

Pate, J.S. 1989. Synthesis, transport, and utilization of products of symbiotic nitrogen fixation. *In Recent Advances in Phytochemistry, Vol. 23: Plant Nitrogen Metabolism.* (Poulton, J.E., Romeo, J.T.,and Conn, E.E. eds.) Academic Press, London. pp. 65-115.

Pate, J.S., Atkins, C.A., Herridge, D.F., and Layzell, D.B. 1981. Synthesis, storage, and utilization of amino compounds in white lupin (*Lupinus albus* L.). *Plant Physiol.* **67**: 37-42.

Patterson, L.W., and LaRue, T.A. 1983. N_2 fixation and ureide content of soybeans: Environmental effects and source sink manipulations. *Crop Sci.* **23**: 819-824.

Peterson, L.W., and Huffaker, R.C. 1975. Loss of ribulose-1,5-bisphosphate carboxylase and increase in proteolytic activity during senescence of detached primary barley leaves. *Plant Physiol.* **55**: 1009-1015.

Peoples, M.B., Beiharz, V.C., Waters, S.P., Simpson, R.S., and Dalling, M.J. 1980. Nitrogen redistribution during grain growth in wheat (*Triticum aestivum* L.).. II. Chloroplast senescence and the degradation of ribulose-1,5-bisphosphate carboxylase. *Planta* **149**: 241-251.

Peoples, M.B., and Dalling, M.J. 1978. Intracellular localization of acid, peptide hydrolases in wheat leaves. *Plant Physiol.* **63** : S-159.

Peoples, M.B., and Dalling M.J. 1988. The Interplay between proteolysis and amino acid metabolism during senescence and nitrogen reallocation. *In Senescence and Aging in Plants.* (Nooden, L.D.,and Leopold, A.C. eds.) Academic Press Inc. New York. pp. 181-217.

Peoples, M.B., Firth, G.J.T., and Dalling M.J. 1979. Proteolytic enzymes in green wheat leaves. IV. Degradation of ribulose-1,5-bisphosphate carboxylase by acid proteinases isolated on DEAE-cellulose. *Plant Cell Physiol.* **20**: 253-258.

Peoples., M.B., Pate, J.S, and Atkins, C.A. 1983. Mobilization of nitrogen in fruiting plants of a cultivar of cowpea. *J. Exp. Bot.* **34**: 563-578.

Peoples, M.B., Pate, J.S., and Atkins, C.A. 1985. The effect of nitrogen source on transport and metabolism of nitrogen in fruiting plants of cowpea (*Vigna unguiculata* (L.) Walp.). *J. Exp. Bot.* **36**: 567-582.

Pfeiffer, N.E., Torres, C.M., and Wagner, F.W. 1983. Proteolytic activity in soybean root nodules. Activity in host cell cytosol and bacteroids throughout physiological development and senescence. *Plant Physiol.* **71**: 792-802.

Phillips, D.A., Center, D.M.,and Jones, M.B. 1983. Nitrogen turnover and assimilation during regrowth in *Trifolium subterraneum* L. and *Bromus mollis* L. *Plant Physiol.* **71**: 472-476.

Phillips, D.R., Horton, R.F., and Fletcher, R.A. 1969. Ribonuclease and chlorophyllase activities in senescing leaves. *Physiol. Plant.* **22**: 1050-1054.

Pladys, D., and Rigaud, J. 1985. Senescence in french-bean nodules: Occurrence of different proteolytic activities. *Physiol. Plant.* **63**: 43-48.

Preston, K.R., and Kruger, J.E. 1976. The nature and role of proteolytic enzymes during early germination. *Cereal Res. Commun.* **4**: 213-218.

Preston, K.R., and Kruger, J.E. 1976. Location and activity of proteolytic enzymes in developing wheat kernels. *Can J. Plant Sci.* **56**: 3217-3223.

Preston, K.R., and Kruger, J.E. 1976. Purification and properties of two proteolytic enzymes with carboxypeptidase activity in germinated wheat. *Plant Physiol.* **58**: 516-520.

Preston, K.R., and Kruger, J.E. 1986. Mobilization of monocot protein reserves during germination. *In Plant Proteolytic Enzymes. Vol. II.* (Dalling,M. J. ed.) CRC Press. Orlando, Florida. pp. 1-18.

Preston, K.R., and Kruger, J.E. 1979. Physiological control of exo - and endoproteolytic activities in germinating wheat and their relationship to storage protein hydrolysis. *Plant Physiol.* **64**: 450-454.

Purvis, A.C., and Barmore, C.R. 1981. Involvement of ethylene in chlorophyll degradation in peel of citrus fruit. *Plant Physiol.* **68**: 854-857.

Ragster, L.E., and Chrispeels, M.J. 1981. Autodigestion in crude extracts of soybean leaves and isolated chloroplasts as a measure of proteolytic activity. *Plant Physiol.* **67**: 104-109.

Robson, R.L., and Postage, J.R. 1980. Oxygen and hydrogen in biological nitrogen fixation. *Annu. Rev. Microbiol.* **34**: 183-207.

Rogers, S., Wells, R., and Rechsteiner, M. 1986. Amino acid sequences common to rapidly degraded proteins: The pest hypothesis. *Science* **234**: 364-368.

Ryan, C.A., and Walker-Simmons, M. 1981. Plant proteinases. In *The Biochemistry of Plants* Vol. VI. (Marcus, A. eds.) Academic Press. New York. pp. 321-350.

Salgo, A., and Feller, U. 1983. Effect of low molecular weight compounds on proteolytic inactivation of glucose-6-phosphate dehydrogenase. *Plant Physiol.* **72** (Suppl.): 35.

Scandalios, J.G., and Espiritu, L.G. 1969. Mutant aminopeptidases of *Pisum sativum*. I. Developmental genetics and chemical characteristics. *Mol. Gen. Genet.* **105**: 101-112.

Schubert, K.R. 1986. Products of biological nitrogen fixation in higher plants: Synthesis, transport, and metabolism. *Annu. Rev. Plant Physiol.* **37**: 539-574.

Shanklin, J., Jabben, M., and Vierstra, R. D. 1987. Red light-induced formation of ubiquitin-phytochrome conjugates: Identification of possible intermediates of phytochrome degradation. *Proc. Natl. Acad. Sci. USA* **84**: 359-363.

Simpson, R.J., and Dalling, M.J. 1981. Nitrogen redistribution during grain gorwth in wheat (*Triticum aestivum* L.). III. Enzymology and transport of amino acids from senescing flag leaves. *Planta* **151**: 447-456.

Simpson, R.J., Lambers, H., and Dalling, M.J. 1983. Nitrogen redistribution during grain growth in wheat (*Triticum aestivum* L.). IV. Development of a quantitative model of the translocation of nitrogen to the grain. *Plant Physiol.* **71**: 7-14.

Sprossler, B., Heilmann, H.D., Gramp. E., and Uhlig, H. 1971. Eigenschaften der Carboxypeptidase C aus Orangnblattern, Hoppe-Seyler's *Z. Physiol Chem.* **352**: 1524-1538.

Storey, R., and Beevers, L. 1977. Proteolytic activity in relationship to senescence and cotyledonary development in *Pisum sativum* L. *Planta* **137**: 37-44.

Streit, L., and Feller, U. 1982. Inactivation of N-assimilating enzymes and proteolytic activities in wheat leaf extracts: Effect of pyridine nucleotides and adenylates. *Experientia* **38**: 1176-1180.

Streit, L., and Feller, U. 1983a. Changing activities and different resistance to proteolytic activity of two forms of glutamine synthetase in wheat leaves during senescence. *Physiol. Veg.* **21**: 103-108.

Streit, L., and Feller, U. 1983b. Nitrogen-metabolizing enzymes from bean leaves (*Phaseolus vulgaris* L.): Stability *"in vitro"* and susceptibility to proteolysis. *Z. Pflanzenphysiol.* **111**: 19-27.

Tajima, S., and Yamamoto, Y. 1975. Enzymes of purine catabolism in soybean plants. *Plant Cell Physiol.* **16**: 271-282.

Tanaka, K., Waxman, L., and Goldberg, A.L. 1984. Vanadate inhibits the ATP-dependent degradation of proteins in reticulocytes without affecting ubiquitin conjugation. *J. Biol. Chem.* **259**: 2803-2809.

Tazaki, K., and Ishikura, N. 1983. Multiple forms of aminopeptidase in *Euonymus* leaves. *Plant Cell Physiol.* **24**: 1263-1268.

Thimann, K.V. 1980. The senescence of leaves. *In Senescence in Plants.* (Thimann, K.V. ed.) CRC Press. Florida, Boca Raton, pp. 85-115.

Thomas, H. 1975. Regulation of alanine aminotransferase in leaves of *Lolium temulentum* during senescence. *Z. Pflanzenphysiol.* **74**: 208-218.

Thomas, H. 1977. Ultrastructure, polypeptide composition and photochemical activity of chloroplasts during foliar senescence of a non-yellowing mutant genotype of *Festuca pratensis* Huds. *Planta* **137**: 53-60.

Thomas, H. 1978. Enzymes of nitrogen mobilization in detached leaves of *Lolium temulentum* during senescence. *Planta* **142**: 161-169.

Thomas, H. 1982. Control of chloroplast demolition during leaf senescence. *In Plant Growth Substances.* (Wareing, P.F. ed.) Academic Press. London. pp. 559-567.

Thomas, H., and Huffaker, R.C. 1981. Hydrolysis of radioactively-labelled ribulose-1,5-bisphosphate carboxylase by an endopeptidase from primary leaf of barley seedlings. *Plant Sci. Lett.* **20**: 251-262.

Thomas, H., and Stoddart, J.L. 1975. Separation of chlorophyll degradation from other senescence processes in leaves of a mutant genotype of meadow fescue (*Festuca pratensis*). *Plant Physiol.* **56**: 438--441.

Thomas, H., and Stoddart, J.L. 1980. Leaf senescence. *Annu. Rev. Plant Physiol.* **31**: 83-111.

Thompson, J.F. 1980. Arginine synthesis, proline synthesis, and related process. *In the Biochemistry of Plants.* (Miflin, B.J. ed.) Academic Press. New York. **5**: 375-402.

Van der Wilden, W., Segers, J.H.L., and Chrispeels, M.J. 1983. Cell walls of *Phaseolus vulgaris* leaves contain the azocoll-digesting proteinase. *Plant Physiol.* **73**: 576-578.

Vance, C.P. 1978. Nitrogen fixation in alfalfa: An overview. *In Proc. 8th Annu. Alfalfa Symp.,* pp. 34-41.

Vance, C.P., Heichel, G.H., Barnes, D.K., Bryan, J.W., and Johnson, L.E. 1979. Nitrogen fixation, nodule development, and vegetative regrowth of alfalfa (*Medicago sativa* L.) following harvest. *Plant Physiol.* **64**: 1-8.

Walker-Simmons, M., and Ryan, C.A. 1977a. Wound-induced peptidase activity in tomato leaves. *Biochem. Biophys. Res. Commun.* **74**: 411-416.

Walker-Simmons, M., and Ryan, C.A. 1977b. Immunological identification of proteinase inhibitors I and II in isolated tomato leaf vacuoles. *Plant Physiol.* **60**: 61-63.

Walker-Simmons, M., and Ryan, C.A. 1980. Isolation and properties of carboxypeptidase from leaves of wounded tomato plants. *Phytochemistry* **19**: 43-47.

Wallace, W., and Shannon, J.D. 1981. Proteolytic activity and nitrate reductase inactivation in maize seedlings. *Aust. J. Plant Physiol.* **8**: 211-219.

Wanger, G.J., Mulready, P., and Cutt, J. 1981. Vacuole/extravacuole distribution of soluble protease in *Hippeastrum* petal and *Triticum* leaf protoplasts. *Plant Physiol.* **68**: 1081-1090.

Watanabe, T., and Kondo, N. 1983. The change in leaf protease and protease inhibitor activities after supplying various chemicals. *Biol. Plant.* **25**: 100-109.

Waters, S.P., and Dalling, M.J. 1979. Distribution and characteristics of aminoacyl β-napthylamidase activities in wheat seedlings. *Aust. J. Plant Physiol.* **6**: 595-606.

Waters, S.P., Nobel, E.R., and Dalling, M.J. 1982. Intracellular localization of peptide hydrolase in wheat (*Triticum aestivum* L.) leaves. *Plant Physiol.* **69**: 575-579.

Waters, S.P., Peoples, M.B., Simpson, R.J., and Dalling, M.J. 1980. Nitrogen redistribution during grain growth in wheat (*Triticum aestivum* L.). 1. Pattern of peptide hydrolase activity and protein breakdown in the flag leaf, glumes and stem. *Planta* **148**: 422-428.

Waxman, L., and Goldberg, A.L. 1982. Protease from *Escherichia coli* hydroyzes ATP and proteins in a linked fashion. *Proc. Natl. Acad. Sci. USA* **79**: 4883-4887.

Wells, J.R.E. 1965. Purification and properties of a proteolytic enzyme from french beans. *Biochem. J.* **97**: 228-235.

Wittenbach, V.A. 1978. Breakdown of ribulose bisphosphate carboxylase and change in proteolytic activity during dark-induced senescence of wheat seedlings. *Plant Physiol.* **62**: 604-608.

Wittenbach, V.A., Lin, W., and Hebert, R.R. 1982. Vacuolar localization of proteases and degradation of chloroplasts in mesophyll protoplasts from senescing primary wheat leaves. *Plant Physiol.* **69**: 98-102.

Woolhouse, H. W. 1982. Leaf senescence. *In The Molecular Biology of Plant Development.* (Smith, H., and Grierson, G. eds.) Blackwell. Oxford, pp. 256-281.

Yao, K., Paliyath, G., Humphrey, R.W., Hallet, R.F., and Thompson, J.E. 1991. Identification and characterization of non-sedimentable lipid-protein microvesicles enriched in phospholipid degradation products. *Proc. Natl. Acad. Sci. USA* **88**: 2269-2273.

Zuber, H., and Matile, Ph. 1968. Acid carboxypeptidases: Their occurrence in plants, intracellular distribution and possible function, *Z. Naturforsch.* **23b**: 663-665.

Zuber, H. 1976. Carboxypeptidase C. *In Methods Enzymol.* **45b**: 561-568.

Nitrogen Nutrition in Higher Plants, 1995
Editors : H.S. Srivastava & R.P. Singh
Associated Publishing Co., New Delhi, India
pp. 367-384.

Light and Nitrogen Assimilation

INEKE STULEN and MARGREET TER STEEGE

I. Introduction

The effect of light on nitrogen metabolism has been studied in many plant species and the stimulatory effect of light on nitrate assimilation has long been recognized (Beevers and Hageman, 1969). Light may exert its influence on nitrogen assimilation in various ways, either directly by (a) affecting the synthesis of the enzymes involved, or indirectly via products of photosynthesis, *viz.* (b) by providing reductant and ATP for functioning of the enzymes, (c) by modulating enzyme activity via ATP/AMP levels and (d) by providing carbon skeletons for accepting the reduced nitrogen. Besides this, light may affect: (a) the transport of inorganic nitrogen into cellular compartments such as vacuoles and chloroplasts; (b) the amount of nitrate accumulated in the vacuole, and hence; (c) the partitioning of nitrogen between inorganic and organic nitrogen fractions.

The role of light on nitrate assimilation, as well as the interaction between carbon and nitrogen metabolism, have both been extensively dealt with in a number of recent review articles (Stulen, 1986; 1990; Hageman and Below, 1990; Sawhney and Naik 1990). Most articles have dealt with the influence of white light mainly; although one included the influence of the spectral quality of light. In the present chapter, the current ideas on the influence of light on nitrogen assimilation in higher plants, including the developments since the advancement of genetical and molecular techniques, are discussed in relation to the compartmentation of the various processes within the leaf cell and in the plant as a whole.

II. Enzyme Synthesis

A. Nitrate Reductase

The first enzyme in the reduction pathway of nitrate to ammonium is nitrate reductase (NR) which has been the object of many studies. It is located in the cytoplasm (Oelmüller et al., 1988; Sopory and Sharma, 1990), although it may be loosely associated with the chloroplast or microbody-like particles (Hewitt, 1975; Vaughn et al., 1984; Sopory and Sharma, 1990). Nitrate reductase is among the most efficient catalysts known (Campbell, 1990; Srivastava, 1992). NADH:NR (EC 1.6.6.1) is the most common form of nitrate reductase in higher plants; it is a homodimer, with two identical subunits of a size between 100 and 120 kD, each containing flavin (FAD), heme, and a molybdenum cofactor. Electrons are transferred between the pyridine nucleotide oxidation site and the nitrite reduction site (Kleinhofs et al., 1989; Caboche and Rouzé, 1990; Callaci and Smarelli Jr. 1991). It is much less stable than the next enzyme in the pathway, nitrite reductase (NiR) (Beevers and Hageman, 1980).

The adaptive nature of NR, *viz.* the induction of the enzyme by its substrate nitrate, is well documented (Beevers and Hageman, 1969). Since the advancement of genetical and molecular techniques, it has become possible to study the induction process of NR, *viz.* the expression of the enzyme protein and the development of the *in vitro* activity, in detail. After the technique for cloning of the NR gene was developed, it has been possible to demonstrate that the induction by nitrate occurs at the level of transcription (Galangau et al., 1988; Daniel-Vedéle et al., 1989; Callaci and Smarelli Jr., 1991). Transfer of nitrate-starved tobacco plants to nitrate resulted within minutes in an accumulation of NR-mRNA (Galangau et al., 1988). Exposure of maize plants to nitrate also resulted in an increase in the steady-state level of NR mRNA. This was followed by an increase in NR protein (NRP) level and finally by the appearance of *in vitro* NR activity (Campbell, 1990). Although the NRP and the appearance of *in vitro* NR activity are both induced by nitrate, there were important differences in the timing and the responses to environmental cues (Melzer et al., 1989; Oaks et al., 1990). Experiments with maize seedlings, grown on nitrate concentrations of varying strength, showed that at very low external nitrate concentrations an inactive form of the NRP (a 116 kD protein) was synthesized, which was activated at higher nitrate concentrations (Oaks et al., 1988; 1990).

In addition to nitrate, light appeared to be necessary for the appearance of the NRP; in maize seedlings the 116 kD NRP was detected in light-grown seedling shoots but not in dark-grown seedling shoots (Oaks et al., 1988). The light require-ment for expression of NR was also shown for tobacco plants. NR mRNA, NRP and *in vitro* NR activity all increased upon illumination of dark-grown tobacco plants (Deng et al., 1990). The exact nature of the stimulatory effect of light on the appear-ance of NR has not been established yet. It has been postulated that light might exert its positive control over the nitrate-mediated induction of NR via the far-red absorbing form of phytochrome (Schuster et al., 1987), possibly in combination with a factor from intact plastids (Rajasekhar and Mohr, 1986; Oelmüller et al., 1988). However, it should be noted that the phytochrome response might exert its influ-ence in a more indirect way, for instance via the release of nitrate from the vacuole (see below), thereby increasing the level of nitrate in the cytoplasm.

A diurnal rhythm of *in vitro* NR activity has been shown for various plant species, with a high activity during the day, and a lower activity during the night (Hageman et al., 1961; Beevers and Hageman, 1969; Gebauer et al., 1984; Lillo and Hen-riksen, 1984; Gebauer, 1990). More detailed experiments revealed that in tobacco leaves the NR mRNA level, the amount of NRP and *in vitro* NR activity all showed diurnal changes (Galangau et al., 1988). Detailed experiments with corn seedlings showed that the light-induced circadian oscillations of NR originate at the nucleic-acid level and are followed by much broader oscillations at the protein and activity level (Lillo, 1989). Experiments with tobacco plants, transferred to continuous light, also indicated that in the expression of the NR gene a circadian rhythm is involved (Deng et al., 1990). In maize shoots transferred to continuous darkness, NR mRNA and *in vitro* NR activity decreased rapidly, and in a parallel trend (Bowsher et al., 1991). In tobacco leaves kept in the dark for a long period, NR mRNA also decreased, but not in parallel with NRP and NR activity. From these experiments the conclu-sion was drawn that light was required for either NR-gene transcription or mRNA stability (Deng et al., 1990).

Rapid changes in *in vitro* NR activity were found in spinach leaves, when rates of net photosynthesis were modulated by varying the CO_2 supply (Kaiser and Brendle-Behnisch, 1991). Further *in vitro* experiments (Kaiser and Spill, 1991) sug-gested that the levels of ATP/AMP in the cytosol might be the central link between *in vitro* NR activity in the cytosol and photosynthesis in the chloroplast. It was pro-posed that phosphorylation/dephosphorylation of NR might be a mechanism for reversible modulation of NR activity by ATP and AMP.

B. Nitrite Reductase

The reduction of nitrite to ammonium is catalyzed by nitrite reductase (NiR, EC 1.7.7.1.), which is located inside the chloroplast in the leaf (Beevers and Hage-man, 1980; Oelmüller et al., 1988; Sopory and Sharma, 1990), and inside a plastid in the root (Beevers and Hageman, 1980; Emes and Fowler, 1983; Oji et al., 1985). The enzyme can be induced by nitrite, as well as by nitrate (Beevers and Hageman, 1969; Rao et al., 1980; Sopory and Sharma, 1990). It is still a matter of debate, however, whether NiR is induced by nitrate itself, or indirectly by nitrite, formed by reduction of nitrate (Sopory and Sharma, 1990). The enzyme is more stable than NR (Beevers and Hageman, 1980). NiR protein is a monomer with a molecular weight between 60 and 70 kD, which contains one siroheme, to which nitrite binds, and one 4Fe-4S center, that functions as the initial electron acceptor (Vega et al., 1980; Guerrero et al., 1981).

The appearance of *in vitro* NiR activity is stimulated by light. Experiments with antibodies against NiR showed that the light-induced increase in *in vitro* NiR activity involved alteration of NiR protein level. It was postulated that light might control the level of transcription or translation of NiR (Gupta and Beevers, 1984; 1985). Elucidation of the mechanism of the nitrate-inducible expression has advanced since the NiR genes have been cloned (Back et al., 1988). Recent experiments have demonstrated that the induction of NiR by nitrate occurs at the level of transcription (Lahners et al., 1988).

No diurnal cycling in *in vitro* NiR activity was found in maize shoots, but the level of NiR mRNA did show considerable diurnal cycling (Bowsher et al., 1991). When maize plants were grown in continuous darkness, NiR activity showed no rapid decrease, but the level of NiR mRNA decreased to an undetectable level after a prolonged period in the dark (Bowsher et al., 1991). In these experiments, therefore, little correlation between mRNA and NiR activity was found. From these experiments the conclusion was drawn that post translational processing must also be involved in maintaining a sufficient level of *in vitro* NiR activity (Bowsher et al., 1991). There is some evidence that a similar phytochrome-mediated process as postulated for the expression of NR, might be involved in the expression of NiR (Sopory and Sharma, 1990; Weber et al., 1990). However, the requirement for a plastid factor for NiR gene-expression appeared to be low, compared to the requirement of NR for this factor (Oelmüller et al., 1988).

Summarizing, there is much experimental evidence that light influences the expression of NR and NiR to a great extent. The exact regulatory mechanism, however, is still unknown, and may involve other (vital) processes, such as the availability of nitrate in the cytoplasm.

III. Enzyme Functioning

A. Nitrate Reductase

The reduction of nitrate to nitrite is markedly stimulated by light. The reductant for the most common form of NR (EC 1.6.6.1.) is NADH, which supplies 2 electrons and 2 H^+, needed for the conversion of nitrate to nitrite.

$$NO_3^- + NADH + H^+ \xrightarrow{NR} NO_2^- + NAD^+ + H_2O$$

It is generally accepted that light stimulates nitrate reduction by increasing the generation and the availability of NADH for functioning of the enzyme. However, there is still some doubt about the question whether the reducing equivalents for functioning of NR originate from within the chloroplast, and/or the mitochondria (Abrol et al., 1983; Stulen, 1986; 1990; Sawhney and Naik, 1990).

From experiments with leaf segments, incubated in the dark under anaerobic conditions (the conditions of the *in vivo* assay of NR), the conclusion was drawn that NADH produced in the glycolytic pathway—the oxidation of triose phosphates by the cytosolic enzyme triose phosphate dehydrogenase—is the source of reductant for the functioning of NR (Klepper et al., 1971). In the light, photosynthetically generated reducing equivalents can be transferred from the chloroplast to the cytosol via shuttle mechanisms, involving transport of triose phosphates or dicarboxylic acids, and generating NADH for the reduction of nitrate in the cytosol. There is also evidence that in the light, when oxidation of NADH in the respiratory chain might be suppressed, the NADH generated in the mitochondria is available for nitrate reduction (Stulen, 1986; 1990; Sawhney and Naik, 1990). Continuous functioning of the TCA cycle in the light, by withdrawal of NADH for NR, would be of great importance to insure a sufficient supply of carbon for accepting the reduced nitrogen (see next section).

The interactions between nitrogen assimilation and respiration during photosynthesis have been studied in detail in the green alga *Selenastrum minutum* (Turpin et al., 1990). During assimilation of nitrate and nitrite in the light, respiratory carbon flow was enhanced in nitrogen-limited *Selenastrum minutum*. It was clearly shown that respiration provided not only carbon for amino acid synthesis, but also the reducing power for functioning of NR and NiR. Respiration, therefore, is important in supporting photosynthetic nitrogen assimilation in this organism (See Turpin et al., 1990 for references). Experiments with barley shoots, in which the interaction between nitrate assimilation, photosynthesis and respiration was also studied, revealed that the alga and the higher plant differ fundamentally in their partitioning of energy to nitrate assimilation. In barley, nitrate assimilation appeared to have little effect on carbon fixation in the light, so that diversion of reductant to nitrate assimilation should not diminish carbon fixation (Bloom et al., 1989).

It is concluded that the supply of reductant for functioning of NR might not be limiting for nitrate assimilation and plant growth at high light intensity. At low light intensities, however, a limited supply of reductant might indeed play a role (Smirnoff and Stewart, 1985; De Kok et al., 1986).

B. Nitrite Reductase

The role of light in nitrite reduction is well understood, since many experiments have clearly shown that in leaves nitrite reduction is a photosynthetic process in which NiR uses Fd_{red} generated by light reactions in the chloroplast to provide the six electrons needed for reduction of nitrite to ammonium (Vega et al., 1980; Guerrero et al., 1981).

$$NO_2^- + 6\ Fd_{red} + 8H^+ \overset{NiR}{\rightarrow} NH_4^+ + 6\ Fd_{ox} + 2H_2O$$

Whether nitrite and CO_2 compete for Fd_{red} in the chloroplast was measured in a reconstituted spinach chloroplast system (Baysdorfer and Robinson, 1985). In this system, competition for Fd_{red} between ferredoxin-NADP reductase and nitrite (added to the system in high concentrations) could indeed be demonstrated. Whether this competition also occurs *in situ* was investigated in later experiments with intact leaf plastids from spinach and mesophyll cells from soybean (Robinson, 1986). The rate of nitrite reduction was measured when the demand for Fd_{red} for CO_2 assimilation was either absent, rate limiting or saturating. In neither of these experiments, a competition between nitrite and CO_2 for Fd_{red} was found. Apparently, at the high light intensities used in these experiments there is sufficient Fd_{red} for functioning of both nitrite reduction and CO_2 assimilation. However, it cannot be ruled out that at low light intensities, or when the photosynthetic apparatus is damaged, as in a polluted atmosphere, a competition for Fd_{red} in the chloroplast might occur (Stulen, 1986).

C. Functioning of NR in Situ

By comparing *in vivo* and *in vitro* activities of NR in relation to the nitrogen flux needed to maintain reduced nitrogen concentration in the plant, the conclusion can be drawn that the level of the enzyme (as determined by an *in vitro* assay) is excessive (Stulen and De Kok, 1993). For *Spinacia oleracea* for instance, *in vitro* activities ranged from 16 - 29 μmol gFW^{-1} h^{-1} (De Kok et al., 1986; Steingröver et al., 1986), while *in vivo* activities in plants, grown under the same experimental conditions, ranged from 1 - 4 μmol g^{1}FW^{1}h^1 (De Kok et al., 1986; Steingröver et al., 1986). The calculated nitrogen flux, needed to maintain the protein level during growth, was approximately 2 μmol g^{1}FW^1 h^1. The same can be concluded from experiments in which the incorporation of ^{15}N nitrate into the reduced nitrogen fraction was compared with the level of NR measured *in vitro* (Stulen et al., 1973). In some cases the *in vivo* assay for NR provided a closer approximation of the accumulation of reduced nitrogen in the plant (Brunetti and Hageman, 1976; Stulen et al., 1981).

The group of Hageman carried out many experiments with wheat and maize, in order to investigate whether a correlation between NR activity *in vitro* or *in vivo*, and protein yield could be established (Hageman, 1990; Hageman and Below, 1990) for references). It was shown that for a given genotype *in vitro* NR activity of the whole plant, integrated over time, was closely associated with the actual amount of protein accumulated by the plant. However, the view that such measurements of *in vitro* nitrate reductase activity could serve as an estimate of the input of reduced nitrogen in the plant in general, was not valid for diverse genotypes. Likewise, no correlation was found between measurements of NR activity *in vivo* and eventual crop protein-nitrogen accumulation. Based on all these experiments the conclusion was drawn that the *in situ* accumulation of reduced nitrogen could also be affected by the amount of nitrate at the induction and assimilation site of NR, and the availability of metabolites for generation of reductant (Hageman, 1990). The influence of light on these processes is dealt with in the next sections.

Summarizing, the conclusion can be drawn that at the biochemical level there are many experimental data to explain the stimulatory effect of light on the functioning of NR and NiR, by increasing the generation of reductant. However, it is unlikely that the supply of either NADH or Fd_{red} for functioning of NR and NiR in the leaf is limiting at normal light intensities.

IV. Synthesis of Organic Nitrogenous Compounds

A. *Amino Acid Synthesis: Synthesis and Functioning of Enzymes*

Ammonia, the product of NiR, is incorporated into amino acids by glutamine synthetase (GS; EC 6.3.1.2) and glutamate synthase (GOGAT), which is either Fd_{red} (EC 1.4.7.1) or NAD(P)H-dependent (EC 1.4.1.13) (Miflin and Lea, 1976; Lea et al., 1992). GS is an octameric protein, with a molecular weight of 350-400 kD (Stewart et al., 1980). In leaves two major isoforms have been isolated: GS_1 in the cytoplasm and GS_2 in the chloroplast (McNally et al., 1983). Fd_{red}-dependent GOGAT is an iron-sulfur protein with a molecular weight from 140-230 kD and is located in the chloroplast; the NADH-dependent form is a monomer, with a molecular weight of 158-200 kD, and is also located in the plastid (see Lea et al., 1992 for references). It has been a matter of long debate whether glutamate dehydrogenase (GDH; EC 1.4.1.3) is also involved in the assimilation of ammonia. However, it has by now been established that in most cases over 95% of the ammonia from nitrate reduction is assimilated via the GS/GOGAT pathway. There is evidence for a catabolic function of GDH, providing carbon for functioning of the TCA cycle (Yamaya and Oaks, 1987; Srivastava and Singh, 1987; Oaks and Yamaya, 1990; Robinson et al., 1991; Lea et al., 1992).

The effect of light, nitrate and ammonium on the appearance of GS (the chloroplast form) was investigated in mustard cotyledons. These experiments showed that light, operating via phytochrome, was the main factor in the process of appearance of *in vitro* GS activity (Schmidt and Mohr, 1989). The *in vitro* activity of Fd_{red}-dependent GOGAT (EC 1.4.7.1) also increased considerably during greening of etiolated leaf tissue (Suzuki et al., 1982; Wallsgrove et al., 1982). Experiments with maize suggested that this increase involved an initial activation of the enzyme, followed by new synthesis of GS protein (Suzuki et al., 1987). Experiments with Scots pine showed that the *in vitro* activity of cytosolic NADH-GOGAT was not significantly affected by light, but that of plastidic Fd_{red}-GOGAT was increased strongly in response to light. It was suggested that the latter increase could be mediated by phytochrome (Elmlinger and Mohr, 1991).

Diurnal variations in *in vitro* activities were found for both GS and Fd_{red}-dependent GOGAT, although less pronounced than for NR (Lillo, 1984). This study showed that *in vitro* NiR and GS activities were about 10 times higher, and GOGAT activity was at least 3 times higher than NR activity, indicating that the level of NiR, GS and GOGAT do not limit the further assimilation of nitrite, generated by NR, either during the day or during the night (Lillo, 1984). It should be pointed out that the main source of ammonia for the GS/GOGAT cycle in the light is ammonia released in the pocess of photorespiration, during the conversion of glycine to serine (Keys et al., 1978; Lea et al., 1992), and that the GS/GOGAT cycle, therefore, is engaged in the recycling of internally produced NH_3 to a great extent.

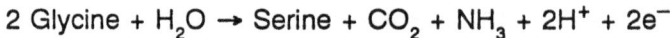

$$2\ \text{Glycine} + H_2O \rightarrow \text{Serine} + CO_2 + NH_3 + 2H^+ + 2e^-$$

The rate of release of NH_3 in photorespiration is at least ten times the rate of NH_3 production by nitrate reduction under normal light conditions (Stulen, 1986; Lea et al., 1992). ^{15}N-nitrate studies with GS-deficient mutants, containing no more than 10% of cytosolic GS activity compared to the wild-type, showed that the remaining GS activity was sufficient to maintain normal growth under non-photorespiratory conditions (Lea et al., 1992).

In the leaf GS is dependent on photosynthetically generated ATP, while GOGAT (EC 1.4.7.1.) is dependent on Fd_{red}, generated in the chloroplast, similar to NiR.

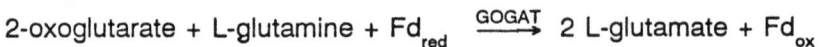

$$\text{L-glutamate} + ATP + NH_3 \xrightarrow{\text{GS}} \text{L-glutamine} + ADP + Pi + H_2O$$

$$2\text{-oxoglutarate} + \text{L-glutamine} + Fd_{red} \xrightarrow{\text{GOGAT}} 2\ \text{L-glutamate} + Fd_{ox}$$

It has been suggested that the enzyme ferredoxin:NADP reductase, which transmits reducing power from ferredoxin to NADP, might have an important function in balancing nitrogen and carbon metabolism by diverting reducing power towards nitrogen assimilation when excess carbon is available (Vennesland and Guerrero, 1979). It has already been pointed out in the previous section that at normal conditions of light and nutrient supply, competition for Fd_{red} by CO_2 assimilation on the one hand, and functioning of NiR and GOGAT on the other seems unlikely (Robinson, 1986; Bloom et al., 1989).

B. Amino Acid Synthesis: Provision of Carbon Skeletons

Carbon skeletons from the citric acid cycle are needed to accept to ammonia generated by NiR. Formation of amino acids, therefore, draws heavily on the citric acid cycle and replenishment of carbon is needed for continuous functioning of the cycle. Malate appears to be an important source of carbon for the anaplerotic operation of the citric acid cycle in the root (Naik and Nicholas, 1984). Experiments with roots suggested that carbon from the citric acid cycle, used for amino acid synthesis, can be replenished by oxaloacetate or malate, generated by phospho-enolpyruvate (PEP) carboxylase (Popp and Summons, 1984). As pointed out in the previous section, there is evidence that GDH might be involved in the replenish-ment of carbon in the citric acid cycle as well (Yamaya and Oaks, 1987; Oaks and Yamaya, 1990; Robinson et al., 1991). Finally, carbon from photorespiration might serve as a carbon source for the citric acid cycle in the light (Singh et al., 1985).

Continuous operation of the citric acid cycle in the light, and continuous respi-ratory carbon flow, therefore, is a prerequisite for incorporation of reduced nitrogen into amino acids. To what extent respiratory carbon flow is influenced by light has been questioned a great deal (Abrol et al., 1983). Experiments with algae and barley have revealed that in the light, under photosynthetic conditions, respiration is impor-tant in providing reductant for nitrate assimilation as well as carbon for amino acid synthesis (Bloom et al., 1989; Turpin et al., 1990).

Addition of ammonia to isolated photosynthesizing leaf cells resulted in an increased flow of carbon into the citric acid cycle due to stimulation of pyruvate kinase, and PEP carboxylase activities and due to an increased flow of 2-oxoglutarate from the citric acid cycle towards glutamine (Platt et al., 1977; Paul et al., 1978). The effect of endogenous ammonia, generated in photorespiration, appeared to be similar to the effect of exogenously added ammonia, thereby suggesting that the latter effect is an increased manifestation of the regulation in the intact plant (Lawyer et al., 1981). Experiments in which malate metabolism was investigated in relation to amino acid metabolism during supply of ammonium to radish cotyledons showed that the increased flux of carbon through the citric acid cycle, due to enhanced gluta-mine synthesis, operated at the expense of malate, derived from the PEP carbo-xylase pathway (Dahlbender and Strack, 1986). When nitrate was the nitrogen source in these experiments, a relationship between the turnover of malate and that of glutamate was also observed. Diurnal changes in leaf malate concentration, similar to those of *in vitro* NR activity were found in tobacco leaves (Deng et al., 1989). It should be noted that overall nitrate reduction uses more H^+ than electrons, and results in an increase in cytoplasmic pH, which in its turn stimu-lates the enzyme PEP carboxylase (Pilbeam and Kirkby, 1990).

C. Protein Synthesis

The effect of light on the synthesis of specific enzymes has been dealt with in previous sections. It should be emphasized that light stimulates overall protein syn-thesis by its effect on the activation of ribosomes, which is phytochrome-mediated (Travis et al., 1974).

V. Partitioning of Nitrogen Compounds within the Cell

A. Nitrate in the Cytoplasm

Light may affect the availability of nitrate within the leaf cell, at the site of induction of NR and NiR, in various ways, *viz.* by affecting the uptake of nitrate by the root and the transport of nitrate to the shoot, and/or by mediating the release of nitrate from the vacuole.

Based on experiments with ^{15}N labelled nitrate, the conclusion was drawn that the rate of nitrate uptake by the root in the light and in the dark did not differ much (Rufty et al., 1984). In a number of experiments the rate of translocation to the shoot was greater in the light than in the dark (Rufty et al., 1984; Mattson et al., 1988). It has long been established that the flux of nitrate to the shoot determines the level of NR in the shoot (Shaner and Boyer, 1976).

A great part of the nitrate present in the cell may be localized in the vacuole, and therefore not available for induction of NR or reduction by NR (Hewitt, 1975; Hewitt et al., 1979). Experiments with barley leaves have shown that in the dark, nitrate stored in the vacuole is not available for the induction of NR, but that upon a dark-light transition nitrate is released and the induction of NR is enhanced (Aslam and Oaks, 1976). From detailed ^{15}N-nitrate labelling studies the conclusion could be drawn that nitrate stored in the vacuole and released at the beginning of the light period was incorporated in proteins (Rufty et al., 1989).

B. Nitrite Transport into the Chloroplast

Nitrite, the product of nitrate assimilation, has to be transported into the chloroplast before it can be reduced by nitrite reductase. To what extent nitrite transport into the chloroplast is influenced by light/dark conditions was investigated with intact pea chloroplasts (Brunswick and Cresswell, 1988 a,b). The experiments showed that nitrite uptake into intact chloroplasts was closely related to stroma pH and nitrite reduction, and suggested a partial dependence on photophosphorylation (photosystem I).

C. Nitrate Accumulation in the Vacuole

Nitrate can be accumulated in leaf vacuoles in high concentrations (Hewitt, 1975; Hewitt et al., 1979). Especially under conditions of a low light intensity and a short day-length nitrate accumulation can be considerable (Steingröver et al., 1982; 1986; Stulen et al., 1990). In the leaf, diurnal changes in nitrate concentration were found. In spinach leaves nitrate concentration increased during the night, and decreased during the day. Compartmental analysis showed that the change in nitrate concentration in the leaf was caused by changes in nitrate accumulation in the vacuole (Steingröver et al., 1986a). From these and other experiments with herbaceous plants the conclusion was drawn that nitrate serves as an osmotic agent in the vacuole when there is a shortage of carbon intermediates for osmotic purposes (Aslam and Oaks, 1976, Veen and Kleinendorst, 1985; Steingröver et al., 1986a). Experiments with cultivars of lettuce showed that accumulation of nitrate in preference to an accumulation of organic compounds can differ between cultivars. The two lettuce cultivars investigated differed in their partitioning of nitrogen and carbon between structural growth and osmotic use (Blom-Zandstra et al., 1988).

The light condition plays a role in the accumulation of nitrate in the vacuole and the release of nitrate from the vacuole. When spinach plants were kept at a very low light intensity during the night period, no increase in the amount of nitrate in the vacuole was found (Steingröver et al., 1982; 1986b). Experiments with barley showed that in the dark nitrate was accumulated in the vacuole, and that light was needed for release of the stored nitrate. It was suggested that a phytochrome effect might play a role in the permeability of the tonoplast to nitrate (Aslam and Oaks, 1976). This might also be the case during a night period with a very low light intensity (Steingröver et al., 1986b), preventing the accumulation of nitrate in the vacuole.

The question now remains how the partitioning of nitrate between protein synthesis, and storage in the vacuole is regulated, and to what extent the transport mechanism of nitrate over the tonoplast is the decisive factor. The mechanism of nitrate transport over the tonoplast has not been elucidated yet (Clarkson, 1986). In order to gain more insight into the regulatory mechanism involved in the partitioning of nitrate within the cell, the concentration of nitrate in the cytoplasm as well as in the vacuole has to be known. Recently, techniques have been developed to measure nitrate concentrations within the cell (Zhen et al., 1991). From experiments with barley root cells the conclusion could be drawn that the gradient of nitrate across the tonoplast *in vivo* is too large to result from a passive process. It is very likely, therefore, that the accumulation of nitrate in to the vacuole is an active process (Miller and Smith, 1992). Knowledge on the efflux of nitrate out of the tonoplast, however, is still lacking at present.

VI. Interaction between Root and Shoot

It is clear that the shoot is dependent on nitrate taken up by the root and that root functioning is completely dependent on carbohydrate translocation from the shoot, and thereby indirectly on light. For root tissue a relationship between soluble sugar and nitrate uptake has been found (Talouizte et al., 1984). A low carbohydrate supply in plants grown at a low light intensity resulted in a reduction in nitrate uptake (Ta and Ohira, 1981). Nitrate reduction in the root is dependent on carbohydrates as well. Reductant for nitrate reduction in non-photosynthetic tissue can be generated in several ways (see Lee, 1980 for a review).

Under normal conditions of light and nutrient supply, plants maintain their protein content within a certain range, around 20% of dry weight (Hewitt et al., 1979). The question can now be raised how the overall uptake and reduction of nitrate is regulated, so that no more reduced nitrogen is accumulated. Experiments with wheat cultivars, differing in nitrate uptake rate, have shown that the differences in uptake rate were correlated with the relative growth rate of the plant (Rodgers and Barneix, 1986), which indicates that differences in uptake rate were indeed related to differences in RGR, and hence plant demand (Clarkson, 1986). The nature of the

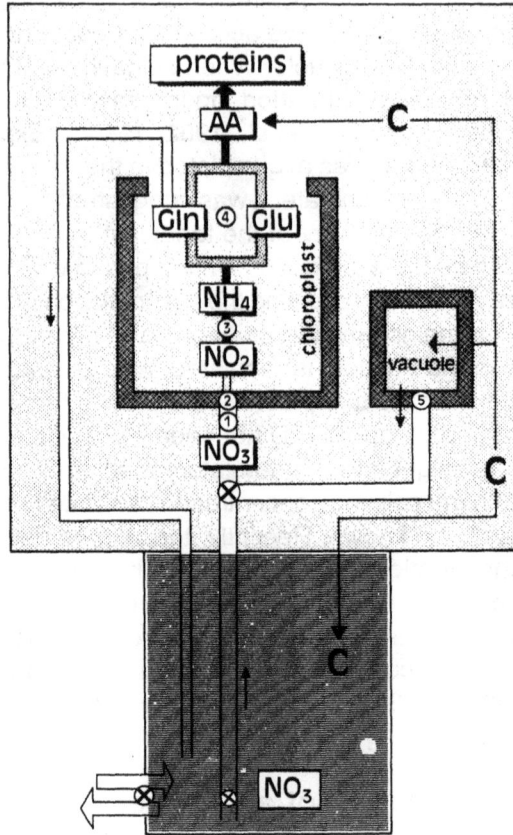

Fig. 1.   Intracellular localization of enzymes and transport processes in assimilatory nitrate reduction
          and some interactions with carbon metabolism. 1-Nitrate reductase; 2-Trar)ort of nitrite
          into the chloroplast; 3-Nitrite reductase; 4-Glutamine synthetase/glutamate synthase;
          5-Transport of nitrate into and out of the vacuole.

signal from shoot to root has not been elucidated yet; it might be a reduced nitrogen compound, as glutamine (Clarkson, 1986) or nitrate itself (Steingröver et al.,
1986a) or both.

VII.   Conclusion and Future Prospects

From the experimental evidence presented in the previous sections it can be concluded that light stimulates overall nitrate reduction and assimilation in several ways.
Its effect on various processes, as enzyme synthesis and the provision of reductant
for functioning of the enzymes is well established at the molecular and biochemical

level. It is concluded that under normal light conditions it is not likely that nitrate reduction is limited by either the level of the enzymes or the supply of reductant or carbon skeletons, but that under low light conditions, the supply of NADH might be the limiting factor in the reduction of nitrate. In this respect, it should be taken into consideration that the level of the enzymes is usually measured with an *in vitro* assay, performed under non-physiological conditions. The coupling of a NR assay system to an NADH-generating system so that NR is measured under low and steady-state concentrations of NADH (Sanchez and Heldt, 1969) might offer a way of measuring NR under more physiological conditions. However, in order to understand the influence of light on the regulation of nitrogen assimilation in the plant as a whole, the compartmentation of the various processes within the cell and in the plant, has to be taken into account as well (Fig. 1). There is evidence that light may influence the transport of nitrate and nitrite over the tonoplast and the chloroplast membrane, respectively. The regulatory mechanism in the cytoplasm, controlling the partitioning of nitrate between protein synthesis and storage in the vacuole, however, is not clear yet. For a full understanding of the influence of light on *in situ* nitrate reduction, the knowledge on nitrate concentration in compartments within the cell, and fluxes of nitrogen compounds, has to be expanded.

Literature Cited

Abrol, Y.P., Sawhney, S.,and Naik, M.S. 1983. Light and dark assimilation of nitrate in plants. *Plant Cell Environ.* **6**: 595-599.

Aslam, M., and Oaks, A. 1976. Effect of light and glucose on the induction of nitrate reductase and on the distribution of nitrate in etiolated barley leaves. *Plant Physiol.* **58**: 588-591.

Back, E., Burkhart, W., Moyer, M., Privalle, L., and Rothstein, S. 1988. Isolation of cDNA clones coding for spinach nitrite reductase: complete sequence and nitrate induction. *Mol. Gen. Genet.* **212**: 20-26.

Baysdorfer, C., and Robinson, J.M. 1985. Metabolic interactions between spinach leaf nitrate reductase and ferredoxin-NADP reductase. *Plant Physiol.* **77**: 318-320.

Beevers, L., and Hageman, R. H. 1969. Nitrate reduction in higher plants. *Annu. Rev. Plant Physiol.* **20**: 485-552.

Beevers, L., and Hageman, R. H. 1980. Nitrate and Nitrite reduction. In *The Biochemistry of Plants*, Vol. 5, (Miflin, B. J. ed.) Academic Press. New York. pp. 115-168.

Blom-Zandstra, M., Lampe, J. E. M., and Ammerlaan, F. H. M. 1988. C and N utilization of two lettuce genotypes during growth under non-varying light conditions and after changing the light intensity. *Physiol. Plant.* **74**: 147-153.

Bloom, A. J., Caldwell, R. M., Finazzo, J., Warner, R. L., and Weissbart, J. 1989. Oxygen and carbon dioxide fluxes from barley shoots depend on nitrate assimilation. *Plant Physiol.* **91**: 352-356.

Bowsher, C. G., Long, D. M., Oaks, A., and Rothstein, S. J. 1991. Effect of light/dark cycles on expression of nitrate assimilatory genes in maize shoots and roots. *Plant Physiol.* **95**: 281-285.

Brunetti, N., and Hageman, R. H. 1976. Comparison of *in vivo* and *in vitro* assays of nitrate reductase in wheat (*Triticum aestivum* L.) seedlings. *Plant Physiol.* **58**: 583-587.

Brunswick, P., and Cresswell, C. F. 1988. Nitrate uptake into intact pea chloroplasts. I. Kinetics and relationship with nitrite assimilation. *Plant Physiol.* **86**: 378-383.

Brunswick, P., and Cresswell, C. F. 1988. Nitrate uptake into intact pea chloroplasts. II. Influence of electron transport regulators, uncouplers, ATPase and anion uptake inhibitors and protein synthesis. *Plant Physiol.* **86**: 384-389.

Caboche, M., and Rouze, P. 1990. Nitrate reductase: a target for molecular and cellular studies in higher plants. *Trends Genet.* **6**: 187-192.

Callaci, J. J., and Smarelli, J. Jr. 1991. Regulation of the inducible nitrate reductase isoform from soybeans. *Biochim. Biophys. Acta.* **1088**: 127-130.

Campbell, W. H. 1990. Purification, characterization and immunochemistry of higher plant nitrate reductase. *In Nitrogen in Higher Plants,* (Abrol, Y.P. ed.) John Wiley & Sons. New York. pp. 65-92.

Clarkson, D. T. 1986. Regulation of the absorption and release of nitrate by plant cells: a review of current ideas and methodology. *In Fundamental, Ecological and Agricultural Aspects of Nitrogen metabolism in Higher Plants,* (Lambers, H.,Neeteson, J. J., and Stulen, I. eds.) Martinus Nijhoff Publishers. Dordrecht. pp. 3-28.

Dahlbender, B., and Strack, D. 1986. The role of malate in ammonia assimilation in cotyledons of radish (*Raphanus sativus* L.). *Planta* **169**: 382-392.

Daniel-Vedele, F., Dorbe, M.-F., Caboche, M., and Rouzè, P. 1989. Cloning and analysis of the tomato nitrate reductase-encoding gene: protein domain structure and amino acid homologies in higher plants. *Gene* **85**: 371-380.

De Kok, L. J., Stulen, I., Bosma, W., and Hibma, J. 1986. The effect of short-term H_2S fumigation on nitrate reductase activity in spinach leaves. *Plant Cell Physiol.* **27**: 1249-1254.

Deng, M.-D., Moureaux, T., and Lamaze, T. 1989. Diurnal and circadian fluctuation of malate levels and its close relationship to nitrate reductase activity in tobacco leaves. *Plant Sci.* **65**: 191-197.

Deng, M.-D., Moureaux, T., Leydecker, M.-T., and Caboche, M. 1990. Nitrate-reductase expression is under the control of a circadian rhythm and is light inducible in *Nicotiana tabacum* leaves. *Planta* **180**: 257-261.

Elmlinger, M. W., and Mohr, H. 1991. Coaction of blue/ultraviolet—A light and light absorbed by phytochrome in controlling the appearance of ferrodoxin-dependent glutamate synthase in the Scots pine (*Pinus sylvestris* L.) seedling. *Planta* **183**: 374-380.

Emes, M. J., and Fowler, M. W. 1983. The supply of reducing power for nitrite reduction in plastids of seedling pea root (*Pisum sativum.* L.). *Planta* **158**: 97-102.

Galangau, F., Daniel-Vedele, F., Moureaux, T., Dorbe,M.F., Leydekker, M. T., and Caboche, M. 1988. Expression of leaf nitrate reductase genes from tomato and tobacco in relation to light-dark regimes and nitrite supply. *Plant Physiol.* **88**: 383-388.

Gebauer, G. 1990. Diurnal changes of nitrate content and nitrate reductase activity in different organs of *Atriplex hortensis* (C_3 plant) and *Amaranthus retroflexus* (C_4 plant). *In Plant Nutrition-Physiology and Applications,* (van Beusichem, M. L. ed.) Kluwer Academic Publishers. Dordrecht. pp. 93-99.

Gebauer, G., Melzer, A., and Rehder, H. 1984. Nitrate content and nitrate reductase activity in *Rumex obtusifolius* L. I. Differences in organs and diurnal changes. *Oecologia* **63**: 136-142.

Guerrero, M. G., Vega, J. M., and Losada, M. 1981. The assimilatory nitrate reducing system and its regulation. *Annu. Rev. Plant. Physiol.* **32**: 169-204.

Gupta, S. C., and Beevers, L. 1984. Synthesis and degradation of nitrite reductase in pea leaves. *Plant Physiol.* **75**: 251-252.

Gupta, S.C., and Beevers, L. 1985. Regulation of synthesis of nitrite reductase in pea leaves: *in vivo* and *in vitro* studies. *Planta* **166**: 89-95.

Hageman, R. H. 1990. Historical perspectives of the enzymes of nitrate assimilation by crop plants and plants and potential for biotechnological application. *In Inorganic Nitrogen in Plants and Microorganisms,* (Ullrich, W. R.,Rigano, C.,Fuggi, A.,and Aparicio, P. J. eds.) Springer Verlag. Berlin. pp. 3-12.

Hageman, R.H., and Below, F. E. 1990. Role of nitrogen metabolism in crop productivity. *In Nitrogen in Higher Plants,* (Abrol, Y. P. ed.) John Wiley & Sons, New York, pp. 313-335.

Hageman, R.H., Flesher, D., and Gitter, A. 1961. Diurnal variation and other light effects influencing the activity of nitrate reductase and nitrogen metabolism in corn. *Crop Sci.* 1: 201-204.

Hewitt, E. 1975. Assimilatory nitrate-nitrite reduction. *Annu. Rev. Plant Physiol.* 26: 73-100.

Hewitt, E., Hucklesby, D. P., Mann, A. F., Notton, B. A., and Rucklidge,G. F. 1979. Regulation of
nitrate assimilation in plants. *In Nitrogen Assimilation of Plants,* (Hewitt, E.,and Cutting, C.
V. eds.) Academic Press, London, pp. 255-288.

Kaiser, W. M., and Brendle-Behnisch, E. 1991. Rapid modulation of spinach leaf nitrate reductase
activity by photosynthesis. I. Modulation *in vivo* by CO_2 availability. *Plant Physiol.* 96:
363-367.

Kaiser, W. M., and Spill, D. 1991. Rapid modulation of spinach leaf nitrate reductase by photosynthesis. II. *In vitro* modulation by ATP and AMP. *Plant Physiol.* 96: 368-375.

Kaiser, W. M., and Spill, D. 1991. Rapid modulation of spinach leaf nitrate reductase by photosynthesis. II. *In vitro* modulation by ATP and AMP. *Plant Physiol.* 96: 368-375.

Kleinhofs, A., Warner, R. L., and Melzer, J. M. 1989. Genetics and molecular biology of higher plant
nitrate reductases. (Romeo, and E.C. Conn eds.), Plenum Press, New York, pp. 117-156.

Klepper, L. A., Flesher, D., and Hageman, R. H. 1971. Generation of reduced nicotinamide adenine
nucleotide for nitrate reduction in green leaves. *Plant Physiol.* 48: 580-590.

Lahners, K., Kramer, V., Back, E., Privalle, L., and Rothstein, S. 1988. Molecular cloning of complementary DNA encoding maize nitrite reductase. Molecular analysis and nitrate reduction.
Plant Physiol. 88: 741-746.

Lawyer, A. L.,Cornwell, K.L., Larsen, P. O., and Bassham, J. A. 1981. Effects of carbon dioxide
and oxygen on the regulation of photosynthetic carbon metabolism by ammonia in spinach
mesophyll cells. *Plant Physiol.* 68: 1231-1236.

Lea, P. J., Blackwell, R. D., and Joy, K. W. 1992. Ammonia assimilation in higher plants. *In Nitrogen
Metabolism of Plants,* (Mengel, K.,and Pilbeam, D. J. eds.) Clarendon Press, Oxford, pp.
153-187.

Lee, R. B. 1980. Sources of reductant for nitrate assimilation in non-photosynthetic tissue: a review.
Plant Cell Environ. 3: 65-90.

Lillo, C. 1984. Diurnal variations of nitrite reductase, transferase and aspartate aminotransferase
in barley. *Physiol. Plant.* 61: 214-218.

Lillo, C. 1989. An unusually rapid light-induced nitrate reductase mRNA pulse and circadian oscillations. *Naturwissenschaften* 76e: 526-528.

Lillo, C., and Henriksen, A. 1984. Comparative studies of diurnal variations of nitrate reductase activity in barley leaves. *Physiol. Plant.* 62: 219-223.

Mattson, M., Lundborg, T., and Larsson, C.-M. 1988. Nitrate utilization in barley: relations to nitrate
supply and light/dark cycles. *Physiol. Plant.* 73: 380-386.

McNally, S. F., Hirel, B., Gadal, P., Mann, A. F., and Stewart, G. R. 1983. Glutamine synthetases
of higher plants. *Plant Physiol.* 72: 22-25.

Melzer, J. M., Kleinhofs, A., and Warner, R. L. 1989. Nitrate reductase regulation: Effects of nitrate
and light on nitrate reductase mRNA accumulation. *Mol. Gen. Genet.* 217: 341-346.

Miflin, B.J., and Lea, P. J. 1976. The pathway of nitrogen assimilation in plants. *Phytochemistry*
15: 873-885.

Miller, A. J., and Smith, S. J. 1992. The mechanism of nitrate transport across the tonoplast of barley
root cells. *Planta* 187: 554-557.

Naik, M. S., and Nicholas, D. J. D. 1984. Origin of NADH for nitrate reduction in wheat roots. *Plant
Sci. Lett.* 35: 91-96.

Oaks, A., Long, D. M., Zoumadakis, M., Li, X.-Z., and Herig, C. 1990. The role of NO_3^- and NH_4^+
ions in the regulation of nitrate reductase in higher plants. *In Inorganic Nitrogen in Plants
and Microorganisms,* (Ullrich, W.R.,Rigano, C.,Fuggi, A.,and Aparicio, P. J. eds.) Springer
Verlag, New York, pp. 165-171.

Oaks, A., Poulle, M., Goodfellow, V. J., Cass, L. A., and Deising H., 1988. The role of nitrate and ammonium ions and light on induction of nitrate reductase in maize leave. *Plant Physiol.* **88**: 1067-1072.

Oaks, A., and Yamaya, T. 1990. Nitrogen assimilation in leaves and roots.—A role for glutamate dehydrogenase. *In Nitrogen in Higher Plants,* (Abrol, Y. P. ed.) John Wiley & Sons, New York, pp. 181-195.

Oelmüller, R., Schuster, C., and Moher, H. 1988. Physiological characterization of a plastidic signal required for nitrate-induced appearance of nitrate and nitrite reductase. *Planta* **174**: 75-83.

Oji, Y., Watanabe, W., Wakiuchi, N., and Okamoto, S. 1985. Nitrite reduction in barley root plastids: Dependence on NADPH coupled with glucose-6-phosphate and 6-phosphoglucotron carrier and a diaphorase. *Planta* **165**: 85-90.

Paul, J. S., Cornwell, K. L., and Bassham, J. A. 1978. Effects of ammonia on carbon metabolism in photosynthesizing leaf discs from *Papaver somniferum* L. *Planta* **142**: 49-54.

Pilbeam, D. J., and Kirkby, E. A. 1990. The physiology of nitrate uptake. *In Nitrogen in Higher Plants,* (Abrol, Y. P. ed.) John Wiley & Sons. New York, pp. 39-65.

Platt, J. S., Plaut, Z., and Bassham, J. A. 1977. Ammonia regulation of carbon metabolism in photosynthesizing leaf discs. *Plant Physiol..* **60**: 739-742.

Popp, M., and Summons, R. E. 1984. Phosphoenolpyruvate carboxylase and amino acid metabolism in roots. *Physiol. Veg.* **21**: 1083-1089.

Rajasekhar, V. K., and Mohr, J. 1986. Appearance of nitrite reductase in cotyledons of mustard (*Sinapis alba* L.) seedling cotyledons as affected by nitrate, phytochrome and photooxidative damage of plastids. *Planta* **168**: 369-376.

Rao, L. V. M., Datta, N., Sopory, S. K., and Guha-Mukerjee, S. 1980. Phytochrome mediated induction of nitrite reductase—a chloroplastic enzyme, in etiolated maize leaves. *Plant Cell Physiol.* **22**: 577-582.

Robinson, J. M. 1986. Carbon dioxide and nitrite photoassimilatory processes do not compete for reducing equivalents in spinach and soybean leaf chloroplasts. *Plant Physiol.* **80**: 676-684.

Robinson, S. A., Slade, A. P., Fox, Phillips, R., Ratcliffe, R. G., and Stewart, G. R. 1991. The role of glutamate dehydrogenase in plant nitrogen metabolism. *Plant Physiol.* **95**: 509-516.

Rodgers, C. O., and Barneix, A. J. 1986. Cultivar differences in the rate of nitrate uptake by intact wheat plants as related to growth rate. *Physiol. Plant.* **72**: 121-126.

Rufty, W. R. Jr., Israel, W. D., and Volk, R. J. 1984. Assimilation of $^{15}NO_3^-$ taken up by plants in the light and in the dark. *Plant Physiol.* **76**: 769-775.

Rufty, T. W., Mackown, C. T., and Volk, R. J. 1989. Effects of altered carbohydrate availability on whole plant assimilation of $^{15}NO_3^-$. *Plant Physiol.* **89**: 442-447.

Sanchez, J., and Heldt, H.W. 1990. On the regulation of spinach nitrate reductase. *Plant Physiol.* **92**: 684-689.

Sawhney, S. K., and Naik, M. S. 1990. Role of light in nitrate assimilation in higher plants. *In Nitrogen in Higher Plants,* (Abrol, Y. P. ed.) John Wiley & Sons, New York. pp. 93-129.

Schmidt, S., and Mohr, H. 1989. Regulation of the appearance of glutamine synthetase in mustard (*Sinapis alba* L.) cotyledons by light, nitrate and ammonium. *Planta* **177**: 526-534.

Schuster, R., Oelmüller, R., and Mohr, H. 1987. Signal storage in phytochrome action on nitrate-mediated induction of nitrate reductase and nitrite reductase in mustard seedling cotyledons. *Planta* **171**: 136-143.

Shaner, D. J., and Boyer, J. S. 1976. Nitrate reductase activity in maize leaves. I. Regulation by nitrate flux. *Plant Physiol.* **58**: 499-504.

Singh, P., Kumar, P. A., Abrol, Y. P., and Naik, M. S. 1985. Photorespiratory nitrogen cycle - A critical evaluation. *Physiol. Plant.* **66**: 169-176.

Smirnoff, N., and Stewart, G. R. 1985. Nitrate assimilation and translocation by higher plants: comparative physiology and ecological consequences. *Physiol. Plant.* **64**: 133-140.

Sopory, S. K., and Sharma, A. K. 1990. Spectral quality of light, hormones and nitrate assimilation. *In Nitrogen in Hihger Plants.* (Abrol, Y. P. ed.) John Wiley & Sons. New York, pp. 129-159.

Srivastava, H. S. and Singh, R. P. 1987. Role and regulation of L-glutamate dehydrogenase activity in higher plants. *Phytochemistry* **26**: 597-610.

Srivastava, H. S. 1992. Multiple functions and forms of higher plant nitrate reductase. *Phytochemistry* **31**: 2941-2947.

Stewart, G. R., Mann, A. F., and Fentem, P. A. 1980. Enzymes of glutamate formation: glutamate dehydrogenase, glutamine synthetase and glutamate synthase. *In The Biochemistry of Plants,* Vol. V, (Miflin, B. J. ed.) Academic Press, New York. pp. 271-327.

Steingröver, E., Oosterhuis, R., and Wieringa, F. 1982. Effect of light treatment and nutrition of nitrate on nitrate accumulation in spinach (*Spinacia oleracea* L.). *Z. Pflanzenphysiol.* **107**: 97-102.

Steingröver, E., Ratering, P., and Siesling, J. 1986. Daily changes in uptake, reduction and storage of nitrate in spinach grown at low light intensity. *Physiol. Plant.* **66**: 550-556.

Steingröver, E., Siesling, J., and Ratering, P. 1986. Effect of one night with "low light" on uptake, reduction and storage of nitrate in spinach. *Physiol. Plant.* **66**: 557-562.

Stulen, I. 1986. Interactions between carbon and nitrogen metabolism in a whole plant context. *In Fundamental, Ecological and Agricultural Aspects of Nitrogen Metabolism in Higher Plants,* (Lambers, H., Neeteson, J. J.,and Stulen, I. eds.) Martinus Nijhoff Publishers, Dordrecht. pp. 261-279.

Stulen, I. 1990. Interactions between carbon and nitrogen metabolism in relation to plant growth and productivity. *In Nitrogen in Higher Plants,* (Abrol, Y. P. ed.) John Wiley & Sons. New York. pp. 297-313.

Stulen, I., and De Kok, L. J. 1993. Whole Plant regulation of sulfur metabolism.—A theoretical approach and comparison with current ideas on nitrogen metabolism. *In Sulfur Nutrition and Assimilation in Higher Plants. Regulatory, Agricultural and Applied Aspects,* (De Kok, L. J., Stulen, I., Rennenberg, H., Brunold, C.,and Rauser, W., eds.) SPB Academic Publishing. The Hague. pp. 77-91.

Stulen, I., Lanting, L., Lambers, H., Posthumus, F., Van de Dijk, S. J., and Hofstra, R. 1981. Nitrogen metabolism of *Plantago major* ssp. *major* as dependent on the supply of mineral nutrients. *Physiol. Plant.* **52**: 108-114.

Stulen, I., Ter Steege, M. W., and Kuiper, P. J. C. 1990. Role of nitrate in growth of higher plants with emphasis on regulation of nitrate accumulation. *In Inorganic Nitrogen in Plants and Microorganism,* (Ullrich, W. R., Rigano, C., Fuggi, A.,and Aparicio, P. J., eds.) Springer Verlag. New York. pp. 336-341.

Stulen, I., Zantinge, J., and Koster, A. 1973. Functioning of nitrate reductase in the light and in the dark in seedlings of *Raphanus sativus. Acta Bot. Neerl.* **22**: 589-596.

Suzuki, A., Audet, C., and Oaks, A. 1987. Influence of light on the ferredoxin-dependent glutamate synthase in maize leaves. *Plant Physiol.* **84**: 578-581.

Suzuki, A., Vidal, J., and Gadal, P. 1982. Glutamate synthase isoforms in rice. Immunological studies in green leaf, etiolated leaf, and root tissues. *Plant Physiol.* **70**: 827-832.

Ta, T.C., and Ohira, K. 1981. Effects of various environmental and medium conditions on the response of Indica and Japonica rice plants to ammonium and nitrate nitrogen. *Soil. Sci. Plant Nutr.* **27**: 347-355.

Talouizte, A., Champigny, M. L., Bismuth, E., and Moyse, A. 1984. Root carbohydrate metabolism associated with nitrate assimilation in wheat previously deprived of nitrogen. *Physiol. Veg.* **22**: 19-27.

Travis. R. L., Key, J. L., and Ross, C. W. 1974. Activation of 80S ribosomes by red light treatment of dark-grown seedlings. *Plant Physiol.* **53**: 28-31.

Turpin, D. H., Weger, H. G., Smith, R. G., Plaxton, W. C., and Lin, M. 1990. Interactions between respiration and nitrogen assimilation during photosynthesis. *In Inorganic Nitrogen in Plants* and Microorganisms (Ullrich, W. R., Rigano, C., Uggi, A.,and Aparicio, P. J., eds.) Springer Verlag. Berlin. pp. 124-130.

Vaughn, K. C., Duke, S. O., and Funkhauser, E. A. 1984. Immunological characterization and localization of nitrate reductase in norflurazon treated soybean cotyledons. *Physiol. Plant.* **62**: 481-484.

Veen, B. W., and Kleinendorst, A. 1985. Nitrate accumulation and osmotic regulation in Italian ryegrass (*Lolium multiflorum* Lam.). *J. Exp. Bot.* **36**: 211-218.

Vennesland, B., and Guerrero, M. G. 1979. Reduction of nitrate and nitrite. *In Encyclopedia of Plant Physiol.* Vol. VI, (Pirson, A.,and Zimmerman, M. H., eds.) Springer Verlag. Heidelberg. pp. 425-444.

Vega, J. M., Cardenas, J., and Losada, M. 1980. Ferredoxin nitrite reductase. *Methods Enzymol.* **69**: 255-270.

Wallsgrove, R. M., Lea, P. J., and Miflin, B.J. 1982. The dependent of NADH-dependent and ferredoxin-dependent glutamate synthase in greening barley and pea leaves. *Planta* **154**: 473-476.

Weber, M., Schmidt, S., Schuster, C., and Mohr, H. 1990. Factors involved in the coordinate appearance of nitrite reductase and glutamine synthetase in the mustard (*Sinapis alba* L.) seedling. *Planta* **180**: 429-434.

Yamaya, T., and Oaks, A. 1987. Synthesis of glutamate by mitochondria - An anaplerotic function for glutamate dehydrogenase. *Physiol. Plant.* **70**: 749-756.

Zhen, R.-G., Koyro, H. W., Tomos, R. A., and Miller, A. J. 1991. Compartmental nitrate concentrations in barley root cells measured with nitrate-selective microelectrodes and by single-cell sap sampling. *Planta* **185**: 356-361.

Nitrogen Nutrition in Higher Plants, 1995
Editors : H.S. Srivastava & R.P. Singh
Associated Publishing Co., New Delhi, India
pp. 385-400.

Nitrogen Nutrition and Flowering

S.M. GRIFFITH and G.M. BANOWETZ

Abbreviations - SDP, short day plants; LDP, long day plants; N, nitrogen

I. Introduction

Nitrogen plays a major role in plant growth, the development of vegetative and reproductive structures, and in determination of crop yield (Hageman and Below, 1990). The effects of nitrogen are both quantitative and qualitative. Both the amount and the form of N nutrition affects plant morphology, rate of development, and final yield in virtually all plants. The biochemical mechanisms by which N affects these growth and development processes are not fully understood. This is particularly true with respect to the role of N nutrition in floral induction, initiation, and sex expression.

The role of nitrogen nutrition in primary floral induction has not received extensive research. Primary floral induction here refers to the set of photoperiod-influenced events that occur in the leaf which result in the transmission of an unidentified signal that influences subsequent development of the shoot apex. After induction occurs, the apex undergoes evocation, a process during which the transition from vegetative to reproductive growth and flowering occurs. Much of the early work on the effect of N on flowering used plants in which flowering had already been induced. Consequently, only N effects on processes that occurred subsequent to floral induction were observed. Furthermore, much of this early work was done before floral induction requirements of the particular species had been identified.

Most of the early studies (and much recent work) concluded that the primary effect of N was on floral development. Conclusive evidence for a direct role in floral induction, floral initiation, and sex expression remains to be discovered.

Research on the interaction of the amount and form of N with phytohormone content and the related effects of these hormones on flowering processes is still being conducted. Some evidence suggests that N may affect phytohormone metabolism and distribution and that these changes may affect both vegetative and reproductive growth. The results to date indicate that the role of N in floral induction and initiation is indirect and relatively complex.

II. Nitrogen and Floral Expression and Development

A.   Kleb's Carbohydrate/Nitrogen Balance Theory

Interest in the effect of N on flowering dates to at least the early 1900's. Because the roles of photoperiod and vernalization in floral induction had not, as yet, been identified, careful interpretation of these early results on the interaction of carbohydrate and N in flowering is required (Cameron and Dennis, 1986). Klebs (1913, 1918) proposed that the flowering condition of plants was controlled by environmental conditions including "nutritive salts" (especially nitrogenous components) and increased light intensity. He attributed the effect of increased light on the formation of organic substances such as carbohydrates. Early field observations in several crops, especially fruit trees, indicated that the amount of vegetative growth was usually inversely related to flowering. In addition, treatments thought to alter the carbohydrate content of the plant, including girdling, drought, or pruning often promoted flowering. These results, and early studies on the effects of nitrogenous fertilizers on vegetative and reproductive growth suggested that the ratio of carbohydrate to N had profound effects on plant growth and development. Additional work of Kraus and Kraybill (1918) with tomato *(Lycopersicon esculentum)* supported this idea.

Their classic work with tomato showed that conditions of high N nutrition favored vigorous vegetative growth and suppressed flowering. When plants grown under high N were switched to moderate N amounts, flowering increased while vegetative growth declined. Their analyses of stem tissue indicated that portions of the plants that contained the most vegetative growth also had the highest nitrate content. Portions of the plants that contained more reproductive tissue had lower nitrate and increased carbohydrate content. They further showed that low N nutrition reduced both vegetative growth and flowering in tomato. Based upon these results, they suggested that higher C/N ratios favored flowering in tomato plants and that these observations were consistent with those of other researchers working with a variety of species.

Cameron and Dennis (1986) reviewed the findings of Kraus and Kraybill and concluded that their "experimental design and analysis were inadequate to establish the correlations for which they have cited". They advised a careful interpretation of similar findings in the literature. The interpretation of C/N ratio measurements made from plant material composed of tissues that differ in age and organ composition is difficult. A detailed examination of specific tissue and cellular sites is required to determine the relationship of carbohydrate and N to any growth process. Subsequent studies with tomato indicated that the role for N in floral induction was less direct than that suggested by Kraus and Kraybill (Fisher, 1969). A more direct role in the development of initiated floral primordia was observed, although interestingly, the greatest flowering response in these plants occurred in the group that received the highest quantity of nitrogen (Wittwer and Teubner, 1957).

Although numerous papers support Kleb's original hypothesis, the idea that a high endogenous C/N ratio is essential for flower induction is not generally accepted (Bernier et al., 1981). Many plants with strict environmental requirements for floral induction do not show a relationship between C/N ratios and flowering. For example, in the SDP 'Biloxi' soybean *(Glycine max L.)*, floral induction and development of floral primordia occur prior to increased C/N ratios (Murneek, 1948). In the SDP *Xanthium*, plants with different endogenous N levels initiated flowers at the same day-length (Naylor, 1941).

Although Klebs's theory concerning regulation of flowering in plants by relative carbohydrate and N content has, for the most part, been dismissed, an important role for N in flowering processes has not been rejected.

B. Floral Induction

After the effects of environmental factors (e.g. vernalization, photoperiodism) on floral induction and development were identified, more precise research on the role of specific nutrients on flowering was possible. It soon became obvious that processes regulating floral induction were quite different from those that affected floral development (Leopold, 1951). Although exceptions can be found, most evidence does not support the hypothesis that floral induction is regulated directly by N nutrition.

Some studies indicate that N nutrition or plant N and carbohydrate status can substitute for or reduce vernalization requirements (Brewster, 1983, 1985). Brewster (1983) demonstrated that low plant N status during cold temperatures promoted inflorescence initiation in onion *(Allium cepa L.)* (Table 1). The low N status also substituted for the long photoperiods required by certain cultivars. Unvernalized winter rye required high N levels for floral induction whereas vernalized plants did not (Gott et al., 1955). Similarly, high nutrition levels may substitute for vernalization of *Dactylis glomerata* (Calder and Cooper, 1961). These results suggest that cold-requiring and possibly photoperiodic species may rely on higher N levels under conditions of marginal induction (e.g., reduced number of degree cold days or higher plant density) (Calder and Cooper, 1961; Bernier et al., 1981).

Nitrogen effects on floral processes are not consistent among plants of similar photoperiodic class (Vince-Prue, 1975). For example, in both SDP and LDP, flowering can be either promoted or delayed by a reduction in available N. In *Cannabis* sativa (Heslop-Harrison, 1964) and strawberry (Guttridge, 1969), low N levels during LD promoted flowering. Further, in *Kalanchoe blossfeldiana* (Rugner, 1961), *Perilla* (Chailakhyan, 1958), and *Xanthium* (Neidle, 1939), high N levels during SD treatment accelerated flowering. In LDP, flowering often is accelerated by N deprivation (Vince-Prue, 1975). An N-starved LDP may even flower in SD before one with a high level of N in LD (Lang, 1965).

Table 1. Effects of photoperiod and nitrogen nutrition on estimated mean time to inflorescence initiation of *Allium cepa* L.[abc]

Cultivar	Photoperiod (h)	Nitrogen level	Estimated mean time to floral initiation (days)
Rijnsburger	8	Normal	86.2
		Low	47.0
	20	Normal	37.8
		Low	32.6
Senshyu	8	Normal	78.0
		Low	56.0
	20	Normal	40.6
		Low	43.0

[a]Adapted from Brewster, 1983

[b]Under low nitrogen conditions, nitrogen content was reduced from 3.2 to 1.4% in the bulb and sheath tissue and from 3.3 to 2.0% in the leaf blade tissue.

[c]Nitrogen was supplied as KNO_3 and $Ca(NO_3)_2$. Normal and low nitrogen levels were 12.0 mM and 1.8 mM, respectively.

Growth of the SDP *Pharbitis* under continuous light and poor nutrient conditions or lower nitrate levels promoted floral-bud formation (Wada, 1974; Wada and Shinozaki, 1985) and shortened the critical dark period (Wada and Shinozaki, 1985). A culture medium with a high C/N ratio favored *in vitro* floral-bud formation in this same species (Ishioka et al., 1991).

In summary, N nutrition has a major influence in determining the number of flowers formed and the rate of their development. The influence of N on processes relating to photoperiodic and vernalized floral induction is less clear (Neidle, 1939; Naylor, 1941; Leopold, 1951; Vince-Prue, 1975; Bernier et al., 1981).

C. Floral Development

Although studies have shown that N nutrition can have a profound effect on floral development, the physiological and biochemical basis for these effects remains unclear. One possible role of N in plant development involves the establishment and maintenance of source and sink regions (Hageman and Below, 1990). Because carbohydrates also may affect source and sink development (Wardlaw, 1990), it is critical to focus on the interaction of N and carbohydrate in these relationships. Environmental conditions, including light and temperature, play a major role in the regulation of plant carbohydrate and N content, and consequently may determine how N will affect flowering. For example, Bunt (1969) and Ryan et al. (1972) demonstrated that winter-grown tomato (day-neutral) had reduced flower development under high N levels. This most likely was a result of carbohydrate deficiency resulting from the shortened photoperiod. High N applied to tomato during the long photoperiod of summer promoted flower development. Onion plants grown under low irradiances, high temperatures and reduced plant carbohydrate levels required more time to flower (Brewster, 1985). These examples illustrate the tight coupling of carbohydrate and N nutrition and metabolism to growth and development processes.

Although the level of N is an important determinant for the expression of the full genetic potential of plant growth and development, N-form can be equally important. Use of different N-forms or different absolute amounts of N may account for contradictory results in previous reports on the effect of N on flowering. In addition, some previous work regarding the effect of nitrate and ammonium nutrition on physiological processes was performed under greenhouse conditions using nutrient solutions and lacking continuous active pH control. Results from these experimental systems may be affected by differences or changes in pH (Hageman, 1980; Griffith, unpublished).

In *Lemna gibba*, ammonium ions decrease the rate of flowering and lengthen the induction time (Oota and Kondo, 1974). These findings suggest that N metabolism, specifically nitrate reductase activity, may influence the flowering response. When nitrate assimilation is suppressed in *Lemna paucicostata* with specific inhibitors, flowering can be induced (Tanaka et al., 1986). This same species also can be induced to flower on a low N or N-deficient medium but floral development proceeds only after plants are transferred to a N-rich medium (Tanaka, 1986; Tanaka et al., 1991). Nitrogen-deficient *Lemna* treated with specific inhibitors of the aminopeptidases, bestatin and elastinal, failed to flower (Umezawa et al., 1976). The authors suggested that exoproteolysis and endoproteolysis may be involved in N deficiency-induced flowering in *Lemna*. This may imply involvement of a low molecular weight polypeptide with florigenic activity (Takeba et al., 1990; Kozaki et al., 1991).

The length of time to flowering can be shortened and the process of floral development accelerated by increasing the levels of N or altering the ratio of nitrate and ammonium nutrition. For instance, ammonium nutrition, especially combined with nitrate (i.e., mixed-N nutrition), compared with nitrate alone, generally leads to early flowering. This phenomenon has been observed in carnation (*Dianthus caryophyllus* L.) (Blake and Harris, 1960; Green et al., 1973), chrysanthemum (*Chrysanthemum morifolium* L.) (Tsujita et al., 1974), China aster (*Callistephus chinensis*) (Haynes and Goh, 1977), and Italian ryegrass (*Lolium multiflorum* L.) (Griffith, 1991), but not with flax (*Linum usitatissimum*) (Abdel-Raouf et al., 1983) (Table 2). Apple (*Pyrus malus*) trees exposed to ammonium formed more flowers than nitrate-fed plants even when ammonium was given only for a short period (Grasmanis and Leeper, 1967; Grasmanis and Edwards, 1974).

Table 2. Effect of nitrate (NO_3^-) and ammonium (NH_4^+) nutrition on plant growth (g fresh weight per plant), time of blooming (number of days from planting to 50% completion of harvest), and total number of flowers per plant in rooted carnation cuttings (*Dianthus caryophyullus* L.)[a]

Nitrogen Source	Plant Growth	Time to Blooming	Total Number of Flowers
100% NO_3^-	267	192	3.25
66% NO_3^-/33% NH_4^+	325	182	3.50

[a]Adapted from Green et al., 1973.

The reason mixed-N nutrition results in earlier flowering is not known. One possible mechanism may involve the interaction between N level and N-form with phytohormone metabolism and distribution (see section III this chapter; Evans, 1971). A more general explanation is that overall plant growth and development proceed at a faster rate and consequently, the plant flowers earlier. Faster growth rates probably result from more efficient utilization of N which provides a greater balance of reduced-N in the plant. The reduced energy requirement for ammonium assimilation may be an important factor in ammonium nutrition-enhanced growth (Haynes and Goh, 1978). Furthermore, the increased photosynthetic capacity that accompanies enhanced shoot growth may help promote earlier flowering. It has been shown that certain plants which receive enhanced ammonium nutrition contain higher amounts of amino acids, amides, and total reduced-N compared to nitrate-grown plants (Hageman, 1980). The elevated levels of these nitrogenous components may favor growth of meristematic zones and their subsequent development. In *Helianthus annuus* L. cv. Hysun, final floret number was a function of the plant's content of reduced-N at the beginning of floret production (Steer and Hocking, 1983). This also may partly explain why higher N levels applied to wheat at the double-ridge stage and ear emergence increased spikelet numbers (Langer and Liew, 1973).

The effect of mixed-N nutrition on flowering processes differs among species, but clearly a mixture of ammonium and nitrate favors growth and protein production (Haynes and Goh, 1978). The optimum ammonium/nitrate ratio may be species-cultivar, or age-dependent (Michael et al., 1970; Uesato, 1974; Haynes and Goh, 1978). In carnation, ammonium nutrition increases vegetative growth and decreases time to flowering, while in apple and aster, ammonium-related early flowering is accompanied by reduced vegetative growth (Haynes and Goh, 1978). Work with apple trees indicates that ammonium promotes a higher ratio of flower buds to total buds, while nitrate promotes vegetative growth (Haynes and Goh, 1978; Grasmanis and Leeper, 1967).

Monocotyledonous plants (in particular, grasses) and dicotyledonous plants respond differently to mixed-N nutrition. Generally the rates of tillering, leaf and stem growth are increased in grasses that receive nitrate enriched with ammonium ions, compared to plants that receive nitrate alone (Spratt and Gasser, 1970; Griffith, 1991). The increased tillering and growth lead to earlier flowering. It is well established that wheat receiving nitrate at heading (boot stage) has increased grain protein compared to plants that receive ammonium (Spratt, 1974).

D. Sex Expression

Changes in sex expression and sex organ development resulting from N nutrition have been reported for monoecious, andromonoecious, and dioecious species (Kinet et al., 1981). Examples include hops, tomato, *Begonia, Cucumis, Aquilegia, Nigella, Xanthium,* and *Cleome.* Generally, plants fed high amounts of N produce more flowers and the ratio of female to male flowers is greater. Nitrogen deficiency has little effect on anther development and pollen viability but does repress the growth of pistils (Kinet et al., 1981). Again caution should be used in interpreting results because of differences in the timing of female and male flower expression and development, light conditions, N-form, and temperature. For example in *Begonia,* male flowers develop prior to the female terminal flowers. N deficiency at the time of female flower expression results in a shift to a maleness condition (Kinet et al., 1981). Many well-designed experiments strongly suggest a role for N in sex expression in plants.

Work with *Cucumis sativus* demonstrated that soil-grown plants fed high amounts of N produced more flowers and showed an increase in female over male flowers (Tiedjens, 1928). Photoperiod can also influence the response in *Cucumis.* For example, under LD and high N conditions, male and female flowers increased 46% and 55%, respectively compared to plants grown under low soil-N conditions. In contrast, plants grown under SD conditions and fed high amounts of N produced an even greater number of female to male flowers, 21% to 3%, respectively. Similar findings were also observed for *C. anguria* and *C. melo* (Tiedjens, 1928; Brantley and Warren, 1960).

Timing of N nutrition can influence sex expression. In the dioecious species hemp (*Cannabis sativa* L.), solution-cultured plants given high N (8x normal) produced only female flowers. Under N-minus conditions, only male flowers developed (Tibeau, 1936). If plants were starved of N for 27 days and then transferred to high N-culture, only femaleness resulted. In contrast, 44-day starvation resulted in only male flowers. This indicates a strong interaction between N and the timing of pistil and stamen developmental expression. For example, Tibeau (1936) found that hemp produced pistillate structures under high N conditions while plants that received little available N were entirely staminate.

III. Nitrogen and the Involvement of Phytohormones

Nitrogen nutrition may affect floral initiation and development indirectly through alteration of phytohormone content. A number of studies have shown that N-form affects both endogenous levels of specific phytohormones and the flowering response of the plant.

In one study, the number of flowering buds and the ratios of flowering buds to total emerged buds were increased in apple rootstocks that received either ammonium or mixed N nutrition (Gao et al., 1992). Xylem sap collected from ammonium-treated rootstocks contained higher levels of both cytokinins (Fig. 1) and gibberellins

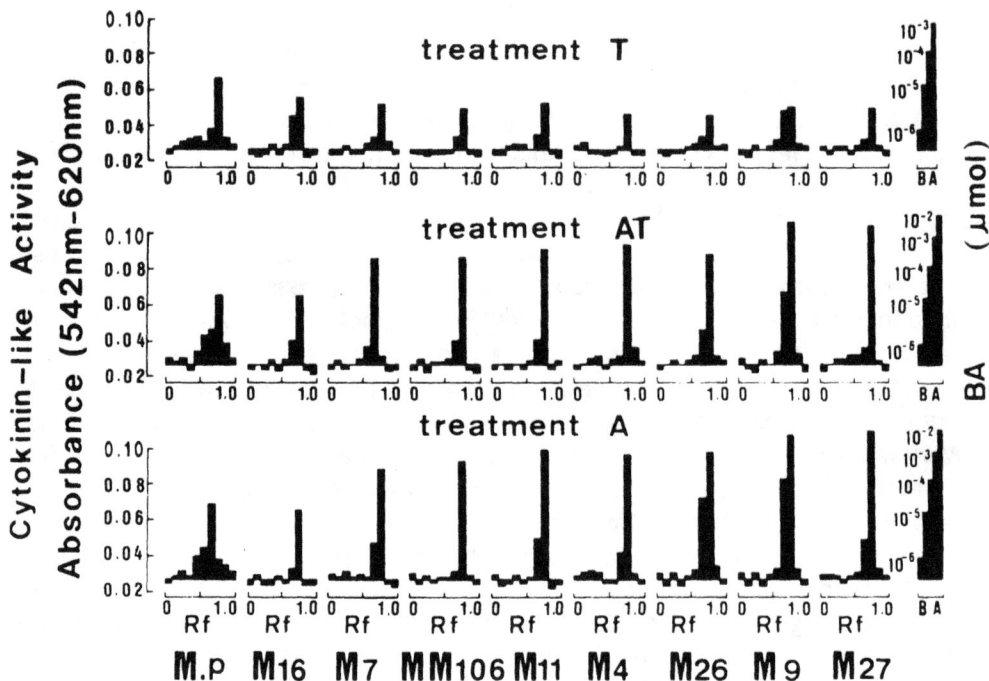

Fig. 1.   Cytokinin-like activity in the xylem sap of 'Fuji' apples grafted on nine rootstocks (M.p, M_{16}, M_7, MM_{106}, M_{11}, M_4, M_{26}, M_9, M_{27}) as affected by nitrate alone[T] ammonium sulphate alone[A] or 1:1 nitrate ammonium sulphate[AT] application. The histograms show the activities in the sap equivalent to 40 g of stem fresh weight. From Gao et al., 1992

(Fig. 2) compared to that measured in sap from nitrate-fed plants. The ammonium-enhanced gibberellin-like activity was specific to higher RF zones on paper chromatography. An earlier study also found increased cytokinin levels in xylem sap from ammonium-supplied rootstocks compared to those which received nitrate (Buban et al., 1978). Because juvenile plants were used in these studies, no flowering data were presented. Their findings agreed with those of Weissman (1972), who found enhanced levels of cytokinins in ammonium-treated soybeans and sunflowers. Other studies also have shown that the form amd concentration of N has profound influences on endogenous cytokinins in *Solanum tuberosum* (Sattelmacher

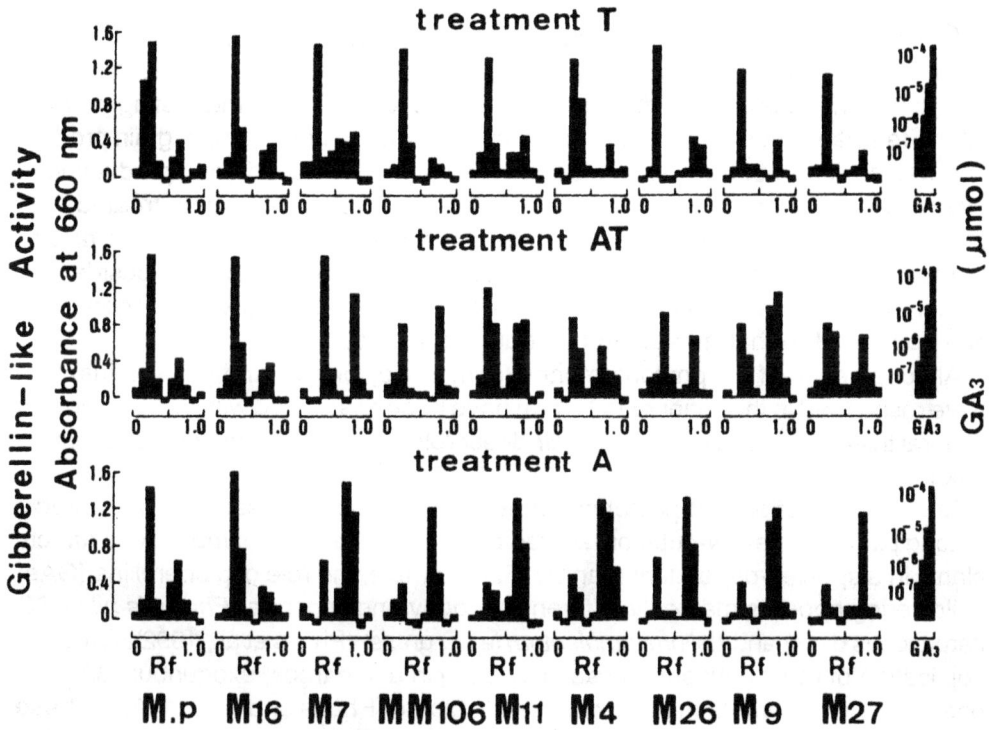

Fig. 2. Gibberellin-like activity in the xylem sap of 'Fuji'l apples grafted on nine rootstocks (M.p, M_{16}, M_7, MM_{106}, M_{11}, M_4, M_{26}, M_9, M_{27}) as affected by nitrate alone[T] ammonium sulphate alone[A], or 1:1 nitrate/ammonium sulphate[AT] application. The histograms show the activities in the sap equivalent to 20 g of fresh weight of stem. From Gao et al., 1992.

and Marschner, 1978), *Betula pendula* and *Acer pseudoplatanus* (Darral and Wareing, 1981), and *Helianthus annuus* (Salama and Wareing, 1979). The last study, in contrast to previous ones, indicated that cytokinin levels were higher in nitrate-treated plants compared to those that received either ammonium or mixed N nutrition. Zeatin or Zeatin riboside were the cytokinins which responded to N nutrition in these studies. In contrast, a study of the effects of N-source on endogenous cytokinins in the orchid (*Epidendrum fulgens*) showed that the cytokinin 2-isopentenyladenine (2-iP) increased in protocorms that received nitrate as the sole N source, while zeatin increased in those that received ammonium (Mercier and Kerbauy, 1991). No 2-iP was detected in protocorms that received ammonium as the sole N source. Bud development was reduced by use of nitrate as the sole N source and enhanced under ammonium nutrition. Maximum protocorm growth occurred with mixed N nutrition, and these protocorms contained the highest levels of both 2-iP zeatin. These data imply that cytokinins are limiting under certain conditions. No attempt was made to determine whether cytokinin treatment could substitute for mixed N nutrition in enhancement of protocorm growth. Results of a recent study also suggest that cytokinins may be limiting in some instances, and that increased grain production in maize plants that received mixed-nitrogen nutrition may involve cytokinins (Smiciklas and Below, 1992). In their study, maize that received mixed nitrogen nutrition produced more grain than nitrate-fed plants. Interestingly, when nitrate-fed plants were treated with the cytokinin, 6-benzylaminopurine, grain production was similar to that of mixed nitrogen-fed plants. Increased grain production in plants that received mixed nitrogen nutrition, or in nitrate-fed plants treated with cytokinin was attributed to enhanced dry matter partitioning to the grain and to reduced kernel abortion. Furthermore, the content of zeatin and zeatin riboside was higher in root tips and xylem exudate of plants that received mixed nitrogen nutrition compared to that measured in nitrate-fed maize.

Although each of the phytohormones (primarily cytokinins and gibberellins) that have responded in this manner to N-form have been associated with certain aspects of floral initiation or development, their direct roles in the flowering process remain unknown.

Because the effects of phytohormones on flowering processes differ markedly among plant species, N-nutrition-altered phytohormone content most likely affects plants in a species-dependent manner. For example, the role of gibberellins (GAs) in flowering appears markedly different in woody angiosperms, *Pharbitis nil* (a SD herbaceous dicot) and *Lolium temulentum* (a LD grass) (Pharis et al., 1992). Although application of most GAs suppresses flowering in apple trees, exogenous GA_4 has been shown to promote the following year's bloom (Pharis and King, 1985). These authors suggested that the reason GA_4 acts differently from other GAs may be due to the relatively short biological "half-life" of GA_4 and GA_7-rich mixtures of $GA_{4/7}$. On the other hand, exogenous GAs, including GA_3, GA_4, GA_5, and GA_{13} may por- mote floral induction in certain SD plants held under non-inductive or marginally inductive conditions (Pharis, 1972; Kohli and Sawhney, 1979; King et al., 1987; Ber- nier, 1988). In the LDP *Lolium temulentum*, applied GA can substitute for a LD treat- ment in floral initiation (Pharis et al., 1987). This same study showed that quantities of endogenous GA-like substances increase following a single LD treatment. Appli- cation of GA_{32} enhanced floral initiation but had little effect on stem elongation. This response closely mimicked that response of the plant to a single LD.

Although mutants that are defective in GA production have been obtained for many species, they have not yet received widespread use in determination of the effects of N nutrition on phytohormone content. Some which may prove useful for this work include the GA-deficient mutants of maize (*Zea mays*), *Arabidopsis*, rice (*Oryza sativa*), tomato (*Lycopersicon esculentum*) and pea (*Pisum sativum*) (Koornneef and van der Veen, 1980; Pharis and King, 1985; Nester and Zeevaart, 1988; Wilson et al., 1992). Some of these GA-deficient plants continue to flower under inductive conditions, although flowers may contain structural defects. The GA-defective mutants of *Brassica rapa* and *Thlaspi arvense* require longer periods to flower compared to wild type plants (Metzger and Hassebrock, 1990; Zanewich et al., 1990). Furthermore, a mutant of red clover will not flower without exogenous GA (Jones, 1990). Recently, studies with GA-deficient mutants that do not flower without exogenous GA suggested that GA was an absolute requirement for floral initiatior in Arabidopsis under SD (Wilson et al., 1992).

One mechanism by which phytohormones, modulated by N nutrition, may affect floral initiation and development is through regulation of gene expression. Sugiharto et al. (1992) presented evidence that cytokinins were involved in the N-regulated expression of phosphoenolpyruvate carboxylase and carbonic anhydrase in detached maize leaves. In their studies, both zeatin, a naturally-occurring cytokinin, and benzyladenine, a synthetic cytokinin, enhanced this N-dependent gene expression. Cytokinins also affect expression of genes which code for the Rubisco small s'bunit (Lerbs et al., 1984), the light-harvesting Chl a/b complex (Flores and Tobin, 1986), and nitrate reductase (Lu et al, 1990; Banowetz, 1992).

Another mechanism by which N-responsive phytohormones may affect floral initiation and development involves polyamines as secondary messengers (Marschner, 1986). In one study, ammonium fertilization stimulated floral development and raised arginine levels in apple trees (Rohozinski et al., 1986). Infiltration of petioles with polyamines caused a similar enhancement of floral development. Because cytokinins and gibberellins can influence polyamine synthesis (Dai et al., 1982; Cho, 1983). It would seem that N-regulated phytohormone levels and the subsequent effect on endogenous polyamines deserves further study.

IV. Conclusions and Future Prospects

This chapter reviewed research dealing with N-nutrition and flowering in higher plants. The hypothesis proposed by Klebs in the early part of this century concerning the role of C/N balance in flowering has not, in many cases, been supported by subsequent studies. Although high C/N ratios prior to or during flowering are not uncommon in many species, a direct role for C/N balance in the induction of floral primordia remains unproven.

Nitrogen level and N-form can have a dramatic effect on plant growth and developmental processes, although the effect appears indirect. Environmental and genetic factors play a more direct role in regulation of flowering. Nitrogen nutrition has a greater influence on floral development than on floral induction. The relative complexity and incomplete understanding of the factors that regulate floral induction continue to complicate efforts to determine the role of N in this process.

It is essential to consider the efforts of both nitrate and ammonium nutrition when evaluating the role of N in plant growth and development. Although different responses occur between species and genotypes, mixed nitrate and ammonium nutrition generally provides plants with an optimum balance of N. When measuring the response of plants to N-form, precautions should be taken to minimize secondary pH effects, especially when the buffering capacity of the growth medium is low (e.g., hydroponics). Much of the data concerning the role of mineral—and reduced-N compounds in the growth and metablism of plants is correlative. Additional studies that provide more direct data are needed. Nitrogen research focused at both cellular and subcellular levels is critical to a better understanding of the molecular regulation of such processes.

Phytohormones play a major role in floral induction and development. Nitrogen nutrition, including both N-level and N-form, can change phytohormone content. These changes may be directly related to molecular events involved in floral initiation and subsequent growth and development. More information is needed to understand the relationship between N and phytohormone metabolism and distribution and the effect of phytohormone content on the initiation and development of floral primordia.

Literature Cited

Abdel-Raouf, M.S., El-Hattab, A.H., Gheith, E.M.S., and Shaban, A.S. 1983. Effect of N level on flowering seed and fiber yield of some introduced and local flax varieties. *Ann. Agric. Sci.* **19**: 69-78.

Banowetz, G.M. 1992. The effects of endogenous cytokinin content on benzyladenine-enhanced nitrate reductase induction. *Physiol. Plant.* **86**: 341-348.

Bernier, G., Kinet, J., and Sachs, R.M. 1981. Control by nutrition and water stress, In *The physiology of Flowering, The initiation of flowers,* Vol. I. CRC Press, Boca Raton, Fla. pp. 13-20.

Bernier, G. 1988. The control of floral evocation and morphogenesis. *Annu. Rev. Plant Physiol.* **39**: 175-219.

Blake, J., and Harris, G.P. 1960. Effects of nitrogen nutrition on flowering in carnation. *Ann. Bot.* **24**: 253-255.

Brantley, B.B., and Warren, G.F. 1960. Sex expression and growth in muskmelon. *Plant Physiol.* **35**: 741-745.

Brewster, J.L. 1983. Effects of photoperiod, nitrogen nutrition and temperature on inflorescence initiation and development in onion (*Allium cepa* L.). *Ann. Bot.* **51**: 429-440.

Brewster, J.L. 1985. The influence of seedling size and carbohydrate status and photon flux density during vernalization on inflorescence initiation in onion (*Allium cepa* L.). *Ann. Bot.* **55**: 403-414.

Buban, V.A., Tromp, J., Knegt, E., and Bruinsma, J. 1978. Effects of ammonium and nitrate nutrition on the levels of zeatin and amino nitrogen in xylem sap of apple rootstocks. *Z. Pflanzenphysiol.* **89**: 289-295.

Bunt, A.C. 1969. Peat-sand substrates for plants grown in containers. I. Effect of base fertilizers. *Plant Soil* **31**: 97-110.

Calder, D.M., and Cooper, J.P. 1961. Effect of spacing and nitrogen-level on floral initiation in cocksfoot (*Dactylis glomerata* L.). *Nature* **191**: 195-196.

Cameron, J.S., and Dennis, F.G. 1986. The carbohydrate-nitrogen relationship and flowering/fruiting: Kraus and Kraybill revisited. *Hort. Sci.* **21**: 1099-1102.

Chailakhyan, M.Kh. 1958. Flowering in different plant species as a response to nitrogenous food. *Dokl. Acad. U.S.S.R.* **47**: 146-149.

Cho, S-C. 1983. Enhancement by putrescine of gibberellin-induced elongation in hypocotyls of lettuce seedlings. *Plant Cell Physiol.* **24**: 305-308.

Dai, Y-R., Kaur-Sawhney, R., and Galston, A.W. 1982. Promotion by gibberellic acid of polyamine biosynthesis in internode of light-grown dwarf peas. *Plant Physiol.* **69**: 103-105.

Darrall, N., and Wareing, P.F. 1981. The effect of nitrogen nutriton on cytokinin activity and free amino acids in *Betula Pendula* Roth. and *Acer pseudoplatanus* L. *J. Exp. Bot.* **32**: 369-379.

Evans, L.T. 1977. Flower induction and the florigen concept. *Annu. Rev. Plant Physiol.* **22**: 365-480.

Fisher, K.J. 1969. Effects of nitrogen supply during propagation of flowering and fruiting of glasshouse tomatoes. *J. Hortic. Sci.* **44**: 407-411.

Flores, S., and Tobin,E.M. 1986. Benzyladenine modulation of the expression of two genes for nuclear-encoded chloroplast proteins in *Lemna gibba*: apparent post-transcriptional regulation. *Planta* **168**: 340-349.

Gao, Y-P., Motosugi, H., and Sugiura, A. 1992. Rootstock effects on growth and flowering in young apple trees grown with ammonium and nitrate nitrogen. *J. Amer. Soc. Hort. Sci.* **117**: 446-452.

Gott, M.B., Gregory, F.G., and Purvis, O.N. 1955. Studies in vernalisation of cereals. XIII. Photoperiodic control of stages in flowering between initiation and ear formation in vernalised and unvernalised Petkus winter rye. *Ann. Bot.* **21**: 87-126.

Grasmanis, V.O., and Edwards, G.R. 1974. Promotion of flower initiation in apple trees by short exposure to ammonium ion. *Aust. J. Plant Physiol.* **1**: 99-105.

Grasmanis, V.O.,and Leeper, G.W. 1967. Ammonium nutrition and flowering of apple trees. *Aust. J. Biol. Sci.* **20**: 761-767.

Green, J.L., Holley, W.D., and Thaden, B. 1973. Effects of the NH_4^+:NO_3^- ratio, chloride, N-serve, simazine on carnation flower production and plant growth. *Proc. Florida State Hort. Soc.* **86**: 385-388.

Griffith, S.M. 1991. Enhanced floral development in hydroponically grown ryegrass receiving mixed nitrate and ammonia nutrition. *Plant Physiol. Suppl.* **96**: 104.

Guttridge, C.G. 1969. Fragaria. In *The Induction of Flowering* (Evans, L.T., ed.) MacMillian, Australia. pp. 247-267.

Hageman, R.H. 1980. Effect of form of nitrogen on plant growth *In Nitrification Inhibitors - Potentials and Limitations,* ASA Spec. Publ., Vol. **38**: ASA, CSSA, SSSA, Madison, WI. pp. 47-117.

Hageman, R.H., and Below, F.E. 1990. Role of nitrogen metabolism in crop productivity. *In Nitrogen in Higher Plants* (Abrol, Y.P., ed.), John Wiley & Sons, Inc., New York. pp. 314-334.

Haynes, R.L.,and Goh, K.M. 1977. Evaluation of potting media for commercial nursery production of container-grown plants. II. Effects of media, fertilizer nitrogen, and a nitrification inhibitor on yield and nitrogen uptake of *Callistephus chinensis* L. Nees 'Pink Princess'. *New Zealand J. Agric. Res.* **20**: 371-381.

Haynes, R.J.,and Goh, K.M. 1978. Ammonium and nitrate nutrition of plants. *Biol. Rev.* **53**: 465-510.

Heslop-Harrison, J. 1964. Sex expression in flowering plants. *Brookhaven Symp. Biol.* **16**: 109-125.

Jones, T.A. 1990. Use of a flowering mutant to investigate changes in carbohydrates during floral transition in red clover. *J. Exp. Bot.* **41**: 1013-1019.

Klebs, G. 1913. Ueber das Verhaltnis der Aussenwelt zur Entwicklung der Pflanzen. Eine theoritische Betrachtung. *Sitz. Heidelb. akad. Wiss. Ser. B* **5**: 1-47.

Klebs, G. 1918. Uber die Blutenbildung bei *Sempervivum*. *Flora* **111/112**: 128-151.

Kinet, J., Sachs, R.M., and Bernier, G. 1981. Control by nutrition. *In The Physiology of Flowering, The development of Flowers*, Vol. III: CRC Press, Boca Raton, Fla. pp. 53-61.

King, R.W., Pharis, R.P., and Mander, L.N. 1987. Gibberellins in relation to growth and flowering in *Pharbitis nil* Chois. *Plant Physiol.* **84**: 1126-1131.

Kohli, R.K.,and Sawhney, S. 1979. Promotory effect of GA_{13} on flowering of *Amaranthus*,—a short day plant. *Biol. Plant.* **21**: 206-213.

Koorneef, M.,and van der Veen, J.H. 1980. Induction and analyses of gibberellin sensitive mutants in *Arabidopsis thaliana* L. Heynh. *Theor. Appl. Genet.* **58**: 257-263.

Kozaki, A., Takeba, G., and Tanaka, O. 1991. A Polypeptide induces flowering in *Lemna paucicostata* at a very low concentration. *Plant Physiol.* **95**: 1288-1290.

Kraus, E.J., and Kraybill, H.R. 1918. Vegetation and reproduction with special reference to the tomato. *Oregon Agr. Expt. Sta. Bul.* **149**: 5-90.

Lang, A. 1965. Physiology of flowering initiation. In *Encyclopedia of Plant Physiology* (Ruhland, W., ed.), Vol. **15**: Springer-Verlag, New York. pp. 1380-1516.

Langer, R.H.M., and Liew, F.K.Y. 1973. Effects of varying nitrogen supply at different stages of the reproductive phase on spikelet and grain production and on grain nitrogen in wheat. *Aust. J. Agric. Res.* **24**: 647-656.

Leopold, A.C. 1951. Photoperiodism in plants. *Quart. Rev. Biol.* **26**: 247-263.

Lerbs, S., Lerbs, W., Klyachdo, N.L., Romanko, E.G., Kulaeva, O.N., Wolligiehn, R., and Parthier, B. 1984. Gene expression in cytokinin- and light-mediated plastogenesis of *Cucurbita* cotyledons: ribulose-1,5-bisphosphate carboxylase/oxygenase. *Planta* **162**: 289-298.

Lu, J-I., Ertl, J.R., and Chen, C-M. 1990. Cytokinin enhancement of the light induction of nitrate reductase transcript levels in etiolated barley leaves. *Plant Mol. Biol.* **14**: 585-594.

Mercier, H., and Kerbauy, G.B. 1991. Effects of nitrogen source on growth rates and levels of endogenous cytokinins and chlorophyll in protocorms of *Epidendrum fulgens. J. Plant Physiol.* **138**: 195-199.

Metzger, J.D., and Hassebrock, A.T. 1990. Selection and characterization of a gibberellin-deficient mutant of *Thlaspi arvense* L. *Plant Physiol.* **94**: 1655-1662.

Michael, G., Martin, P., and Owassia, I. 1970. The uptake of ammonium and nitrate from labelled ammonium nitrate in relation to the carbohydrate supply of the roots, In *Nitrogen Nutrition of the Plant* (Kirkby, E.A. ed.), Waverley Press, Leeds. pp. 22-29.

Naylor, A.W. 1941. Effect of nutrition and age upon rate of development of terminal staminate inflorescences of *Xanthium pennsylvanicum. Bot. Gaz.* **103**: 342-353.

Neidle, E.K. 1939. Nitrogen nutrition in relation to photoperiodism in *Xanthium pennsylvanicum. Bot. Gaz.* **100**: 607-618.

Nester, J.A., and Zeevaart, J.A.D. 1988. Flower development in normal tomato and a gibberellin-deficient (ga-2) mutant. *Am. J. Bot.* **75**: 45-55.

Oota, Y., and Kondo, T., 1974. Removal by cyclic AMP of the inhibition of duckweed flowering due to ammonium and water-treatment. *Plant Cell Physiol.* **15**: 403-411.

Pharis, R.P. 1972. Flowering of *Chrysanthemum* under non-inductive long day by gibberellins and N6-benzyladenine. *Planta* **105**: 205-212.

Pharis, R.P., Evans, L.T., King R., and Mander, L.N. 1987. Gibberellins, endogenous and applied, in relation to flower induction in the long day plant *Lolium temulentum, Plant Physiol.* **84**: 1132-1138.

Pharis, R.P., Evans, L.T. King, R., and Mander, L. 1992. Gibberellins and flower initiation/early differentiation: Evidence for differences in structure and dose, relative to efficacy in shoot elongation. *Proc. Plant Growth Regul. Soc. Amer.*: 17th Annual Meeting, pp. 94-105.

Pharis, R.P., and King, R. 1985. Gibberellins and reproductive development in seed plants. *Annu. Rev. Plant Physiol.* **36**: 517-566.

Rohozinki, J., Edwards, G.R., and Hoskyns, P. 1986. Effects of brief exposure to nitrogenous compounds on floral initiation in apple trees. *Physiol. Veg.* **24**: 673-677.

Runger, W. 1961. Uber den Einfluss der Stickstoffernahrung und der Temperatur wahrend der Langtag-und Kurztagperioden auf die Blutenbildung von *Dalanchoe blossfeldiana Planta* **56**: 517-529.

Ryan, E., Smillie, G.W., and McAleese, D.M. 1972. Effect of natural light conditions on the growth of tomato plants propagated in peat. 1. Growth and flowering responses. *Ir. J. Agric. Res.* **11**: 295-303.

Salama, A.M.S.,and Wareing, P.F. 1979. Effects of mineral nutrition on endogenous cytokinins in plants of sunflower (*Helianthus annuus* L.). *J. Exp. Bot.* **30**: 971-981.

Sattelmacher, B.,and Marschner, H. (1978). Nitrogen nutrition and cytokinin activity in *Solanum tuberosum*. *Physiol. Plant.* **42**: 185-189.

Smiciklas, K.D.,and Below, F.E. 1992. Role of cytokinin in enhanced productivity of maize supplied with NH_4^+ and NO_3^-. *Plant and Soil* **142**: 307-313.

Spratt, E.D. 1974. Effect of ammonium and nitrate forms of fertilizer-N and their time of application on utilization of N by wheat. *Agron. J.* **66**: 57-61.

Spratt, E.D.,and Gasser, J.K.R. 1970. The effect of ammonium sulphate treated with nitrification inhibitor, calcium nitrate, on growth and N-uptake of spring wheat, ryegrass and kale. *J. Agric. Sci.* **74**: 111-117.

Steer, B.T., and Hocking, P.J. 1983, Leaf and floret production in sunflower (*Helianthus annuus* L.) as affected by nitrogen supply. *Ann. Bot.* **52**: 267-277.

Sugiharto, B., Burnell, J.N., and Sugiyama, T. 1992. Cytokinin is required to induce the nitrogen-dependent accumulation of mRNAs for phosphoenolpyruvate carboxylase and carbonic anhydrase in detached maize leaves. *Plant Physiol.* **100**: 153-156.

Takeba, G. Nakajima, Y., Kozaki, A. Tanaka, O., and Kassai, Z. 1990. A flower inducing substance of high molecular weight from higher plants. *Plant Physiol.* **94**: 1677-1681.

Tanaka, O. 1986. Flower induction by nitrogen deficiency in *Lemna paucicostata* 6746. *Plant Cell Physiol.* **27**: 875-880.

Tanaka, O., Horikawa, W., Nishimura, H., and Nasu, Y. 1986. Flower induction by suppression of nitrate assimilation in *Lemna paucicostata* 6746. *Plant Cell Physiol.* **27**: 127-133.

Tanaka, O., Yamamoto, T., Nakaymam, Y., Ozaki, T., and Takeba, G. 1991. Flowering induced by nitrogen deficiency in *Lemna paucicostata* 151. *Plant Cell Physiol.* **32**: 1173-1177.

Tibeau, M.E. 1936. Time factor in utilization of mineral nutrients by hemp. *Plant Physiol.* **11**: 731-774.

Tiedjens, V.A. 1928. Sex ratios in cucumber flowers as affected by different conditions of soil and light. *J. Agric. Res.* **36**: 721-746.

Trewavas, A.J. 1983. Nitrate as a plant hormone. In *Interactions between Nitrogen and Growth Regulations in the Control of Plant Development*, Monograph **9** (Jackson, M.B. ed.), British Plant Growth Regulator Group, Wantage. pp. vii-111.

Tsujita, M.J., Kiplinger, D.C., and Tayama, H.K. 1974. The effects of nitrogen nutrition, temperature, and light intensity on the growth, flowering, quality, and chemical composition of 'Indianapolis yellow' chryanthemum. *Hort. Sci.* **9**: 294.

Uesato, K. 1974. Effects of different forms of nitrogen sources in the culture media on the growth of *Dendrobium nobile* seedlings. *Science Bulletin of the College of Agriculture*, University of Ryukyus, Okinawa **21**: 73-81.

Umezawa, H., Aoyagi, T., Suda, M. Hamada, M., and Takeuchi, T. 1976. Bestatin, an inhibitor of aminopeptidease B, production by actinomycetes. *J. Antibiotics.* **26**: 97-99.

Vince-Prue, D. 1975. The biochemistry of flowering. In *Photoperiodism in Plants* (Wilkins, M.B. ed.) McGraw-Hill, England. pp. 213.

Wada, K. 1974. Floral initiation under continuous light in *Pharbitis nil*: a typical short-day plant. *Plant Cell Physiol.* **15**: 381-384.

Wada, K.,and Shinozaki, Y. 1985. Flowering response in relation to C and N contents of *Pharbitis nil* plants cultured in nitrogen-poor media. *Plant Cell Physiol.* **26**: 525-535.

Wardlaw, I.F. 1990. The control of carbon partitioning in plants. *New Phytol.* **116**: 341-381.

Weissman, G.S. 1982. Influence of ammonium and nitrate nutrition on the pyridine and adenine nucleotides of soybean and sunflower. *Plant Physiol.* **49**: 142-145.

Wilson, R.N., Heckman, J.W., and Somerville, C.R. 1992. Gibberellin is required for flowering in *Arabidopsis thaliana* under short days. *Plant Physiol.* **100**: 403-408.

Wittwer, S.H., and Teubner, F.G. 1957. The effect of temperature and nitrogen nutrition on flower formation in the tomato. *Am. J. Bot.* **44**: 125-129.

Zanewich, K.P., Rood, S.B., and Williams, P.H. 1990. Growth and development of *Brassica* genotypes differing in endogenous gibberellin content. I. Leaf and reproductive development. *Physiol. Plant.* **79**: 673-678.

Nitrogen Nutrition in Higher Plants, 1995
Editors : H.S. Srivastava & R.P. Singh
Associated Publishing Co., New Delhi, India
pp. 401-416.

Regulation of Nitrogen Assimilation by Plant Growth Regulators

P.K. JAIWAL and R.P. SINGH

Abbreviations: ABA, abscisic acid; IAA; indole-3-acetic acid; IPA, indole pyruvic acid; GAs; gibberellins; GDH, glutamate dehydrogenase; GS, glutamine synthetase; GOGAT, glutamate synthase; NR, nitrate reductase; NiR, nitrite reductase; NAA, napthalene acetic acid; PGR, plant growth regulator; SA, salicylic acid.

I. Introduction

The plant growth regulators affect most phases of growth and development of plants. These endogenous hormones and synthetic growth regulators are specific in nature and affect the plant metabolism which subsequently coordinate the growth and development of tissues and organs. The mechanisms by which PGRs mediate their effects are still obscure. They are synthesized in different tissues in the same plant, they elicit different effects in different cells, and in some cases the effect of one PGR overlaps those of the other (Rogers and Rogers, 1992). The PGRs bind with their receptors and change the membrane permeability and/or effect the transcription of mRNA or its translation during protein synthesis or even modulate the folding and structural configuration of proteins. These processes lead to the change in metabolism which subsequently affect the growth and development of plants.

Though ammonium and other forms of inorganic nitrogen may also be available to the plants, nitrate is the predominant form which is reduced to ammonium before its entry into the organic pool in most of the plants. In nitrogen fixing prokaryotes

karyotes and some symbiotically associated higher plants, however, a major input of nitrogen comes through biological fixation of dinitrogen gas. Several PGRs are known to regulate the nitrogen assimilation in higher plants (Luckwill, 1968). Simultaneously, the metabolism of some of the endogenous hormones like auxins and cytokinins which contain nitrogen is also regulated by the availability of nitrogen at the site of their synthesis/degradation. The effects of PGRs on nitrogen metabolism in general (Luckwill, 1968), on NR (Srivastava, 1980) and GDH activities (Srivastava and Singh, 1987) and on photo-regulation of nitrate assimilation (Sopory and Sharma, 1990) have been reviewed. However, the recent approach to understand regulation of plant processes at molecular level have elucidated several new lines of research to explain the nitrogen assimilation in plants. The present review is planned to analyse the work reported on regulation of this important plant process by PGRs and also to find out future perspectives.

II. Regulation of Nitrate and Ammonia Assimilation

The process of nitrate assimilation is governed primarily by the availability of substrate at the site of NR activity. A perusal of literature reveals that very few studies have been conducted to elucidate the role of PGR in the regulation of nitrate and ammonia assimilation and it appears that PGR have their specific effects on these processes.

A. Auxins and Cytokinins

The exogenous application of IAA does not affect the induction of NO_3^- uptake in the roots of nitrogen depleted bean seedlings, although, kinetin causes a concentration dependent suppression of the NO_3^- uptake (Hanisch ten Cate and Breteler, 1982). This study did not explain the inhibitory effect of kinetin or anything related to the mechanism of the NO_3^- uptake. Indoleacetic acid has no effect on NR activity in bean roots (Hanisch ten Cate and Breteler, 1982) but inhibits it slightly in maize leaves (Rao et al., 1984). Cytokinins, on the other hand, increase NR activity in fenugreek cotyledons, (Rijven and Parkash, 1971; Parkash, 1972), maize leaves (Rao et al., 1984), wheat seedlings (Banowetz, 1992) in the presence of nitrate and in excised embryos of *Agrostemma githago* (Borriss, 1967; Kende et al., 1971; Hirschberg et al., 1972), cucumber cotyledons, (Knypl, 1973; Kuznetsov et al., 1985) and in bean roots, (Hanisch ten Cate and Breteler, 1982) in its absence. The induction of enzyme activity by cytokinins is, however, negligible in the absence of nitrate in wheat (Lu et al., 1983) and maize (Sharma and Sopory, 1987). Thus, there appears to be differences in the response of NR activity to cytokinins in different species and the response also seems to be modified by the absence or presence of nitrate in some species (Lu et al., 1983; Banowetz, 1992) (Table-1). Further, the effect of cytokinins on NR activity has been shown to be dependent upon the plant organ and the nature and concentration of the cytokinin also in *Leucaena leucocephala*

(Pandey and Srivastava, 1992). The effects of NO_3^- and cytokinins on NR activity have been reported to be either additive (Kende et al., 1971; Kuznetsov et al., 1979; Hanisch ten Cate and Breteler, 1982) or synergistic (Parkash, 1972; Knypl, 1973). Schmerder and Borriss (1986) have indicated that the presence of ethylene is necessary for induction of NR activity by cytokinins in *Agrostemma githago* embryos. Inhibitors and density labelling studies have demonstrated that the induction of NR by cytokinins is due to *de novo* synthesis of the enzyme (Hirschberg et al., 1972; Kende and Shen, 1972). Further, anticytokinins inhibit cytokinin induced NR activity

Table 1. Factors influencing the Induction of nitrate reductase activity by cytokinins

Plant Material	Factors			Effect	Reference
	Nitrate	Light/Dark	Other PGR		
Agrostemma githago embryo	-	-	-	Increase	Borriss, 1967, Kende et al., 1971; Hirschberg et al., 1972.
A. githago embryo	-	-	+ABA	ABA reversed NO_3^- and cyto-kinin-induced enhancement of NR activity	Lu et al., 1992.
A githago embryo	-	-	+ Ethylene	Increase	Schmerder and Borriss, 1986.
Cucumber cotyledon	-	-	-	Increase	Knypl, 1973; Kuznetsov et al., 1985.
Bean roots	-	-	-	Increase	Hanisch ten Cate and Breteler, 1982.
Fenugreek cotyledon	+	Light	-	Increase	Rijven and Parkash, 1971.
Barley leaves	+	Light	-	Increase	Keanov and Pavlovo, 1976.
Maize leaves	+	Light	-	Increase	Rao et al., 1984.
Maize leaves	+	Dark	-	Increase	Rao et al., 1984.
Wheat seedlings	+	Dark	-	Increase	Banowetz, 1992.
Barley seedlings	+	Dark	-	No effect	Guadinova, 1990.
	-	Light	-	Increase	Guadinova, 1990.
Tobacco	-	-	+GAs	Increase (GA can substitute light)	Roth-Bejerano and Lips, 1970.
Barley seedlings	-	-	+ABA	Increase	Kende et al., 1971.

(Kulaeva, 1980), but have no effect on nitrate-induced activity suggesting that cytokinins and nitrate stimulate NR activity through different mechanisms, although some physico-chemical and kinetic properties of nitrate-induced and cytokinin-induced NR are the same (Dilworth and Kende, 1974; Pandey and Srivastava,

1992). This indicates that nitrate and cytokinins enhance the expression of same genes. However, the identity of nitrate - and cytokinin - induced NR is yet to be compared. Lu et al. (1983) have noted that cytokinin enhancement of NR induction in etiolated wheat seedlings is directly related to the amount of light the plants received prior to the cytokinin treatment. They suggested that availability of higher photosynthetically derived energy may be responsible for this dependence. However, in etiolated barley leaves, Lu et al. (1990) showed using NR c-DNA probe that requirement of light for cytokinin dependent NR induction was, indeed, mediated through increased transcription of NR specific m-RNA. Further, it is noted that the effects of exogenous cytokinin on NR gene expression in wheat seedlings are dependent upon the endogenous levels of cytokinins (Banowetz, 1992). A low induction of NR by exogenous cytokinin in those tissues, where endogenous level of the PGR is already high may be due to the saturation of putative cytokinin binding sites or receptors by high endogenous cytokinin level (Banowetz, 1992).

Light seems to be interacting with PGR, as far as the effect of NR is concerned. Kinetin and red light together have additive effects on the induction of NR by NO_3^- suggesting that the phytochrome and kinetin acted independently to increase the induction of NR in maize tissue (Rao et al., 1984). On the other hand, kinetin in combination with gibberellin replaces the requirement of light for the induction of NR by NO_3^- in tobacco leaves (Lips and Roth-Bejerano, 1969) but no beneficial effect of the application of mixtures of gibberellic acid and kinetin on synthesis of NR in the dark was observed in green leaves of rice and light was still required for the induction of NR in the presence of GA and kinetin (Gandhi and Naik, 1974).

The effect of cytokinins on NiR varies according to the cytokinin and the plant species. Benzyladenine and kinetin do not affect NiR activity in *Agrostemma* (Dilworth and Kende, 1974; Kuznetsov and Litonova, 1985). On the other hand, kinetin stimulates NiR activity in excised maize leaves. The stimulatory effect of kinetin on NiR activity is inhibited by abscisic acid (Sopory and Sharma, 1990).

Kinetin has no effect on GS activity in *Agrostemma* (Kuznetsov and Litonova, 1985). Glutamine synthetase is considered to be the primary enzyme for assimilation of ammonia in higher plants incoupling with GOGAT (Lea et al., 1990). Studies regarding hormonal regulation of GS-GOGAT cycle are rare and hence this area should be explored. Indole-acetic acid and other auxins (NAA, IPA etc.) increase GDH activity in excised pea roots (Sahulka and Gaudinova, 1975). On the other hand, the effect of cytokinins on GDH activity is variable, to some extent, according to their nature and concentration, to the seedling age and to plant species. Kinetin has either no influence on GDH activity as in maize leaves (Garg and Srivastava, 1992) or inhibits it as in wheat leaves (Kar and Feierabend, 1984) or increases the enzyme activity as in pearl millet seedlings (Huber and Sankhla, 1974). However, benzyladenine increases GDH activity with increase in its concentration and age of the leaves in maize seedlings (Garg and Srivastava, 1992).

B.   Gibberellins

The gibberellic acid has no effect on the uptake of NO_3^- (Hanisch ten Cate and Breteler, 1982). Plant growth regulators of this group do not affect induction of NR in *Hordeum vulgare* aleurone layers (See, Srivastava, 1980) and in maize leaves (Rao et al., 1984) but they inhibit NR acitivity in rice (Aleshin et al., 1983). On the other hand, GA induces substantial NR activity in bean roots (Hanisch ten Cate and Breteler, 1982). Further, GA in combination with a cytokinin not only increased NR activity but also substituted light for maintenance of NR activity in tobacco leaves (Roth-Bejerano and Lips, 1970).

Gibberellins promote the activity of GS is lettuce seeds (Tasrba, 1984). However, the effect of GAs on NiR, GDH and GOGAT have rarely been studied.

C.   Ethylene

Ethephon, an ethylene releasing substance does not affect NO_3^- uptake but it stimulates the induction of NR by NO_3^- in bean roots (Hanisch ten Cate and Breteler, 1982). Similar stimulation of induction of NR by ethylene has also been reported in roots and stems of potato (Palmer, 1985). However, it inhibits NR in potato leaves (Palmer, 1985). Further, induction of NR by cytokinins in *Agrostemma* requires the presence of ethylene (Schmerder and Borriss, 1986) and ethylene and cytokinin probably act on the same NR-induction mechanism (Schmerder and Borriss, 1986). It appears that different organs of the plant have different regulatory mechanisms for NR activity in the presence of ethylene. Ethephon, at concentration range of 60-480 mgl^{-1} stimulates GDH activity in *Pennisetum* seedlings (Sankhla and Huber, 1974). The exact mechanism of enzyme induction is yet to be elucidated.

D.   Abscisic acid

No effect of ABA on NO_3^- uptake has been observed in bean roots (Hanisch ten Cate and Breteler, 1982). Abscisic acid has no effect on NR activity in bean cotyledons (Sankhla and Huber, 1975) but inhibited the same in roots, and hypocotyl (Higgins et al., 1974). It also inhibits NR activity in *Agrostemma* (Kende et al., 1974) and NO_3^- induced NR in bean roots whereas cytokinin induced NR of bean roots is stimulated (Hanisch ten Cate and Breteler, 1982).

Abscisic acid inhibits NRA in potato tuber slices in the presence of exogenous NO_3^- during the initial stages of incubation. However, pretreatment with hormone for a period of upto 24 h in the absence of exogenous NO_3^-, significantly enhanced NR activity and increased endogenous NO_3^- content (Palmer, 1981). Inhibitor studies reveal that ABA response requires protein synthesis but may be independent of RNA transcription and enhancement of NR activity by ABA is possibly by suppressing the production of inhibitory substances (Palmer, 1982).

Studies related to the effect of ABA on other enzymes are rare. In maize leaves,

ABA inhibited kinetin induced NiR completely but phytochrome mediated NiR induction was partially reduced (Sopory and Sharma, 1990). Glutamate dehydrogenase activity in pearl millet is also inhibited by ABA whereas GS activity in the same plant is increased (Huber, 1974; Huber et al., 1977; Eder and Huber, 1977). Since, ABA in known to accumulate under stress conditions (Boucaud and Ungar, 1976; Henson 1984; Mohapatra et al., 1988), its role in regulation of ammonia assimilation is yet to be explored.

E. Phenolics

A differential inhibition of uptake of NO_3^- and NH_4^+ by ferulic acid was noted in maize seedlings, although, the acid did not affect the NO_3^- reduction (Bergmark et al., 1992). Salicylic acid at 0.01 to 0.1 mM increases NR activity indirectly in maize seedlings by protecting the enzyme from inactivation but the acid at higher concentration inhibits the enzyme activity (Jain and Srivastava, 1981a). Asthana and Srivastava (1978) have reported increase in NR activity in endosperm, embryo and primary leaves of maize in the presence of SA and ascorbic acid either alone or in combination. The NiR activity is hardly affected and GDH activity is stimulated slightly by the lower concentration (0.01 mM) of SA whereas the higher concentrations (upto 1-5 mM) inhibit these enzymes (Jain and Srivastava, 1981b). The inhibitory effect of SA on GDH activity seems to be at the activity level rather than at synthesis level of the enzymes as it accelerates the *in vitro* inactivation of the enzyme which is restored to a great extent by divalent cations such as Ca^{2+} Mg^{2+}, Mn^{2+} and Zn^{2+} (Jain and Srivastava, 1981b). It appears, therefore, that SA inhibits the enzyme activity by chelating one or more of these ions. Alternatively, the acid may inhibit the enzyme activity by changing the physico-chemical nature of the enzyme molecule.

Effect of phenolic compounds have not been studied on GS activity whereas NADH-GOGAT is stimulated by 0.001 to 1 mM SA in the leaves and by 0.05 to 0.1 mM SA in the roots of maize (Singh and Srivastava, 1987b). Higher concentrations of SA, however, inhibit the enzyme activity in both the tissues. It is suggested that, SA increases GOGAT activity by inducing synthesis of the enzyme as well as by some kind of direct modulation of the enzyme molecule (Singh and Srivastava, 1987b).

Of various phenyl propanoids and their derivatives, only transcinnamic acid and coumarin inhibit the induction of NR in maize seedlings (see Srivastava, 1980).

F. Others

Triadimenol, a triazole derivative increases NR activity and nitrate accumulation in the leaves of canola and at least part of the increased enzyme activity is independent of nitrate accumulation (Srivastava and Fletcher, 1992).

Cell-free extracts of maize and cucumber showed an increased and reduced levels of NR respectively, as a result of treatment of the plants with 2, 4-D (Luckwill, 1968).

The apparent differences in the enzyme activity of the two species may be due to differences in the endogenous levels of 2, 4-D related compounds. Luckwill (1968) has suggested that the effect may be due to modification of NR activity through intramolecular changes in the protein sulphydryl-groups.

Application of 2, 4-D also increases GDH activity in pea and maize roots at low concentrations (0.002 to 0.02 mM) but inhibits the same at higher concentrations (Tkemaladze, 1981). The pea shoot GDH is also stimulated at lower concentrations of the herbicide although the enzyme from maize is unaffected (Tkemaladze, 1981).

Herbicides such as triazines, urea, diazines, bipyridyliums and benzonitriles inhibit NR acitivity in duckweed but dinitroanilines and amides (excluding propanil) have no effect on the activity (see, Srivastava, 1980). On the other hand, simazine and atrazine increase NR activity in maize and barley.

Foliar applications of 2-chloro-4, 6-bis (ethylamino-5-triazine, 2-methylmercapto-4-ketyl-amino-6-isobutylamino-5-triazine and 2-methoxy-4-isopropylamino-s-triazine and 2-methoxy-4-isopropylamino-s-triazine and 2-methoxy-4-isopropylamino-6-butylamino-S-triazine (2 mg L^{-1}) increase NADH-GDH levels along with NR and protein in pea and maize leaves (See Srivastava and Singh, 1987). They (Srivastava and Singh, 1987) postulate that sublethal concentration of S-triazine stimulate general nitrogen assimilation and protein synthesis. Inhibition of specific activities of NADH-dependent GDH and GOGAT in maize leaves have been reported by DCMU which is possibly through interference to photosynthetic NADH-production (Singh, 1984; Singh and Srivastava, 1987a). This is evidenced from the observation that the same compound does not affect enzyme activities in maize roots, (Singh and Srivastava, 1987a).

Table 2. The enzymes of nitrogen assimilation pathway showing a positive response to plant growth regulator, in the absence of any other inducer.

Enzyme	Plant, growth regulator	Reference
NR	Cytokinin	Kende et al., 1971; Rijven and Parkash, 1972; Gandhi and Naik, 1974; Hanisch ten Cate and Breteler, 1982; Cuello and Sabater, 1982; Rao et al., 1984.
	Gibberellins	Harisch ten Cate and Breteler, 1982.
	Ethylene	Harisch ten Cate and Breteler, 1982.
	Abscisic acid	Sankhla and Huber, 1975b; Palmer, 1982.
NiR	Cytokinin	Sharma, 1986.
GDH	Auxin	Sahulka and Gaudinova, 1975.
	Cytokinin	Huber and Sankhla, 1974.
	Ethylene	Palmer, 1985.
GS	Gibberellins	Tasrba, 1984.
	Abscisic acid	Huber, 1974.

III.   Regulation of Nodulation and N_2 Fixation

Root nodules are the site of N_2- fixation in most of the legumes. Initiation, development and efficiency of the symbiotic root nodules are regulated by the PGRs, (Dart, 1977).

High levels of IAA have been found in root nodules of pea (Pate, 1958) and lupine (Dullaart, 1970) which indicate its importance in the nodule development. Other auxins like IPA, indole-3-acetic acid, indole-3-propionic acid, indole-3-butyric acid and indole-3-carboxylic acid have also been detected in pea root nodules (Badenoch-Jones et al., 1984). Auxin levels in the nodules of pea, bean, soybean and lupine are much higher than in the roots (Pate, 1958; Dullaart, 1970 Badenoch-Jones et al., 1984).

The major cytokinins detected in root nodules of pea are zeatin, zeatin riboside and dihydroxy zeatin riboside (Syono and Torrey, 1976; Syono et al., 1976; Badenoch-Jones et al., 1987). The isopentenyladenine and its riboside are also detected in the noduels of bean (Puppo et al., 1974) and blackgram (Jaiswal et al., 1981). High levels of cytokinin activity have been found in root nodules of non-leguminous plants too (Henson and Wheeler, 1976). Cytokinin levels in nodule tissue were higher than in the root tissue (Badenoch-Jones et al. 1987). Cytokinin contents of the nodules declined with age but there was no major qualitative change in nodule cytokinins with age (Badenoch-Jones et al., 1987). In pea, cytokinin concentration is the highest in young nodule and is positively correlated with mitotic indices (Syono et al., 1976). The nodule cytokinins, zeatin and zeatin riboside stimulate bacterial growth and nitrogenase activity (Jaiswal et al., 1982). However, no clear correlation between nodule cytokinin levels and the effectiveness of nodule in nitrogen fixation was found (Badenoch-Jones et al., 1987). Involvement of auxin and cytokinin in the development of infection and formation of root nodules have been suggested by several workers (Libbenga et al., 1973; Syono et al., 1976). Root hairs of clover were not curled but infection was stimulated by adding IAA and tryptophan (1.0 pM to 1.0 nM) (See, Dart, 1977).

Exogenous application of IAA shows variable effects on nodulation as it is ineffective in cowpea (Mishra and Mohanty, 1967), inhibitory in chickpea (Jaiwal, 1984) or promotory in alfalfa (Gruodiene and Zvironaite, 1971; Sabel' Nikova et al., 1971). This is probably due to the differences in the endogenous levels of auxins in nodules of different species. Application of NAA induces lesser number of nodules which are larger in size, in alfalfa (Kefford et al., 1960) whereas it increases the number as well as nodule weight in groundnut (Srinivasan and Gopal Krishna, 1977).

Tri-iodobenzoic acid, an antiauxin, increases nodulation in pea (Bond, 1948) and mungbean (Jaiwal, 1984). The other antiauxins, 1-naphthoxyacetic acid and (p-chlorophenoxy) - isobutyric acid also stimulate the nodules in the excised bean roots at lower and inhibit the same at higher concentrations (Cartwright, 1967).

Exogenous supply of lower concentrations of cytokinin do not affect nodulation, but higher concentrations inhibit the growth of main root and its branches which subsequently lead to poor nodulation (Kefford et al., 1960). Cytokinins, however, have been found to be beneficial for nodulation in alfalfa (Sabel' Nikova et al., 1970) and pea (Nandwal and Bharti, 1982). Cytokinins also increase the N_2-fixing efficiency of nodules in pea (Nandwal and Bharti, 1982).

Increased amounts of GA have been detected in root nodules of bean and pea (Radley, 1961) and lupine (Dullaart and Duba, 1970). More GA are found in nodules than in the roots (Radley, 1961; Dullaart and Duba, 1970). Addition of GA to the agar root medium or foliar spray of the hormone reduces nodulation of several plants e.g. bean (Thurbur et al., 1958), alfalfa (Galston, 1959; Kefford et al., 1960); clover (Prakash, 1966) blackgram and mungbean (Kandasamy and Prasad, 1976; Jaiwal, 1984). However, foliar spray of GA_3 at 10 µM, increases the number, fresh and dry weights of nodules and nitrogenase activity in peanut (Bishnoi and Krishnamoorthy, 1990). On the other hand, application of GA to rooting medium enhances nodulation in *Cicer arietinum* (Swaraj and Garg, 1970a) and sunnhemp (Kandaswamy and Prasad, 1976). These conflicting results are probably due to the varying endogenous GA levels in different plants. The effects of GA on nodulation varies with the time of its application also (Mes, 1960). If it is applied at an early stage of they plant, it usually reduces nodulation. Radley (1961) suggested that GA could be the substance formed by existing nodules that restricted the development of further nodules.

In pea, application of 1.9 µ ABA to the rooting medium inhibits *Rhizobium* growth and also the formation of nodules, apparently by inhibiting the cortical cell divisions (Phillips, 1971). Abscisic acid also induces pseudonodules in the roots of chickpea (Jaiwal and Gulati, 1989). Ethylene and ethylene releasing compounds such as ethrel (2-chloroethyl-phosphonic acid) also inhibit root growth and nodulation in several species (Grobbelaar et al., 1971; Drennan and Norton, 1972; Jaiwal, 1984). The inhibition is observed even when ethylene is supplied at a very low concentration.

Benzimidazol and B-9 either separately or in combination with IAA have inhibitory effects on the size and the number of nodules in cowpea (Mishra and Mohanty, 1967). On the other hand, CCC increases nodulation in clover (Prakash, 1966) and soybean and bean (Chailakhyan, 1973). This effect is attributed to the inhibitory action of CCC on the production and activity of GA (Prakash, 1966). B-9, suppresses early nodulation but has no effect on the total number of nodules (Lazan, 1976).

Ascorbic acid at higher concentrations suppresses nodulation (Tonzig and Bracci, 1951) while its lower concentrations enhances the initiation and number of nodules as well as nitrogen content of plants in chickpea (Swaraj and Garg, 1969, 1970b); clover (Garg, 1972); peanut (Goswami, 1974) and mungbean (Jaiwal, 1984). The increase in nodulation by ascorbic acid is due to the enhancement of cell divisions (Garg, 1966).

Morphactin reduces the rhizosphere and rhizoplane microflora of chilli plants (Rao et al., 1972a) and inhibits nodulation in chickpea (Rao et al., 1972b) and mungbean (Jaiwal, 1984). The reduction in the number of lateral roots and nodulation is attributed either to the decrease in the endogenous levels of IAA in the roots resulting from the morphactin mediated stimulation of IAA oxidase activity (Jaiwal, 1984) or to the inhibition of downward transport of IAA to the roots from the apical buds (Ganganias and Berg, 1977). Morphactin also induces pseudonodule formation in the roots of chickpea (Jaiwal and Gulati, 1989).

Root exudates, leaf leachates and decaying organic matters, contain phenolic substances, which in turn, inhibit nitrogen fixation (Rice, 1971). Contrary to this, Blum and Rice (1969) have reported to promotory effect of a few phenols on nitrogen fixation. Salicylic acid increases the quantity of natural auxins and subsequently the nodulation (Arutyunyan et al., 1975). Similar increase in the number and dry weight of nodules in chickpea with SA (Jaiwal, 1984) and nodule production and nitrogen fixation in pigeonpea with chlorogenic acid have been reported (Garg et al., 1989).

IV. Conclusions and Future Prospects

The effect of PGR on the uptake of NO_3^- and NH_4^+ have been studied in a few plants. These studies could not reveal, however, any important regulatory role of the PGR. Nitrate reductase being the rate limiting enzyme in the nitrate assimilation pathway, has been extensively studied in the presence of PGR. Cytokinins increase NR activity in the presence or absence of NO_3^-. The regulation of NR by cytokinins and nitrate may be at the transcriptional level and the nitrate and the cytokinins enhance perhaps independently the expression of the same or different NR genes. The effect of cytokinins on other enzymes of nitrate assimilation is variable and to some extent depends on the nature and concentration of the hormone and also on their endogenous levels. The other PGR also affect NR activity variably and their effects depend upon the organ, plant species, endogenous hormone level and some other factors. The role of PGRs in regulation of NiR, GDH and GS-GOGAT have rarely been studied and these aspects should be explored intensively using the modern techniques of molecular biology.

Higher levels of auxins, cytokinins and even GAs have been detected in root nodules than in the roots. The PGRs regulate initiation, development and N_2-fixing efficiency of root nodules. The effect of PGRs, however, has not been studied on nitrogenase system. The expression and regulation of host and bacterial genes and gene products by PGR may elucidate some interesting findings which will help in monitoring this process for crop improvements.

Acknowledgement

Authors are grateful to Dr. Anju Gulati for her kind assistance in preparation of this MS.

Literature Cited

Aleshin, N.E., Avakyan, E.R., and Aleshin, E.R. 1983. Effect of gibberellic acid on nitrate reductase activity in rice. *Sel'sko Khozyaiztvennaya Biologiya,* O: 55-57.

Arnison, P.G., and Boll, W.G. 1978. The Effect of 2,4-D and kinetin on the activity and isoenzyme pattern of various enzymes in cotyledon cell suspension cultures of bush bean. *Can. J. Bot.* **56**: 2185-2195.

Arutyunyan, R.S.H., Stepanyan, M.D., Karapatyan, N.A., and Chailakhyan, M.K.H. 1957. Effect of growth regulators on plant acceptance of root nodule bacteria and an content of auxin like and phenolic substance in pea plant. *Dok lady Academic Nauk Armyanskoi SSR* **61**: 250-256.

Asthana, J.S., and Srivastava, H.S. 1978. Effect of pre-sowing treatment of maize seeds with ascorbic and and salicylic acid on seed germination, seedling growth and nitrate assimilation in the seedlings. *Indian J. Plant Physiol.* **21**: 150-155.

Badenoch-Jones, J., Summons, R.E., Rolfe B.G., and Letham, D.S. 1984. Phytohormones, *Rhizobium* mutants and nodulation in legumes IV. Auxin metabolites in pea root nodules. *J. Plant Growth Regul.* **3**: 23-29.

Badenoch-Jones, J., Parker, C.W., and Letham, D.S. 1987. Phytohormones, *Rhizobium* mutants and nodulation in legumes. VII. Identification and quantification of cytokinins in effective and ineffective pea root nodules using radio immunoassay. *J. Plant Growth Regul.* **6**: 97-111.

Banowetz, G.M. 1992. The effect of endogenous cytokinin content on benzyladenine enhanced nitrate reductase induction. *Physiol. Plant.* **86**: 341-348.

Bergmark, C.L., Jackson, W.A., Volk R.J., and Bhem, U. 1992. Differential inhibition by ferulic acid of nitrate and ammonium uptake in *Zea mays*. *Plant Physiol.* **98**: 639-645.

Bishnoi, N.R., and Krishnamoorthy, H.N. 1990. Effect of water logging and gibberellic acid on nodulation and nitrogen fixation on peanut. *Plant Physiol. Biochem.* **28**: 663-666.

Blum, U., and Rice, E.L. 1969. Inhibition of symbiotic nitrogen fixation by gallic and tannic acid and possible roles in old-field succession. *Bull. Torrey Bot. Club.* **96**: 531-544.

Bond, L. 1948. Origin and developmental morphology of root nodules of *Pisum sativum*. *Bot Gaz.* **109**: 411-434.

Borriss, H. 1967. Untersuchungen uber die steuerung der Enzymakti vital in pylanzlichen Embryonen durch cytokinine. *Wiss Z. Univ. Rostock, Math-Naturwiss Reihe* **16**: 629-639.

Boucaud, J., and Ungar, I.A. 1976. Hormonal control of germination under saline conditions of three halophytic taxa in the genus *Suaeda*. *Physiol. Plant.* **37**: 143-148.

Cartwright, P.M. 1967. The effect of growth regulators on the growth and nodulation of excised roots of *Phaseolus vulgaris* L. Wiss. *Z. Univ. Rostock, Math-Naturwiss, Reihe* **16**: 537-538.

Chailakhyan, M. K.H., Arutyunyan, R.S.H., Stepanyan, M.D.,and Karapetyan, N.A. 1973. Effect of growth retardants CCC on the growth of leguminous plants and nodulation under different methods of application. *Doklady Akademic Naud. Armyan-stoic SSR* **56**: 182-187.

Cuello, J.,and Sabater, B. 1982. Control of some enzymes of nitrogen metabolism during senescence of detached barley (*Hordeum vulgare* CV. Moezoncillo) leaves. *Plant Cell Physiol.* **23**: 561-566.

Dart, P. 1977. Infection and development of leguminous nodules, In *A Treatise on Dinitrogen Fixation*. Section III - Biology. (Hardy, R.W.F. and Silver, W.S., eds.): John Wiley and Sons, New York. pp. 367-472.

Day, J.M., Dart, P.J., and Percy, R. 1971. *Report of Rothamsted Experimental Station*. Part 1 p. 99.

Dilworth, M.F., and Kende, H. 1974. Control of nitrate reductase activity in excised embryos of *Agrostemma githago*. *Plant Physiol.* **54**: 826-828.

Drennan, D.S.H., and Norton, C. 1972. The effect of ethrel on nodulation in *Pisum sativum* L. *Plant Soil* **36**: 53-57.

Dullaart, J. 1970a. The auxin contents of root nodules of *Alnus glutinosa* L. vill, *J. Exp. Bot.* **21**: 975-984.

Dullaart J. 1970b. The bioproduction of indole-3-acetic acid and related compounds in root nodules and roots of *Lupinus luteus* L. and by its rhizobial symbiont. *Acta Bot. Neerl.* **19**: 513-515.

Dullaart, J., and Duba L.I. 1970. Presence of gibberellin like substances and their possible role in auxin bioproduction in root nodules and roots of *Lupinus luteus* L. *Acta Bot. Neerl.* **19**: 877-881.

Eder, A., and Huber, W. 1977. About the effect of abscisic acid and kinetin on biochemical changes in *Pennisetum typhoides* during stress conditions. *Z. Pflanzenphysiol.* **84**: 303-312.

Fletcher, W.W., Aleorn, J.W.S.,and Raymond, J.C. 1958. Effect of GA on nodulation of white clover (*Trifolium repens*). *Nature* **182**: 1319-1320.

Ganganias, A.A., and Berg, A.R. 1977. The effect of morphactin (methyl 2-chloro 9-hydroxy fluorene 9-carboxylate) on basipetal transport of indole-3 acetic acid in hypocotyl sections of *Phaseolus vulgaris* L. *Ann. Bot.* **41**: 1135-1148.

Galston, A.W. 1959. Gibberellins, and nodulation. *Nature* **183**: 545.

Gandhi, A.P., and Naik, M.S. 1974. Role of roots, hormones and light in the synthesis of nitrate reductase and nitrite reductase in rice seedlings. *FEBS Lett.* **40**: 343-345.

Garg, B.K. 1972. Effect of applied ascorbic acid and thymine, a DNA base on nodulation and nitrogen fixation in berseem. *M.Sc. Thesis,* C.C.S. H.A.U. Hisar. India

Garg, O.P. 1966. Promotion of lateral bud development: A New Concept. *Curr. Sci.* **135**: 155-156.

Garg, N., Garg, O.P., and Dua, I.S. 1989. Symbiotic nitrogen fixation in relation to mono; di-or polyphenols in *Cajanus cajan*. *Indian J. Plant Physiol.* **32**: 86-89.

Garg, S., and Srivastava, H.S. 1992. Seedlings age and cytokinin effects on glutamate dehydrogenase activity and nitrogen assimilation in maize leaves. *Biol. Plant.* **34**: 152-157.

Goswami C.L. 1974. The effect of applied ascorbic acid and adenine on nodulation and nitrogen fixation in leguminous crops. *Ph.D. Thesis,* C.C.S. H.A.U. Hisar. India

Grobbelaar, N., Clarke, B., and Hough, M.C. 1971. The nodulation and nitrogen fixation of isolated roots of *Phaseolus vulgaris* L. III. The effect of carbon dioxide and ethylene. *Plant Soil* (special Vol): 215-233.

Gruodien, J., and Zvironaite, V. 1971. Effect of IAA on growth and synthesis of N compounds in lucerne. *Luk. TSR Aukstuju Mosklo Darbai Biologia* **17**: 77-87.

Guadinova, A. 1990. The Effect of cytokinins on nitrate reductase activity. *Biol. Plant.* **32**: 89-96.

Gunther, G. 1974. The influence of kinetin on the induction of nitrate reductase activity in tissue cultures. *In Plant Growth Substances* (Skoog, F., ed.) Hirakawa, Tokyo. pp. 671-674.

Hanisch ten Cate, C.H., and Breteler, H. 1982. Effect of Plant growth regulators on nitrate utilization by roots of nitrogen depleted dwarf bean. *J. Exp. Bot.* **33**: 37-46.

Henson, I.E. 1984. Effects of atmospheric humidity on abscisic acid accumulation and water status in leaves of rice (*Oryza sativa* L.) *Ann. Bot.* **54**: 569-582.

Henson, I.A., and Wheeler, G.T. 1976. Hormones in plants bearing nitrogen fixing nodules: The distribution of cytokinins in *Vicia faba* L. *New Phytol* **76**: 433.

Higgins, T.J. V., Goodwin P.B., and Carr, D.J. 1974. The inhibition of nitrate reductase in mungbean seedlings. *Aust. J. Plant Physiol.* **1**: 1-8.

Hirschberg, K., Hubner, G., and Borris, H. 1972. Cytokinin induced *de novo* synthesis of nitrate reductase in isolated embryos of *Agrostemma githago*. *Planta* **108**: 333-337.

Hofmann, K. 1967. Wntersuchungen uber die Induktion der Nitrate reduktase in pflanzlichen embryonen. *M. Sc Thesis, University of Greifswald.*

Huber, W. 1974. The influence of sodium chloride and abscisic acid treatment on protein metabolism and some further enzymes of amino acid metabolism in seedlings of *Pennisetum typhoides*. *Planta* **121**: 225-236.

Huber, W., and Sankhla N. 1974. Abscisic-kinetin interaction in growth and activities of enzymes of amino acid metabolism in *Pennisetum typhoides* seedlings. *Z. Pflanzenphysiol.* **73**: 160-166.

Huber, W., Kreutmeier, F., and Sankhla, N. 1977. Ecophysiological studies on Indian arid zone plants VI. Effects of sodium chloride and abscisic acid on amino acid and protein metabolism in leaves of *Phaseolus aconitifolius*. *Z. Pflanzen Physiol.* **81**: 234-247.

Jain, A., and Srivastava, H.S. 1981a. Effect of salicylic acid on nitrate reductase activity in maize seedlings. *Physiol. Plant.* **51**: 339-342.

Jain, A., and Srivastava, H.S. 1981b. Effect of salicylic acid on nitrite reductase and glutamate dehydrogenase activities in maize roots. *Physiol. Plant.* **53**: 285-288.

Jaiwal, P.K. 1984. Studies on the effects of certain growth regulating substances on the morphology of some important leguminous crops. *Ph. D.Thesis.* Kurukshetra Univ. Kurukshetra, India.

Jaiwal, P.K., and Gulati, A. 1989. Morphactin and abscisic acid induced pseudonodule formation in the roots of *Cicer arietinum* L. *Proc Nat. Acad. Sci. India* **59**: 463-477.

Jaiswal, V., Rizvi J.H., Mukherjee, D., and Mathur, S.N. 1981. identification of cytokinins present in root nodules of *Phaseolus mungo* L. *Biol. Plant.* **23**: 342-344.

Jaiswal, V., Rizvi, S.J.H., Mukherji, D., and Mathur, S.N. 1982. Nitrogenase activity in root nodules of *Vigna mungo*. The role of nodule cytokinins. *Angew. Bot.* **56**: 143-148.

Kandaswamy, D.,and Prasad, N.N. 1976. Foliar spray of growth regulators on nodulation in three plants and its relation to *Rhizobium* population. *Indian J. Exp. Biol.* **14**: 737-739.

Kar, M., and Feierabend, J. 1984. Changes in the activities of enzymes involved in amino acid metabolism during the senescence of detached wheat leaves. *Physiol. Plant.* **62**: 39-44

Keanov, E., and Pavlovo, A. 1976. Effect of some cytokinin analog N-allyl-N'-phenyl-thiourea derivatives on the inhibition of chlorophyll decomposition and the nitrate reductase activity in young leaves of *Hordeum vulgare*, *Biochem. Physiol. Pflanzen.* **170**: 479-485.

Kefford, N.P., Zwar, J.A., and Bruce, M.I. 1960. The symbiotic synthesis of auxin by legume and nodule bacteria and its role in nodule development. *Aust. J. Biol. Sci.* **13**: 456-467.

Kende, H., Hahn, H., and Kays, S.E. 1971. Enhancement of nitrate reductase activity by benzyladenine in *Agrostemma githago*. *Plant. Physiol.* **48**: 702-706.

Kende, H., and Shen, T.C. 1972. Nitrate reductase in *Agrostemma githago*: Comparison of the inductive effects of nitrate and cytokinins. *Biochim. Biophys. Acta,* **286**: 118-125.

Kende, H., M. Fukuyama D., and De Zacks, R. 1974. On the control of nitrate reductase by nitrate and benzylalenine in *Agrostemma githago* embryos. *In Plant Growth Substances.* (Skoog, F. ed.) Hirokawa, Tokyo, pp. 675-682.

Knypl, J.S. 1973. Synergistic induction of nitrate reductase activity by nitrate and benzylaminopurine in detached cucumber cotyledons. *Z. Pflanzenphysiol.* **70**: 1-11.

Kulaeva, O.N. 1980. Cytokinin action on enzyme activities in plants. *In: Plant Growth Substances* (Skoog F. ed.) Springer-Verlag, Berlin. pp. 119-129.

Kuznetsov, V.V.,and Litonova, G.N. 1985. Cytokinin regulation of the process of nitrate nitrogen assimilation in corn cockle embryos. *Fiziol. Rast.* **32**: 494-500.

Kuznetsov, V.V. Kuznetsov V, and Kulaeva O.N. 1979. Influence of nitrate and cytokinin on nitrate reductase activity in isolated *Agrostema githago* seed buds, *Biokhimiya* **44**: 529-545.

Lazan, H. 1976. Effect of growth regulators and combined nitrogen on growth and nodulation of soybean. *Malaysian Agri. Res.* **5**: 9-17.

Lea, P.J, Robinson, S.A., and Stewart, G.R. 1990. The enzymology and metabolism of glutamine, glutamate and asparagine, In the Biochemistry of plants : A comprehensive Treatise Intermediary, Nitrogen Metabolism (Miflin B.J. and Lea P.J. eds) Vol. 16 Academic Press Inc., New York. pp. 121-159.

Libbenga, K.R., Van Iren, F., Bogers, R.J.,and Schraag-Lamers, M.F. 1973. The role of hormones and gradients in the initiation of cortex proliferation and nodule formation in *Pisum sativum* L. *Planta,* **114**: 29-39.

Lips, S.H., and Roth-Bejerano, N. 1969. Light and hormones: an interchangeability in the induction of nitrate reductase. *Science* **166**: 109-110.

Lu, J.L., Ertl, J.R., and Chen, C.M. 1992. Transcriptional regulation of nitrate reductase m-RNA levels by cytokinin-abscisic acid interactions in etiolated barley leaves. *Plant Physiol.* **98**: 1255-1260.

Lu, J.L., He, W-Z., Chen, W., Jhang Y.H.,and Teng Y.W. 1983. Studies on nitrate reductase II. Effect of benzyladenine and light on the induction of nitrate reductase in wheat seedlings. *Acta Phytophysiol. Sinica* **9**: 41-49.

Lu J.L., Extl, J.R., and Chen C.M. 1990. Cytokinin enhancement of the light induction of nitrate reductase transcript levels in etiolated barley leaves. *Plant Mol. Biol.* **14**: 585-594.

Luckwill, L.C. 1968. Relations between plant growth regulators and nitrogen metabolism. *In Recent Aspects of Nitrogen Metabolism in Plants* (Hewitt, E. J. and Cutting, C.V. eds.) Academic Press, London. pp. 189-199.

Mes, M.G. 1960. Influence of gibberellic acid and photoperiod on growth, nodulation and nitrogen assimilation of *Vicia villosa*. *Nature* **184**: 2035-2036.

Mishra, D., and Mohanty, B. 1967. External effect of IAA, benzymidazole and B-9 on nodulation of cowpea. *Nature* **214**: 320-321.

Miura, G.A., and Miller, C.O. 1969. 6- (γ-γ-Dimethyl-allylamino) purine as a precursor of zeatin. *Plant Physiol.* **44**: 372-376.

Mohapatra, S.S., Poole, R.J., and Dhindsa, R.S. 1988. Abscisic acid regulated gene expression in relation to feezing tolerance in alfalfa. *Plant Physiol.* **87**: 468-473.

Nandwal, A.S., and Bharti, S. 1982. The effect of kinetin and IAA on growth yield and nitrogen fixing efficiency of nodules in pea. *Indian J. Plant Physiol.* **25**: 358-363.

Palmer, C.E. 1981. Influence of abscisic acid on nitrate accumulation and nitrate reductase activity in potato tuber slices. *Plant Cell Physiol.* **22**: 1541-1551.

Palmer, C.E. 1982. Abscisic acid stimulation of nitrate reductase activity in potato tuber slices. Effect of inhibition of protein and nucleic acid synthesis. *Plant Cell Physiol.* **23**: 301-308.

Palmer, C.E. 1985. Influence of ethephon on nitrate reductase, protein, amino nitrogen and nitrate content of *Solanum tuberosum* L. *Plant Cell Physiol.* **26**: 407-417.

Pandey, S., and Srivastava, H.S. 1992. Cytokinin effects on nitrate reductase activity in *Leucaena leucocephala* seedlings in the presence of different nitrogenous salts. *Indian J. Plant Physiol.* **35**: 64-72.

Parkash, V. 1972. Synergism between cytokinins and nitrate in induction of nitrate reductase activity in fenugreek colyledons. *Planta* **102**: 105-111.

Pate, J.S. 1958. Studies of the growth substances of legume nodules using paper chromatography. *Aust. J. Biol. Sci.* **11**: 516-526.

Phillips, D.A. 1971. Abscisic acid inhibition of root nodule initiation in *Pisum sativum*. *Planta* **100**: 181-190.

Phillips, D.A., and Torrey, J.G. 1972. Studies on cytokinin production of *Rhizobium*. *Plant Physiol.* **49**: 11-15.

Prakash, V. 1966. Effect of 2-Chloroethyl trimethyl ammonium chloride, trimethyl ammonium chloride and gibberellic acid on nodulation in *Trifolium alexandrinum*. *Indian J. Exp. Biol.* **4**: 251.

Puppo, A. Rigaud, J., and Barthe, P. 1974. Sur La presence de cytokinines dons les nodules de *Phaseolus vulgaris* L. *C.R. Acad. Sc. Ser. D.* **279**: 2029-2032.

Radley, M.E. 1961. Gibberellin-like substances in plants. *Nature* **191**: 684-685.

Rao U.M., Datta, N., Mahadevan, M. Guha-Mukherjee, S., and Sopory, S.K. 1984. Influence of cytokinins and phytochrome on nitrate reductase activity in etiolated leaves of maize. *Phytochemistry* **23**: 1875-1879.

Rao, V.R., Jayakar, M., Sharma, K.R., and Mukherjee, K.G. 1972a. Effect of foliar spray of morphactin on funginin the root zone of *Capsicum annuum*. *Plant Soil* **37**: 179-182.

Rao, V.R. Subba Rao, N.S., and Mukherjee, K.G. 1972b. Inhibition of nodulation in gram (*Cicer arietinum*) by morphactin. *Indian J. Microbiol.* **12**: 264-266.

Rice, E.L. 1971. Inhibition of nodulation of inoculated legumes by leaf leachates from pioneer plant sp. from abandoned fields. *Am. J. Bot.* **58**: 368-371.

Rijven A.H.G.C.,and Parkash, V. 1971. Action of kinetin on cotyledons of fenugreek. *Plant Physiol.* **47**: 59-64.

Rogers, J.C., and Rogers, S.W. 1992. Definition and functional implication of gibberellin and abscisic acid cis-acting hormone response complexes. *The Plant Cell* **4**: 1443 - 1451.

Roth-Bejerano, N., and Lips, S.H. 1970. Hormonal regulation of nitrate reductase activity in leaves. *New Phytol.* **69**: 165-169.

Sabel, Nikova, V.T., Zhighina, A.S.,and Voloshova, M.M. 1971. Effect of growth substances on infection of legumes by nodule bacteria. *Izvestiya Adak\demii Nauk Moldavshi SSR Biologicheskishi Khimieheskish Nauk* **3**: 48-51.

Sankhla, N., and Huber, W. 1974. Activities of enzymes of amino acid metabolism in *Pennisetum* seedlings grown in the presence of 2-chloroethyl-phosphonic acid. *Phytochemistry* **13**: 1319-1321.

Sankhla, N., and Huber, W. 1975. Effect of salt and abscisic acid on *in vivo* activity of nitrate reductase in seedlings of *Phaseolus aconitifolius*. *Z. Pflanzenphysiol.* **76**: 467-470.

Sahulka, J., and Gaudinova A. 1975. Regulation of glutamate dehydrogenase level in excised pea roots by some exogenously supplied compounds with auxin activity. *Biol. Plant.* **17**: 228-230.

Schmerder, B., and Borriss, H. 1986. Induction of nitrate reductase by cytokinins and ethylene in *Agrostemma githago* L. embryos. *Planta* **169**: 589-593.

Sharma A.K. 1986. Phytochrome regulation of nitrate reductase and nitrate reductase in maize. *Ph.D Thesis,* Jawaharlal Nehru Univ., New Delhi.

Sharma A.K., and Sopory S.K. 1987. Effect of phytochrome and kinetin on nitrate reductase in *Zea mays* L. *Plant Cell Physiol.* **28**: 447-454.

Singh, R.P. 1984. Assimilation of inorganic nitrogen and primary amination reactions in maize seedlings. *Ph. D. Thesis,* D.A. Univ. Indore, India.

Singh, R.P., and Srivastava, H.S. 1987a. In vivo effect of some metabolic inhibitors on glutamate dehydrogenase and glutamate synthase activities in excised maize tissues. *Curr. Sci* **56**: 93-94.

Singh, R.P., and Srivastava, H.S. 1987b. Effect of salicylic acid on NADH glutamate synthase activity in root and leaf tissue of maize seedlings. *Indian J. Plant Physiol.* **30**: 60-65.

Sopory, S.K., and Sharma A.K. 1990. Spectral quality of light, hormones and nitrate assimilation. *In Nitrogen in higher Plants* (Y.P. Abrol ed) Research Studies Press Ltd. England. pp. 129-157.

Srinivasan, P.S., and Gopal Krishna, S. 1977. Effect of planofix-an NAA formulation on groundnut var. TMU-7. *Curr Sci.* 119-120.

Srivastava, H.S. 1980. Regulation of nitrate reductase activity in higher plants. *Phytochemistry* **19**: 725-733.

Srivastava, H.S., and Singh, R.P. 1987. Role and regulation of L-glutamate dehydrogenase activity in higher plants. *Phytochemistry* **26**: 597-610.

Srivastava, H.S., and Fletcher, R.A. 1992. Triadimenol increase nitrate levels and nitrate reductase activity in canola leaves. *J. Exp. Bot.* **43**: 1267-1271.

Subba Rao, N.S. 1975. *Soil Microorganisms and Plant Growth,* Oxford and IBH Publ., New Delhi.

Swaraj, K., and Garg, O.P. 1969. Effect of treatment of growing apices of *Cicer arietinum* with IAA and GA_3 on nodulation and nitrogen fixation. *Indian J. Plant Physiol.* **12**: 38-47.

Swaraj, K., and Garg, O.P. 1970a. The effect of gibberellic acid (GA_3) when applied to the rooting medium on nodulation and nitrogen fixation in gram (*Cicer arietinum*). *Physiol. Plant.* **23**: 747-754.

Swaraj, K., and Garg, O.P. 1970b. The effect of ascorbic acid when applied to the rooting medium on nodulation and nitrogen fixation in gram (*Cicer arietnum*) L. *Physiol. Plant.* **23**: 889-897.

Syono, K., and Torrey, J.G. 1976. Identification of cytokinins in root nodules of garden pea (*Pisum sativum* L.). *Plant Physiol.* **57**: 602-606.

Syono, K., Newcomb, W., and Torrey J.G. 1976. Cytokinin production in relation to the development of pea root nodules. *Can J. Bot.* **54**: 2155-2162.

Tasrba, G. 1984. Effect of gibberellic acid on glutamine synthetase activity in two varieties of lettuce seeds New York 515 and grand raphids. *Plant Cell Physiol.* **25**: 293-297.

Thurber, G.A. Douglas, J.R., and Galston, A.W. 195 8. Inhibitory effect of gibberellins on nodulation in dwarf bean (*Phaseolus vulgaris*). *Nature* **181**: 1082-1083.

Tkemaladze, G.S.H. 1981. Effect of 2, 4-D on activity of glutamate and malic dehydrogenase in pea and corn seedlings. *Soviet Plant Physiol.* **28**: 1013-1021.

Tonzing, S., and Bracci, L. 1951. Ricerche sulla fisiologic dell, acido ascorbico VI. Attirit, tubercol-igene del *Rhizobium Leguminosarum* acido aseorbico. *Nuovo Gior. Bot. Ital.* **58**: 2558-270.

Wallsgrove, R.M., Lea, P.J., and Miflin, B.J. 1979. Distribution of the enzymes of nitrogen assi-milation within the pea leaf cell. *Plant Physiol.* **63**: 232-236

White, R., and Kende, H. 1992. Regulation of nitrate reductase by cytokinins. *Plant Physiol.* S-470.

Wu, M.T., Singh, B., and Salunkhe, D.K. 1971. Influence of S-triazines on some enzymes of car-bohydrate and nitrogen metabolism in leaves of pea (*Pisum sativum*) and sweet corn (*Zea mays*) L. *Plant Physiol.* **48**: 517-520.

Nitrogen Nutrition in Higher Plants, 1995
Editors : H.S. Srivastava & R.P. Singh
Associated Publishing, Co., New Delhi, India
pp. 417-430.

Assimilation of Nitrogen Dioxide by Plants and its Effects on Nitrogen Metabolism

H.S. SRIVASTAVA, D.P. ORMROD and B. HALE

Abbreviations: GDH, L-glutamate dehydrogenase; GOGAT, glutamine oxoglutarate aminotransferase (glutamate synthase); GS, glutamine synthetase; NiR, nitrite reductase; NR, nitrate reductase; PGA, phosphoglyceric acid; RUBP, ribulose 1-5, bisphosphate.

I. Introduction

Nitrogen dioxide (NO_2), in the gaseous state and as a component of photo-chemical smog and acid rain, has been recognized as a major air pollutant in industrialized and urbanized regions. It also helps generate the secondary pollutants ozone and peroxyacyl nitrates (PAN) in the atmosphere through reactions with hydrocarbons, under the influence of ultra-violet radiation. The important anthropogenic sources of the pollutant are motor vehicles, power plants, industries, incinerators and household activities where high temperatures are created. Another oxide of nitrogen, nitric oxide, is often co-generated along with NO_2, but in a warm atmosphere, it is rapidly oxidised to NO_2 by gas phase reaction (Finlayson-Pitts and Pitts, 1986). In most industrialized countries, the annual emission of NO_2 increased steadily until the early 1980's when it became relatively constant, largely because of the control of source emissions. Total global emission of NO_2, from anthropogenic sources, was estimated to be more than four times greater in 1980 than in 1930 (Dignon and Hameed, 1989). Global emissions are likely to increase further due to the accelerated pace of industrialization and urbanization in most developing countries. However, even at the current rate of emission, the toxic effects of the gas are a matter of great concern (Wellburn, 1990). Besides restricting the emission of NO_2, efforts to identify and expand the potential of various types of sinks are also needed.

Plants are able to absorb this pollutant through their exposed leaf surfaces (Hanson and Lindberg, 1991), although the principal route of entry into the plant is through the stomata, as is the case with other gases (Mansfield and Majernik, 1970). After entering the leaf through the stomata, NO_2 may react with apoplastic water and with the cell sap to produce equal amounts of nitrate and nitrite ions, which can be reduced and assimilated :

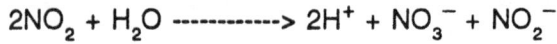

$$2NO_2 + H_2O \text{------------>} 2H^+ + NO_3^- + NO_2^-$$

However, the reactive dissolution of the gas does not seem to be the rate limiting step in the absorption of NO_2, as the amount of NO_3^- accumulated inside the leaf is several times higher (Rogers, et al., 1979) than expected from simple dissolution at that partial pressure (Lee and Schwartz, 1981). The mesophyll or biochemical resistance ($\tau_m NO_2$), indicating the extent of 'fixation' of NO_2 inside the plant cell, seems to be the most important factor governing the uptake of NO_2 in certain situations (Srivastava et al., 1975 a, b). Grennfelt et al. (1983) have also suggested that there must be a biochemical reaction involved in the assimilation of NO_2 by Scots pine (*Pinus sylvestris*) needles. Since most plants have the ability to assimilate nitrate and nitrite (Beevers and Hageman, 1969), a role of NO_2 gas as a source of nutrient nitrogen has been suggested by several investigators (Amundson and Maclean, 1982; Cowling and Koziol, 1982; Rowland et al., 1985; Srivastava et al., 1989). Thus, besides its general effect on plant growth and metabolism, the pollutant is expected to specifically interact with the usual processes of inorganic nitrogen assimilation. This aspect is examined in this chapter.

II. Nitrate and Nitrite Reduction

Assimilation of acquired nitrate is initiated by its reduction to nitrite by the enzyme NADH-nitrate reductase (NR:E.C. 1.6.6.1), a rate limiting and regulatory enzyme (Beevers and Hageman, 1969; Srivastava, 1980). Exposure to NO_2 affects NR activity, the effect ranging from an increase to a decrease in the activity, depending on the species, cultivar, tissue examined, dosage of the pollutant, soil nitrogen status and possibly environmental factors as well (Table 1). In most species, a low dose of NO_2 increases NR activity, especially when the plants have been raised on a nitrogen deficient medium (Srivastava and Ormrod, 1984; Rowland et al., 1987). In the bryophytes, *Clenidium molluscum, Homalothecium sericeum, Pleurozium schreberi* and *Hylocomium splendens* an exposure to 35 nmol mol^{-1} NO_2 for 24 h increases NR activity, although the same level of NO decreases activity (Morgan et al., 1992). Among various environmental factors, the photosynthetic photon flux (PPF) density seems to be an important factor influencing the increase or decrease in enzyme activity (Srivastava et al., 1990). In bean leaves, the effect of NO_2 on NR activity (increase or decrease) is much more pronounced at a continuous PPF density of 300 than at 100, 500 or 700 μmol m^{-1} s^{-1} (Fig. 1).

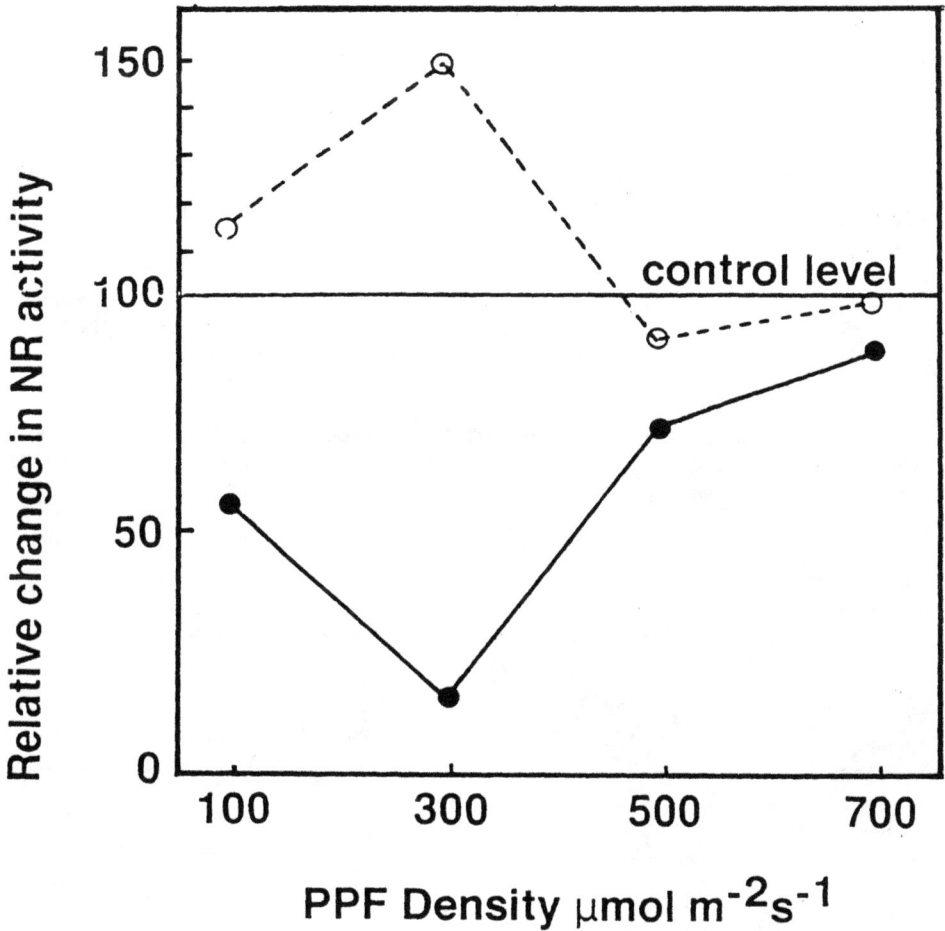

Fig. 1. Influence of photosynthetic photon flux on the increase or decrease in nitrate reductase activity by NO_2 in the presence or absence of nitrate. Values relative to respective controls ($-NO_2 \pm NO_3^-$) are given._____$-NO_3^-$,_____ $+ NO_3^-$. The control line represents $-NO_2 -NO_3^-$ for $-NO_3^-$ and $-NO_2 + NO_3^-$ for $+ NO_3^-$.

[Based on data from Srivastava, et al., 1990]

Table 1.   Effect of nitrogen dioxide on nitrate reductase activity

Species and tissues	NO$_2$ dosage (μmol mol^{-1}h)	Soil N Status	Assay	Effect	Reference
Capsicum annuum cv. Bell Boy					
- leaves	27	Compost soil	*in vitro*	No effect	Murray and Wellburn (1985).
- roots	"	" "	" "		
cv. Rumba					
- leaves	"	" "	" "		
Cucurbita maxima cv. Duch					
- cotyledons	4, 8, 16	Soil with Supplementary N	*in vivo* and *in vitro*	Inhibition	Hisamatsu et al. (1988).
"	12, 24 and 48	" "	" "	"	Takeuchi et al. (1985).
- first leaves	"	" "	" "	No effect	"
Hordeum vulgare cv. Patty					
- roots	60.6	0.01 and 0.1mM NO$_3^-$	*in vivo*	Increase at 0. 1mM but decrease at 0.01 mM NO$_3^-$	Rowland et al. (1987).
- shoots	"	" "	" "	Increase	"
cv. Steptoe (non mutant wild type); narla nar2a (nitrate reductase deficient mutants)	64.8	5mM NO$_3^-$	*in vivo*	No effect (Steptoe) Inhibition (nar1a, nar 2a)	Rowland-Bamford et al. (1989).
- roots	"	5mM NH$_4$Cl	" "	Inhibition	"
Lycopersicon esculentum cv. Ailsa Craig					
- roots	27	Compost soil	*in vitro*	No effect	Murray and Wellburn (1985).
- leaves	"	" "	" "	Increase	" "
cv. Eurocross BB					
- roots	"	" "	" "	No effect	" "
- leaves	"	" "	" "	" "	" "
Phaseolus vulgaris cv. Kinghorn Wax					
- leaves	2.4 to 59.4	1 to 5mM NO$_3^-$	*in vitro*	Increase	Srivastava and Ormrod (1984).
		10 to 20mM NO$_3^-$	" "	No effect	" "
	0.5 to 12	No N	*in vivo*	Increase	Srivastava and Ormrod (1989).
		10mM NO$_3^-$	" "	No effect	" "
Picea abies	1.4	Commercial soil	*in vitro*	Increase	Thoene et al. (1991).
- needles	4.3	" "	" "	No effect	" "
- roots	1694 (500 ppb for 11 weeks)	0.9 mM	*in vivo*	Decrease	Tischner, et al. (1988).

Contd....

Species and tissues	NO$_2$ dosage (μmol mol^{-1}h)	Soil N Status	Assay	Effect	Reference
- shoots	1694	0.9 mM	*in vivo*	Increase	Tischner, et al. (1988).
Picea rubens					
- leaves	1.8 to 12.6	Promix-A	*in vivo*	Increase	Norby et al. (1989).
Pinus sylvestris					
- needles	2.04 to 20.4	8.5 or 17 mMNO$_3^-$	*in vivo*	Increase	Wingsle et al. (1987).
Lycopersicon esculentum cv. Ailsa Craig					
- leaves	4 to 12	Ammonium grown Plants	*in vivo*	Increase	Zeevart (1974).

An increase in NR activity has also been observed with exposure to nitric acid vapour in *Picea rubens* (Norby et al., 1989). Nitrogen penta-oxide, another oxide of nitrogen, also causes a small increase in enzyme activity in the needles of *Picea abies* seedlings (Brown and Roberts, 1988). In the same species, NO$_2$, in the presence of O$_3$ and O$_3$ + SO$_2$ either decreases or increases NR activity, depending upon the age of the plant (Klumpp et al., 1989).

Increase in NR activity is apparently due to the accumulation of nitrate in the exposed plants, which acts both as an inducer and as a substrate for the enzyme. Exposure to NO$_2$ increases nitrate content in most cases (Zeevart, 1976; Kaji et al., 1980; Srivastava and Ormrod, 1984, 1989). The flux of NO$_2$ into plants however, does not seem to depend upon the level of NR activity (Rowland-Bamford et al., 1989) and besides the rate of reduction of nitrate to nitrite, some other 'internal' factors also seem to govern the absorption of NO$_2$ (Saxe, 1986; Rowland-Bamford and Drew, 1988). The inhibition of NR activity may be either another manifestation of the generalized toxic effect of NO$_2$, or due to some specific effect of the pollutant on the formation and function of the enzyme. Several hypotheses can be advanced regarding the specific inhibition of NR activity by NO$_2$.

(1) There is a rapid conversion of nitrate and nitrite to ammonia in NO$_2$-exposed plants in the light (Zeevart, 1976), which may reduce the reductant (NAD(P)H) supply which is an essential factor for both nitrate and nitrite reductases. However, NADH content of *Cucurbita maxima* cotyledons does not change with NO$_2$ fumigation, although the treatment inhibits both *in vivo* and *in vitro* NR activity (Takeuchi et al., 1985).

(2) It has also been suggested that the assimilatory products of NO$_2$, such as amino acids, might inhibit NR activity (Takeuchi et al., 1985). Although as

potential end products of nitrate assimilation amino acids are expected to inhibit NR activity, the effects of exogenously supplied amino acids on NR activity vary considerably, and in some cases some amino acids even increase the activity (Srivastava, 1980). The situation with the endogenous pool of amino acids, however, might be different.

(3)   The pollutant may inhibit the transfer of nutrient derived nitrate to the metabolic pool, as a consequence of its own rapid assimilation. It has been demonstrated that a metabolic pool of nitrate is essential for the induced synthesis of NR (Renner and Beck, 1988) and that the enzyme rapidly turns over in the absence of nitrate (Guerrero et al., 1981).

Nitrite reductase (NiR, E.C. 1.7.7.1) is the second enzyme in the pathway of nitrate assimilation, reducing nitrite to ammonia, and as such it has to be active to prevent a toxic accumulation of nitrite. Increase in NiR activity by NO_2 has been reported in several cases (Yoneyama et al., 1979; Wellburn et al., 1981; Wellburn, 1982; Rowland et al., 1987). In another study, nitrite reductase activity increases in spinach but not in kidney bean, when both species are exposed to 8.0 μmol mol^{-1} NO_2 for 24h in the light (Yu et al., 1988). Exposure to NO_2 in the dark however, activates the enzyme in both species. The effect of NO_2 on NiR appears to be tissue specific as well; in barley, while NO_2 increases NiR activity in the shoot, it has no effect on the root enzyme (Rowland et al., 1987). Like NR, the increase in NiR activity is light intensity dependent. The enzyme increases substantially with NO_2 exposure at 250 μmol m^{-2} s^{-1}, while there is no effect at lower photosynthetic photon flux densities (Murray and Wellburn, 1985). Increased NiR activity has been observed with NO exposure as well (Wellburn et al., 1980 a,b). In tomato, the increased activity in the sensitive cultivar Ailsa Craig is much more apparent than in the resistant cultivar Sonato (Wellburn et al., 1980b).

Besford and Hand (1989) by employing an immuno-blotting technique have demonstrated that the increase in NiR activity by nitrogen oxides (NO_x) at high CO_2 levels in lettuce (*Lactuca sativa* L.) is due to net synthesis of the enzyme. They have also suggested that the increased levels of NiR may be the result of NO_x stimulating the flow of nitrogen through the nitrogen metabolizing pathway, thereby raising the levels of protein in general. Consistent with this hypothesis they found an increase in the activities of some enzymes of carbohydrate metabolism.

III.   Primary Amination

Ammonium, either taken from the soil or generated from the reduction of nitrate/ nitrite, is assimilated primarily to glutamic acid through either the GDH or the GS/GOGAT pathway (Miflin and Lea, 1976). Among various factors determining the operation of either the GDH or the GS/GOGAT pathway, the source of inorganic nitrogen seems to be the most important (Srivastava and Singh, 1987). In studies on the effects of NO_2 on the enzymes of ammonium assimilation, Wellburn, et al. (1980a) found that GDH increases with 0.2 μmol mol^{-1} NO$_2$, but is unaffected by 0.1 μmol mol^{-1}. In the same study, GS activity was unaffected by both levels of NO$_2$. In spruce (*Picea*) needles also GS activity is unaffected by 1.4 to 9.8 μmol mol^{-1} x h NO$_2$ (Theone et al., 1991). In *Picea abies* seedlings, GDH activity in the roots is little affected, although GS activity in the shoots increases substantially with NO$_2$ exposure (Tischner et al., 1988).

The activities of the transaminating enzymes glutamate pyruvate transaminase and glutamate oxaloacetate transaminase are also unaffected by 0.1 and 1.0 μmol mol^{-1} NO$_2$ in pea seedlings (Horsman and Wellburn, 1975). However, NADH-GOGAT in bean leaves is substantially increased by exposure to NO$_2$ ranging from 0.02 to 0.5 μmol mol^{-1} (Srivastava and Ormrod, 1984). The increase is more pronounced in the absence of nutrient nitrate than in its presence. Since the other enzyme of the ammonium assimilation pathway, i.e. NADH-GDH, does not show any consistent response to NO$_2$, it has been suggested that NO$_3^-$/NO$_2^-$ generated from NO$_2$ is assimilated through the GS/GOGAT pathway (Srivastava and Ormrod, 1984). Experimental proof of this suggestion, involving the use of the ^{13}N or ^{15}N isotope of nitrogen has yet to be obtained.

IV. Amino Acid and Protein Content

Variable effects of NO$_2$ exposure on protein and amino acid contents (the eventual products of inorganic nitrogen assimilation) have been observed. Increased contents of the soluble amino group, as measured by the ninhydrin reaction, and of protein content have been reported for many species (Zeevart, 1976). Increased nitrite, ammonium and rapidly metabolized amino acids such as aspartate, glutamate and alanine have been recorded in squash seedlings (Takeuchi et al., 1985). Amino acid content of barley roots and shoots is also increased substantially by a 9-day exposure to 0.3 μ mol mol^{-1} NO$_2$ (Rowland 1986). On the other hand, 85 nmol mol^{-1} NO$_x$ for 10 days decreases amino acid content in pine needles, glutamine and arginine decreasing more than many other amino acids (Wingsle et al., 1987). The authors have suggested that the decrease in arginine is due to either its increased turnover or its transport to other organs of the seedlings. In bilberry (*Vaccinium myrtillus*) tissues, levels of NH$_4^+$, free amino acids and total protein content are higher in areas polluted with nitrogenous gases, compared with unpolluted areas (Lahdesmaki et al., 1990). In this species, the protein profile is also altered due to nitrogenous pollutants (Pietila et al., 1990). In *Picea abies* seedlings, exposure to 500 n mol mol^{-1}NO$_2$ for 11 weeks increases root and shoot protein contents by about 10% (Tischner et al., 1988). Clearly, the effects of NO$_2$ on amino acid and protein contents vary among species.

Alterations in the levels of individual amino acids in response to NO$_2$ have also been examined in a few cases. In tomato (*Lycopersicon*) leaves, increases in glutamine, asparagine and serine account for most of the increase in total amino acids (Murray and Wellburn, 1985). In kidney bean also, increase in total amino acid

content consists mainly of glutamine in the early stages of exposure and asparagine in the later stages (Ito et al., 1986). Preferential increases in some specific
amino acids have also been demonstrated using $^{15}NO_2$, although in most cases
there is invariably an increase in the amides, glutamine and asparagine (Yoneyama
and Sasakawa, 1979; Kaji et al., 1980; Okano et al., 1987). The amides are key
amino donors for many other amino acids and their rapid synthesis during NO_2
exposure may perhaps be aimed towards detoxifying the NH_4^+ formed from the
reduction of NO_3^- and NO_2^-, the cellular products of NO_2. However, in NO_x-
exposed scots pine seedlings, glutamine content is little affected while glycine is
increased substantially (Nasholm et al., 1991). These differences among studies
might be due to differences in the soil nitrogen status.

V. Symbiotic Nitrogen Fixation

Although symbiotic fixation is an important source of nitrogen, specially for
legumes, the effects of air pollutants on the process have rarely been examined.
In studies with bean (*Phaseolus vulgaris*), exposing 8-day old seedlings for 6 h each
day for 15 days to NO_2 concentrations ranging from 0.02 to 0.5 μmol mol^{-1} decreases total nodule weight at 0 and 1 mol m^{-3} nutrient nitrate, and nitrogenase activity at 0 and 20 mol m^{-3} nutrient nitrate (Srivastava and Ormrod, 1986). However,
exposing soybean plants at the flowering stage to low levels of NO_2, causes some
increase in nitrogenase activity in the nodules (Gupta and Narayanan, 1992).

VI. Assimilation of NO_2 and its Role as a Nutrient

Although NO_2 is known to suppress plant growth and productivity (Taylor et al.,
1975), its stimulatory effects on plant growth at relatively low concentrations have
also been reported in several studies (Spierings, 1971; Faller, 1972; Capron and
Mansfield, 1977; Kress et al., 1982; Whitmore and Freer-Smith, 1982; Marie and
Ormrod, 1984). This stimulation is believed to be due to the assimilation of NO_2
which provides additional nitrogen, and it has been suggested that NO_2 can act
as a source of nutrient nitrogen (Amundson and Maclean, 1982; Cowling and Koziol,
1982; Rowland et al., 1985). This suggestion is supported by the following evidence:

(1) As described earlier, most of the enzymes involved in NO_2 assimilation are
     activated by NO_2.
(2) There is an increase in total organic nitrogen content following NO_2 expo-
     sure (Spierings, 1971; Zeevart, 1976; Yoneyama and Sasakawa, 1979; Sriv-
     astava and Ormrod, 1984; Ito et al., 1985; 1986; Rowland, 1987).
(3) The contribution to total organic nitrogen in plants by NO_2 is greater in plants
     grown under conditions of low soil nitrogen than in high soil nitrogen (Sriv-
     astava and Ormrod, 1984; Okano and Totsuka, 1986; Rowland-Bamford and
     Drew, 1988).

(4) Nitrogen dioxide may suppress the alternate mode of nitrogen acquisition in bean i.e. symbiotic nitrogen fixation, even in plants raised in low soil nitrogen (Srivastava and Ormrod, 1986).

(5) Finally, direct measurements of ^{15}N have shown that the N of $^{15}NO_2$ is incorporated into organic nitrogen (Rogers et al., 1979). In 3 h, about 97% of the absorbed NO_2 is incorporated into organic nitrogen.

Some estimates have been made about the extent of N input from NO_2 pollution in larger ecosystems. Hanson et al. (1989) estimate N inputs between 0.04 and 1.9 kg ha^{-1} yr^{-1} for natural forests and inputs up to 12 kg ha^{-1} yr^{-1} for forests in urban environments at prevailing NO_2 levels. Hill (1971) estimates an input of 100 kg NO_2 ha^{-1} yr^{-1} in the southern California landscape at an average ambient atmospheric concentration of 60 nmol mol^{-1} NO_2. It has been realized, however, that the role of NO_2 as a source of nutrient nitrogen is limited and it can not act as an alternative source of nitrogen to that in soil (Srivastava and Ormrod, 1984; Wellburn, 1990; Taylor and Bell, 1992). Further, in some situations, although the pollutant is able to increase total organic nitrogen, it suppresses the overall growth of the plant (Srivastava and Ormrod, 1986). Considerable speculation focuses on the phytotoxicity of nitrite formed from nitrogen oxides (Wellburn, 1990; Shimazaki et al., 1992). These generalizations can be made about the role of low levels of NO_2 in plant metabolism:

(1) Plants may assimilate NO_2 as a reaction to toxic accumulation of nitrite and possibly of ammonia and protons (H^+) in the cell. Experiments involving inhibitors of nitrogen assimilation pathways indicate that the assimilation of NO_2 is faster than that of nitrate taken up from the soil (Srivastava and Ormrod, 1989). The rapid assimilation of the pollutant may spare the plants from its toxicity in two ways: (a) it may reduce the toxic levels of nitrite and ammonium and (b) it may neutralize the acidity (H^+) caused by the dissolution of NO_2 in the cell sap. The assimilation of nitrate is believed to cause alkalization of the cell sap, as per the following equation, suggested by Raven (1988) :

$$3NO_3^- + 45CO_2 + 37H_2O \rightarrow C_{45}H_{72}N_3O_{32} + 51O_2 + 2OH^-$$

(2) Apparent stimulation of plant growth by low levels of NO_2 appears to be an indirect consequence of the detoxification processes, rather than a direct effect of NO_2 assimilation. In this connection, it is important to mention that NO_2 increases the activity of many non-nitrogen pathway enzymes as well, such as RUBP carboxylase, 3-PGA phosphokinase, PEP carboxylase (Besford and Hand, 1989), and root malate dehydrogenase (Tischner et al., 1988). It also increases chlorophyll and carotenoid contents in some cases (Horsman and Wellburn, 1975; Srivastava and Ormrod, 1984). As far as enzymes and chlorophyll are concerned, one may argue that they increase because

the assimilation of NO_2 provides the amino acids for induced synthesis of these compounds. But the increase in the non-nitrogenous pigment, carotenoid, indicates that the pollutant stimulates general biogenesis in the chloroplasts. Another oxidant, O_3, also stimulates plant growth in some cases (Bennett et al., 1974). Brown and Roberts (1988) however, suggest that the growth stimulation is due to nitrogen oxides produced as a contaminant in ozonators supplied with air rather than oxygen.

The biochemical and molecular mechanisms involved in the stimulation of plant growth by NO_2 are yet to be fully elucidated. Obviously the search must be linked with that for the mechanisms of its toxicity. There is a threshold dose above which plant growth is suppressed instead of increased. In this context, increasing NO_2 with the aim of achieving higher organic nitrogen and growth could be counterproductive. One must also look for the limitations on the optimum assimilation of NO_2, despite the fact that the biochemical machinery for the process is generally stimulated during exposure to the pollutant. It is likely that some of the reduced NO_2 escapes from the plant in a volatile form. Rogers et al. (1979) have shown hat only about 65% of the absorbed NO_2 is accounted for in the plant tissues. Alternatively, some of the reduced and assimilatory products of NO_2 might be compartmentalized in an organelle, where they are unavailable for plant metabolism and growth.

VII. Conclusions and Future Prospects

Nitrogen dioxide (NO_2), as an atmospheric pollutant and as a possible source of nutrient nitrogen, influences almost every step of nitrogen metabolism from nitrate reduction and assimilation to synthesis of organic nitrogen. Exposure to NO_2 increases nitrate and nitrite reductase activity in most species, particularly those grown on nitrogen deficient media. Total organic nitrogen content is increased following NO_2 exposure, particularly when little soil nitrogen is available. Nitrogen dioxide may suppress symbiotic nitrogen fixation, even when there is little soil nitrogen present. Finally, ^{15}N studies have demonstrated that the N from NO_2 is incorporated into organic nitrogen. Despite these experimental indications that NO_2 may be a significant source of nutrient nitrogen, we suggest that the assimilation of the pollutant itself may be aimed at its detoxification. Therefore, in spite of the fact that its own contribution to organic nitrogen is significant, NO_2 appears to serve a role as nutrient nitrogen only to a limited extent.

Literature Cited

Amundson, R.G., and Maclean, D.C. 1982. Influence of oxides of nitrogen on crop growth and yield. *In Air Pollution of Nitrogen Oxides* (T. Schneider, and L. Grants, eds.) Elsevier Scientific, Amsterdam. pp. 501-510.

Beevers, L., and Hageman, R.H. 1969. Nitrate reduction in higher plants. *Ann. Rev. Plant Physiol.* **20**: 495-522.

Bennett, J.P., Resh, H.M., and Runeckles, V.C. 1974. Apparent stimulation of plant growth by air pollutants. *Can. J. Bot.* **52**: 35-41.

Besford, R.T., and Hand, D.W. 1989. The effects of CO_2 enrichment and nitrogen oxides on some Calvin cycle enzymes and nitrite reductase in glasshouse lettuce. *J. Exp. Bot.* **40**: 329-336.

Brown, K.A., and Roberts, T.M. 1988. Effects of ozone on foliar leaching in Norway spruce (*Picea abies* L. Karst.): Confounding factors due to NO_x production during ozone generation. *Environ. Pollut.* **55**: 55-73.

Capron, T.M., and Mansfield, T.A. 1977. Inhibition of growth in tomato by air polluted with nitrogen oxides. *J. Exp. Bot.* **28**: 112-116.

Cowling, D.W., and Koziol, M.J. 1982. Mineral nutrition and plant response to air pollutants. *In Effects of Gaseous Air Pollution in Agriculture and Horticulture* (Unsworth, M.H. and Ormrod, D.P. eds.) Butterworths Scientific, London. pp. 349-375.

Dignon, J., and Hameed, S. 1989. Global emissions of nitrogen and sulfur oxides from 1860 to 1980. *J. Air Pollut. Contr. Assoc.* **39**: 180-186.

Faller, V.N. (1972). Schwefeldioxid, schwefelwasserstoff, nitrose gase und ammonoiak als ausschliessblich S-bzue, N-quellen der hoheren pflanze. *Zeitschrift Pflanzenernahrung Bodenkunde* **131**: 120-130.

Finlayson-Pitts, B.J., and Pitts, J.N. Jr. 1986. *Atmospheric Chemistry: Fundamentals and Experimental Techniques.* John Wiley, New York. pp. 405-588.

Grennfelt, P., Bengston, C.S., and Skarby, L. 1983. Dry deposition of nitrogen dioxide to Scots pine needles. *In Precipitation, Scavenging, Dry Deposition and Resuspension*, (Pruppacher, H.R., Semonin, R.G. and Slinn, W.G.N. eds.) Vol. 2. Elsevier, New York. pp. 753-761.

Guerrero, M.G, Vega, J.M., and Losada, M. 1981. The assimilatory nitrate reducing system and its regulation. *Ann. Rev. Plant Physiol.* **32**: 169-204.

Gupta, G., and Narayanan, R. 1992. Nitrogen fixation in soybean treated with nitrogen dioxide and molybdenum. *J. Environ. Quality* **21**: 46-49.

Hanson, P.J., Rott, K., Taylor, G.E. Jr., Gunderson, C.A., Lindberg, S.E., and Ross-Todd, B.M. 1989. NO_2 deposition to elements representative of a forest landscape. *Atmos. Environ.* **239**: 1783-1794.

Hanson, P.J., and Lindberg, S.E. 1991. Dry deposition of reactive nitrogen compounds: A review of leaf, canopy and non-foliar measurements. *Atmos. Environ.* **25**: 1615-1634.

Hill, A.C. 1971. Vegetation: a sink for atmospheric pollutants. *J. Air Pollut. Contr. Assoc.* **21**: 341-346.

Hisamatsu, S., Nihira, J., Takeuchi, Y., Satoh, S., and Kondo, N. 1988. NO_2 suppression of light induced nitrate reductase in squash cotyledons. *Plant Cell Physiol.* **9**: 395-401.

Horsman, D.C., and Wellburn, A.R. 1975. Synergistic effects of SO_2 and NO_2 polluted air upon enzyme activity in pea seedlings. *Environ. Pollut.* **8**: 123-133.

Ito, O., Okano, K., Kuroiwa, M., and Totsuka, T. 1985. Effects of NO_2 and O_3 alone and in combination on kidney bean plants (*Phaseolus vulgaris* L.). Growth partitioning of assimilates and root activities. *J. Exp. Bot.* **36**: 652-662.

Ito, O., Okano, K., and Totsuka, T. 1986. Effects of NO_2 and O_3 alone or in combination on kidney bean plants. II. Amino acid pool size and composition. *Soil Sci. Plant Nutr.* **32**: 351-363.

Kaji, M., Yoneyama, T., Totsuka, T., and Iwaki, H. 1980. Absorption of atmospheric NO_2 by plants and soils. VI. Transformation of NO_2 absorbed in the leaves and transfer of nitrogen through the plants. *In "Studies on the Effects of Air Pollutants on Plants and Mechanism of Phytotoxicity".* Res. Rep., Nat. Inst. Environ. Studies, Japan. **11**: 51-58.

Klumpp, A., Kuppers, K., and Guderian, R. 1989. Nitrate reductase activity of needles of Norway spruce fumigated with different mixtures of ozone, sulfur dioxide and nitrogen dioxide. *Environ. Pollut.* **58**: 261-271.

Kress, L.W., Skelly, J.M., and Kinkelmann, K.H. 1982. Growth impact of O_3, NO_2 and/or SO_2 on *Platanus occidentalis*. *Agric. Environ.* **7**: 265-274.

Lahdesmaki, P., Pakonen, T., Saari, E., Laine, K., and Havas, P. 1990. Environmental factors affecting basic nitrogen metabolism and seasonal levels of various nitrogen fractions in tissues of bilberry, *Vaccinium myrtillus*. *Holarctic Ecol.* **13**: 19-30.

Lee, Y.N., and Schwartz, S.E. 1981. Evaluation of the rate of uptake of nitrogen dioxide by atmospheric and surface liquid water. *J. Geophys. Res.* **86**: 11971-11983.

Mansfield, T.A., and Majernick, O. 1970. Can stomata play a part in protecting plants against air pollutants? *Environ. Pollut.* **1**: 149-154.

Marie, B.A., and Ormrod, D.P. 1984. Tomato plants grown with continuous exposure to sulfur dioxide and nitrogen dioxide. *Environ. Pollut.* **33**: 257-265.

Miflin, B.J., and Lea, P.J. 1976. The pathway of nitrogen assimilation in plants. *Phytochemistry* **15**: 875-885.

Morgan, S.M., Lee, J.M., and Ashenden, T.W. 1992. Effects of nitrogen oxides on nitrate assimilation in bryophytes. *New Phytol.* **120**: 89-92.

Murray, A.J.S., and Wellburn, A.R. 1985. Differences in nitrogen metabolism between cultivars of tomato and pepper during exposure to glasshouse atmosphere containing oxides of nitrogen. *Environ. Pollut.* **39**: 303-316.

Nasholm, T., Hogberg, P., and Edfast, A.B. 1991. Uptake of NO_x by mycorrhizal and nonmycorrhizal scots pine seedlings: quantities and effects of amino acid and protein concentrations. *New Phytol.* **119**: 83-92.

Norby, R.J., Weerasuriya, Y., and Hanson, P.J. 1989. Induction of nitrate reductase activity in red spruce needles by NO_2 and HNO_3 vapour. *Can. J. Forest Res.* **19**: 889-896.

Okano, K., Tatsumi, J., Yoneyama, T., Kono, Y., and Totsuka, T. 1987. Comparison of the fates of $^{15}NO_2$ and $^{14}CO_2$ absorbed through a leaf of rice plants. *Res. Rep., Nat. Inst. Environ. Studies, Japan.* **66**: 59-66.

Okano, K., and Totsuka, T. 1986. Absorption of nitrogen dioxide by sunflower plants grown at various levels of nitrate. *New Phytol.* **102**: 551-562.

Pietila, M., Lahadesmaki, P., Pakonen, T., Laine, K., Saari, E., and Havas, P. 1990. Effects of nitrogenous air pollutants on changes in protein spectra with the onset of winter in the leaves and shoots of bilberry (*Vaccinium myritillus*, L.). *Environ. Pollut.* **66**: 103-116.

Raven, J.A. 1988. Acquisition of nitrogen by the shoots of land plants: Its occurrence and implications for acid-base regulation. *New Phytol.* **109**: 1-20.

Renner, U., and Beck, E. 1988. Nitrate reductase activity of photoautotrophic suspension culture cells of *Chenopodium rubrum* is under the hierarchical regime of NO_3^-, NH_4^+ and light. *Plant Cell Physiol.* **29**: 1123-1131.

Rogers, H.H., Campbell, J.C., and Volk, R.J. 1979. Nitrogen-15 dioxide uptake and incorporation by *Phaseolus vulgaris* L. *Science* **206**: 333-335.

Rowland, A.J. 1986. Nitrogen uptake, assimilation and transport in barley in the presence of atmospheric nitrogen dioxide. *Plant Soil.* **91**: 353-356.

Rowland, A.J., Drew, M.C., and Wellburn, A.R. 1987. Foliar entry and incorporation of atmospheric nitrogen dioxide into barley plants of different nitrogen status. *New Phytol.* **107**: 357-371.

Rowland, A.J., Murray, A.J.S., and Wellburn, A.R. 1985. Oxides of nitrogen and their impact upon vegetables. *Rev. Environ. Health* **5**: 295-342.

Rowland-Bamford, A.J., and Drew, MC. 1988. The influence of plant nitrogen status on NO_2 uptake, NO_2 assimilation and on the gas exchange characteristics of barley plants exposed to atmospheric NO_2. *J. Exp. Bot.* **39**: 1287-1297.

Rowland-Bamford, A.J., Lea, P.J., and Wellburn, A.R. 1989. NO_2 flux into leaves of nitrate reductase deficient barley mutants and corresponding changes in nitrate reductase activity. *Environ. Exp. Bot.* **29**: 439-444.

Saxe, H. 1986. Stomatal dependent and stomatal independent uptake of NO_x, *New Phytol.* **103**: 199-205.

Shimazaki, K., Yu, S.-W., Sakaki, T., and Tanaka, K. 1992. Differences between spinach and kidney bean plants in terms of sensitivity to fumigation with NO_2. *Plant Cell Physiol.* **33**: 267-273.

Spierings, F.H.G.C. 1971. Influence of fumigation with NO_2 on growth and yield of tomato plants. *Netherland J. Plant Pathol.* **77**: 194-200.

Srivastava, H.S. 1980. Regulation of nitrate reductase activity in higher plants. *Phytochemistry* **19**: 725-733.

Srivastava, H.S., Jolliffe, P.A., and Runeckles, V.C. 1975a. Inhibition of gas exchange in bean leaves by NO_2. *Can. J. Bot.* **53**: 466-474.

Srivastava, H.S., Jolliffe, P.A., and Runeckles, V.C. 1975b. Effects of environmental conditions on the inhibition of gas exchange by NO_2 in bean leaves. *Can. J. Bot.* **53**: 475-482.

Srivastava, H.S., and Ormrod, D.P. 1984. Effects of nitrogen dioxide and nitrate nutrition on growth and nitrate assimilation in bean leaves. *Plant Physiol.* **76**: 418-423.

Srivastava, H.S., and Ormrod, D.P. 1986. Effects of nitrogen dioxide and nitrate nutrition on nodulation, nitrogenase activity, growth and nitrogen content of bean plants. *Plant Physiol.* **81**: 737-741.

Srivastava, H.S., and Ormrod, D.P. 1989. Nitrogen dioxide and nitrate nutrition effects on nitrate reductase activity and nitrate content of bean leaves. *Environ. Exp. Bot.* **29**: 433-438.

Srivastava, H.S., Ormrod, D.P., and Marie, B.A. 1989. Methods for uptake and assimilation studies of nitrogen dioxide. *In Modern Methods of Plant Analysis* New Series, Vol. 9 (Linskens, H.F. and Jackson, J.F. eds.) Springer-Verlag, Berlin. pp. 213-226.

Srivastava, H.S., Ormrod, D.P., and Hale-Marie, B.A. 1990. Photosynthetic photon flux effects on bean response to nitrogen dioxide. *Environ. Exp. Bot.* **30**: 463-467.

Srivastava, H.S., and Singh, R.P. 1987. Role and regulation of L-glutamate dehydrogenase activity in higher plants. *Phytochemistry.* **26**: 597-610.

Takeuchi, Y., Nihira, J., Kondo, N., and Tezuka, T. 1985. Change in nitrate reducing activity in squash seedlings with NO_2 fumigation. *Plant Cell Physiol.* **26**: 1027-1035.

Taylor, H.J., and Bell, J.N.B. 1992. Tolerance to SO_2, NO_2 and their mixture in *Plantago major* L. populations. *Environ. Pollut.* **59**: 19-24.

Taylor, O.C., Thompson, C.R., Tingey, D.T., and Reinert, R.A. 1975. Oxides of nitrogen. *In Responses of Plants to Air Pollution* (Mudd, J.B. and Kozlowski, T.T. eds.) Academic Press, New York. pp. 121-139.

Thoene, B., Schroder, P., Papen, H., Egger, A., and Rennenberg, H. 1991. Absorption of atmospheric NO_2 by spruce (*Picea abies* L. Karst) trees I. NO_2 influx and its correlation with nitrate reduction. *New Phytol.* **117**: 575-585.

Tischner, R., Peuke, A., Godbold, D.L., Feig, R., Merg, G., and Hutterman, A. 1988. The effect of NO_2 fumigation on aseptically grown spruce seedlings. *J. Plant Physiol.* **133**: 243-246.

Wellburn, A.R. 1982. Effects of SO_2 and NO_2 on metabolic functions. *In Effects of Gaseous Air Pollution in Agriculture and Horticulture* (Unsworth, M.H. and Ormrod, D.P. eds.) Butterworths Scientific, London. pp. 169-187.

Wellburn, A.R. 1990. Why are atmospheric oxides of nitrogen usually phytotoxic and not alternative fertilizers? *New Phytol.* **115**: 395-429.

Wellburn, A.R., Capron, T.M., Chan, H.S., and Horsman, D.C. 1980a. Biochemical effects of atmospheric pollutants on plants. *In Effects of Air Pollutants on Plants.* (Mansfield, T.A. eds.) Cambridge Univ. Press, Cambridge. pp. 105-114.

Wellburn, A.R., Wilson, J., and Aldridge, P.H. 1980b. Biochemical responses of plants to nitric oxide polluted atmosphere. *Environ. Pollut.* **22**: 219-228.

Wellburn, A.R., Higginson, C., Robinson, D., and Walmsley, C. 1981. Biochemical explanation of more than additive inhibitory effects of atmospheric levels of sulfur dioxide plus nitrogen dioxide upon plants. *New Phytol.* **88**: 223-237.

Whitmore, M.E., and Freer-Smith, P.H. 1982. Growth effects of SO_2 and/or NO_2 on woody plants and grasses during spring and summer. *Nature* **300**: 55-57.

Wingsle, G., Nasholm, T., Lundmark, T., and Ericsson, A. 1987. Induction of nitrate reductase in needles of Scots pine seedlings, by NO_x and NO_3^-. *Physiol. Plant.* **70**: 399-403.

Yoneyama, T., and Sasakawa, H. 1979. Transformation of atmospheric NO_2 absorbed in spinach leaves. *Plant Cell Physiol.* **20**: 263-266.

Yoneyama, T., Sasakawa, H., Ishizuka, S., and Totsuka, T. 1979. Absorption of atmospheric NO_2 by plants and soils. II. Nitrite accumulation, nitrite reductase activity and diurnal changes of NO_2 absorption in leaves. *Soil Sci. Nutrit.* **25**: 267-76.

Yu, S., Li, L., and Shimazaki, K.I. 1988. Response of spinach and kidney bean plants to nitrogen dioxide. *Environ. Pollut.* **55**: 1-13.

Zeevart, A.J. 1974. Induction of nitrate reductase by NO_2. *Acta Botanica Netherlandica* **23**: 345-346.

Zeevart, A.J. 1976. Some effects of fumigating plants for short periods with NO_2. *Environ. Pollut.* **11**: 97-108.

Nitrogen Nutrition in Higher Plants, 1995
Editors : H.S. Srivastava & R.P. Singh
Associated Publishing Co., New Delhi, India
pp. 431-445.

Nitrogen Uptake and Assimilation in Halophytes

D.V. AMONKAR and S.M. KARMARKAR

Abbreviations: ESP, Exchangeable sodium percentage; GDH, glutamate dehydrogenase; GOGAT, glutamine oxoglutarate amino transferase (glutamate synthase); GS, glutamine synthetase; NiR, nitrite reductase; NR, nitrate reductase; SAR, sodium absorption ratio.

I. Introduction

Higher plant species have developed different strategies for the assimilation of nitrogen, but among them perhaps the most remarkable adaptive mechanism for nitrogen uptake as well as its utilisation occurs in the halophytes. This is apparent from the fact that halophytes inhabit environments which show wide fluctuations not only in the levels of salinity but also in the structure of the soil and its mineral composition. Of particular interest in the present review is the fact that saline soils by and large have a poor nitrogen content or are deficient in nitrogen, but surprisingly halophytic species inhabiting such areas are known to possess a vigorous nitrogen metabolism, and such plants usually have a higher amino acid and protein content. In addition, it is now known that most halophytes synthesize a variety of nitrogen rich compounds which till very recently were regarded as a means for storage of nitrogen, but of late have been shown to function as osmoregulants. This review however, is confined to a discussion on the present status of our understanding of nitrogen uptake and its assimilation in halophytes. Much attention, therefore, is being focussed on the nutritional status of the soil and the role of salinity in regulating ion uptake and the concomitant effects on metabolic processes in halophytes vis-a-vis other plant species under the influence of salinity stress.

II. The Saline Habitat

Total soluble salts in saline soils mostly include sulphates, nitrates, chlorides and bicarbonates of sodium, potassium, calcium and magnesium. Soils containing more than 0.1% salts can be termed as saline. In coastal regions the soil is continuously bathed in sea water, hence the higher soluble salt content of these soils is due to chlorides of sodium and magnesium. During the formation of saline soils, the deposition of sediments is accompanied by a physico-chemical precipitation of dissolved substances caused by a mixing of fresh and salt water in the estuarine or marine regions (Jackson, 1958). Carbonates, phosphates, nitrates etc. are thereby added to the soil, leading to a further increase in its salt content. In coastal regions carbonates also occur on account of the presence of molluscan shells, arthropod appendages and other calcareous material deposited by marine organisms.

Saline soils are characterized by a high sodium-absorption ratio (SAR), which tends to limit the availability of calcium, magnesium and potassium ions from the soil. Such a soil is also charcterized by a poor aeration, low water availability. An increased exchangeable sodium percentage (ESP) in saline soils results in a decreased uptake of calcium, magnesium and potassium. A high ESP is generally accompanied by a high SAR. This reduces the availability of water to the plant and halophytes generally overcome this difficulty by increasing leaf succulence.

Potassium plays a vital role in plant metabolism and especially in halophytes, it ranks next to sodium. Saline soils are usually rich in sodium but poor in potassium. In spite of this fact, plants of saline habitats absorb this element selectively and, in fact, it is generally believed that the salt tolerance of halophytes depends to a large extent on their capacity to absorb potassium ions from a sodium rich environment (Eppley et al. 1969; Neales and Sharkey, 1981). Halophytes in general, therefore have a high potassium content on account of its preferential absorption.

Saline soils are rich in chloride ions; although halophytes absorb more sodium than chlorides, they have a greater affinity for chloride ions (Waisel and Eshel, 1971; Waisel, 1972). Sodium and chloride ions are taken up by roots of halophytes in non-equivalent quantities; sodium uptake is greater even in a chloride rich environment (Waisel, 1972).

In view of the variety of adaptations in halophytic species to overcome the problems posed by their habitats which are generally unfavourable, it would be of interest to examine one area on which little attention has been focussed so far, namely the strategies adopted by this plant group for the acquisition of nitrogen and its consequent assimilatory processes.

III. Nitrogen Availability and Its Acquisition

A saline environment, whether it be the soil or water, is generally deficient in nitrogen and as is the case with other plant groups faced with the similar problem (low availability of nitrogen), halophytes adapt to such an environment through the development of efficient mechanisms not only for the uptake of nitrogen but also for their assimilation. In addition, halophytes also have the ability to conserve nitrogen and recycle it within the plant.

Halophytes utilize nitrogen mostly in the form of nitrates, but it is well known that saline soils have a nitrate content which is insufficient to meet the requirements of the plants. Nitrate availability is further reduced because this anion is readily leached by rain into deeper layers of the soil where roots cannot penetrate, or often nitrates are washed away with the tidal water. In sea water, the concentration of nitrate nitrogen is as low as 0.001 mg to 0.6mg per litre (Swerdrup et al., 1961). Eppley et al., (1969) and Eppley and Thomas (1969) have studied the absorption kinetics of nitrate and ammonium ions in several species of marine phytoplankton and they have observed a simple hyperbolic relationship between external NO_3^-/ NH_4^+ concentration and their rates of absorption. It has generally been observed that increase in soil salinity interferes with NO_3^- uptake by halophytes (Neales and Sharkey, 1981).

Several species of soil microorganisms are involved in the mineralization of organic matter, as a result of which nitrogen is made available to the plants. Thus, nitrogen metabolism of plants is to a large extent dependent on the availability of nitrogen in the soil and also the microbial activity within the soil. The soil nitrogen undergoes nitrification through bacterial action and these processes are dependent on several environmental factors such as soil pH, salinity, temperature, aeration of soil, etc. Optimum nitrification occurs in a neutral soil or where the pH is alkaline, but nitrification is markedly affected at a pH less than 6.0 (Mahasneh et al., 1984). In the soils supporting *Suaeda* and *Arthrocnemum*, the pH was observed to fluctuate between 7.4 and 8.5 and this has little effect on nitrification (Mahasneh et al., 1984). Earlier, Warrick (1960) observed that the mangrove soils of Bombay show pH values between 7.4 and 8.0. In soils supporting *Clerodendron inerme*, a borderline mangrove, the pH is the highest during the rainy season but it declines during the dry season (Minotti et al., 1968). Chittar (1971) and Amonkar (1977) have observed a seasonal variation in the pH of soils supporting *Suaeda maritima* and *Salvadora persica*, respectively. In the latter case, the soil pH varies between 8.1 and 8.3. In general, the pH of most saline soils does not act as a limiting factor in nitrification.

Another important factor influencing the microbial activity and consequently the process of nitrification is the factor of soil salinity. A salinity of 0.44% is sufficient for completely inhibiting nitrification (Laura, 1977).

Despite the fact that NO_3^- and Cl^- in the growth medium/soil environment do not exhibit any competition among themselves for their uptake by the root systems of plant species (Smith, 1973), the high levels of Cl^- ions, inside the tissues have been shown to inhibit NO_3^- uptake (Cram, 1973). However, halophytes seem to show a different response; in spite of the high levels of chlorides in their tissues they are able to accumulate noticeable levels of NO_3^- (Stewart et al., 1974); that too from a soil environment, known to be poor or deficient in nitrogen content.

On the other hand, the effects of salinity on ammonification are less pronounced. While increasing salinity does tend to retard ammonification, its effect is only partial; ammonification is never completely suppressed (Laura, 1977). The changes in the rates of nitrification and ammonification which accompany changes in the levels of soil salinity, can also be correlated with the increase-decrease in the NH_4^+ and NO_3^- content of the soil as was observed by Doddema et al. (1986) in the soils supporting *Arthrocnemum fruticosum* (L.) Moq. It has also been reported that in natural saline soils, nitrification occurs as long as aeration is sufficient (Stewart et al., 1979). Stewart et al. (1972) examined the utilization of nitrate by the soil marsh plants as well as the variation in nitrogen supply within a salt marsh ecosystem. Apparently, NO_3^- is the most important form of nitrogen available on the marsh; the nitrogen content in the marsh varies from a high nitrogen content in the lower marsh to a low content in the upper marsh. Thus competition for nitrogen in the upper marsh limits the growth of certain species. Stewart et al. (1972) further observed that annuals such as *Suaeda* and *Salicornia* and perennials such as *Puccinellia*, *Spartina* and *Spergularia* are less successful than *Glaux*, *Limonium* and *Triglochin* in competing for the available NO_3^- leading them to suggest that other factors might be involved in this competition for NO_3^-

Conflicting evidence has been presented regarding the form of nitrogen utilized by salt marsh plants. While halophytic species of baltic shore meadows exhibit growth responses to NH_4Cl rather than to $NaNO_3$, suggesting an adaptation to the preferential utilization of NH_4^+ (Tyler, 1967), estuarine marsh species like *Suaeda maritima* are able to utilize NO_3^- (Stewart et al., 1972).

A. Site of Nitrate Reduction

Nitrates may be utilised directly in the roots or may be translocated to the shoots. In the majority of the halophytes examined, it is sufficiently clear that NO_3^- assimilation occurs in roots as well as shoots. e.g. *Atriplex hastata, Cochleria officinalis, Honkenya peploides, Plantago maritima, Salicornia europaea, Suaeda maritima* and *Triglochin maritima* (Stewart et al., 1973), *Arthrocnemum fruticosum* (Saad-Eddin and Doddema, 1986). Thus the current concept is that NO_3^- reduction occurs in the roots, as well as in shoots. Even today it is difficult to assess the role or contribution in exact terms of the root and shoot to nitrogen metabolism as a whole in the entire plant.

B. Nitrate Reductase

Nitrate reductase from the halophytic alga, *Dunaliella* is sensitive to KCl and NaCl (Heimer, 1973). In *Suaeda maritima*, NaCl inhibits the *in vitro* activity of NR.

Similarly in the salt tolerant species, *Salicornea europaea* NR activity gradually declines as NaCl concentrations are raised above 25mM (Austenfeld, 1974). From kinetic studies on NR activity, it has been established that the inhibition caused by NaCl is uncompetitive in nature as is evident from the fact that inhibition in the reduction of NO_3^- is neither dependent on the concentration of the substrate or the cofactor involved (Austenfeld, 1974). This observation is in contrast with those of Flowers et al. (1977) and Boucaud and Billard (1978) who reported that many enzymes of halophytes, particularly the dehydrogenases exhibit a competitive inhibition in response to increment of salt in the external medium.

In *Suaeda maritima* the NR activity is located almost exclusively in the shoots. The root tissues do indicate the presence of NR acitvity but it is as low as 1% of total plant NR activity on a fresh weight basis (Billard and Boucaud, 1982). These workers have further observed that the activity of NR in the root constitutes less than 5% of the activity in the leaf tissue. As the NaCl concentration in the growth medium is raised from O to 129 mM NaCl, NR activity is enhanced by 50% after 25 days, and after 45 days NR activity showed a further increase of 20% over that observed after 25 days, suggesting that the beneficial effect of NaCl on enzyme activity is enhanced depending upon the length of time the plant is in contact with the salt (Billard and Boucaud, 1982). In a growth medium devoid of NaCl, NR activity declines by 70% after 45 days in *S. maritima*. Billard and Boucaud (1982) have compared this to a situation similar to that arising through conditions akin to water loss and high temperature in crop plants like wheat and barley. This is further exemplified by the fact that NaCl *in vivo* (long term treatment) tends to stimulate NR activity while absence of NaCl leads to a 6% decline in enzyme activity. In wheat and barley also NR activity declines by 6% in the absence of water. It has thus been suggested that in *S. maritima* NR activity is due to the net effect of NaCl which simultaneously inhibits the catalytic activity of the enzyme while stimulating the synthesis of enzymatic protein (Billard and Boucaud, 1982).

In another series of experiments conducted with *Arthrocnemum fruticosum* grown in the greenhouse it was observed that plants adapt to their changed environment and the *in vivo* activity of NR is unaffected by salinity levels upto 2%. In fact after one week of salt treatment, the NR activity is the highest in plants not exposed to NaCl in the growth medium, and a negative correlation exists between NR activity and salt concentration. However, prolonged salt treatment (1 month or more) does affect NR activity to the extent that in the roots maximum enzyme activity is noted in the presence of 2% NaCl in the growth medium while in the shoots a similar effect is produced only at 5% level of NaCl (Saad - Eddin and Doddema, 1986).

Interestingly, as NaCl concentrations are raised the NR activity tends to shift from the roots to the shoots. While the soil water potentials influence the levels of NR in the roots during the vegetative phase, in the shoots NR activity is to a large extent dependent on the NO_3^- supply from the roots (Saad-Eddin and Doddema, 1986).

As noted earlier, species from the strand line and lower marsh exhibit a higher N content as well as higher NR activity than those from the upper marsh; this activity runs parallel to the decrease in N content as one proceeds from the lower marsh to the upper marsh (Stewart et al., 1973). This has been explained as being due to the addition of NO_3^- to the lower marsh when tidal inundations occur.

As in most other species, the NR activity in the halophytes is also nitrate inducible (Stewart et al., 1972). Any site within a marsh has an equal potential for inducting NR, inspite of this, *S. maritima* in the lower marsh contains fifty times higher level of activity than NR from plants growing in the upper marsh. This can in fact be correlated with the higher NO_3^- content of the lower marsh as compared to that of the upper marsh (Stewart et al., 1972).

The enzyme basically has a requirement for NADH and hence NR activity with NADPH never exceeds 5% of the level of activity with NADH. Further, the salt marsh plants (*Atriplex hastata, Cochleria officinalis, Spergularia media, Suaeda maritima, Triglochin maritima, Aster tripolium* and *Puccinellia maritima*) respond favourably to sugars and other glycolytic intermediates as reported for other species by Klepper et al. (1971). Stewart et al. (1973) have, therefore, suggested that reducing power necesssary for NO_3^- reduction is made available through generation of NADH during glycolysis. On the basis of the Michaelis constants, optimum pH, etc. they have further concluded that the properties of NR from the salt marsh plants do not in any way differ from the enzyme from other higher plants.

Earlier studies have indicated that plants can be distinguished into those in which reduction of NO_3^- occurs in the roots and others in which the roots do not have the ability to reduce NO_3^- (Bollard, 1957; Wallace and Pate, 1967). In the eight salt marsh plants examined by Stewart et al. (1973), NR activity was detected in their roots. But, inspite of NR being functional in the roots, the shoot represented the main site for NO_3^- reduction, the only exception being *Triglochin*.

High salt concentration in the environment tends to inhibit NR activity in most mesophytes. While Heimer (1973) noted that NaCl inhibited NR activity in halophytes and a halophytic alga, *Dunaliella parva*, Barber et al. (1989) observed that in spinach it is the Cl^- which was responsible for inhibiting NR activity possibly through the involvement of the Mo center.

Few reports have also appeared on the stimulation of NR activity in response to salt stress (Joshi, 1987; Sudhakar and Veeranjaneyula, 1988). In the two mangrove species, *Aegiceras corniculatum*, an excreter, and *Sesuvium portulacastrum*, an accumulator, the higher NR activity has been correlated to the adaptation of these species to salinity stress (Bhosale and Shinde, 1982). Panikkar (1992) has undertaken a detailed investigation of nitrogen metabolism in the halophyte, *Acanthus ilicifolius*. Her data indicate that NaCl concentrations in the range 0-25 mM enhance the *in vivo* activity of NR in the leaves while at higher concentration of salt (50-100 mM) there is a gradual loss of enzyme activity. On the other hand, an *in vitro* assay of the enzyme indicates that NR from the young and mature leaves shows an initial stimulation at 30 mM NaCl but enzyme activity declines at higher level of salinity (60 and 90 mM NaCl). In the senescent leaves, NR activity is much below that observed in young and mature leaves; in the absence of additional salt and even in the presence of salt it continues to decline at all NaCl levels. Apparently, NR from the young and mature leaves is more sensitive to salt than the enzyme from the senescent leaves.

C. Nitrite Reductase

The enzyme NiR catalyses the second step in the reduction of NO_3^- whereby the NO_2^- is reduced to NH_3. The enzyme is affected by various types of stresses but compared to NR, it is less sensitive to water and salt stress. Heuer et al. (1979) could not observe any effect of salinity on NiR from intact or excised wheat leaves. It has been observed that in the halophyte *Arthrocnemum fruticosum*, NiR activity declines under the influence of salinity although the enzyme is less sensitive to salt than NR (Doddema et al., 1986).

Nitrite reductase from the leaves of *Acanthus ilicifolius* responds to NaCl in a manner similar to NR. The salt stimulates NiR activity at 25 mM concentration but the enzyme activity is somewhat inhibited at 50 mM, although the activity is maintained at a higher level than in the control. At 100 mM NaCl in the growth medium NiR activity more or less equals the enzyme activity in the control (Panikkar, 1992).

IV. Assimilation of Ammonium

Soil nitrogen is also made available to the roots of higher plants in the form of NH_4^+. However, within the plant tissues NH_4^+ concentrations are kept at extremely low levels, thereby protecting the tissues from the ill effects of NH_4^+ toxicity. This is made possible through the ability of plants to assimilate NH_4^+ efficiently through two enzymatic pathways namely glutamine synthetase - glutamate synthase pathway, and the glutamate dehydrogenase pathway.

A. Glutamine synthetase - Glutamate synthase (GS-GOGAT) Pathway

The enzymes of the halophyte, *Salicornia herbaceae* are equally or even more resistant to salt than the corresponding enzymes from glycophytes such as pea, bean and rice (Rakova et al., 1978a,b). While increasing NaCl concentrations inhibit GS activity in pea, bean and *Salicornia* roots, the same is stimulated to a considerable extent in maize roots. Even at a concentration of about 1.0 M NaCl the GS activity remains at the level of the control, indicating thereby, that in this species the enzyme surpasses the same enzyme from *Salicornia* in its resistance to salt. Further, the addition of Na_2SO_4 to enzyme preparation stimulates GS activity in maize roots but proves inhibitory in pea roots. Rakova et al. (1978a) observed that the response of GS to the *in vitro* addition of salt was similar to the effect of *in vivo* salt supply. In *Salicornia* for example, in the presence of 470 mM NaCl *in vitro* GS activity was inhibited by 5% while activity during *in vivo* salt supply was stimulated by 8%.

Suaeda maritima var. macrocarpa when grown in optimal saline conditions (129-mM NaCl) exhibits a high GS activity (Boucaud and Billard, 1979). Variations in the levels of GS activity have been recorded in *S. maritima* grown in the lower marsh, drift line and upper marsh (Stewart and Rhodes, 1978). While plants located in the lower marsh exhibit maximum GS activity, the enzyme activity is at a lower level in the drift line zone and at the lowest level in plants of the upper marsh. In the various halophytes examined (*Aster tripolium, Halimione portulacoides, Honkenya peploides, Plantago maritima, Puccinellia maritima, Salicornia europaea, Spartina anglica, Spergularia media, Suaeda maritima, Triglochin maritima*) the levels of GS and GDH decrease in the roots in response to an increase in external salinity. On the other hand, in the shoot tissues, an increase in external salinity brings about a corresponding increase in the level of GS activity. According to Stewart and Rhodes (1978) such changes possibly reflect the tendency for the shoot to play a greater role in nitrogen assimilation under saline conditions.

Along with GS and GDH, glutamate synthase (GOGAT) has also been reported in halophytes (Stewart and Rhodes, 1978). The enzyme in roots is NAD(P)H specific and while it is dependent on glutamine and 2-oxoglutarate, it is inhibited by azaserine. No glutamate synthesis occurs when NAD(P)H is replaced by Fd or when NH_3^- or asparagine substitute for glutamine. In *Triglochin*, however, GOGAT system in the root tissues is active with reduced Fd as well as with the NAD(P)H (Stewart et al., 1979). In general, the level of GOGAT present in the root tissues of the halophytes is much lower than those of GS and GDH. When the four halophytic species, *Aster tripolium Atriplex hastata, Plantago coronocarpus* and *Triglochin maritima* were subjected to changes in salinity in the external medium varying from O to 200 mM (*A. hastata* and *P. coronocarpus*) or from 0-400 mM NaCl (*A. tripolium* and *T. maritima*), the level of GOGAT activity from the shoots, was unaffected while the enzyme from the roots exhibited a slight change in its activity. According to Stewart and Rhodes (1978) the fact that GOGAT in the roots has the ability to utilize NADPH and NADH as well as reduced Fd is indicative of the presence of two forms of the enzyme. However, in the absence of sufficiently purified enzyme samples, their characterization remains incomplete and therefore, the multiple enzyme forms of GOGAT in halophytes are yet to be identified.

Billard and Boucaud (1980) compared the effect of NaCl on the activity of GOGAT from the halophyte, *Suaeda maritima* with that on the GOGAT from the glycophyte, *Phaseolus vulgaris*. While addition of salt in the nutrient medium lowered the activity of GOGAT in the glycophyte, it had no effect on the activity of the enzyme in the halophyte (Billard and Boucaud, 1980).

Studies on the effect of age on the *in vivo* and *in vitro* activities of GS, GOGAT and GDH in the halophyte, *Acanthus ilicifolius* have indicated that the levels of activity of GS and GOGAT are the highest in the young leaves of *Acanthus* and the activities of these enzyme decline in the senescent leaves (Panikkar, 1992). However, ageing pea leaves seem to have enough potential for the assimilation of glutamine derived from protein hydrolysis (Storey and Beevers, 1978). In *A. ilicifolius* a decrease in the activity of GS-GOGAT is accompanied by a simultaneous increase in the activity of GDH, as senescence progresses.

GS from the leaves of *A. ilicifolius* responds favourably to the *in vivo* NaCl treatment; the activity of the enzyme is stimulated at lower NaCl concentrations (0-25mM) while higher concentrations (50-100mM) are inhibitory, the inhibition being more pronounced in the senescent leaves than in young leaves (Panikkar, 1992). In contrast, GOGAT activity declines in response to increments in NaCl in the growth medium.

Both GS and GOGAT from young and mature leaves of *Acanthus* show similar responses to NaCl *in vitro*; the enzymes exhibit a slight stimulation in their activities at lower levels of salinization (30mM NaCl) followed by a gradual decrease in the enzyme activities as NaCl concentrations are raised to 60 and 90 mM. In the senescent leaves, NaCl inhibits the activities of GS and GOGAT (Panikkar, 1992). The decline is more pronounced in the case of GOGAT as compared to GS. These observations are in contrast to the data reported by Boucaud and Billard (1981) where *in vivo* NaCl treatment (0-500 mM) had no effect on GS activity at saturated substrate concentrations in *Suaeda maritima*, while *in vitro* treatment of the enzyme with 25-300 mM NaCl did inhibit GOGAT activity in *Phaseolus* and *Suaeda* (Billard and Boucaud, 1980).

B.   Glutamate Dehydrogenase Pathway.

The effects of salinity on GDH have been demonstrated in several laboratories (See Srivastava and Singh, 1987). An increase in GDH activity may result on account of the accumulation of NH_4^+ and amides under salinity stress, probably due to the GS-GOGAT route becoming inoperative under such stress conditions (Billard and Boucaud, 1980; Srivastava and Singh, 1987). As stated earlier, the addition of NaCl in the nutrient medium is an absolute requirement for the complete development of *Suaeda maritima* var. macrocarpa and optimal growth occurs when NaCl in the growth medium has a concentration of 129 mM; under such conditions GDH in the leaves shows minimum activity in contrast to GOGAT which functions at optimal levels (Billard and Boucaud, 1980). In *Suaeda*, therefore GDH lacks a positive role in the assimilation of NH_4^+ in the leaves, but unfavourable conditions such as high or low salinities lead to the changes in activity in the roots (Billard and Boucaud, 1980).

A specific control mechanism is believed to operate in regulating these activities of the enzymes NR, GS and GDH. While a low NH_3 supply serves to simultaneously maintain the activity of GS at a high level and that of GDH at a low level, enhanced supplies of NH_3 are capable of reversing this trend, in as much that GDH activity is enhanced with a simultaneous depression of GS activity. The experiments of Stewart and Rhodes (1978) have demonstrated that halophytes do have the potential to assimilate NH_3 via the GS-GOGAT as well as GDH pathways. Such a machinery becomes necessary in view of the fact that there could be fluctuations in the availability of NH_3 within the plant tissues on account of limitation in nitrogen supply in a nitrogen poor environment and also because, as mentioned earlier, salinity levels also determine which of the two pathways must remain operative at any point of time.

An analysis of the action of different concentrations of NaCl on the activity of NAD-specific GDH from plants with different degrees of salt resistance (pea, maize cotton and *Salicornia herbaceae*) has provided interesting data on the variability of the responses. While 100 mM NaCl in the medium results in a decline in GDH activity to 87%, 85%, and 74% in pea, maize and *Salicornia* plants respectively, in the cotton plant the degree of inhibition is as much as 43% of the control. An increases in the NaCl concentration upto about 700 mM results in a gradual loss of enzyme activity in the four plant species (Rakova et al., 1978b). The authors thus, concluded that there was no direct correlation between salt resistance of the plant and salt resistance of the enzymes in an *in vitro* system. These observations are in support of the findings of Osmond and Greenway (1972) who noted that PEPcase from the salt resistant halophyte, *Atriplex spongiosa*, was more sensitive to the presence of salt than the same enzyme from the maize plant.

On the other hand, NAD specific GDH activity increased by 38% over that of the control with an *in vivo* salt treatment in *Salicornia*, but was inhibited by almost the same degree in pea and cotton (41% and 35% respectively) and by a slightly lower magnitude (28%) in maize (Rakova et al., 1978b).

A comparative study on the *in vitro* and *in vivo* levels of GDH activity from the young, mature and senescent leaves of the halophyte, *Acanthus ilicifolius*, has been undertaken by Panikkar (1992). Data obtained indicate that the enzyme responds favourably to the presence of salt in the growth medium. The activity of the enzyme increases in response to 0-50 mM NaCl. On the other hand, when the enzyme extract is treated with NaCl solution (*in vitro* effect) the activity in young and senescent leaves is increased by 30 mM NaCl but the activity declines at 60 and 90 mM NaCl in the assay medium. In the mature leaves, GDH activity exhibits a steady decline with increasing concentrations of NaCl in the assay medium (Panikkar, 1992).

From the correlation of data on kinetic characteristics of the enzymes, aspartate aminotransferase (AspAT), alanine amino transferase (AlaAT) and GDH from the leaves of the mangrove, *Sonneratia apetala*, it has been observed that there is a characteristic coordination between the activities of the above noted enzymes, as far as 2-oxoglutarate utilization is concerned (Geeta, 1988). In the mature leaves the enzyme AspAT functions efficiently at low concentrations of 2-oxoglutarate, the Km being 0.35 mM 2-oxoglutarate. Concentrations of 2-oxoglutarate above 1 mM tend to inhibit the activity of AspAT and at 6 mM level of 2-oxoglutarate, AspAT activity declines by 66%. Apparently the AspAT system is uncompetitively inhibited at this concentration and 2-oxoglutarate thus becomes available for utilization by other enzyme systems. The enzyme AlaAT then takes over the function of AspAT. AlaAT is able to utilize 2-oxoglutarate efficiently at concentrations approximating 6mM oxoglutarate. AlaAT attains maximum velocity when the concentration of

2-oxoglutarate in the medium is 6.2 mM, the Km at this stage being 16 mM 2-oxo-glutarate. As the concentration of this substrate is increased to 7.5 mM, the AlaAT system is inhibited (Geeta and Amonkar, 1989).

The two step interaction between GDH and the substrate 2-oxoglutarate suggests that beyond 8 mM 2-oxoglutarate, the utilization of this substrate is through GDH. GDH of the mature leaf thus has the ability to tolerate higher concentrations of 2-oxoglutarate and this favours glutamate synthesis (Geeta, 1988).

In the senescent leaves of *S. apetala,* both AspAT and AlaAT show peak activity at low concentrations of 2-oxoglutarate (1mM). Higher concentrations prove inhibitory for AspAT and AlaAT activity but do not interfere with GDH activity. The enzyme activity (GDH) is sustained even as 2-oxoglutarate concentrations are raised upto 25 mM. It seems that the GDH system in the senescent leaves is more effective than in the mature leaves of *S. apetala* (Geeta and Amonkar, 1989). During senescence therefore, the enzyme GDH efficiently prevents the accumulation of NH_4^+ which otherwise could prove toxic and, coupled with AspAT and AlaAT, it aids in amination reactions.

V.   Nitrogenous Compounds as Osmoregulants; Adaptive strategies

Although halophytes have been found to accumulate different soluble nitrogenous compounds to very high levels, they do exhibit differences in their capacity to accumulate these compounds. While some halophytes accumulate proline, others accumulate glycine betaine or both proline and glycine betaine or none of these compounds (Stewart et al., 1978). Interestingly, the accumulation of β-alanine is restricted to halophytes from the family Plumbaginaceae (Larher, 1976), the accumulation of glycine betaine is characteristic of members of the family Chenopodiaceae, although it also does accumulate in other plant species (Storey et al., 1977). Further, halophytic species which accumulate proline have been shown to contain low levels of proline when grown in the absence of NaCl; proline levels do however, increase in response to enhanced salinity levels (Stewart and Lee, 1974; Treichel, 1975; Stewart et al., 1978). In contrast, high levels of glycine betaine occur in glycine betaine accumulating species when grown in the absence of NaCl. This is also true for the accumulation of β-alanine betaine in species which accumulate this compound.

In addition to what has been stated above, several halophytes, particularly those which accumulate proline, are known to accumulate amino acids, amides and other amino acids, and the concentrations of the amino acids, pipecolic acid and 5-hydroxypipecolic acid are known to be influenced by changes in the level of external salinity (Goas et al., 1970; Larher, 1976). Besides the accumulation of glutamine, asparagine and serine along with proline in *Agrostis stolonifera* (Ahmad, 1978), glutamine and asparagine + glutamine have also been found to accumulate in the halophyte, *Limonium vulgare.* These observations are indicative of a very efficient transaminating system in halophytes.

Apparently the discovery of the presence of compounds such as proline, glycine betaine and other quaternary ammonium compounds in halophytic tissues, their cytoplasmic location and the fact that they do not inhibit or interfere with enzyme activity has resolved the controversy regarding the role of salts in influencing enzyme activity. It has thus been suggested that such compounds function as compatible solutes and serve in intracellular osmotic adjustments (Flowers et al., 1977). In short, in spite of the high ionic concentration in halophytic tissues, metabolic function is not altered or affected due to compartmentalisation of the salts in the vacuoles of cells; under such circumstances the nitrogen compounds such as proline and glycine betaine take over the function of osmotic adjustment.

It is thus evident that halophytic species have developed a variety of strategies for their survival in habitats which by no means are congenial for the growth and development of most plant species. In addition to several other physiological adaptations, the development of an efficient mechanism for the uptake of nitrogen and its assimilation is most striking; more so, when one recalls that halophytic soils are usually poor in nitrogen content. Essentially, this becomes feasible through the coordinated functioning of the three main enzymes involved in nitrogen metabolism, namely NR, GS and GDH, which perform functions under a wide range of salinity, and are capable of responding in a concerted manner to the fluctuating salt levels. As a result of the efficient turnover of these enzymes, maintained at different levels, depending on the level of salinity, a vigorous amino acid metaboliism is ensured. This, in turn, leads to the synthesis and accumulation of a variety of nitrogen rich compounds, some of which are now known to serve as osmoregulants. At the cellular level it is thus apparent that the compartmentalization of salts in the vacuoles, allows a free play of the metabolic machinery to the extent that the high salt levels within cell and tissues of halophytes do not interfere with enzyme function. Under such circumstances, while a perfect balance is struck between the enzyme systems involved in nitrogen metabolism, it also ensures that variations of salts in the environment do not damage the enzyme systems, which otherwise could lead to a virtual breakdown of metabolic systems.

VI. Conclusions and Future Prospects

Saline habitats characteristically have a low nitrogen content but the levels of salinity fluctuate from time to time as well as from one environment to another. Such conditions are generally known to have an adverse influence on nitrogen uptake and its assimilation in plant species, but surprisingly halophytes have developed strategies which ensure an efficient nitrogen metabolism.

Nitrate assimilation in the majority of halophytes examined occurs in the roots as well as in shoots. While NR activity is influenced by changes in the levels of salinity, the enzyme response is variable, affected positively or negatively, in different halophytic species.

The NH_4^+ generated within the plant tissues is effectively utilised on account of both the GS-GOGAT as well as GDH pathways being operative. Thus NH_4^+ toxicity is avoided. This is further ensured by the fact that when there is an excesssive accumulation of NH_4^+ or amides, the GDH pathway functions efficiently, more so when the GS-GOGAT pathway is subdued.

An efficient nitrogen metabolism results in the synthesis and accumulation of several nitrogen rich compounds such as proline, glycine-betaine and several other compounds which are believed to function as compatible solutes, thus serving as osmoregulants.

Acknowledgements

The authors are grateful to their research students, particularly Ms. Nishigandha Misal, Ms. Aruna Rai and Mrs. Padma Panikkar for their involvement at various stages during the preparation of this manuscript.

Literature Cited

Ahmad, N, 1978. Aspects of glycine-betaine phytochemistry and metabolic functions in plants. *Ph.D Thesis*, Univ. of Wales Cardiff.

Amonkar, D.V. 1977. Physiological studies in halophytes: Studies in *Salvadora persica* Linn. *Ph.D Thesis*, Univ. of Bombay, India.

Austenfeld, F.A. 1974. Der Einflu* des NaCl and andere Alkalisalze auf die Nitratreduktase aktivitat von *Salicornia europaea* L. *Z. Pflanzenphysiol.*, 71: 288-296.

Barber, M.J., Notton, B.A., Kay, C.J., and Solomonson, L.P. 1989. Chloride inhibition of spinach NR. *Plant Physiol.*, 90: 70-94.

Bhosale, L.J., and Shinde, L.S. 1982. Photosynthetic products and enzymes in mangrove *Aegiceras corniculatum* (L.) Blanco and a halophyte *Sesuvium portulacastrum* L. *Photosynthetica*, 17: 59-63.

Billard, J.P., and Boucaud, J. 1980. Effect of NaCl on the activities of glutamate synthase from a halophyte, *Suaeda maritima* and from a glycophyte, *Phaseolus vulgaris*. *Phytochemistry*, 19: 1939-1942.

Billard, J.P., and Boucaud, J. 1982. Effect of NaCl on the nitrate reductase of *Suaeda maritima* var macrocarpa. *Phytochemistry*, 21: 1225-1228.

Bollard, E.G. 1957. Nitrogenous compounds in the tracheal sap of woody members of the family rosaceae. *Aust. J. Biol. Sci.*, 10: 288-291.

Boucaud, J., and Billard, J.P. 1978. Characterisation 'de la glutamate d' eshydroge'nase chez un halophyte obligatorie. le *Suaeda maritima* var. macrocarpa. *Physiol. Plant.*, 44:31-37.

Boucaud, J., and Billard, J.P. 1979. Etude comparec des activities glutamate dehydrogenasique et de glutamine synthelasique dans les recines et les parties aeriennes d'un halophyte: obligatorie la *Suaeda maritima* var. macrocarpa et d'un glycophyte: la *Phaseolus vulgaris* cultives en presence de differentes concentration on NaCl. *C.R. Hebd. Seances Acad. Sci.*, 289: 599-602.

Boucaud, J., and Billard, J.P. 1981. La glutamine synthetase du *Suaeda maritma* action *in vivo* et *in vitro* du NaCl. *Physiol. Plant.*, 53: 558-564.

Chittar, V.M. 1971. Physiological studies in halophytes: Studies in *Suaeda maritima* Dumort. *M. Sc. Thesis*, Univ. of Bombay, India.

Cram, W.J. 1973. Internal factors regulating nitrate and chloride influx in plant cells. *J. Exp. Bot.,* **24**: 328-341.

Doddema, H., Saad-Eddin, R., and Mahasneh, A. 1986. Effects of seasonal changes of soil salinity and soil nitrogen on the N-metabolism of the halophyte *Arthrocnemum fruticosum* (L.) Moq. *Plant and Soil,* **92**: 279-294.

Eppley, R.W., Rogers, J.N., and McCarthy, J.J. 1969. Half saturation constants for uptake of nitrate and ammonium by marine phytoplankton. *Limnol. and Oceanog.,* **14**: 912-920.

Eppley, R.W., and Thomas, W.M. 1969. Comparison of half saturation constants for growth and nitrate uptake of marine phytoplankton. *J. Phycol.,* **5**: 375-379.

Flowers, T.J., Troke, P.F., and Yeo, A.R. 1977. The mechanism of tolerance of halophytes. *Annu. Rev. Plant Physiol.,* **28**: 89-121.

Geeta, M. 1988. Physiology of halophytes: Studies on senescence in *Sonneratia apetala* Buch-Ham. *Ph. D. Thesis.* Univ. of Bombay, India.

Geeta, M., and Amonkar, D.V. 1989. Study of aminating and transaminating enzymes in senescent leaves of *Sonneratia apetala* Buch-Ham. *Strat. Physiol. Regul. Plant Productivity, Proc. Natl. Seminar ISPP, Bombay (Dec. 1989)* pp. 123-127.

Goas, M., Larher, F., and Goas, G.. 1970. Mise en evidence des acides pipecolique et 5-hydroxy pipecolique dans certains halophytes. *C.R. Acad. Sci., Paris.* **271**: 1368.

Heimer, Y.M. 1973. The effects of sodium chloride, potassium chloride and glycerol on the activity of nitrate reductase of a salt tolerant and two non-tolerant plants. *Planta.* **113**: 279-281.

Heuer, B., Plant, Z., and Federman, E. 1979. Nitrate and nitrite reduction in wheat leaves as affected by different types of water stress. *Physiol. Plant.* **46**: 318-323.

Jackson, M. L. 1958. *Soil Chemical Analysis.* Prentice Hall, Englewood Cliffs,N.J. Page 498.

Joshi, S. 1987. Effect of soil salinity on nitrogen metabolism in *Cajanus cajan. Indian. J. Plant Physiol.,* **30**: 223-225.

Klepper, L., Flesher, D., and Hageman, R.H. 1971. Generation of reduced NAD for nitrate reduction in green leaves. *Plant Physiol.,* **48**: 580-590.

Larher, F. 1976. Sur quelques particularites due metabolisme azote dune halophyte : *Limonium vulgare. These Doct. Sc Nat. Rennes.*

Laura, R.D. 1977. Salinity and nitrogen mineralization in soil. *Soil Biol. Biochem.,* **9**: 333-336.

Mahasneh, A., Budour, S., and Doddema, H. 1984. Nitrification and seasonal changes in bacterial populations in the rhizosphere of *Suaeda* and *Arthrocnemum* species growing in saline soils. *Plant and Soil* **82**: 149-154.

Minotti, P.L., Williams, D.C., and Jackson, W.A. 1968 . Nitrate uptake and reduction as affected by calcium and potassium. *Soil Sci. Soc. Amer. Proc.,* **32**: 692-698.

Neales, T.F., and Sharkey, P.J. 1981. Effect of salinity on growth and on mineral and organic constituents of the halophyte, *Disphyma australe* (Solard). *Aust. J. Plant Physiol.,* **8**: 165-179.

Osmand, C.B., and Greenway, H. 1972. Salt response of carboxylating enzymes from species differing in salt tolerance. *Plant Physiol.,* **42**: 656-658.

Panikkar, P. 1992. Physiology of halophytes : Stuides on some aspects of senescence in *Acanthus ilicifolius,* Linn. *Ph. D. Thesis,* Univ. of Bombay, India.

Rakova, N.M., Klyshev, L.K., and Kasymbekov, B.K. 1978a. Effect of Na_2SO_4 and NaCl on activity of the enzymes of primary ammonium nitrogen assimilation in plant roots. *Sov. Plant Physiol.,* **25**: 26-30.

Rakova, N.M., Klyshev, L.K., and Kasymbekov, B.K. 1978b. Effects of sodium salts on the activity of nitrogen metabolism enzymes isolated from plants with different salt resistance. *Sov. Plant Physiol.,* **25**: 585-588.

Saad-Eddin, R., and Doddema, H. 1986. Effects of NaCl on the nitrogen metabolism of the halophyte *Arthrocnemum fruticosum* (L). *Plant and Soil* **92**: 373-385.

Smith, F.A. 1973. The internal control of nitrate uptake into excised barley roots with differing salt contents. *New Phytol.* **72**: 769-782.

Srivastava, H.S., and Singh, R.P. 1987. Role and regulation of L-glutamate dehydrogenase activity in higher plants. *Phytochemistry.* **26**: 597-610.

Stewart, G. R., Lee, J.A., and Orebamjo, T.O. 1972. Nitrogen metabolism of halophytes. I. Nitrate reductase activity in *Suaeda maritima. New Phytol.* **71**: 263.

Stewart, G. R. Lee, J. A., and Orebamjo, T. O. 1973. Nitrogen metabolism of halophytes II. Nitrate availability and utilization. *New Phytol.* **72**: 539.

Stewart, G.R., Lee, J.A., Orrebamjo, T.O., and Havill, D.C. 1974. *In Mechanisms of regulation of plant growth.* (Bieleski, R.L. Ferguson, A.R. and Cresswell, M. M. eds.) Bulletin 12, The Royal Society of New Zealand, Wellington. pp. 41-47.

Stewart, G.R., and Lee, J.A. 1974. The role of proline accumulation in halophytes. *Planta* **120**: 279.

Stewart, C. R., and Rhodes, D. 1978. Nitrogen metabolism of halophytes. III. Enzymes of ammonia assimilation. *New Phytol.,* **80**: 307-316.

Stewart, G.R., Larher, F., Ahmed, I., and Lee, J. A. 1978. Nitrogen metabolism and salt tolerance in halophytes. *In Ecological Processes in Coastal Environments. I European Ecological Symposium.* (Jefferis, R.J. and Davy, J.A. eds.) Blackwell Scientific Publ., Oxford. pp. 42-49.

Stewart, G.R., Larher, F., Ahmad, I., and Lee, J.A. 1979. Nitrogen metabolism and salt tolerance in higher plant halophytes. *In: Ecological processes in coastal environments.* (Jeffries, R.L. and Davy, A. J. eds.) Backwell Scientific Publ. Oxford. pp. 211-227.

Storey, R., Ahmad, N., and Wyn Jones, R. G. 1977. Taxonomic and ecological aspect of the distribution of glycinebetaine and related compounds in plants. *Oecologica,* **27**: 319-332.

Storey, R., and Beevers, L. 1978. Enzymology of glutamine metabolism related to senescence and development in pea (*Pisum sativum* L.). *Plant Physiol.,* **61**: 393-500.

Sudhakar, C., and Veeranjaneyula, K. 1988. Effect of salt stress on some enzymes of nitrogen metabolism is horsegram (*Dolichos biflorus* L.) subjected to salt stress *Ind. J. Exp. Biol.,* **25**: 479-482.

Swerdrup, H.V., Johnson, M.W., and Fleming, R. H. 1961. The Oceans - their physics, chemistry and general biology. Asia Publishing House, Bombay. pp. 181.

Treichel, S. 1975. Effect of NaCl on concentration of proline in different halophytes. *Z. Pflanzen-Physiol.,* **76**: 56-68.

Tyler, G. 1967. On the effect of phosphorus and nitrogen supplied to Baltic shore meadow vegetation. *Bot. Notiser,* **120**: 443.

Waisel, Y. 1972. *Biology of Halophytes.* Academic Press. New York and London. pp. 90

Waisel, Y. and Eshel, A. 1971. Localisation of ions in the mesophyll cells of the succulent halophyte *Suaeda monoica* Forssk. by X-ray microanalysis. *Experientia,* **27**: 230-232.

Wallace, W., and Pate, J.S. 1967. Nitrate assimilation in higher plants with special reference to the cocklebur (*Xanthium pennsylvanicum* Wallr.). *Ann. Bot.* **31**: 213.

Warrick, R.P. 1960. Physiological and ecological studies on halophytes. *Ph. D. Thesis.* Univ. of Bombay. India.